PEROXIDASES AND CATALASES

CONTENTS

PREFACE

Peroxidases and Catalases is a sequel, and more, to *Heme Peroxidases* published in 1999. The book has been completely rewritten, covers important historical discoveries and the exciting research of the last 10 years. *Peroxidases and Catalases* has been strengthened by chapters contributed by authorities in their respective fields: Catalases by Peter Jones of the University of Newcastle Upon Tyne; Biomedical Aspects of Myeloperoxidase by Jeffrey Henderson, Washington University, St. Louis, and Jay Heinecke, University of Washington, Seattle; Vanadium Peroxidases by Rokus Renirie, Wageningen University and Ron Wever, University of Amsterdam. Ron Wever is also the author of the chapter on Application of Peroxidases. A complete chapter on theoretical considerations has been added.

The organization of the book is as follows. After a brief historical summary and discussion of widely used experimental techniques in Chapter 1, the plant, fungal, and bacterial peroxidase superfamilies and their organization are described in Chapter 2, a summary of the pioneering work of Karen Gjessing Welinder. This is followed by chapters on individual members of the plant superfamily. Horseradish peroxidase, the most widely studied and historically most important member, requires three chapters (3 to 5) to do it justice. The first peroxidase crystal structure was that of yeast cytochrome *c* peroxidase and its structure and reactions are described in Chapters 6 and 7. Care is taken in Chapter 6 to document that yeast cytochrome *c* peroxidase is an abnormal peroxidase in terms of its reaction with small reducing substrates. It does not react via the usual cycle established for typical peroxidases. Its reaction with its natural substrate, the macromolecule cytochrome *c* appears well understood due largely to the work of Joseph Kraut and coworkers (Chapter 7).

Chapters 8 to 10 were written with two objectives: first to introduce spectroscopic and theoretical techniques with which one may not be familiar, and second to document a large number of accomplishments in these areas which can easily be grouped together.

The three classes of plant, fungal, and bacterial peroxidases are returned to in Chapters 11 to 14. Class I peroxidases, according to the Welinder classification, are described in Chapters 11 and 12. Ascorbate peroxidase in Chapter 11 has rightly been described as a better model for Class I than is yeast cytochrome *c* peroxidase. The remarkable catalase-peroxidases are the topic of Chapter 12 along with a discussion of *Mycobacterium tuberculosis* and the recurrence of a once-thought-banished disease. Class II fungal peroxidases, those that can degrade lignin, are the topic of Chapter 13. Other well studied Class III plant peroxidases are discussed in Chapter 14.

The catalases, widely distributed throughout the plant and animal kingdoms, are the topic of Chapter 15. A total shift to mammalian peroxidases is made in Chapters 16 to 20. The enzymology of myeloperoxidase, perhaps the best understood of mammalian enzymes, is described in Chapter 16, and remarkable in vivo and model system medical discoveries involving myeloperoxidase are described in Chapter 17. Chapter 18 is devoted to prostaglandin H synthase, the first enzyme in the arachidonic cascade, with care devoted to delineating what is known and what is still controversial. Perhaps the least understood mammalian peroxidase, but one of paramount importance, is thyroid peroxidase (Chapter 19). Lacto- and salivary peroxidases are covered in Chapter 20.

The heme chloroperoxidase from the slime mold *Caldariomyces fumago* (Chapter 21) is amazingly versatile, and could lay better claim to the name versatile peroxidase, which has been bestowed on a member of Class II with characteristics of both lignin and manganese peroxidases. Digression is made to other important elements, selenium enzymes in Chapter 22, and vanadium peroxidases in Chapter 23. Chapter 24 concludes the discussion of heme proteins. Last but not least is the application of peroxidases in Chapter 25.

I am indebted to the contributing authors described above. My editor at John Wiley & Sons, Anita Lekhwani, her assistant Rebekah Amos, and James Terner at Virginia Commonwealth University gave invaluable advice, and my information technology wheels were kept turning by Nurhan Dunford. Figures were donated by Michael Gajhede, Andrea Gumiero, Anette Henriksen, Peter Loewen, Thomas Poulos, Emma Lloyd Raven, and James Terner. My thanks to all.

H. BRIAN DUNFORD

Stillwater, Oklahoma

CONTRIBUTORS

Peter Jones, University of Newcastle Upon Tyne, UK (contributor of Chapter 15)

Jeffrey Henderson, Department of Internal Medicine at Washington University, St. Louis, MO, USA (contributor of Chapter 16)

Jay Heinecke, Department of Medicine, University of Washington, Seattle, WA, USA (contributor of Chapter 16)

Rokus Renirie, University of Wageningen, The Netherlands (contributor of Chapter 23)

Ron Wever, University of Amsterdam, The Netherlands (contributor of Chapter 23 and Chapter 25)

CHAPTER 1

HISTORICAL: PIONEERING WORK ON HORSERADISH AND YEAST CYTOCHROME *c* PEROXIDASES

INTRODUCTION

Although examples of what is now known as enzyme action were described in the early nineteenth century (see Paul, 1963; Saunders et al., 1964), the beginning of the modern era of peroxidase research may be defined as starting with the studies of Bach and Chodat (1903). They demonstrated that their enzyme isolated from horseradish was a true peroxidase, not an oxygenase.

Peroxidase research was almost exclusively on the horseradish enzyme until the 1940s. Today the annual research publications on horseradish peroxidase (HRP) still outnumber the total for all other peroxidases. Discovery dates for some other well-known peroxidases are yeast cytochrome *c* peroxidase and chloroperoxidase, 1940 (although further work did not begin for another 25 years); myeloperoxidase, 1941; and lactoperoxidase, 1947. The majority of the known peroxidases have been discovered since 1970.

In the 1920s and 1930s a controversy raged about the true nature of enzymes, with Richard Willstätter insisting upon the role of metal ions, from his work with horseradish peroxidase, and James Sumner claiming that enzymes were pure proteins, from his work on jackbean urease and other enzymes. Thus, from the controversy between a Nobel Prize winner and a future winner, it was established that there are metalloenzymes and also pure proteins that are enzymes.

The classic work of Michaelis and Menten (1913) interpreted observed enzyme kinetics in terms of an enzyme–substrate complex, a lock-and-key hypothesis. This became the the accepted explanation for rectangular hyperbolic kinetics. The asymptotic limit was ascribed to saturation of the enzyme binding site by substrate.

Peroxidases and Catalases: Biochemistry, Biophysics, Biotechnology, and Physiology, Second Edition.
By H. Brian Dunford

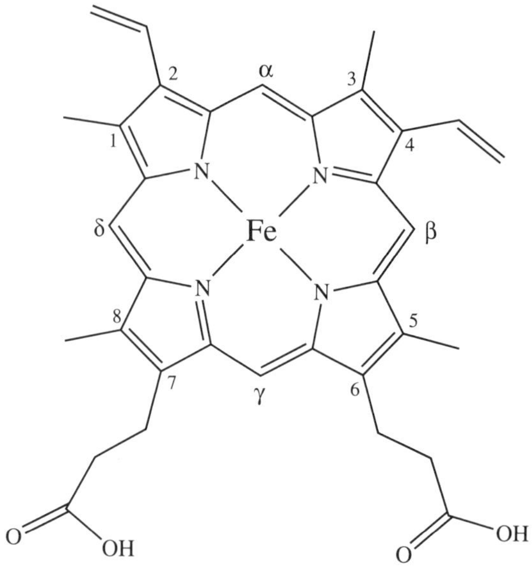

Figure 1.1. Iron protoporphyrin IX (heme *b*).

Therefore, it took time and many investigations, as exemplified by the work of Philip George (1952, 1953a, 1953b), to establish that horseradish peroxidase formed covalent compounds between its heme iron and fragments of hydrogen peroxide, not loose complexes with the entire hydrogen peroxide molecule. What is now known as Compound I was discovered by Hugo Theorell (1941). Theorell was awarded the Nobel Prize in Medicine in 1955 for his work on oxidative enzymes. The discovery of Compound II occurred earlier (Keilin and Mann, 1937). Compound II of HRP is the first enzyme intermediate to have been identified.

Free radical formation in peroxidase reactions violated another perception about enzymes, namely that they were designed to prevent free radical formation, all deemed to be physiologically harmful. Heme peroxidases, their compounds and complexes, are highly colored, especially in the near ultraviolet (Soret) region of the spectrum, a property that offers many advantages over colorless proteins in designing experimental work.

Both yeast cytochrome *c* peroxidase and horseradish peroxidase have ferriprotoporphyrin IX, also called heme *b*, as their prosthetic group, a simpler heme than is found in most mammalian peroxidases (Fig. 1.1). The intensity of research on these two enzymes makes them ideal topics for an introduction to peroxidase and catalase enzymes.

TECHNIQUES AND INSTRUMENTATION

The following is a brief discussion of dominant techniques used in peroxidase research. In addition to standard biochemical procedures, many of the techniques used in inorganic chemistry provide essential information about metalloenzymes.

Chromatography

Although botanists utilized chromatographic techniques as early as 1900, it was the development of a host of gels, resins, and polymers with a wide range of properties that made chromatography the key to successful enzyme isolation and purification. Thus, the late Karl-Gustav Paul (1958) used carboxymethylcellulose chromatography to both simplify and improve the purification of horseradish peroxidase. Paul described enzymology in terms of the chromatographic era and the prechromatographic era.

Amino Acid Sequencing

Frederick Sanger won the Nobel Prize in Chemistry in 1958 for determining the primary amino acid sequence of insulin. His pioneering work was followed rapidly by the sequencing of many proteins. The first primary sequence of a peroxidase was that of horseradish (Welinder, 1976). This was followed by the determination of the yeast cytochrome *c* peroxidase sequence (Takio et al., 1980; Bosshard et al., 1991). Today amino acid sequences are often deduced indirectly from nucleic acid sequences using the genetic code.

Site-Directed Mutagenesis

The 1993 Nobel Prize in Chemistry was awarded to Kary Mullis and Michael Smith for development of the techniques required for site-directed mutagenesis. Yeast cytochrome *c* peroxidase was one of the first enzymes to be cloned and undergo mutagenesis. A mutant enzyme is a new molecule, for which all the techniques that were applied to the wild-type enzyme may be required for characterization.

UV-Visible Spectrophotometers

In today's biochemical laboratories use of sophisticated spectrophotometers is routine. At the end of World War II the best commercial spectrophotometer was the Beckman DU using a car battery as the most stable power supply. One can see the limitations placed on earlier researchers. A paper that I found especially valuable when I started research on horseradish peroxidase was from David Keilin's laboratory where they used a Zeiss microspectroscope and a Hartridge reversion spectroscope (Keilin and Hartree, 1951). Spectrophotometry is especially relevant to peroxidase research because of the intense and different coloring of native enzyme, Compound I, and Compound II. Indeed, transformations in peroxidase reactions provide a direct measure of the number of electrons being transferred.

Spectroscopic Techniques Developed by Physicists

The following four spectroscopic techniques invented by physicists are described in some detail in Chapters 8 and 9. Of the four, only Mössbauer spectroscopy has remained largely in the hands of physicists.

Nuclear Magnetic Resonance (NMR). Discovered by Konrad E. Bloch and Edward M. Purcell in 1946, it led to their Nobel Prize in Physics in 1952. All modern chemical and biochemical laboratories have NMR facilities, although full utilization of the power of NMR requires the knowledge of specialists. In peroxidase research these include Gerd La Mar, Isao Morishima, James Satterlee, and Nigel Veitch. Nobel Prizes in Chemistry were awarded to Richard R. Ernst (1991) for the development of two-dimensional NMR, and to Kurt Wüthrich (2002) for his contributions to three-dimensional protein structural determinations in solution by NMR.

Electron Spin Resonance (ESR). Described by E. Zavoisky in 1945. At an anniversary celebration in Russia, both American and Russian scientists had words with a Nobel Prize committee member saying that Zavoisky deserved a prize, unfortunately to no avail. ESR spectroscopy is widely used, but the work of Ronald Mason deserves special mention.

Mössbauer Spectroscopy. Named after its discoverer, Rudi Mössbauer, who published a description in 1957 and received the Nobel Prize in Physics in 1961. Use of this technique relevant to peroxidases has been conducted by Peter Debrunner, Vincent Boi Han Huynh, Eckard Münck, and Alfred Trautwein.

Resonance Raman Spectroscopy. Raman spectroscopy is named after Chandrasekhara Venkata Raman who discovered it in 1928 and received the Nobel Prize in Physics in 1930. Like infrared spectroscopy, Raman spectroscopy measures transitions between vibrational levels. Resonance Raman spectroscopy is a post-World War II development. In it a laser is used to cause electronic as well as vibrational transitions with enhancement of the latter. Specialists include Teizo Kitagawa, Giulietta Smulevich, Thomas Spiro, and James Terner.

Crystal Structures

The use of X-ray diffraction to elucidate the structures of simple salt crystals and crystals of small molecules was developed by physicists and led to Nobel Prizes in Physics to Max von Laue in 1914 and to the father-and-son team of William Henry Bragg and William Lawrence Bragg in 1915. However the structures of macromolecules were too complicated for the techniques developed by von Laue and the Braggs. The use of heavy metal ions to provide a marker and the determination of primary amino acid sequences provided the necessary breakthroughs, and chemists led the way. The first protein crystal structure determined by X-ray diffraction was that of myoglobin (Nobel Prize in Chemistry to John Kendrew and Max Perutz in 1962).

The first published peroxidase crystal structure was that of yeast cytochrome *c* peroxidase (Poulos et al., 1980). That of horseradish peroxidase was not published until much later (Gajhede et al., 1997), and was aided by comparisons to the crystal structure of peanut peroxidase (Schuller et al., 1996). X-ray analysis of peroxidase crystal structures is now widespread. The groups headed by Thomas Poulos and Ignacio Fita deserve special mention because of the extent of their contributions.

Rapid Reaction Kinetics

Steady-state enzyme kinetic studies are often performed in a conventional spectrophotometer where the dead time (time elapsed between start of the reaction and experimental observations) is one second or greater. By using a plunger to drive the solutions in two syringes through a mixing chamber, the dead time can be reduced to milliseconds. The first example was an accelerated flow apparatus (Chance, 1940a, 1940b), in which the two syringes were pushed manually. A mechanical stop was introduced, thus creating a stopped flow apparatus (Gibson, 1954). For many years the Durrum–Gibson stopped flow apparatus was the commercial instrument of choice. Reaction traces could be obtained only at one wavelength at a time.

I had long desired a stopped flow with rapid spectral scan capabilities. A commercial instrument became available that had a rapidly rotating prism. I wrote to Jui Wang at Yale, who possessed one of the instruments, asking his opinion. His reply, dated 1969, said that the apparatus was fine for preliminary work but was unsatisfactory for quantitative studies. I waited eight more years until the Union Giken stopped flow with rapid scan became available, and possessed the first instrument sold outside Japan. The instrument uses a fixed grating for spectral measurements. Now there are several reliable commercial instruments that combine stopped flow, rapid scan, and sequential mixing capabilities.

The time required to mix two solutions uniformly limits the time at which accurate measurements can be started. The ingenious relaxation techniques, developed principally by Manfred Eigen, eliminated the mixing step. Eigen shared the Nobel Prize in Chemistry in 1967. If a reversible reaction has a finite enthalpy of reaction, then a temperature jump will shift the equilibrium position. The rate of the shift is a function of both the forward and reverse rate constants, and the time required to reach the new equilibrium position is called the relaxation time. Discharge of a high voltage capacitor through the solution can cause the temperature rise. Reaction rates in the microsecond range can be measured using the temperature jump technique (Czerlinski and Eigen, 1959). In principle, the relaxation times, if not the individual rate constants, of all of the steps in a reversible enzyme reaction could be measured in a single T-jump experiment. To the best of my knowledge this has not yet been accomplished. With the use of lasers, reaction times in the femtosecond range and faster can now be measured.

Mass Spectrometry

Developments in mass spectrometry have led to its use as a valuable tool for analysis of biopolymers (McLafferty, 2008). The technique appears to be underused by peroxidase researchers; there are few examples (Donald et al., 2003; Choudhary et al., 2006).

SUMMARY AND CONCLUSIONS

Nearly all of the contributions outlined above were made after World War II. One can only marvel at the patient accomplishments of earlier scientists who paved the way.

The peroxidases are extremely potent catalysts of oxidation. One of the first stopped flow apparatuses I used was machined from a block of polypropylene. Horseradish peroxidase and its substrates started to eat it. There is a recent report that horseradish peroxidase will degrade single-walled carbon nanotubes (Allen et al., 2008). The potency of peroxidase catalysis deserves more emphasis.

A property that has perhaps been overemphasized is the nonspecificity of peroxidase catalysis. The requirements for an oxidizing substrate are highly specific: a molecule of the general formula ROOH that enters the heme cavity to donate an oxygen atom to iron(III). Specific reducing substrates appear to choose one of three routes to donate electrons: at the δ-heme edge, the γ-heme edge, or through the heme propionate side chains. Look at the natural reducing substrates of specific peroxidases. Three of them are included in the name: ascorbate peroxidase, manganese peroxidase, and yeast cytochrome *c* peroxidase. Lignin peroxidase utilizes veratryl alcohol, chloroperoxidase from *C. fumago* favors monochlorodimedone. Catalases utilize hydrogen peroxide as both oxidizing and reducing substrate. Salivary peroxidase utilizes thiocyanate. *Mycobacterium tuberculosis* catalase-peroxidase reacts very well with the antituberculosis drug isoniazid (but perhaps the past tense is more appropriate). Myeloperoxidase reacts with chloride but subsequent reactions may occur through chloramines; taurine is a favored substrate. Similarly, thyroid peroxidase reacts with iodide but further reaction may occur via sulfenyl iodides. There appears to be a physiological symbiosis between enzyme (peroxidase or catalase) and reducing substrate. Peroxidases continue to evolve and new ones to be discovered. Chlorite dismutase is a recent example. A detailed story of the peroxidases and catalases follows, not only of heme peroxidases, but glutathione peroxidases and the vanadium peroxidases.

REFERENCES

Allen BL, Kichambare PD, Gou P, Vlasova II, Kapralov AA, Konduru N, Kagan VE, Star A. 2008. Biodegradation of single-walled carbon nanotubes through enzymatic catalysis. *Nano Lett.* **8**: 3899–3903.

Bach A, Chodat R. 1903. Untersuchungen über die rolle der peroxyde in der chemie der lebenden zelle. IV. Ueber peroxydase. *Ber.* **36**: 600–605.

Bosshard HF, Anni H, Yonetani T. 1991. Yeast cytochrome *c* peroxidase. In: Everse J, Everse KE, Grisham MB, editors, *Peroxidases in chemistry and biology*, Vol. II, CRC Press, Boca Raton, FL, pp. 51–84.

Chance B. 1940a. The accelerated flow method for rapid reactions. Part I. Analysis. *J. Franklin Inst.* **229**: 455–476.

Chance B. 1940b. The accelerated flow method for rapid reactions. Part II. Design, construction, and tests. *J. Franklin Inst.* **229**: 737–766.

Choudhary K, Spicer VL, Donald LJ, Duckworth HW, Ens W, Loewen PC, Standing KG. 2006. Method for estimating the isotopic distributions of metabolically labeled proteins by MALDI-TOFMS: application to NMR samples. *Anal. Chem.* **78**: 5419–5423.

Czerlinski G, Eigen M. 1959. Eine temperatursprungmethode zur untersuchung chemischer relaxation. *Z. Elektrochem.* **63**: 652–661.

Donald LJ, Krokhin OV, Duckworth HW, Wiseman B, Deemagarn T, Singh R, Switala J, Carpena X, Fita I, Loewen PC. 2003. Characterization of the catalase-peroxidase KatG from *Burkholderia pseudomallei* by mass spectrometry. *J. Biol. Chem.* **278**: 35687–35692.

Gajhede M, Schuller DJ, Henriksen A, Smith AT, Poulos TL. 1997. Crystal structure of horseradish peroxidase C at 2.15 Å resolution. *Nature Struct. Biol.* **4**: 1032–1038.

George P. 1952. Chemical nature of the secondary hydrogen peroxide compound formed by cytochrome *c* peroxidase and horseradish peroxidase. *Nature* **169**: 612–613.

George P. 1953a. The chemical nature of the secondary hydrogen peroxide compound formed by cytochrome *c* peroxidase and horseradish peroxidase. II. Formation and decomposition. *Biochem. J.* **55**: 220–230.

George P. 1953b. Intermediate compounds formed when horseradish peroxidase reacts with potassium molybdicyanide and potassium chloroiridate. *Science* **117**: 220–221.

Gibson Q. 1954. Stopped-flow apparatus for the study of fast reactions. *Disc. Faraday Soc.* **17**: 137–139.

Keilin D, Hartree EF. 1951. Purification of horseradish peroxidase and comparison of its properties with those of catalase and methaemoglobin. *Biochem. J.* **49**: 88–104.

Keilin D, Mann T. 1937. On the haematin compound of peroxidase. *Proc. R. Soc. (London)* **122B**: 119–133.

McLafferty FW. 2008. Mass spectrometry across the sciences. *Proc. Natl. Acad. Sci. USA* **105**: 18088–18089.

Michaelis L, Menten LM. 1913. Die kinetik der invertinwirkung. *Biochem. Z.* **49**: 333–369.

Paul KG. 1958. Die isolierung von meerretichperoxydase. *Acta Chem. Scand.* **12**: 1312–1318.

Paul KG. 1963. Peroxidases. In: Boyer PD, Lardy H, Myrback K, editors, *The Enzymes*, Vol. 8, Part B, 2nd ed., Academic Press, New York, pp. 227–231.

Poulos TL, Freer ST, Alden RA, Edwards SL, Skogland UK, Takio K, Eriksson B, Xuong N, Yonetani T, Kraut J. 1980. The crystal structure of cytochrome *c* peroxidase. *J. Biol. Chem.* **259**: 13027–13036.

Saunders BC, Holmes-Siedle AG, Stark BP. 1964. *Peroxidases*. Butterworths, London.

Schuller DJ, Ban N, van Huystee RB, McPherson A, Poulos TL. 1996. The crystal structure of peanut peroxidase. *Structure* **4**: 311–321.

Takio K, Titani K, Ericsson LH, and Yonetani T. 1980. Primary structure of yeast cytochrome c peroxidase. II. The complete amino acid sequence. *Arch. Biochem. Biophys.* **203**: 615–629.

Theorell H. 1941. Crystalline peroxidase. *Enzymologia* **10**: 250–252.

Welinder KG. 1976. Covalent structure of the glycoprotein horseradish peroxidase (EC 1.11.1.7). *FEBS Lett.* **72**: 19–23.

CHAPTER 2

HEME PEROXIDASE AND CATALASE FAMILIES

PLANT, FUNGAL, AND BACTERIAL PEROXIDASES

Until crystal structures became available deductions about peroxidase heme environments were largely guesswork. An exception was a photooxidation study of apo-horseradish peroxidase plus protoporphyrin IX (Mauk and Girotti, 1974), which provided evidence that a histidyl residue is coordinated to the heme in native horseradish peroxidase.

Publication of the crystal structure of yeast cytochrome *c* peroxidase in 1980 was followed by the structure of beef catalase (Murthy et al., 1981). There was a 12-year hiatus before another peroxidase crystal structure was published. In the meantime the plant, fungal, and bacterial peroxidase were classified into families by Karen Welinder (Welinder, 1992; Welinder and Gajhede, 1993). Using primary amino acid sequence data, and initially the crystal structure of yeast cytochrome *c* peroxidase as a reference, she was able to separate α-helical regions and the connecting strands between them. This despite the fact that sequence homologies are often not very high, and the surface loops connecting helices are highly variable.

Each peroxidase in the superfamily consists of ten α-helices, labeled A to J, all in the same orientation. Heme binding occurs between helices B and F (Fig. 2.1). Table 2.1 shows examples of conserved catalytic distal groups and the wide diversity of other amino acid residues in B helices. The work of Welinder filled a huge gap in knowledge about heme environments and peroxidase structures, and has been confirmed as correct with the wide range of crystal structures now available. Her three classes of plant, fungal, and yeast heme peroxidases are described below. Another of her findings is that a gene duplication event must have occurred because of the

Peroxidases and Catalases: Biochemistry, Biophysics, Biotechnology, and Physiology, Second Edition.
By H. Brian Dunford

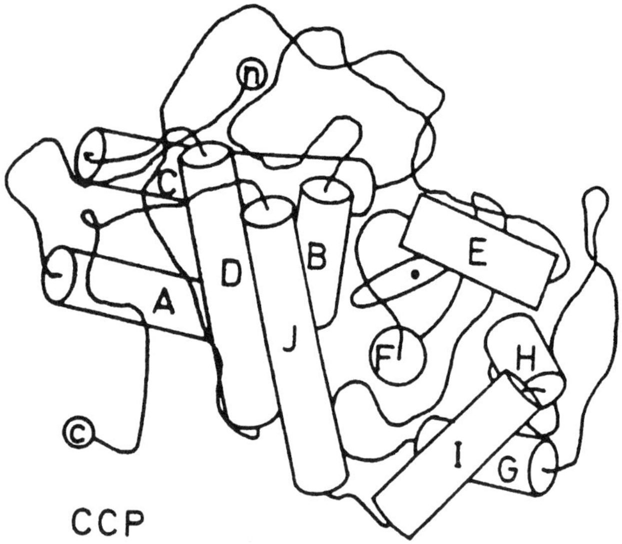

Figure 2.1. Helical structures of yeast cytochrome *c* peroxidase. The heme *b* group is located between helices B and F. (Reproduced with permission from Weliinder KG. 1992. In: Penel C, Gaspar T, Greppin H, editors, University of Geneva Press, Geneva, pp. 1–24.)

similarities of helices 1 to 5 (A to D) compared to helices 6 to 10 (E to J; Fig. 2.2). Those peroxidases that bind calcium ions do so in the same positions in the distal and proximal sides.

Class I

Yeast cytochrome *c* peroxidase, ascorbate peroxidases, and bacterial catalase-peroxidases are members of Class I. Class I peroxidases have a conserved proximal tryptophan residue. They have no carbohydrate, no cystine bridges, no calcium ion, and no signal peptide for secretion. Ascorbate peroxidases in cytosol are dimers; those found in chloroplasts are monomers. The bacterial catalase-peroxidases have roughly double the molecular weight of other Class I peroxidases. It appears that further gene duplication occurred with heme bound to only one half of the molecule. See Chapters 6, 7, 11, and 12.

TABLE 2.1. Conserved Catalytic Distal His and Arg Residues in B Helices

		48		52	
Group I Yeast C*c*P	YGPVLV	**R**	LAW	**H**	TS
		42		46	
Group II Mn Perox.	EVI	**R**	LTF	**H**	DA
		38		42	
Group III HRP	SIL	**R**	LHF	**H**	DC

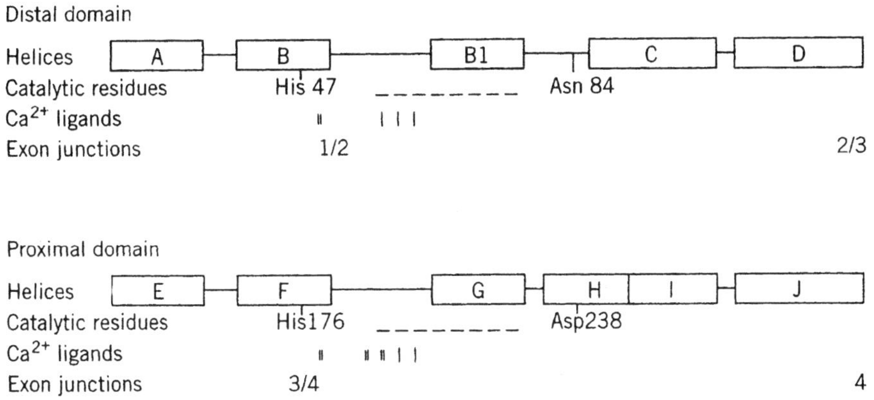

Figure 2.2. Correlation between the five helices of the distal domain and the five helices of the proximal domain of any peroxidase of Classes I to III. (Reproduced with permission from Welinder KG, Gajhede M. 1993. In: Welinder KG, Rasmussen SK, Penel C, Greppin H, editors, *Plant Peroxidases. Biochemistry and Physiology.* University of Geneva Press, pp. 35–42.)

Class II

Class II peroxidases are secreted by fungal organisms. Examples are the lignin and manganese peroxidases from *Phanerochaete chrysosporium*, and the peroxidase from ink cap mushrooms, known by either of the names *Coprinus cinereus* or *Arthromyces ramosus*. Class II peroxidases have about 5% carbohydrate, two calcium ions, and four conserved disulfide bridges. They have an amino-terminal signal peptide for secretion through the endoplasmic reticulum. See Chapter 13.

Class III

Class III consists of the classical secretory plant peroxidases. Examples are the peroxidases from horseradish, peanut, and barley. Class III peroxidases have a wide range of carbohydrate content, ranging from 0 to 25%. They possess two calcium ions and four disulfide bridges in different positions from those in Class II peroxidases. They have an N-terminal signal peptide for secretion, and some have a C-terminal peptide that may target them for vacuoles. Horseradish peroxidase is described in detail in Chapters 3 to 5, and other Class III peroxidases in Chapter 15.

Chloroperoxidase found in the slime mold *Caldariomyces fumago* does not fit into the above classifications (see Chapter 22). Neither do the peroxidases with two hemes bound to a single polypeptide chain (see Chapter 25).

MAMMALIAN PEROXIDASES

The mammalian peroxidases appear to belong in the same family. These include eosinophil peroxidase, lactoperoxidase, myeloperoxidase, prostaglandin H synthase, salivary peroxidase, and thyroid peroxidase. Most have hemes that have covalent attachments to the apoprotein in addition to binding of heme iron to a proximal amino acid residue. An exception is prostaglandin H synthase, which has a rather loosely bound heme *b* prosthetic group. However, large portions of prostaglandin H synthase and myeloperoxidase have very similar structures. Details are in Chapters 17 to 21.

CATALASES

Catalases are found in bacteria, fungi, animals, and plants. They are tetramers with heme prosthetic groups. There are three main families of catalases, which have a common origin. They are much different from the peroxidases, with considerable pleated sheet structure, and usually with a bound NADPH molecule. They have no relationship with the catalase-peroxidases. See Chapter 15 for details.

REFERENCES

Mauk MR, Girotti AW. 1974. The protoporphyrin-apoperoxidase complex: Photooxidation studies. *Biochemistry* **13**: 1757–1763.

Murthy MRN, Reid TJ, Sicignano A, Tanaka N, Rossmann MG. 1981. *J. Mol. Biol.* **152**: 465–499.

Welinder KG. 1992. Superfamily of plant, fungal and bacterial peroxidases. *Curr. Opin. Struct. Biol.* **2**: 388–393.

Welinder KG, Gajhede M. 1993. Structure and evolution of peroxidases. In: Welinder KG, Rasmussen SK, Penel C, Greppin H, editors, *Plant Peroxidases: Biochemistry and Physiology*. University of Geneva Press, Geneva, pp. 35–42.

CHAPTER 3

HORSERADISH PEROXIDASE. I. THE NATIVE ENZYME, COMPOUNDS I AND II, THEIR STRUCTURES, AND THEIR CYCLE

INTRODUCTION

"The classic heme peroxidase and catalase intermediates, known as Compounds I and II, have held the key to understanding the mechanisms of cytochrome oxidase, other heme enzymes, and non-heme iron enzymes; and currently serve as important hypothetical models for cytochrome P-450 intermediates that are yet to be fully characterized" (Terner et al., 2006). Of all the peroxidases that obtained from horseradish (*Armoracia rusticana*) stands out as the one most commonly used. It is readily available, its intermediates are among the most stable, and, unlike yeast cytochrome *c* peroxidase, it has a conventional Compound I structure. The discussion in this chapter is largely limited to horseradish peroxidase isoenzyme C (HRP C). For reviews on horseradish peroxidase see Dunford (1991) and Veitch and Smith (2001).

THE CLASSIC PEROXIDASE CYCLE

The classic peroxidase cycle, common for most heme peroxidases, will be described using horseradish peroxidase (HRP) as an example. HRP is oxidized by hydrogen peroxide to form Compound I (HRP-I) in a two-electron reaction. A wide range of organic hydroperoxides of the general formula ROOH will also form Compound I. Reducing substrates, typically phenols or aromatic amines, reduce Compound I in a

Peroxidases and Catalases: Biochemistry, Biophysics, Biotechnology, and Physiology, Second Edition.
By H. Brian Dunford

one-electron step to Compound II, and Compound II back to native enzyme in a sequential one-electron step, as shown in Figure 3.1.

Free radicals are formed from the substrates, leading to a rich chemistry. A reducing substrate typically reacts with Compound I about an order of magnitude faster than with Compound II. Further details of the reaction cycle are discussed later.

All of the enzyme species in the cycle are highly colored in a unique manner, so that they can be identified by optical spectroscopy. Optical spectra are shown in Figure 3.2. As observed in a test tube, native enzyme is brown, Compound I is green, and Compound II is red.

Accurate molar absorptivities provide the tool for the simplest way of measuring concentrations of peroxidase species. The maximum molar absorptivity of native horseradish peroxidase is the reference from which all other molar absorptivities are calculated. The concentration of native peroxidase is equated to the concentration of its heme. The latter is measured by extracting the heme *b* in pyridine and comparing the spectra to that known for the pyridine–heme *b* complex (Paul et al., 1953). Thus, from absorbance measured on native enzyme, and concentration determined using the pyridine hemochromogen method, the molar absorptivity of native enzyme is obtained using Beer's law.

The two most accurate measurements of the maximum absorbance of native horseradish peroxidase at 403 nm (the Soret region maximum) are in agreement: $1.02 \times 10^5\ M^{-1}\ cm^{-1}$ (Schonbaum and Lo, 1972; Ohlsson and Paul, 1976). This value provides the reference point for all spectra of horseradish peroxidase isoenzme C compounds and complexes.

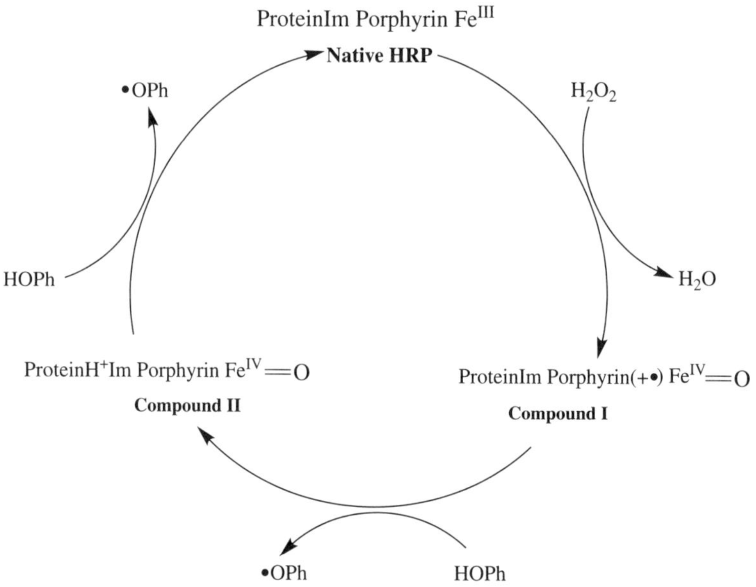

Figure 3.1. The classic peroxidase cycle using hydrogen peroxide and a phenol as oxidizing and reducing substrates.

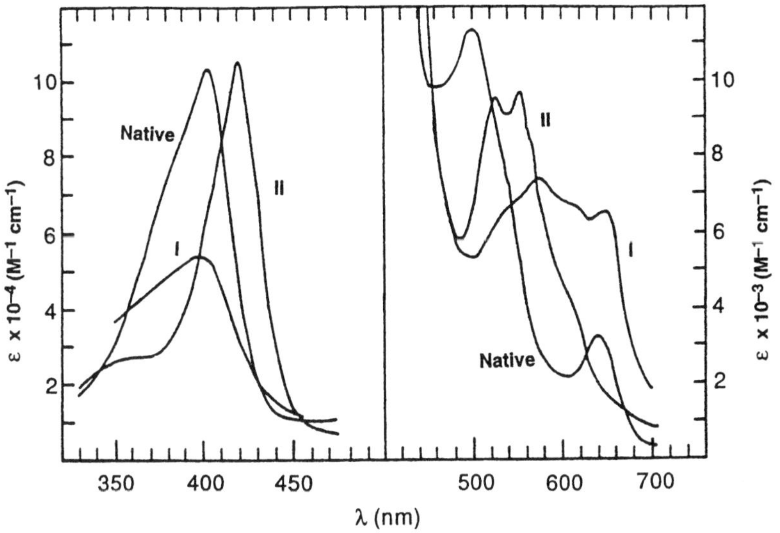

Figure 3.2. Soret and visible region optical spectra of horseradish peroxidase and its Compounds I and II. (Reproduced from Dunford HB. 1982. *Adv. Inorg. Biochem.* **4**: 41–68, with permission from Elsevier Science.)

Preparation of Compound I

Care must be taken in preparing and handling Compounds I and II. Compound I in particular is unstable. A lengthy and confusing literature exists concerning endogenous (internal) and exogenous (external) electron donors that will reduce the compounds. If more than a slight excess of hydrogen peroxide is required to prepare Compound I it is an indication of an impure enzyme preparation. The best preparations of HRP-I will last for an hour or more, indicating that little or no electron donor exists external to the enzyme. Another indication of purity of a native HRP preparation is its Reinheitzahl (RZ; in English, purity number). RZ values, widely quoted in the literature, are the ratio of absorbances at 403 nm and 280 nm where aromatic amino acid residues absorb light. An RZ value in the range 3.0 to 3.4 is one indicator of a pure HRP preparation. Some investigators have used preincubation of peroxidase with some hydrogen peroxide to remove exogenous electron donors, a procedure not favored in my research group. A wide variety of organic hydroperoxides of the general formula ROOH will also form Compound I.

Photolysis of Compounds I and II A clue to the nature of an endogenous donor is given by studies of metal ion-peptide complexes by radiation chemists. Depending upon relative reduction potentials, electron transfer from peptide to metal ion can occur. In HRP-I electron transfer is accelerated by light of ~400 nm corresponding to the maximum absorbance of the heme group (Stillman et al., 1975; Van Wart and Zimmer, 1985). Thus care must be taken that a preparation of Compound I is not photochemically reduced. The source of the electron to reduce Compound I must be within the protein, and this was proven in electron spin resonance

(ESR) studies of the photolysis-induced product (McIntosh and Stillman, 1977; Nadezhdin and Dunford, 1979). Photolysis of Compound I appears to produce a new species with unpaired electrons on both the protein and porphyrin ring and the iron reduced from an oxidation state of IV to III.

Compound II can also be reduced photochemically at room temperature. The reduction will not occur at 10 K, apparently because the necessary conformation change from $Fe^{IV}{=}O$ heme to Fe^{III} heme cannot occur in the frozen solution (Stillman et al., 1975). Photolysis of Compound II occurs most strongly with light of wavelength shorter than 300 nm (Nadezhdin and Dunford, 1979). Again, a free radical is generated on the protein and iron is reduced to an oxidation state of III. These photochemical processes might be called light-directed mutagenesis. One might expect more severe damage by radiation of higher energy than UV light, and this indeed does occur (Berglund et al., 2002).

Preparation of Compound II

Preparation of pure Compound II requires knowledge of the pH dependence of its reactions. Compound I must first be prepared, then reduced to Compound II. An electron donor that reduces Compound I is also capable of reducing Compound II. The necessary trick, using as reductant a phenol such as *p*-cresol, for example, is to carry out the Compound I reduction at pH 9 or slightly higher, where Compound II does not react. A 100% yield of Compound II can be obtained, and the pH can then be adjusted to a desired value (Critchlow and Dunford, 1972). Thus the first spectrum of pure Compound II was obtained, as shown in Figure 3.1. In the absence of light Compound II preparations can last for up to 24 hours.

Intimate details of the three steps in the reaction cycle are now known. Reduction potentials of the Compounds I–II, and Compound II–native enzyme couples have been accurately measured as a function of pH in a classic paper (Hayashi and Yamazaki, 1979). At pH 6 they are essentially the same for both couples ($E_o' =$ 1.1 V). The E_o' values were calculated directly from equilibrium data and also from rate constants for forward and reverse reactions of K_2IrCl_6 and K_3IrCl_6. Thus, horseradish peroxidase reactions can be reversed by reagents with the right redox potentials. A unique reversal (native HRP to Compound II, and Compound II to Compound I) was obtained using a photochemically generated oxidizing agent $[Ru(bpy)_3]^{3+}$ where bpy = 2,2′-bipyridine (Berglund et al., 1997).

The accepted wisdom that HRP reactions are irreversible fits the observed kinetics for many reactions. However, two chemical reactants have to come together before they can react. The inability to detect enzyme–substrate intermediates in the HRP → HRP-I → HRP-II → HRP cycle means that complex formation in each step is rate controlling and the complex reacts faster than it can be detected. When an intermediate complex is detected it simply means complex formation is no longer totally rate controlling.

The Heme Structures of Compounds I and II

The accepted wisdom about enzyme mechanisms was that reaction occurred after formation of a Michaelis–Menten type of complex. The pioneer of the concept of compound formation, not complex formation, in peroxidases was Philip George, who studied reactions of one-electron redox reagents potassium molybdicyanide and potassium chloriridate (George 1952a, 1952b, 1953). Inorganic chemistry met biochemistry.

The Ferryl ($Fe^{IV}{=}O$) Group and the Porphyrin π-Cation Radical If Compound I was a complex of intact hydrogen peroxide with the enzyme, then cyanide should be able to displace the peroxide. The peroxidase catalysis of ferrocyanide oxidation by hydrogen peroxide was inhibited by cyanide binding to the native enzyme only, not to Compounds I and II, showing the iron position was blocked (Cotton et al., 1973). More direct evidence for the ferryl group in Compound I was obtained by reacting *m*-nitroperbenzoic acid with HRP and detecting *m*-nitrobenzoate as product. Similarly when ethyl hydrogen peroxide ethanol was reacted ethanol was produced. The results strongly indicate one oxygen atom is added to the heme iron (Schonbaum and Lo, 1972). The disappearance of *m*-chloroperbenoic acid and the appearance of *m*-perbenzoate was followed kinetically, showing a 1:1 correlation between the amount of perbenzoic acid reacted and the amount of Compound I formed (Adediran and Dunford, 1983).

Porphyrin π-cation radical formation in a model system was demonstrated in a famous paper (Dolphin et al., 1971). The elusive porphyrin free radical of HRP Compound I was finally detected by ESR spectroscopy, with the strength of the free radical signal correlating with the amount of Compound I present (Aasa et al., 1975). The iron(IV) oxidation state was established for both Compounds I and II by Mössbauer spectroscopy, and the existence of the porphyrin π-cation radical confirmed by ESR (Schulz et al., 1979). Spin coupling between the iron(IV) and the free radical electron weakens the ESR signal. The $Fe^{IV}{=}O$ vibrational frequency of HRP compounds was detected by resonance Raman spectroscopy (Terner et al., 1985, 2006; Hashimoto et al., 1986). Thus, spectroscopic techniques and crystal structure data have nailed down the total heme structures of Compounds I and II. For details on the results of spectroscopic techniques see Chapters 8 and 9, including a discussion of the current controversy over whether the $Fe^{IV}{=}O$ group of Compound II is protonated.

Some Details of the Reaction Cycle

One would envisage that hydrogen peroxide reacts directly with the heme to form a formal ferryl $Fe^{V}{=}O$ group. Then an electron is abstracted from the porphyrin ring by the iron so the net result is a $Fe^{IV}{=}O$ ferryl group and a π-cation radical on the porphyrin. On a stopped flow time scale the $Fe^{V}{=}O$ state cannot be detected. The reacting hydrogen peroxide is reduced to a water leaving group. Compound I formation thus may be represented as follows.

$$\underset{\text{Native enzyme}}{\text{PorphyrinFe}^{\text{III}}} + H_2O_2 \longrightarrow \underset{\text{Compound I}}{\text{Porphyrin}(+^{\bullet})\text{Fe}^{\text{IV}}{=}O} + H_2O$$

As we shall see later the above depiction summarizes a series of highly coordinated, complicated steps.

Formation of Compound II occurs in a reaction that directly involves the protein as well as the heme. The reducing substrate, a phenol for example, donates an entire hydrogen atom, proton plus electron, but they go to different locations. The electron is added to the porphyrin ring negating the cation radical; and the proton is added to the imidazole side chain of distal His42.

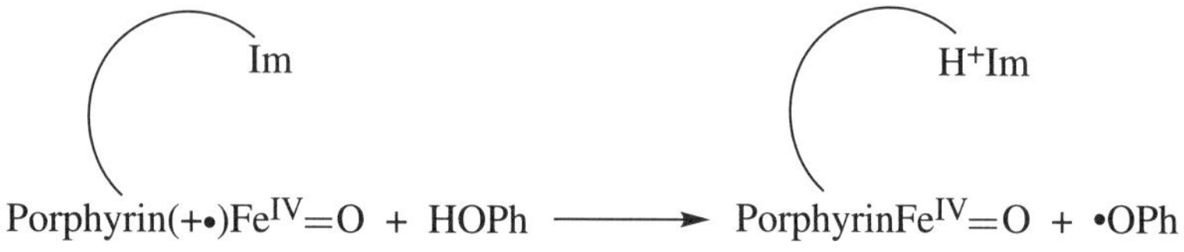

Compound I Compound II

Thus the substrate is oxidized to a free radical.

Reduction of Compound II to native enzyme is more complicated since the oxidation state and coordination number of the iron are changed. A ternary complex is formed between HOPh, H^+Im, and $Fe^{IV}{=}O$, from which the products are $^{\bullet}$OPh, Im, Fe^{III} and a water leaving group.

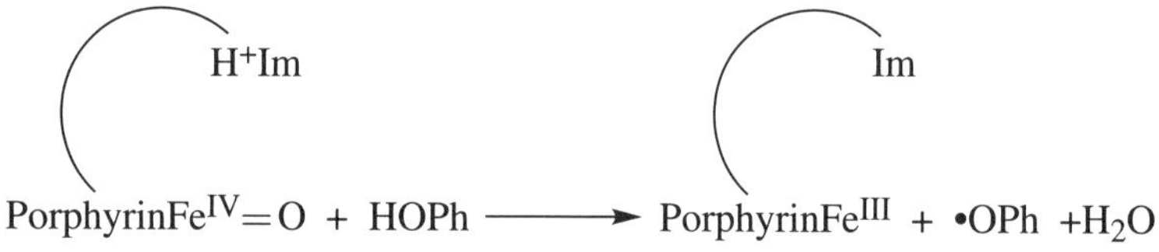

Compound II Native HRP

STRUCTURE AND PROPERTIES OF NATIVE HORSERADISH PEROXIDASE C

The p*I* value for horseradish peroxidase isoenzyme C is 8.8 (Marklund et al., 1974). Its primary sequence was the first to be published for any peroxidase (Welinder, 1976, 1979, 1985). The molecular weight of the single polypeptide chain is 33.89 kDa. In addition it contains a heme *b* prosthetic group, two calcium ions, and eight neutral carbohydrate side chains. Assuming an average composition of $(GlcNAc)_2$, Man_3, Fuc, Xyl for each carbohydrate side chain, the total molecular weight is ~44 kDa. The difficulty in crystallizing HRP appears to be caused by microheterogeneity of carbohydrate side chains, greater than in other Class III plant peroxidases. It is possible to remove most carbohydrate side chains by mild chemical deglycosylation and obtain a fully active enzyme. Solubility in salt solution is greatly reduced (Tams and Welinder, 1995).

Structural Studies and Conformational Changes

Leaching of Ca^{2+} from HRP at low pH by anions was investigated (Wright and English, 2001). The effects of heme substitution, Ca^{2+} removal, and reduction of disulfide bonds were examined as a function of pressure using Fourier transform infrared spectroscopy (Smeller et al., 2003). Conformational changes are induced by changes in temperature and pH (Chattopadhyay and Mazamdar, 2000; Pina et al., 2001). Guanidinium chloride denaturation/renaturation of apo- and holo-HRP were followed using tryptophan fluorescence and circular dichroism (Pappa and Cass, 1993; Chakrabarti and Basak, 1996). Effects on HRP structure of low concentrations of the denaturant urea were studied (Haque et al., 1999). Reagents specific for arginine and tyrosine residues caused loss of activity (Adak et al., 1996). Effects of temperature and heme removal were probed by fluorescence spectroscopy using Nile red (Hungerford et al., 2005).

Synthetic Genes, Crystallization, and Structural Determination

cDNA structures for some of the horseradish peroxidase isoenzymes have been determined (Fujiyama et al., 1988, 1990; Bartonek-Roxå et al., 1991). A synthetic gene for horseradish peroxidase C (the enzyme designated as HRP C*), was expressed in *Escherichia coli* (Smith et al., 1990). No carbohydrate side chains are present. The crystal structure of HRP-C* was finally obtained (Gajhede et al., 1997), aided by comparison to the crystal structure of peanut peroxidase. It took two decades from completion of the primary sequence of HRP to crystal structure determination, whereas only six years elapsed for peanut peroxidase, another Class III peroxidase (Buffard et al., 1990; Ban et al., 1992; Schuller et al., 1996). The stereo ribbon structure of HRP C is shown in Figure 3.3 and another representation in Figure 3.4.

Later a baculovirus expression system was used successfully so carbohydrate side chains were incorporated into the horseradish peroxidase structure, probably different from that found in the native enzyme (Hartmann and Ortiz de Montellano, 1992). Preserved features of Classes I, II, and III peroxidases shown in the HRP structure are the distal His and Arg residues with the His in particular essential for catalytic activity; and the proximal His residue coordinated to the heme iron. Unique features of Class III peroxidases, shown in HRP, are the extended sequences forming helices F′ and F″, and a ring of three peripheral Phe residues, 142, 68, and 179, guarding the approach to the exposed heme edge.

Crystal structures of HRP C* complexes rapidly followed, that with benzhydroxamic acid (Henriksen et al., 1998), and binary and ternary cyanide and ferulic acid complexes (Henriksen et al., 1999). Benzhydroxamic acid occupies a distal hydrophobic pocket with little rearrangement of the heme crevice structure. The pocket is provided by residues H42, F68, G69, A140, P141, F179, and heme C18, C18-methyl, and C20. The structural determinations shed further light on extensive thermodynamic data accumulated on aromatic donor binding (Paul and Ohlsson, 1978), and binding of aromatic hydroxamic acid analogs (Aitken et al., 2001).

Ferulic acid enhances the strength of plant cell walls by forming cross-links and is a known peroxidase substrate. It reacts very rapidly with Compounds I and II. Using the cyanide complex as a model for Compound 0, an intermediate between native enzyme and Compound I, a mechanism for reaction of ferulic acid with Compounds I and II was proposed, in which distal Arg plays a more important role than in the

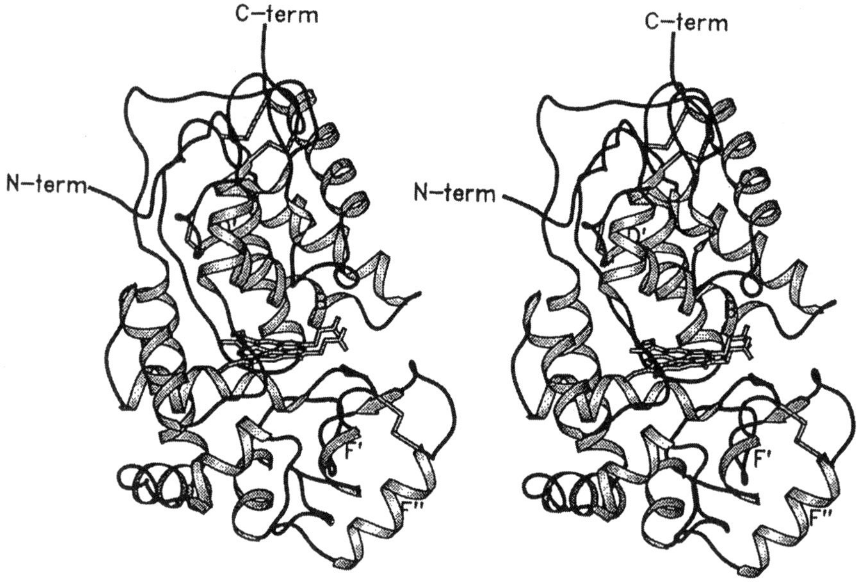

Figure 3.3. Stereo ribbon structure of horseradish peroxidase C showing the unique F′ and F″ helices found in Class III peroxidases. (Kindly supplied by Drs. M. Gajhede and A. Henriksen.)

Poulos–Kraut mechanism for yeast cytochrome *c* peroxidase. This is supported by the 500-fold decrease in rate of Compound I formation in the Arg38Lys mutant (Smith et al., 1992). Ferulic acid hydrogen bonds to Arg38. Protonated distal His42 moves 0.5 Å towards CN^-, providing a strong hydrogen bond.

Reaction of HRP with hydrogen peroxide and ferulic acid in the presence of a tyrosine-containing tripeptide led to many cross-linked products, detected by electrospray ionization (tandem) mass spectrometry (Oudgenoeg et al., 2002). The results on this model system provide striking evidence for the role of ferulic acid in plant cell wall structure.

The crystal structures of Compounds I and II of HRP C* have been published, as well as that of Compound III, the analog of oxymyoglobin (see Chapter 4; Berglund et al., 2002). Great care was taken to ensure that structures were determined before oxidation states were changed by electrons released by the x-ray beam. The experiments were conducted on a multitude of crystals from which the average structure was captured, a technique similar to that used for the determination of single molecule behavior. Compound I was produced by reacting peracetic acid with crystals of HRP C*. The iron–oxygen distance in Compound I is 1.7 Å. The ferryl oxygen forms a 3.0 Å hydrogen bond with the Nε of Arg38, and a 2.7 Å hydrogen bond with a water molecule which in turn is hydrogen bonded to both His42 (2.8 Å) and Arg38 (2.9 Å).

Compound I is unstable and decays spontaneously to Compound II. After disappearance of the green color the crystals were flash frozen. A hydrogen bonded network involving the ferryl oxygen of Compound II was very similar to that in

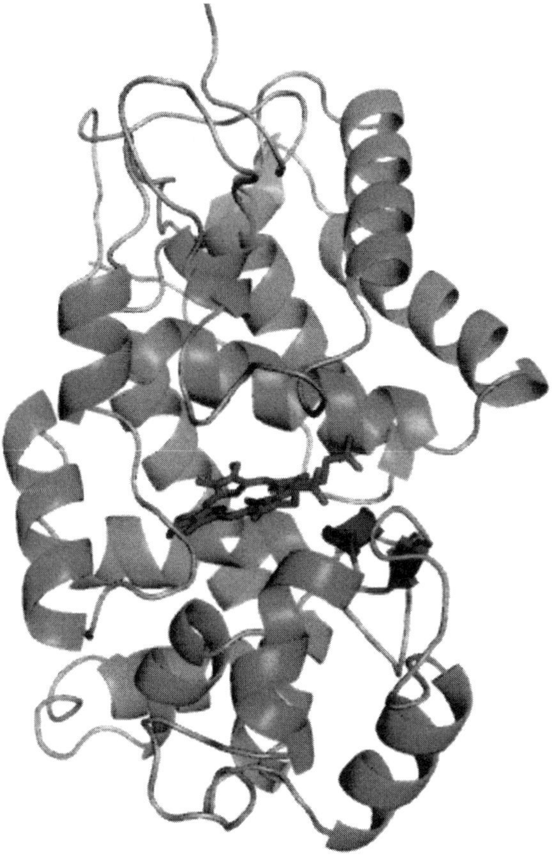

Figure 3.4. Another view of recombinant horseradish peroxidase. (Kindly supplied by Dr. Tom Poulos.)

Compound I. The iron–oxygen distance is slightly longer in Compound II: 1.8 Å, interpreted incorrectly to indicate protonation of the ferryl oxygen. The slightly longer bond length can be accounted for by a hydrogen bond to the ferryl oxygen (see Chapter 8).

The more recent crystal structure of HRP C* (Berglund et al., 2002) confirms that the iron(III) is five coordinate (Lanir and Schejter, 1975). Structures of its complexes with acetate and formate, the ferrous form and its CO complex have also been published (Berglund et al., 2002; Carlsson et al., 2005). Acetate occupies a position above the heme occupied by two water molecules in the acetate-free form (Gajhede et al., 1997). Formate binds directly to the iron. CO binds perpendicularly to the heme and appears to form hydrogen bonds with His42 and Arg38.

The Alkaline Transition

An alkaline transition occurs at pH 11.0, ionic strength 0.1, and temperature 25°C, resulting in a unique optical spectrum. Resonance Raman spectral studies showed that a hydroxyl group binds to five-coordinate Fe^{III} converting it into a six-coordinate Fe^{III}—OH species (Sitter et al., 1988; Feis et al., 1994). The rate constant for the forward reaction in the alkaline transition of horseradish peroxidase

$$Fe^{III} + OH^- \rightleftharpoons Fe^{III}\text{—}OH^-$$

is $(7.8 \pm 0.5) \times 10^4\ M^{-1}\ s^{-1}$ and the reverse $(78 \pm 9)\ s^{-1}$, obtained from pH-jump experiments (Job and Dunford, 1978). Turnip peroxidase P_1 behaves in a similar fashion in its alkaline transition.

Ligand Binding

The binding of ligands to Fe^{III} are the simplest reactions of native HRP. From a large number of pH-dependent studies a clear and consistent picture emerges for the mechanism of these reactions. Initially it was assumed, based on inorganic models, that the anionic form of a ligand would bind, not the protonated ligand. Wrong! HF binds to native HRP, not fluoride ion, and HCN binds, not cyanide ion. These reactions provide mimics of the first step in Compound I formation from native enzyme and hydrogen peroxide. Using fluoride as an example, the proton from the ligand is transferred to the imidazole side chain of the distal His42 residue as fluoride ion binds to the iron(III).

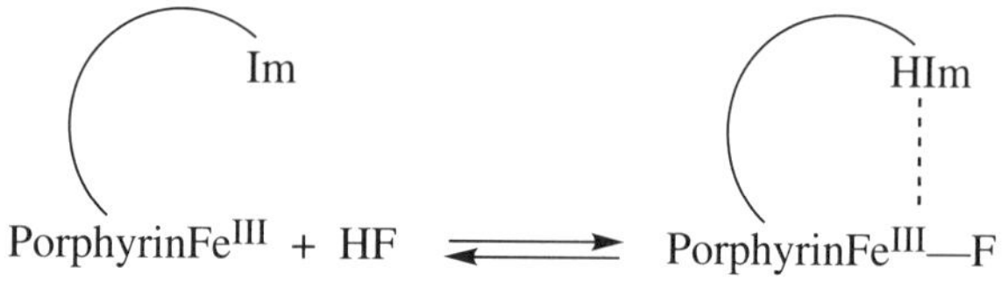

The positive charge on $^{+}$HIm and the negative charge on F^- are partially offset by linking hydrogen bonds, now known to involve an intervening water molecule. The pK_a value of HF is 3, so as the pH is increased over the range 4 to 8, the log of the concentration of HF decreases and that of fluoride ion increases linearly. Over this range the log of the "on" rate constant decreases linearly with increasing pH as expected for HF binding. The necessary proton for the "off" reaction is contained within the enzyme–ligand complex, so fluoride dissociation would be expected to be pH independent, as observed (Dunford and Alberty, 1967).

The pK_a of HCN is 9, so over the range of pH 4 to 8, all of the ligand is in its acid form. Thus the "on" rate constant is expected to be pH independent, as observed. Again the necessary proton for the dissociation reaction is contained within the enzyme–ligand complex, leading to a pH-independent "off" rate constant (Ellis and Dunford, 1968; Dunford et al., 1978). Cyanide binding mimics the first steps in Compound I formation (Fig. 3.5).

Fluoride ion provides a weak ligand field, leading to a high-spin six-coordinate iron(III) complex, whereas cyanide binding brings a strong ligand field so the six-coordinate complex is low spin (see Chapter 8).

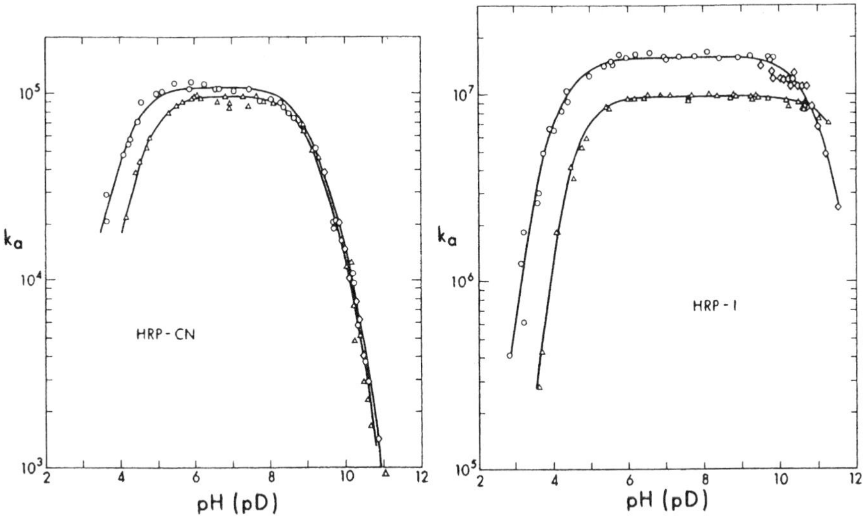

Figure 3.5. pH dependence of cyanide binding and Compound I formation of horseradish peroxidase in H_2O and D_2O. (Reproduced with permission from Dunford HB, Hewson WD, Steiner H. 1978. *Can. J. Chem.* **56**: 2844–2852.)

A possible kinetic ambiguity could occur if an anion bound to a protonated enzyme; in other words if the reacting enzyme is protonated, not the ligand. This alternative mechanism for ligand binding was eliminated in a study of azide binding to chloro- and horseradish peroxidase. The azide binding is so fast that the reaction had be studied using a laser temperature-jump apparatus capable of measurements in nanoseconds. The binding of HN_3 to neutral peroxidase fit the data, and the binding of azide ion to protonated peroxidase could be eliminated because it required a rate constant exceeding the diffusion-controlled limit (Holzwarth et al., 1988).

HORSERADISH PEROXIDASE COMPOUND I (HRP-I)

Kinetics of HRP-I Formation

The kinetics of the conversion of native HRP into HRP-I has been investigated in many laboratories, with good agreement on the results (Dunford et al., 1978; Kato et al., 1984; Ohlsson et al., 1984; Rodriguez-Lopez et al., 1996b). At room temperature a plot of observed rate constant versus $[H_2O_2]$ is linear with zero intercept, indicating that any intermediate species is too short lived to be detected by optical spectroscopy. The rate constant for Compound I formation k_1, obtained from the slope of the linear plot, is $(1.8 \pm 0.2) \times 10^7\ M^{-1}\ s^{-1}$ at 25°C over a wide range of pH (Fig. 3.4). The alkaline transition at pH 11 stops Compound I formation as does an active site protonation at ~pH 2.5. Originally the pK_a value appeared to be 4.0, not 2.5. At pH values below 4 the rate of Compound I formation decreased. It was found that both acetate buffer and nitrate anion used to maintain constant ionic strength were inhibiting Compound I formation at low pH. Nitrate anion binds to an acid form of native HRP (Araiso and Dunford, 1980). Acetate also binds (Carlsson et al., 2005), and is a substrate for reaction with Compound I at low pH (Huang et al., 2004). It appears that the carboxyl group is oxidized to a carboxylic radical which adds to the heme δ-meso position causing enzyme inactivation.

The huge range of the pH-independent rate of formation of Compound I, from pH 2.5 to 11 is, to the best of our knowledge, a record in enzymology. The pK_a value of 2.5 for the distal imidazole ($^+$HIm) of His42 excludes a mechanism for Compound I formation involving $^+$HIm in the resting enzyme (Filizola and Loew, 2000; Rodriguez-López et al., 2001). The Im group is the proton acceptor from H_2O_2 over the entire pH range for Compound I formation.

Tests were made using viscous solvents to determine whether Compound I formation is diffusion controlled. When hydrogen peroxide is used it is not. With the faster reacting *m*-chloroperbenzoic acid the rate of diffusion of reactants is partially rate controlling (Dunford and Hewson, 1977).

Formation of Compound 0 The first evidence for an intermediate species was obtained by measuring Compound I formation as a function of temperature so that the activation energy could be determined (Marklund et al., 1974; Hewson and Dunford, 1975). It is lower than that for a diffusion-controlled reaction, which is impossible for a simple bimolecular reaction. It can be explained in terms of the reversible formation of a precursor complex.

$$\text{HRP} + \text{H}_2\text{O}_2 \underset{k_\text{p}}{\overset{k_\text{p}}{\rightleftarrows}} \text{Precursor complex} \xrightarrow{k_\text{ox}} \text{HRP-I}$$

Thus the overall rate constant k_1 is equal to $K_\text{p} \cdot k_\text{ox}$, where $K_\text{p} = k_\text{p}/k_{-\text{p}}$. It follows that the overall activation energy is given by

$$E_1^\ddagger = \Delta H_\text{p} + E_\text{ox}^\ddagger$$

and a negative value for ΔH_p accounts for the overall low activation energy. Thus physical evidence for the existence of an intermediate was obtained (Jones and Dunford, 1977). At high pH values, in the region where the alkaline transition of HRP occurs, saturation rather than linear kinetics are observed, indicating formation of an intermediate species prior to Compound I formation (Job and Dunford, 1978). Similar evidence was obtained from studies of mutants (Rodriguez-Lopez et al., 1996, 2001).

By working at low temperature in mixed methanol-DMSO-water solvent, kinetic evidence for an intermediate was obtained (Balny et al., 1987). At temperatures below −16°C in methanol-water solvent saturation kinetics were observed and a new species with absorption bands at 330 and 410 nm was detected, labeled Compound 0 (Baek and Van Wart, 1989). Similar precursors of Compound I were observed at low temperatures with use of other hydroperoxides (Baek and Van Wart, 1992). Compound 0 was also produced by the low temperature radiolysis of Compound III (oxyperoxidase; Denisov et al., 2002).

Mechanism of HRP-I Formation

With the establishment of the essential features of the mechanism for Compound I formation in yeast cytochrome *c* peroxidase (Poulos and Kraut, 1980; see Chapter 6), the determination of the crystal structure of recombinant horseradish peroxidase isoenzme C*, and the similarities among all the plant, fungal, and bacterial peroxidases (Welinder, 1985), it was clear that distal His42 and Arg38, and proximal His170 were important catalytic groups for horseradish peroxidase Compound I formation. There is general agreement emerging on the following picture (Hiner et al., 2002). The overall reaction can be broken down, rather arbitrarily, into two steps.

In H—O—O—H the oxygen atom that approaches closer to iron(III) is labeled α and the farther one β.

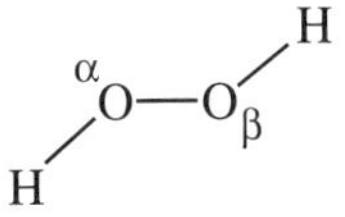

Step 1 is the reversible formation of Compound 0 from hydrogen peroxide and native enzyme, consisting of proton transfer from the α oxygen to the imidazole side chain of distal His42 and binding of the α oxygen to iron(III). Catalysis is therefore promoted by a neutral distal imidazole side chain. If it is protonated, which occurs at pH ~2.5, no Compound I formation can occur.

In step 2 a complex electronic and atomic concerted rearrangement occurs. Transfers of electrons from porphyrin and iron facilitate double bond formation with the α oxygen. Thus the ferryl $Fe^{IV}{=}O$ group and a porphyrin π-cation radical are formed, and the single αO—Oβ bond is broken. The residual ^{-}OH group accepts the proton from distal imidazole forming a water leaving group. A theory of electron density shifts from one side of the heme to the other then back again may account for the efficiency of Compound I formation (see Chapter 10). Step 2 has been assumed to be irreversible but ^{18}O kinetic isotope effects indicate that this is not the case. ^{18}O is incorporated into hydrogen peroxide when it is reacted with horseradish peroxidase (Roth and Cramer, 2008).

Horseradish peroxidase will react stereochemically with racemic secondary hydroperoxides to produce pure *S* hydroperoxides and pure *R* alcohols. Poor enzyme recognition is observed for hydroperoxides with branched aliphatic chains, and none with tertiary hydroperoxides (Adam et al., 1995).

Hydrogen peroxide is generated in the glucose oxidase-glucose reaction. Spectral changes in HRP can be used to quantitate the amount of glucose present (Sanz et al., 2005).

Reactions of HRP-I

Transient state investigations of rates of reduction of HRP-I to HRP-II by a variety of substrates are well documented (Dunford, 1999). In principle they are easier to study than Compound II reactions because preparation of a high yield of HRP-I is simpler. Some of these reactions are so fast that they push the limits of a stopped-flow apparatus. The speed of HRP-I reductions can be correlated by the electron donating ability of the substrate. Thus a linear $\rho\sigma$ Hammett plot is obtained for a series of phenols and aromatic amines (Job and Dunford, 1976). Phenols react by phenolic hydrogen atom donation (electron plus proton) so phenolic radicals are formed. This behavior is observed with vitamin E and its analog Trolox C (Nakamura and Hayashi, 1992). A more complicated explanation for rate correlations assigns a role for enzyme–substrate dissociation constants and reorganization energy for electron transfer (Candeias et al., 1997).

Phenoxyl radicals formed in HRP reactions can dimerize, trimerize, and in the case of *p*-cresol form Pummerer's ketone (Hewson and Dunford, 1976a, 1976b). Assays based on tetraguaiacol formation are unreliable because radicals obtained from guaiacol can undergo many reactions. They will also inactivate HRP (Huang et al., 2005).

Reactions with Tyrosine and Derivatives Tyrosine and diiodotyrosine are oxidized by both Compounds I and II by the same mechanism as other phenols (Ralston and Dunford, 1978, 1980). Chemiluminescence occurs (Totsune et al., 1999). For iodination of tyrosine by HRP, see Chapter 4. A novel di-isodityrosine appears to be important as a plant cell wall cross-link (Brady et al., 1996). Horseradish peroxidase will nitrate tyrosine and tryptophan derivatives in a reaction involving nitrite and hydrogen peroxide (Monzani et al., 2004; Sala et al., 2004). Horseradish peroxidase can oxidize tyrosine residues in other proteins, apparently by transfer of an oxidizing equivalent from the heme edge (Østdal et al., 1999). Oxidation by horseradish peroxidase and subsequent polymerization of tyrosine-containing peptides has been examined (Michon et al., 1997).

HORSERADISH PEROXIDASE COMPOUND II (HRP-II)

The reduction of Compound II to native enzyme, a one-electron reaction as is the conversion of Compound I to Compound II, appears disarmingly simple. Closer examination soon leads to respect for a complicated process. The distal protonated imidazole group donates one proton and the substrate donates another (by an unknown mechanism) to the ferryl oxygen, which is converted into a water leaving group. Substrates that can reduce HRP-I can also reduce HRP-II, but at a rate up to 10-fold slower. Is the slower rate caused by a different site of attack or is it caused by the more complex rearrangement occurring in the heme cavity; reduction of iron and conversion of the ferryl oxygen into a water leaving group? At the present time both explanations may be valid. Reduction rates of HRP-II, as of HRP-I, by families of substrates can be correlated by linear ρσ Hammett plots (Dunford and Adeniran, 1986; Sakurada et al., 1990).

The free radicals formed by substrate oxidations can form colored products. At longer times these can interfere with measurements of interconversion of enzyme species. A clever way of avoiding this problem is to use ascorbate, which is a poor substrate for horseradish peroxidase, as a free radical scavenger (Rodriguez-López et al., 2000). Ascorbate, its free radical monodehydroascorbate, and the two-electron oxidation product dehydroascorbate are all colorless. The reaction of HRP-II with 2,2′-azino-bis-[3-ethylbenzthiazzoline-6-sulfonic acid] (ABTS) was shown to occur via formation of a precursor complex.

SOME DIVERSE APPROACHES TO AN UNDERSTANDING OF HORSERADISH PEROXIDASE

Chemical Modification and Replacement of Heme

There are alternative reaction sites that offer possible mechanisms of reaction: the heme cavity and an exposed heme edge. Hydrazine derivatives are known inhibitors of peroxidases (Ebihara, 1939). Spin trapping showed formation of the hydralazyl radical ($RNHNH^{\bullet}$) (Sinha, 1983). When horseradish peroxidase is reacted with hydrogen peroxide and phenylhydrazine its heme is converted into δ-meso-phenyl and 8-hydroxymethyl derivatives and enzyme activity is greatly reduced (Ator and Ortiz de Montellano, 1987). Similar results were obtained with alkylhydrazines (Ator et al., 1987). Studies on the phenylhydrazine reaction were extended to distal and heme-access residue mutants (Gilfoyle et al., 1996). The results appear to indicate that reducing substrates cannot pass the δ-meso edge of the heme (Ortiz de Montellano, 1992). However, the crystal structures of benzhydroxamic and ferulic acid complexes show the heme cavity is readily accessible (Henriksen et al., 1998, 1999). The suicide inactivation by hydrazine derivatives appears to involve unique reactions initiated by the porphyrin π-cation radical of Compound I, indicating the direction of approach of other substrates but not necessarily their stopping point. Benzhyrazide was also shown to be a peroxidase inhibitor (Aitken et al., 2003). Reaction of HRP with hydrogen peroxide and carboxylic acid at acid pH also leads to inactivation. Carboxylic acid radicals are formed that attach mainly at the δ-meso carbon (Huang et al., 2004). Horseradish peroxidase was reconstituted with 8-(hydroxymethyl) and 8-formylheme derivatives in an attempt to convert the enzyme into a species similar to the mammalian peroxidases (Harris et al., 1993).

The heme *b* was replaced with a heme containing the dimethyl ester of the propionates, leading to a different heme orientation and loss of activity (Adak and Banerjee, 1998). Heme *b* was replaced with mesoporphyrin IX (Herenyi et al., 1998), and naphthalene derivatives (Lasagna et al., 1996) so that fluorescence measurements could be used to investigate porphyrin–protein interactions.

Mutants and Other Peroxidases

The expression of the HRP C gene in *E. coli* (no carbohydrate attachments) by the group of Andrew Smith, and in a baculovirus system (with carbohydrate) by the group of Paul Ortiz de Montellano, opened the way for preparation and study of site-directed mutants. The enzyme expressed in *E. coli* is designated as HRP C*. Enzyme stability decreases in the order: HRP C > HRP C* > HRP C* mutants (Hiner et al., 1995). The horseradish peroxidase gene has also been expressed in transgenic tobacco (Pellegrineschi et al., 1995).

The Phe41Val mutant decreased the rate of reaction with hydrogen peroxide but rates of reaction of *p*-aminobenzoic acid with Compounds I and II were slightly increased (Smith et al., 1992). Ferrous horseradish peroxidase His42Arg, His42Lys, and Arg38Lys mutants, and carbon monoxide and cyanide complexes were studied using laser photolysis to study recombination rates (Meunier et al., 1995). His42Ala, His42Val, and Phe41Ala mutants showed the essential role of His42 in catalysis (Newmyer and Ortiz de Montellano, 1995). The His42Leu and Arg38Leu mutants also showed greatly reduced activity (Rodriguez-López et al., 1996a). Detailed kinetic studies showed that reversibly formed intermediates could be detected in Compound II reactions of distal mutants (Rodriguez-Lopez et al., 2000), something that was observed in Compound II reactions of *p*-cresol at high pH (Critchlow and Dunford, 1972), and of di-(*N*-acetyl-L-tyrosine) (Wang et al., 1999). Compound 0 could be detected in distal mutants (Rodriguez-López et al., 1996a, 2001).

Compound I with a porphyrin cation radical was subsequently converted to protein radical in the Phe172Tyr mutant (Miller et al., 1995). Degradation of the heme occurred when one attempted to form Compound I in the proximal His-170 → Ala mutant, but activity was restored upon addition of saturating amounts of imidazole (Newmyer et al., 1996). Similarly, the catalytic activity of the distal His42Ala mutant could be restored by added imidazole (Newmyer and Ortiz de Montellano, 1996). Some activity remains after replacement of His42 by Glu, a substitution suggested by the active site structure of chloroperoxidase (Tanaka et al., 1997).

The role of the Asn70 histidine hydrogen bond was probed using Asn70Val and Asn70Asp mutants. Reaction rates with phenols were greatly reduced, whereas ABTS reactivity was increased (Nagano et al., 1996). Tryptophanless HRP has been constructed using a Trp117Phe mutant. Folding and stability were affected (Gazaryan et al., 1999). The decrease in reactivity of the His42Ala mutant could be partially offset by a concurrent Asn70Asp mutation (Savenkova and Ortiz de Montellano, 1998).

Rates of sulfoxidation of thioanisole and styrene epoxidation are increased in Arg38 mutants, particularly in the combined Arg38His/His42Val mutant (Savenkova et al., 1998). The heme group in mammalian peroxidases is covalently bonded to the apoprotein in an autocatalytic reaction, which was duplicated by introducing aspartate and glutamic acid residues into horseradish peroxidase. No covalent bonds formed until hydrogen peroxide was added (Colas and Ortiz de Montellano, 2004). Formation of a single heme-protein covalent bond in horseradish peroxidase was sufficient to protect its porphyrin vinyl groups from attack by thiocyanate, bromide, and chloride oxidation products, mimicking the protection that exists in mammalian peroxidases (Huang et al., 2006). Peroxidase activity was demonstrated in a heme-antibody complex (Cochran and Schultz, 1990; Yin et al., 2004) and in heme proteins obtained from a designed combinatorial library (Moffett et al., 2000).

Reactions of Unusual Substrates

HRP was used to help unravel cytochrome P-450 reaction mechanisms. An *N*-cyclopropylamine and substrates containing an *N*-carboxymethyl group were reacted with the HRP-hydrogen peroxide system. *N*-dealkylation occurred and aminium ion intermediates were formed (Shaffer et al., 2001; Totah and Hanzlik, 2001). *N*-isopropyl-*N*-phenyl glycine is decarboxylated in its peroxidase reaction (Totah and Hanzlik, 2002).

Capsaicinoids, the pungent ingredient of hot pepper, react with HRP in a mechanism in which Compound II reduction is the rate-determining step. Ascorbate scavenged colored free radical products in a reaction that may occur in the *Capsicum* fruits (Goodwin and Hertwig, 2003).

Chlorpromazine as a peroxidase substrate stimulates attack on isoniazid, an example of redox mediation (Goodwin et al., 1997). Promethazine is oxidized by free radicals generated from the tetramethylbenzidine reaction with HRP, another example of redox mediation (Olorunniji et al., 2000).

Catharanthine and vindoline are indole alkaloids that are precursors of dimeric anticancer drugs. Their reactions with horseradish peroxidase have been studied (Sottomayor et al., 1997). Deuteroporphyrin IX is oxidized into chlorines by horseradish peroxidase (Dayan et al., 1998). Hydroxyurea is used in an approved treatment for sickle cell anemia and a number of cancers. It reacts with HRP to form nitric oxide (J. Huang et al., 2002). Glycosylated flavonols are found in vegetables and fruits. Oxidation of the flavanol quercetin by horseradish peroxide was studied. Entirely different products were obtained in the presence and absence of glutathione (Awad et al., 2000).

α-Tocopherol was reacted with horseradish peroxidase and the initiation of low-density lipoprotein peroxidation by the α-tochoperoxyl radical was studied (Witting et al., 1997). Peroxyl radicals generated by peroxidases have been claimed to cause oxidative damage to DNA (Adam et al., 2000), a claim that has been contradicted (Shane and Ingold, 2002).

Reactions in Different Solvent Systems and on Electrodes

The versatility of horseradish peroxidase is demonstrated by its reactivity in reversed micelles (Gazaryan et al., 1997; Holzwarth et al., 1991). Aqueous mixtures with dioxane, methanol, and acetonitrile cause partial but not complete denaturation of HRP (Ryu and Dordick, 1992). Methanol, acetone, and ethylene glycol greatly reduced the rate of Compound II reaction with *p*-cresol, but Compound I formation and reaction was only moderately affected (Sato et al., 1995). Horseradish peroxidase encapsulated in sol-gel material is catalytically active and can be reused, providing a solid-state catalytic material (Smith et al., 2002). Polyethylene glycol was covalently bound to HRP and the modified enzyme was reacted in organic solvents. Compound 0 is stabilized in chlorobenzene at –20°C (Ozaki et al., 1998). The use of aqueous sucrose and glycerol solutions to slow diffusion rates has already been discussed (Dunford and Hewson, 1977).

Plants in an environment of high salt concentration undergo osmotic stress. Hydrogen peroxide concentration and peroxidase activity are increased. Osmoprotectants include zwitterions, sugar alcohols, and quaternary ammonium compounds. The protective effect of choline on salt-induced superactivity of horseradish peroxidase has been examined (Pinna et al., 2005).

Use of enzymes in electrode reactions have been studied intensively in the laboratory of Lo Gorton. As one example, immobilization of horseradish peroxidase on gold electrodes is cited (Abad et al., 2002). Another example is entrapment of horseradish peroxidase in a wired polymer matrix (Calvente et al., 2003). Caffeic acid on modified gold electrodes was polymerized by horseradish peroxidase and hydrogen peroxide (Xu et al., 2005).

Reactions of Single Peroxidase Molecules

By use of microfabricated arrays of femtoliter chambers, reactions of single molecules of horseradish peroxidase have been studied (Rondelez et al., 2005; Gorris and Walt, 2009). The substrate Amplex Red reacts as do other phenols with horseradish peroxidase. Its oxidation leads to free radicals that react in a nonenzymatic reaction to form a flouresecent product. The steady-state results of this reaction showed clearly that individual molecules of horseradish peroxidase react at different rates. A distribution pattern of reactivities is observed (Gorris and Walt, 2009). Thus bulk kinetic measurements of an elementary process produce an averaged rate constant, and we shall have to learn to live with the fact that not a single mechanism, but a family of mechanisms, is required to account for enzyme behavior (Benkovic et al., 2008; Hammes, 2008).

REFERENCES

Aasa R, Vänngård T, Dunford HB. 1975. EPR studies on Compound I of horseradish peroxidase. *Biochim. Biophys. Acta* **391**: 259–264.

Abad JM, Vélez M, Santamaria C, Guisán JM, Matheus PR, Vásquez L, Gazaryan I, Gorton L, Gibson T, Fernández VM. 2002. Immobilization of peroxidase glycoprotein on gold electrodes modified with mixed epoxy-boronic acid monolayers. *J. Am. Chem. Soc.* **124**: 12845–12853.

Adak S, Banerjee RK. 1998. Haem propionates control oxidative and reductive activities of horseradish peroxidase by maintaining the correct orientation of the haem. *Biochem. J.* **334**: 51–56.

Adak S, Mazmunder A, Banerjee RK. 1996. Probing the active site residues in aromatic donor oxidation in horseradish peroxidase: involvement of an arginine and tyrosine residue in aromatic donor binding. *Biochem. J.* **314**: 953–991.

Adam W, Hoch U, Lazarus M, Saha-Möller CR, Schreier P. 1995. Enzyme-catalyzed asymmetric synthesis: kinetic resolution of racemic hydroperoxides by enantioselective reduction to alcohols with horseradish peroxidase. *J. Am. Chem. Soc.* **117**: 11898–11901.

Adam W, Kurz A, Saha-Möller CR. 2000. Peroxidase-catalyzed oxidative damage of DNA and 2′-deoxyguanosine by model compounds of lipid hydroperoxides: involvement of peroxyl radicals. *Chem. Res. Toxicol.* **13**: 1199–1207.

Adediran SA, Dunford HB. 1983. Structure of horseradish peroxidase Compound I. Kinetic evidence for the incorporation of one oxygen atom from the oxidizing substrate into the enzyme. *Eur. J. Biochem.* **132**: 147–150.

Aitken SM, Ouellet M, Percival MD, English AM. 2003. Mechanism of horseradish peroxidase inactivation by benzhydrazide: a critical evaluation of arylhydrazides as peroxidase inhibitors. *Biochem. J.* **375**: 613–621.

Aitken SM, Turnbull JL, Percival MD, English AM. 2001. Thermodynamic analysis of the binding of aromatic hydroxaamic analogues to ferric horseradish peroxidase. *Biochemistry* **40**: 13980–13989.

Araiso T, Dunford HB. 1980. Horseradish peroxidase. XLI. Complex formation with nitrate and its effect upon Compound I formation. *Biochem. Biophys. Res. Commun.* **94**: 1177–1182.

Ator MA, David SK, Ortiz de Montellano PR. 1987. Structure and catalytic mechanism of horseradish peroxidase. *J. Biol. Chem.* **262**: 14954–14960.

Ator MA, Ortiz de Montellano PR. 1987. Protein control of prosthetic heme reactivity. *J. Biol. Chem.* **262**: 1542–1551.

Awad HM, Boersma MG, Vervoort J, Rietjens IMCM. 2000. Peroxidase-catalyzed formation of quercetin quinine methide-glutathione adducts. *Arch. Biochem. Biophys.* **378**: 224–233.

Baek HH, Van Wart HE. 1989. Elementary steps in the formation of horseradish peroxidase Compound I: direct observation of Compound 0, a new intermediate with a hyperporphyrin structure. *Biochemistry* **28**: 5714–5719.

Baek HH, Van Wart HE. 1992. Elementary steps in the reaction of horseradish peroxidase with several peroxides: kinetics and thermodynamics of formation of Compound 0 and Compound I. *J. Am. Chem. Soc.* **114**: 718–725.

Balny C, Travers F, Barman T, Douzou P. 1987. Thermodynamics of the two step formation of horseradish peroxidase Compound I. *Eur. Biophys. J.* **14**: 375–383.

Ban N, van Huystee RB, Day J, Greenwood A, Larson S, Esnault R, McPherson A. 1992. Preliminary crystallographic study of peanut peroxidase. *Acta Cryst. B* **48**: 109–111.

Bartonek-Roxå E, Eriksson H, Mattiasson B. 1991. The cDNA sequence of a neutral horseradish peroxidase. *Biochim. Biophys. Acta* **1088**: 245–250.

Benkovic SJ, Hammes GG, Hammes-Schiffer S. 2008. Free-energy landscape of enzyme catalysis. *Biochemistry* **47**: 3317–3321.

Berglund GI, Carlsson GH, Smith AT, Szöke H, Henriksen A, Hajdu J. 2002. The catalytic pathway of horseradish peroxidase at high resolution. *Nature* **417**: 463–468.

Berglund J, Pascher T, Winkler JR, Gray HB. 1997. Photoinduced oxidation of horseradish peroxidase. *J. Am. Chem. Soc.* **119**: 2464–2469.

Brady JD, Sadler IH, Fry SC. 1996. Di-isodityrosine, a novel tetrameric derivative of tyrosine in plant cell wall proteins: a new potential cross-link. *Biochem. J.* **315**: 323–327.

Buffard D, Breda C, van Huystee RB, Asemota O, Pierre M, Ha DBD, Esnault R. 1990. Molecular cloning of complementary DNAs encoding two cationic peroxidases from cultivated peanut cells. *Proc. Natl. Acad. Sci. USA* **87**: 8874–8878.

Calvente JJ, Narváez A, Dominguez E, Andreu R. 2003. Kinetic analysis of wired enzyme electrodes: application to horseradish peroxidase entrapped in a redox polymer matrix. *J. Phys. Chem. B* **107**: 6629–6643.

Candeias LP, Folkes LK, Wardman P. 1997. Factors controlling the substrate specificity of peroxidases: kinetics and thermodynamics of the reaction of horseradish peroxidase Compound I with phenols and indole-3-acetic acids. *Biochemistry* **36**: 7081–7085.

Carlsson GH, Nicholls P, Svistunenko D, Berglund GI, Hajdu J. 2005. Complexes of horseradish peroxidase with formate, acetate, and carbon monoxide. *Biochemistry* **44**: 635–642.

Chakrabarti A, Basak S. 1996. Structural alterations of horseradish peroxidase in the presence of low concentrations of guanidinium chloride. *Eur. J. Biochem.* **241**: 462–467.

Chattopadhyay K, Mazumdar S. 2000. Structural and conformational stability of horseradish peroxidase: effect of temperature and pH. *Biochemistry* **39**: 263–270.

Cochran AG, Schultz PG. 1990. Peroxidase activity of an antibody-heme complex. *J. Am. Chem. Soc.* **112**: 9414–9415.

Colas C, Ortiz de Montellano PR. 2004. Horseradish peroxidase mutants that autocatalytically modify their heme prosthetic group. Insights into mammalian peroxidase heme protein covalent bonds. *J. Biol. Chem.* **279**: 24131–24140.

Cotton ML, Dunford HB, Raycheba MT. 1973. Studies on horseradish peroxidase. XIII. The kinetic effect of cyanide on the oxidation-reduction cycle. *Can. J. Biochem.* **51**: 627–631.

Critchlow JE, Dunford HB. 1972. Studies on horseradish peroxidase. IX. Kinetics of the oxidation of *p*-cresol by Compound II. *J. Biol. Chem.* **247**: 3703–3713.

Dayan FE, Duke SO, Faibis V, Jacobs JM, Jacobs NJ. 1998. Horseradish peroxidase-dependent oxidation of deuteroporphyrin IX into chlorines. *Arch. Biochem. Biophys.* **351**: 27–34.

Denisov IG, Makris TM, Sligar SG. 2002. Formation and decay of hydroperoxo-ferric heme complex in horseradish peroxidase studied by cryoradiolysis. *J. Biol. Chem.* **277**: 42706–42710.

Dolphin D, Forman A, Borg DC, Fajer J, Felton RH. 1971. Compounds I of catalase and horse radish peroxidase: π-cation radicals. *Proc. Natl. Acad. Sci. USA* **68**: 614–618.

Dunford HB. 1991. Horseradish peroxidase: structure and kinetic properties. In: Everse J, Everse KE, Grisham MB, editors, *Peroxidases in Chemistry and Biology*, Vol. II. CRC Press, Boca Raton, FL, pp. 1–24.

Dunford HB. 1999. *Heme Peroxidases*. Wiley-VCH, New York.

Dunford HB, Adeniran AJ. 1986. Hammett $\rho\sigma$ correlation for reactions of horseradish peroxidase Compound II with phenols. *Arch. Biochem. Biophys.* **251**: 536–542.

Dunford HB, Alberty RA. 1967. The kinetics of fluoride binding to ferric horseradish peroxidase. *Biochemistry* **6**: 447–451.

Dunford HB, Hewson WD. 1977. Effect of mixed solvents on the formation of horseradish peroxidase Compound I: the importance of diffusion-controlled reactions. *Biochemistry* **16**: 2949–2957.

Dunford HB, Hewson WD, Steiner H. 1978. Horseradish peroxidase. XXIX. Reactions in water and deuterium oxide: cyanide binding, Compound I formation, and reactions of Compounds I and II with ferrocyanide. *Can. J. Chem.* **56**: 2844–2852.

Dunford HB, Ralston IM. 1983. On the mechanism of iodination of tyrosine. *Biochem. Biophys. Res. Commun.* **116**: 639–643.

Ebihara T. 1939. Untersuchungen über ascorbinsäure-oxidase. II. Reinigung und eigenschaften der ascorbinsäure-oxidase. *J. Biochem. (Tokyo)* **29**: 199–215.

Ellis WD, Dunford HB. 1968. The kinetics of cyanide and fluoride binding by ferric horseradish peroxidase. *Biochemistry* **7**: 2054–2062.

Feis A, Marzocchi MP, Paoli M, Smulevich G. 1994. Spin state and axial ligand bonding in the hydroxide complexes of metmyoglobin, methemoglobin, and horseradish peroxidase at room and low temperatures. *Biochemistry* **33**: 4577–4583.

Filizola M, Loew GH. 2000. Role of protein environment in horseradish peroxidase Compound I formation: molecular dynamics simulations of horseradish peroxidase-HOOH complex. *J. Am. Chem. Soc.* **122**: 18–25.

Fujiyama K, Takemura H, Shibayama S, Kobayashi K, Choi J-K, Shinmyo A, Takano M, Yamada Y, Okada H. 1988. Structure of the horseradish peroxidase isoenzyme C genes. *Eur. J. Biochem.* **173**: 681–687.

Fujiyama K, Takemura H, Shinmyo A, Okada H, Takano M. 1990. Genomic DNA structure of two new horseradish-peroxidase-encoding genes. *Gene* **89**: 163–169.

Gajhede M, Schuller DJ, Henriksen A, Smith AT, Poulos TL. 1997. Crystal structure of horseradish peroxidase C at 2.15 Å resolution. *Nature Struct. Biol.* **4**:1032–1038.

Gazaryan IG, Chubar TA, Ignatenko OV, Mareeva EA, Orlova MA, Kapeliuch YL, Savitsky PA, Rojkova AM, Tishkov VI. 1999. Tryptophanless recombinant horseradish peroxidase: stability and catalytic properties. *Biochem. Biophys. Res. Commun.* **262**: 297–301.

Gazaryan IG, Klyachko NL, Dulkis YK, Ouporov IV, Levashov AV. 1997. Formation and properties of dimeric recombinant horseradish peroxidase in a system of reversed micelles. *Biochem. J.* **328**: 643–647.

George P. 1952a. Chemical nature of the secondary hydrogen peroxide compound formed by cytochrome *c* peroxidase and horseradish peroxidase. *Nature* **169**: 612–613.

George P. 1952b. The specific reactions of iron in some hemoproteins. *Adv. Catalysis* **4**: 367–428.

George P. 1953. Intermediate compounds formed when horseradish peroxidase reacts with potassium molybdicycanide and potassium chloriridate. *Science* **117**: 220–221.

Gilfoyle DJ, Rodriguez-Lopez JN, Smith AT. 1996. Probing the aromatic-donor-binding site of horseradish peroxidase using site-directed mutagenesis and the suicide substrate phenylhydrazine. *Eur. J. Biochem.* **236**: 714–722.

Goodwin DC, Grover TA, Aust SD. 1997. Role of efficient substrates in enhancement of peroxidase-catalyzed oxidations. *Biochemistry* **36**: 139–147.

Goodwin DC, Hertwig KM. 2003. Peroxidase-catalyzed oxidation of capsaicinoids: steady-state and transient-state kinetic studies. *Arch. Biochem. Biophys.* **417**: 18–26.

Gorris HH, Walt DR. 2009. Mechanistic aspects of horseradish peroxidase elucidated through single molecule studies. *J. Am. Chem. Soc.* **131**: 6277–6282.

Hammes GG. 2008. How do enzymes really work? *J. Biol. Chem.* **283**: 22337–22346.

Haque ME, Debnath D, Basak S, Chakrabarti A. 1999. Structural changes of horseradish peroxidase in presence of low concentrations of urea. *Eur. J. Biochem.* **259**: 269–274.

Harris RZ, Liddell PA, Smith KM, Ortiz de Montellano PR. 1993. Catalytic properties of horseradish peroxidase reconstituted with 8-(hydroxymethyl)- and 8-formylheme derivatives. *Biochemistry* **32**: 3658–3663.

Hartmann C, Ortiz de Montellano P. 1992. Baculovirus expression and characterization of catalytically active horseradish peroxidase. *Arch. Biochem. Biophys.* **297**: 61–72.

Hashimoto S, Tatsuno Y, Kitagawa T. 1986. Resonance Raman evidence for oxygen exchange between the $Fe^{IV}=O$ heme and bulk water during enzymic catalysis of horseradish peroxidase and its relation with the heme-linked ionization. *Proc. Natl. Acad. Sci. USA* **83**: 2417–2421.

Hayashi Y, Yamazaki I. 1979. The oxidation-reduction potentials of Compound I/Compound II and Compound II/ferric couples of horseradish peroxidases A_2 and C. *J. Biol. Chem.* **254**: 9101–9106.

Henriksen A, Schuller DJ, Meno K, Welinder KG, Smith AT, Gajhede M. 1998. Structural interactions between horseradish peroxidase C and the substrate benzhydroxamic acid determined by X-ray crystallography. *Biochemistry* **37**: 8054–8060.

Henriksen A, Smith AT, Gajhede M. 1999. The structures of horseradish peroxidase C-ferulic acid and the ternary complex with cyanide suggest how peroxidases oxidize small phenolic substrates. *J. Biol. Chem.* **274**: 35005–35011.

Herenyi L, Suialu A, Mauring K, Kis-Petik K, Fidy J, Kikas J. 1998. Variety in the coupling of mesoporphyrin IX to apohorseradish peroxidase C studied by energy selected fluorescence excitation and vibronic hole burning spectroscopy. *J. Phys. Chem. B.* **102**: 5932–5940.

Hewson WD, Dunford HB. 1975. Horseradish peroxidase. XVIII. The Arrhenius activation energy for the formation of Compound I. *Can. J. Chem.* **53**: 1928–1932.

Hewson WD, Dunford HB. 1976a. Oxidation of *p*-cresol by horseradish peroxidase Compound I. *J. Biol. Chem.* **251**: 6036–6042.

Hewson WD, Dunford HB. 1976b. Stoichiometry of the reaction between horseradish peroxidase and *p*-cresol. *J. Biol. Chem.* **251**: 6036–6042.

Hiner ANP, Hernández-Ruíz J, García-Cánovas F, Smith AT, Arnao MB, Acosta M. 1995. A comparative study of the inactivation of wild-type, recombinant and two mutant horseradish peroxidase isoenzymes C by hydrogen peroxide and *m*-chloroperbenzoic acid. *Eur. J. Biochem.* **234**: 506–512.

Hiner ANP, Raven EL, Thorneley RNF, Garcia-Cánovas F, Rodriguez-López JN. 2002. Mechanisms of Compound I formation in heme peroxidases. *J. Inorg. Biochem.* **91**: 27–34.

Holzwarth J, Jobe D, Dunford HB. 1991. Kinetics of azide binding to chloroperoxidase and horseradish peroxidase in water and reversed micelles. *Czech. J. Phys.* **41**: 293–295.

Holzwarth JF, Meyer F, Pickard M, Dunford HB. 1988. Mechanism of azide binding to chloroperoxidase and horseradish peroxidase: use of an iodine laser temperature jump apparatus. *Biochemistry* **27**: 6628–6633.

Huang J, Sommers EM, Kim-Shapiro DB, King SB. 2002. Horseradish peroxidase catalyzed nitric oxide formation from hydroxyurea. *J. Am. Chem. Soc.* **124**: 3473–3480.

Huang L, Colas C, Ortiz de Montellano PR. 2004. Oxidation of carboxylic acids by horseradish peroxidase results in prosthetic heme modification and inactivation. *J. Am. Chem. Soc.* **126**: 12865–12873.

Huang L, Wojciechowski G, Ortiz de Montellano PR. 2006. Role of heme-protein covalent bonds in mammalian peroxidases: protection of the heme by a single engineered heme-protein link in horseradish peroxidase. *J. Biol. Chem.* **281**: 18983–18988.

Huang Q, Huang Q, Pinto RA, Griebenow K, Schweitzer-Stenner R, Weber WJ Jr. 2005. Inactivation of horseradish peroxidase by phenoxyl radical attack. *J. Am. Chem. Soc.* **127**: 1431–1437.

Hungerford G, Rei A, Ferreira IC. 2005. Studies on the interaction of Nile red with horseradish peroxidase in solution. *FEBS J.* **272**: 6161–6169.

Job D, Dunford HB. 1976. Substituent effect on the oxidation of phenols and aromatic amines by horseradish peroxidase Compound I. *Eur. J. Biochem.* **66**: 607–614.

Job D, Dunford HB. 1977. The alkaline transition of turnip peroxidases. *Arch. Biochem. Biophys.* **179**: 95–99.

Job D, Dunford HB. 1978. Horseradish peroxidase. XXVIII. Formation and reactivity of the alkaline form. Evidence for an enzyme-substrate complex in Compound I formation. *Can. J. Chem.* **56**: 1327–1334.

Jones P, Dunford HB. 1977. On the mechanism of Compound I formation from peroxidases and catalases. *J. Theor. Biol.* **69**: 457–470.

Kato M, Aibara S, Morita Y, Nakatani H, Hiromi K. 1984. Comparative studies on kinetic behavior of horseradish peroxidase isoenzymes. *J. Biochem.* **95**: 861–870.

Lanir A, Schejter A. 1975. Nuclear magnetic resonance evidence for the absence of iron coordinated water in horseradish peroxidase. *Biochem. Biophys. Res. Commun.* **62**: 199–203.

Lasagna M, Vargas V, Jameson DM, Brunet JE. 1996. Spectral properties of environmentally sensitive probes associated with horseradish peroxidase. *Biochemistry* **35**: 973–979.

Marklund S, Ohlsson P-I, Opara A, Paul K-G. 1974. The substrate profiles of the acidic and slightly basic horseradish peroxidases. *Biochim. Biophys. Acta* **350**: 304–313.

McIntosh AR, Stillman MJ. 1977. Electron-paramagnetic resonance studies on a photochemically produced species of horseradish peroxidase compound I. *Biochem. J.* **167**: 31–37.

Meunier B, Rodriguez-Lopez JN, Smith AT, Thorneley RNF, Rich PR. 1995. Laser photolysis behavior of ferrous horseradish peroxidase with carbon monoxide and cyanide: effects of mutations in the distal heme pocket. *Biochemistry* **34**: 14687–14692.

Michon T, Chenu M, Kellershon N, Desmadril M, Guéguen J. 1997. Horseradish peroxidase oxidation of tyrosine-containing peptides and their subsequent polymerization: a kinetic study. *Biochemistry* **36**: 8504–8513.

Miller VP, Goodin DB, Friedman AE, Hartmann C, Ortiz de Montellano PR. 1995. Horseradish peroxidase $Phe^{172} \rightarrow Tyr$ mutant: sequential formation of Compound I with a porphyrin radical cation and a protein radical. *J. Biol. Chem.* **270**: 18413–18419.

Moffet DA, Certain LK, Smith AJ, Kessel AJ, Beckwith KA, Hecht MH. 2000. Peroxidase activity in heme proteins derived from a designed combinatorial library. *J. Am. Chem. Soc.* **122**: 7612–7613.

Monzani E, Roncone R, Galliano M, Koppenol WH, Casella L. 2004. Mechanistic insight into the peroxidase catalyzed nitration of tyrosine derivatives by nitrite and hydrogen peroxide. *Eur. J. Biochem.* **271**: 895–906.

Nadezhdin AD, Dunford HB. 1979. On the mechanism of photolysis of Compounds I and II of horseradish peroxidase at 77 K. *Photochem. Photobiol.* **29**: 899–903.

Nagano S, Tanaka M, Ishimori K, Watanabe Y, Morishima I. 1996. Catalytic roles of the distal asparagine-histidine couple in peroxidases. *Biochemistry* **35**: 14251–14258.

Nakamura M, Hayashi T. 1992. Oxidation mechanism of vitamin E analogue (Trolox C, 6-hydroxy-2,2,5,7,8-pentamethylchroman) and vitamin E by horseradish peroxidase and myoglobin. *Arch. Biochem. Biophys.* **299**: 313–319.

Newmyer SL, Ortiz de Montellano PR.1995. Horseradish peroxidase His-42 → Ala, His-42 → Val, and Phe-41 → Ala mutants. Histidine catalysis and control of substrate access to the heme iron. *J. Biol. Chem.* **270**: 19430–19438.

Newmyer SL, Ortiz de Montellano PR.1996. Rescue of the catalytic activity of an H42A mutant of horseradish peroxidase by exogenous imidazoles. *J. Biol. Chem.* **271**: 14891–14896.

Newmyer SL, Sun J, Loehr TM, Ortiz de Montellano PR. 1996. Rescue of the horseradish peroxidase His-170 → Ala mutant activity by imidazole: importance of proximal ligand tethering. *Biochemistry* **35**: 12788–12795.

Ohlsson P-I, Paul K-G. 1976. The molar absorptivity of horseradish peroxidase. *Acta Chem. Scand.* **B30**: 373–375.

Ohlsson P-I, Paul K-G, Wold S. 1984. The formation of Compound I of lactoperoxidase and horseradish peroxidase. A comparison. *Acta Chem. Scand.* **B38**: 853–859.

Olorunniji FJ, Malomo SO, Adediran SA, Odutuga AA. 2000. Promethazine oxidation by redox mediation in peroxidase reactions. *Arch. Biochem. Biophys.* **380**: 251–256.

Ortiz de Montellano PR. 1992. Catalytic sites of hemoprotein peroxidases. *Annu. Rev. Pharmacol. Toxicol.* **32**: 89–107.

Østdal H, Andersen HJ, Davies MJ.1999. Formation of long-lived radicals on proteins by radical transfer from heme enzymes: a common process? *Arch. Biochem. Biophys.* **362**: 105–112.

Oudgenoeg G, Dirksen E, Ingemann S, Hilhrst R, Gruppen H, Boeriu CG, Piersma SR, van Berkel WJH, Laane C, Voragen AGJ. 2002. Horseradish peroxidase-catalyzed oligomerization of ferulic acid on a template of a tyrosine-containing peptide. *J. Biol. Chem.* **277**: 21332–21340.

Ozaki S, Inada Y, Watanabe Y. 1998. Characterization of polyethylene glycolated horseradish peroxidase in organic solvents: generation and stabilization of transient catalytic intermediates at low temperature. *J. Am. Chem. Soc.* **120**: 8020–8025.

Pappa HS, Cass AEG. 1993. A step towards understanding the folding mechanism of horseradish peroxidase: tryptophan fluorescence and circular dichroism equilibrium studies. *Eur. J. Biochem.* **212**: 227–235.

Paul K-G, Ohlsson P-I. 1978. Equilibria between horseradish peroxidase and aromatic donors. *Acta Chem. Scand. B.* **32**: 395–404.

Paul K-G, Theorell H, Åkeson A. 1953. The molar light absorption of pyridine ferroprotoporphyrin (pyridine haemochromogen). *Acta Chem. Scand.* **7**: 1284–1287.

Pellegrineschi A, Kis M, Dix I, Kavanagh TA, Dix PJ. 1995. Expression of horseradish peroxidase in transgenic tobacco. 1995. *Biochem. Soc. Trans.* **23**: 247–250.

Pina DG, Shnyrova AV, Gavilanes F, Rodriguez A, Leal F, Roig MG, Sakharov IY, Zhadan GG, Villar E, Shnyrov VL. 2001. Thermally induced conformational changes in horseradish peroxidase. *Eur. J. Biochem.* **268**: 120–126.

Pinna MC, Bauduin P, Touraud D, Monduzzi M, Ninham BW, Kunz W. 2005. Hofmeister effects in biology: effect of choline addition on the salt-induced super activity of horseradish peroxidase and its implication for the salt resistance of plants. *J. Phys. Chem. B* **109**: 16511–16514.

Poulos TL, Kraut J. 1980. The stereochemistry of peroxidase catalysis. *J. Biol. Chem.* **255**: 8199–8205.

Ralston I, Dunford HB. 1978. Horseradish peroxidase. XXXII. pH dependence of the oxidation of L-(−)-tyrosine by Compound I. *Can. J. Biochem.* **56**: 1115–1119.

Ralston I, Dunford HB. 1980. Horseradish peroxidase. XLII. Oxidations of L-tyrosine and 3,5-diiodo-L-tyrosine by Compound II. *Can. J. Biochem.* **58**: 1270–1276.

Rodriguez-López JN, Gilabert MA, Tudela J, Thorneley RNF, Garcia-Cánovas F. 2000. Reactivity of horseradish peroxidase Compound II toward substrates: kinetic evidence for a two-step mechanism. *Biochemistry* **39**: 13201–13209.

Rodriguez-López JN, Lowe DJ, Hernández-Ruiz J, Hiner ANP, Garcia-Cánovas F, Thorneley RNF. 2001. Mechanism of reaction of hydrogen peroxide with horseradish peroxidase: identification of intermediates in the catalytic cycle. *J. Am. Chem. Soc.* **123**: 11838–11847.

Rodriguez-López JN, Smith AT, Thorneley RNF. 1996a. Role of arginine 38 in horseradish peroxidase. A critical residue for substrate binding and catalysis. *J. Biol. Chem.* **271**: 4023–4030.

Rodriguez-López JN, Smith AT, Thorneley RNF. 1996b. Recombinant horseradish peroxidase isoenzyme C: the effect of distal haem cavity mutations (His42 → Leu and Arg38 → Leu) on Compound I formation and substrate binding. *J. Biol. Chem.* **271**: 4023–4030.

Rondelez Y, Tresset G, Tabata KV, Arata H, Fujita H. 2005. Microfabricated arrays of femtoliter chambers allow single molecule enzymology. *Nature Biotech.* **23**: 361–365.

Roth JP, Cramer CJ. 2008. Direct examination of H_2O_2 activation by a heme peroxidase. *J. Am. Chem. Soc.* **130**: 7802–7803.

Ryu K, Dordick JS. 1992. How do organic solvents affect peroxidase structure and function? *Biochemistry* **31**: 2588–2598.

Sakurada J, Sekiguchi R, Sato K, Hosoya T. 1990. Kinetic and molecular orbital studies on the rate of oxidation of monosubstituted phenols and anilines by horseradish peroxidase Compound II. *Biochemistry* **29**: 4093–4098.

Sala A, Nicolis S, Roncone R, Casella L. Monzani E., 2004. Peroxidase catalyzed nitration of trytophan derivatives: mechanism, products and comparison with chemical nitrating reagents. *Eur. J. Biochem.* **271**: 2841–2852.

Sanz V, de Marcos S, Castillo JR, Galbán J. 2005. Application of molecular absorption properties of horseradish peroxidase for self-indicating enzymatic interactions and analytical methods. *J. Am. Chem. Soc.* **127**: 1038–1048.

Sato K, Hasumi H, Tsukidate A, Sakurada J, Nakamura S, Hosoya T. 1995. Effect of mixed solvents on the three elementary steps in the reactions of horseradish peroxidase and lactoperoxidase. *Biochim. Biophys. Acta* **1253**: 94–102.

Savenkova MI, Kuo JM, Ortiz de Montellano PR. 1998. Improvement of peroxygenase activity by relocation of a catalytic histidine within the active site of horseradish peroxidase. *Biochemistry* **37**: 10828–10836.

Savenkova MI, Ortiz de Montellano PR. 1998. Horseradish peroxidase: partial rescue of the His-42 → Ala mutant by a concurrent Asn-70 → Asp mutation. *Arch. Biochem. Biophys.* **351**: 286–293.

Schonbaum GR and Lo S. 1972. Interaction of peroxidases with aromatic peracids and alkyl peroxides. Product analysis. *J. Biol. Chem.* **247**: 3353–3360.

Schuller DJ, Ban N, Van Huystee RB, McPherson A, Poulos TL. 1996. The crystal structure of peanut peroxidase. *Structure* **4**: 311–321.

Schulz CE, Devaney PW, Winkler H, Debrunner PG, Doan N, Chiang R, Rutter R, Hager LP. 1979. Horseradish peroxidase Compound I: evidence for spin coupling between the heme iron and a "free" radical. *FEBS Lett.* **103**: 102–105.

Shaffer CL, Morton MD, Hanzlik RP. 2001. *N*-Dealkylation of an *N*-cyclopropylamine by horseradish peroxidase: fate of the cyclopropyl group. *J. Am. Chem. Soc.* **123**: 8502–8508.

Shane RA, Ingold KU. 2002. Cleavage of supercoiled DNA by horseradish peroxidase plus *tert*-butyl hydroperoxide is not due to *tert*-butylperoxyl radicals. *Chem. Res. Toxicol.* **15**: 1324–1329.

Sinha BK. 1983. Enzymatic activation of hydrazine derivatives. *J. Biol. Chem.* **258**: 796–801.

Sitter AJ, Shifflett JR, Terner, J. 1988. Resonance Raman spectroscopic evidence for heme iron-hydroxide ligation in peroxidase alkaline forms. *J. Biol. Chem.* **263**: 13032–13038.

Smeller L, Meersman F, Fidy J, Heremans K. 2003. High-pressure FTIR study of the stability of horseradish peroxidase. Effect of heme substitution, ligand binding, Ca++ removal, and reduction of the disulfide bonds. *Biochemistry* **42**: 553–561.

Smith AT, Sanders SA, Greschik H, Thorneley RNF, Burke JF, Bray RC. 1992. Probing the mechanism of horseradish peroxidase by site-directed mutagenesis. *Biochem. Soc. Trans.* **20**: 340–345.

Smith AT, Santama N, Dacey S, Edwards M, Bray RC, Thorneley RNF, Burke JF. 1990. Expression of a synthetic gene for horseradish peroxidase C in *Escherichia coli* and folding and activation of the recombinant enzyme with Ca^{2+} and heme. *J. Biol. Chem.* **265**: 13335–13343.

Smith K, Silvermail NJ, Rodgers KR, Eigren TE, Castro M, Parker RM. 2002. Sol-gel encapsulated horseradish peroxidase: a catalytic material for peroxidation. *J. Am. Chem. Soc.* **124**: 4247–4252.

Sottomayor M, DiCosmo F, Barceló AR. 1997. *Enz. Microb. Tech.* **21**: 543–549.

Stillman JS, Stillman MJ, Dunford HB. 1975. Photochemical reactions of horseradish peroxidase Compounds I and II at room temperature and 10°K. *Biochemistry* **14**: 3183–3188.

Tams JW, Welinder KG. 1995. Mild chemical deglycosylation of horseradish peroxidase yields a fully active, homogeneous enzyme. *Anal. Biochem.* **228**: 48–55.

Tanaka M, Ishimori K, Mukai M, Kitagawa T, Morishima I. 1997. Catalytic activities and structural properties of horseradish peroxidase His42 → Glu or Gln mutant. *Biochemistry* **36**: 9889–9898.

Terner J, Palaniappan V, Gold A, Weiss R, Fitzgerald MM, Sullivan AM, Hosten CM. 2006. Resonance Raman spectroscopy of oxoiron(IV) porphyrin π-cation and oxoiron(IV) hemes in peroxidase intermediates. *J. Inorg. Biochem.* **100**: 480–501.

Terner J, Sitter AJ, Reczek CM. 1985. Resonance Raman characterizations of horseradish peroxidase: observation of the $Fe^{IV}{=}O$ stretching vibration of Compound II. *Biochim. Biophys. Acta* **828**: 73–80.

Totah RA, Hanzlik RP. 2001. Detection of aminium ion intermediates: *N*-cyclopropyl versus *N*-carboxymethyl groups as reporters. *J. Am. Chem. Soc.* **123**: 10107–10108.

Totah RA, Hanzlik RP. 2002. Non-oxidative decarboxylation of glycine derivatives by a peroxidase. *J. Am. Chem. Soc.* **124**: 10000–10001.

Totsune H, Ohno C, Kambayashi Y, Nakano M, Ushijima Y, Tero-Kubota S, Ikegami Y. 1999. Characteristics of chemiluminescence observed in the horseradish peroxidase-hydrogen peroxide-tyrosine system. *Arch. Biochem. Biophys.* **369**: 233–242.

Van Wart HE, Zimmer J. 1985. Resonance Raman studies of the photoreduction of horseradish peroxidase Compounds I and II. *J. Am. Chem. Soc.* **107**: 3379–3381.

Veitch NC, Smith AT. 2001. Horseradish peroxidase. *Adv. Inorg. Chem.* **51**: 107–162.

Wang W, Noël S, Desmadril M, Guéguen J, Michon T. 1999. Kinetic evidence for the formation of a Michaelis–Menten-like complex between horseradish peroxidase Compound II and di-(N-acetyl-L-tyrosine). *Biochem. J.* **340**: 329–336.

Welinder KG. 1976. Covalent structure of the glycoprotein horseradish peroxidase (EC 1.11.1.7). *FEBS Lett* **72**: 19–23.

Welinder KG. 1979. Amino acid sequence studies of horseradish peroxidase. Amino and carboxy termini, cyanogen bromide and tryptic fragments, the complete sequence, and some structural characteristics of horseradish peroxidase C. *Eur. J. Biochem.* **96**: 483–502.

Welinder KG. 1985. Plant peroxidases. Their primary, secondary and tertiary structures, and relation to cytochrome *c* peroxidase. *Eur. J. Biochem.* **151**: 497–450.

Witting PK, Upston JM, Stocker R. 1997. Role of α-tocopheroxyl radical in the initiation of lipid peroxidation in human low-density lipoprotein exposed to horse radish peroxidase. *Biochemistry* **36**: 1251–1258.

Wright PJ, English AM. 2001. Buffer-anion-dependent Ca^{2+} leaching from horseradish peroxidase at low pH. *J. Biol. Inorg. Chem.* **6**: 348–358.

Xu P, Uyama H, Whitten JE, Kobayashi S, Kaplan DL. 2005. Peroxidase-catalyzed in situ polymerization of surface orientated caffeic acid. *J. Am. Chem. Soc.* **127**: 11745–11753.

Yin J, Mills JH, Schultz PG. 2004. A catalysis-based selection for peroxidase antibodies with increased activity. *J. Am. Chem. Soc.* **126**: 3006–3007.

CHAPTER 4

HORSERADISH PEROXIDASE. II. TWO-ELECTRON REACTIONS, FERROUS PEROXIDASE, COMPOUND III, THE FIVE OXIDATION STATES, OXYGEN EVOLUTION, AND INACTIVATION

INTRODUCTION

Compound I formation has long been understood as a two-electron oxidation of native enzyme by hydrogen peroxide and other hydroperoxides and peroxy acids of the general formula ROOH. The participation of Compound I in direct two-electron oxidations of substrates is a more recent observation. Horseradish peroxidase exhibits oxidation states other than those of native enzyme, Compounds I and II. It can be reduced to an iron(II) ferrous state which can bind molecular oxygen to form Compound III (oxidation state VI). Thus, oxidation states from II to VI exist, as do interesting pathways between them. Finally, enzyme inactivation can be caused by substrates that oxidize as well as by free radical products. These provide the topics covered in this chapter.

TWO-ELECTRON OXIDATIONS BY COMPOUND I

The first well-documented example is oxidation of iodide by HRP-I (Roman and Dunford, 1972). From pH 5 to 10 the log of the rate constant decreases linearly

Peroxidases and Catalases: Biochemistry, Biophysics, Biotechnology, and Physiology, Second Edition. By H. Brian Dunford with increasing pH, not your typical bell-shaped curve! At lower pH the as-yet-not-understood interference of acetate buffer occurred. Thus it is clear that iodide reacts with an acid form of HRP. The acid form is now known to contain the $^{+}$HIm side chain of distal His42, with a pK_a value of ~2.5. The simplest explanation is that iodide ion accepts the ferryl oxygen of Compound I and the proton from $^{+}$HIm to form hypoiodous acid.

$$\text{Porphyrin}(+\bullet)\text{Fe}^{IV}{=}\text{O} \quad {}^{+}\text{HIm} \quad \text{I}^{-} \longrightarrow \text{PorphyrinFe}^{III} \quad \text{Im} \quad \text{HOI}$$

Hypoiodous acid has a pK_a value of 10.6 so it remains undissociated.

The iodide results are in striking contrast to those obtained when the substrate is a phenol such as *p*-cresol. Not only is a two-electron substrate oxidation occurring in the iodide reaction, but the catalyst is the acid form of Compound I, not the basic form. The overall steady-state reaction is

$$\text{Native HRP} + H_2O_2 \xrightarrow{k_1} \text{Compound I} + H_2O$$

$$\text{Compound I} + \text{Iodide} \xrightarrow{k_2} \text{Native HRP} + \text{HOI}$$

$$\frac{[\text{Native HRP}]}{[\text{Compound I}]} = \frac{k_2[\text{Iodide}]}{k_1[H_2O_2]}$$

By having large [H_2O_2] and small [Iodide] it is possible to make [Native HRP] approach zero. Therefore, a steady-state reaction can be observed where all of the enzyme is in the form of Compound I. Thus an accurate Compound I spectrum was obtained in the steady state (Roman and Dunford, 1972). When the substrate is a one-electron reductant, then Compound II is formed from Compound I, and Compound II will always dominate in steady-state concentration over Compound I because it is a slower reactant. In the same year another accurate determination of the Compound I spectrum was published (Schonbaum and Lo, 1972). The maximum molar absorptivity of Compound I occurs at 400 nm. Its value is 5.43×10^4 M^{-1} cm^{-1} using the value of 1.02×10^5 M^{-1} cm^{-1} for native HRP as reference (Roman and Dunford, 1972). Two identical values of 5.4×10^4 M^{-1} cm^{-1} for Compound I at 400 nm are estimated from published spectra (Schonbaum and Lo, 1972; Roman and Dunford, 1973). The latter spectrum was obtained in a steady-state study of sulfite oxidation (see below). As for the molar absorptivity of native HRP, three significant figures appear justified, so the best value for Compound I appears to be 5.43×10^4 M^{-1} cm^{-1} at 400 nm.

The kinetic results on iodide oxidation have stood the test of time. They do not provide evidence about the binding site of iodide. Work with horseradish peroxidase reconstituted with δ-modified hemes and meso-hemes (where the two vinyls are replaced with ethyl groups) has shed valuable light (Harris et al., 1993). It appears likely that iodide approaches the δ-methyne bridge of the heme.

Iodination of Tyrosine

The oxidation of tyrosine by HRP-H_2O_2 was described in Chapter 3. When iodide is introduced into the HRP-H_2O_2-tyrosine system oxidation is replaced by iodination (Dunford and Ralston, 1983; Sun and Dunford, 1993). Iodination of tyrosine occurs as rapidly as iodide is oxidized by Compound I so hypoiodous acid, HOI, either free or enzyme-bound, is the iodinating reagent. If the reaction is run long enough in the steady state then iodination of the enzyme occurs (Björkstén, 1968), but this is not observed in the transient-state experiments. The HRP system appears to be an appropriate model for the halogenation reactions of the mammalian peroxidases, but an iodinating enzyme intermediate is detected in the lactoperoxidase iodination of tyrosine, which is not present in the HRP system (Sun and Dunford, 1993). Coupling of diiodotyrosine catalyzed by horseradish peroxidase was studied (Doerge et al., 1994).

Iodide can also be oxidized by Compound II so that a free radical product is formed (Roman et al., 1971). The reaction is an artifact, in the sense that in a steady-state reaction of iodide and hydrogen peroxide catalyzed by horseradish peroxidase no Compound II is generated. Transient-state kinetic experiments show the reaction of Compound II with iodide has the same linear relation of log (rate constant) versus pH as the Compound I-iodide reaction. However, the Compound II reaction is slower by a factor of about 400. This rules out the possibility that iodide might be oxidized in two subsequent steps, with the second faster than the first (Harris et al., 1993). Interestingly, acetate buffer does not affect the Compound II-iodide reaction at low pH.

$$\text{Porphyrin Fe}^{IV}\!=\!\text{O}\ (^{+}\text{HIm})\quad \text{I}^{-} \longrightarrow \text{PorphyrinFe}^{III}\ (\text{Im})\quad \text{HOI}$$

Cyanate and thiocyanate are known as pseudo-halogens because of the similarities in their reactions to that of the halides. Thiocyanate is oxidized by horseradish peroxidase, but in a much slower reaction compared to lactoperoxidase. Inhibition by cyanide generated in the HRP reaction appears to be the cause (Adak et al., 1997).

With the success of the investigation of iodide oxidation, a search was made for other substrates that would reduce Compound I directly to native enzyme. Sulfite and nitrite were tested (Roman and Dunford, 1973; Araiso et al., 1976). Sulfite reacts with an acid form of both Compounds I and II, so that both two- and one-electron substrate oxidations occur. Sulfite reductions of Compound II occur about as rapidly as Compound I, an exception to the general rule about relative reactivities of the two compounds. A steady-state spectrum of pure Compound I was obtained in the reaction of hydrogen peroxide with sulfite catalyzed by horseradish peroxidase. This rules out the possibility of two subsequent one-electron reactions in the steady state. The Compound I direct reduction to native enzyme by sulfite is indicated.

$$\underset{\text{Compound I}}{\text{Porphyrin}(+\bullet)\text{Fe}^{IV}\!=\!\text{O}}\quad {}^{-}\text{O}_2\text{SOH} \longrightarrow \underset{\text{Native HRP}}{\text{PorphyrinFe}^{III}}\quad {}^{-}\text{O}_3\text{SOH}$$

Nitrite reacts by one-electron reactions only, again with acid forms of oxidized HRP.

The reaction scheme proposed for the nitrite reaction was

$$\text{HRP-I} + NO_2^- \longrightarrow \text{HRP-II} + NO_2$$

$$\text{HRP-II} + NO_2^- \longrightarrow \text{HRP} + NO_2$$

$$2NO_2 + H_2O \xrightarrow{\text{fast}} NO_3^- + NO_2^- + 2H^+$$

A water molecule would be produced in the Compound II reduction. Therefore, the source of oxygen in the oxidation of nitrite is indicated to be water. The authors were unaware of the interest in reactions of NO, NO_2, nitrite, and peroxynitrite, that would occur a generation later. To the best of my knowledge, iodide and sulfite remain as the only substrates known to reduce Compound I directly to native enzyme. When the active site is modified by replacing Phe41 and His42 with smaller amino acids, then styrene can be epoxidized (Ortiz de Montellano et al., 1995).

One of the myths of halide ion oxidation has been that horseradish peroxidase oxidizes iodide only, lactoperoxidase iodide and bromide, and myeloperoxidase iodide, bromide, and chloride. This myth, started by respected experimentalists who performed experiments only at pH $\sim$7, has only recently been shattered.

At low pH both bromide and chloride can be oxidized by horseradish peroxidase Compound I (Huang et al., 2005), again indicating that protonated distal imidazole is a catalytic group. The resultant HOBr (pK_a 8.7) and HOCl (pK_a 7.5) react with the heme vinyl groups. The results led to the query of how myeloperoxidase escapes inactivation by HOCl, assumed to be the bactericidal reagent in white blood cells. The vinyl groups can be protected against attack by a mutant that leads to a protein-heme cross-link (Huang et al., 2006). The assumption that free HOCl is generated by myeloperoxidase is almost certainly another myth (Dunford, 2005).

OXYGEN TRANSFER BY ONE-ELECTRON MECHANISMS

It is possible to transfer the oxygen of the $Fe^{IV}{=}O$ group to substrate without direct two-electron reduction of Compound I. Thioanisole and substituted thioanisoles (Kobayashi et al., 1987; Ortiz de Montellano et al., 1987; Pérez and Dunford, 1990a, 1990b; Harris et al., 1993; Goto et al., 1999), phenothiazine derivatives, including promazine and chlorpromazine (Kelder et al., 1994), and water-soluble aromatic sulfides (Baciocchi et al., 1996) are examples. Oxidation of thioanisole by Compound I and Compound II results in formation of a sulfur cation radical. A competing and unique Compound II reaction may be caused by electrophilic attack of the ferryl oxygen by the sulfur cation radical, resulting in hydroxyl radical formation, which in turn attacks the sulfur (Pérez and Dunford, 1990b), as shown in Figure 4.1. Sulfoxidation of aryl alkyl sulfides is enhanced in the Phe41Leu mutant (Ozaki and Ortiz de Montellano, 1994).

Aminopyrine (Griffin and Ting, 1978) and *N,N*-dimethylaniline (Kedderis and Hollenberg, 1983; Chen and Nobe, 1993) undergo one-electron oxidations leading to demethylation and formaldehyde formation, apparently with water as the source of oxygen in formaldehyde.

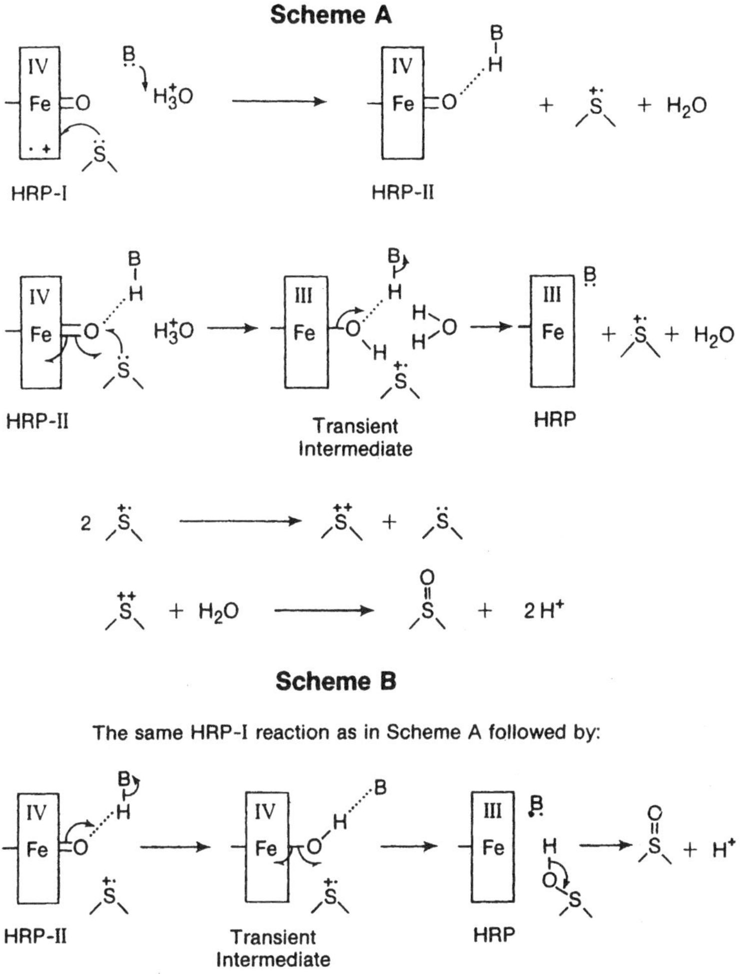

Figure 4.1. Proposed mechanisms for the one-electron oxidations of Compounds I and II by thioanisoles. (Reproduced with permission from Pérez U, Dunford HB. 1990. *Biochim. Biophys. Acta* **1038:** 98–104.)

Oxidation of 9-alkylanthracenes by horseradish peroxidase was studied in a comparison with cytochrome P-450 and a model system. Oxidation in the HRP-H_2O_2-anthracene system occurs in a one-electron oxidation to a free radical followed by reaction with O_2. Labeling experiments showed no added oxygen was derived from H_2O_2 (Anzenbacher et al., 1996).

FERROUS HORSERADISH PEROXIDASE AND COMPOUND III

The iron in native (ferric) horseradish peroxidase can be reduced to its ferrous state using sodium dithionite $Na_2S_2O_4$ (Keilin and Mann, 1937). The brown color of native enzyme is replaced by a bright red. For a stable preparation, an oxygen-free system is required. In the presence of oxygen the ferrous peroxidase was oxidized to its ferric form and no intermediate was detected in initial experiments. The thermodynamics of the Fe^{3+}/Fe^{2+} couple in HRP has been explored by measuring the temperature dependence (Battistuzzi et al., 2002).

Ferroperoxidase binding of carbon monoxide was quantitated (Keilin and Hartree, 1951; Kertesz et al., 1965). Dissociation of the CO complex was studied by laser-induced optoacoustic spectroscopy (Feis and Angeloni, 2001).

With purer preparations of ferrous HRP than used earlier, it was shown that oxygen will bind to form Compound III but not cleanly (Wittenberg et al., 1967). Some oxidation of the ferrous enzyme occurs and there are complications from excess dithionite, which react directly with hydrogen peroxide. The Compound III preparation was shown to have a visible region spectrum close to that observed when HRP is reacted with excess hydrogen peroxide. Ferrous horseradish peroxidase and Compound III are the peroxidase analogs of myoglobin and oxymyoglobin. Compound III, also called oxyperoxidase, is more reactive than oxymyoglobin. It has been suggested that the oxygen in oxyperoxidase is more reactive because of a stronger iron-His binding causing more π^* iron-to-oxygen back donation (Van Wart and Zimmer, 1985). Oxygen binding to ferrous HRP has been studied as a function of selected mutants (Rodriguez-López et al., 1997). The oxygen binding rate is increased by two orders of magnitude in replacement of Arg38 by Lys, Ser, or Gly. His42 and Arg38 replacements produced a less stable oxyperoxidase, which decomposes to the ferric enzyme. In a comparison to the behavior of myoglobin and hemoglobin, the binding and dissociation rates of *tert*-butyl isocyanide to ferrous HRP were measured. Binding is extremely slow. Escape to form a secondary caged pair occurs (Bandyopadhyay et al., 1996).

Just as Compound I is formed from the two-electron oxidation of native peroxidase by hydrogen peroxide, Compound II can be formed by the two-electron oxidation of ferrous horseradish peroxidase in a bimolecular reaction. Quantitative results were obtained by using only the minimum required amount of dithionite to form ferrous HRP; or by using less than the stoichiometric amount of dithionite so that a mixture of ferric and ferrous HRP is obtained. The three reactions that occur are on vastly different time scales and can be separated kinetically. The fastest is Compound I formation from ferric HRP and hydrogen peroxide. Next is Compound II formation from ferrous HRP and hydrogen peroxide. Slowest is the conversion of Compound II to Compound III again by reaction with hydrogen peroxide. The rate constant for oxidation of ferrous HRP to Compound II is $10^5\ M^{-1}\ s^{-1}$ at 20°C and pH 7, 100-fold slower than the rate constant for Compound I formation from the ferric enzyme. (Noble and Gibson, 1970).

$$\underset{\text{Ferrous HRP}}{Fe^{II} + H_2O_2} \longrightarrow \underset{\text{Compound II}}{Fe^{IV}{=}O + H_2O}$$

$$\underset{\text{Compound II}}{Fe^{IV}{=}O + H_2O_2} \longrightarrow \underset{\text{Compound III}}{Fe^{II}{-}O{-}O + H_2O}$$

The work of Noble and Gibson did not establish the mechanism(s) of Compound III formation from Compound II. Two important papers provide details (Nakajima and Yamazaki, 1987; Adediran and Lambeir, 1989). After Compound I formation a series of reactions initiated by a small amount of one-electron reductions by hydrogen peroxide led to the following mechanism.

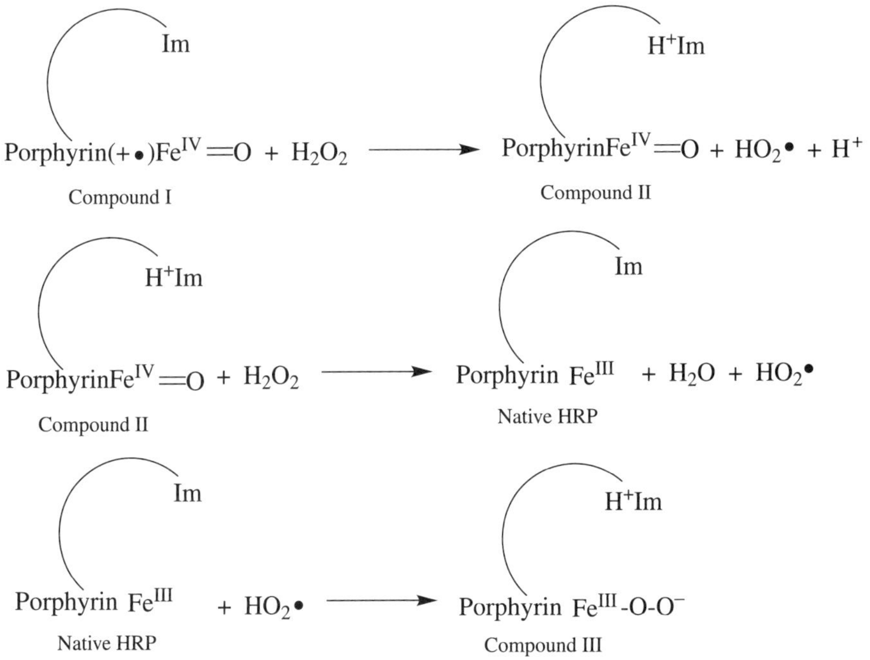

A two-step reaction of hydrogen peroxide with Compound II (called ferryl peroxidase in their paper) yielded Compound III (Nakajima and Yamazaki, 1987). Starting with Compound II a mechanism similar to the last two steps was proposed, although hydrogen peroxide rather than hydroperoxyl radical was the reactant in the last step (Adediran and Lambeir, 1989). The rate constant for conversion of Compound II to III is 20 $M^{-1}\ s^{-1}$.

Twin peaks occur at 543 and 577 nm in the Compound III spectrum, compared to 526 and 554 in Compound II (Wittenberg et al., 1967; Nakajima and Yamazaki, 1987; Adediran and Lambeir, 1989). The difference in spectra in the visible region is the easiest way to spectrally distinguish Compounds II and III, since their Soret region spectra are very similar with peaks in the 416 to 420 nm region.

Still another pathway from Compound II to Compound III is possible (Nakajima and Yamazaki, 1987). Exchange of the ferryl oxygen with water (Hashimoto et al., 1986) means that some free Fe^{IV} exists, which will react with hydrogen peroxide to form Compound III directly. Hydroperoxyl (superoxide) radical scavengers were employed. Superoxide dismutase was unsatisfactory because it is inactivated by hydrogen peroxide. Tetranitromethane (TNM) was ideal. It shut down the first mechanism involving hydroperoxyl radical, but not the second, dependent on hydrogen peroxide.

Compound III can therefore be formed by oxygen addition to the ferrous enzyme, and by oxidation of Compound II by hydrogen peroxide. It can also be formed by superoxide addition to the ferric enzyme.

$$\mathrm{Fe^{III} + O{-}O^- \longrightarrow Fe^{III}{-}O{-}O^-}$$

The formulae $\mathrm{Fe^{II}{-}O{-}O}$ and $\mathrm{Fe^{III}{-}O{-}O^-}$ are different electronic representations (resonance structures) of the same Compound III species.

Structure of Compound III

Compound III was produced by reaction of excess hydrogen peroxide with crystals of ferric HRP C* with structural determination of the resultant crystals. "The dioxygen species ligates the heme iron in a bent configuration with the OA atom 1.8 Å away from the iron. ... The OB atom forms hydrogen bonds with the Nε2 atom of His42 (2.9 Å), the Nε atom of Arg38 (2.9 Å) and with a water molecule (2.9 Å)" (Berglund et al., 2002, p. 467).

Compound III as a Terminal Oxidase Model

Since Compound III can be represented as $\mathrm{Fe^{II}\text{-}O_2}$, as is the bound oxygen in cytochrome *c* oxidase, it has been suggested to be a model of a terminal oxidase (Phelps et al., 1974; Berglund et al., 2002). Crystals of Compound III prepared from HRP C* are extremely sensitive to x-ray radiation and rapidly decay to ferrous peroxidase with formation of two molecules of water (Berglund et al., 2002).

$$\mathrm{Fe^{II}{-}O_2 + 4H^+ + 4e^- \longrightarrow Fe^{II} + 2H_2O}$$

Whether the reaction is a good model of a terminal oxidase would require knowledge about the source of the four electrons and four protons, and their method of transfer.

THE FIVE OXIDATION STATES OF HORSERADISH PEROXIDASE

The establishment of a route from ferrous horseradish peroxidase to Compound II to Compound III in separate discrete reactions helped to define the interrelationship between the five oxidation states of horseradish peroxidase and the pathways between them. These are depicted in Figure 4.2.

There are electron changes and sometimes proton changes that occur when one oxidation state is changed into another. These are shown in Figure 4.2. The addition of both a proton and electron by so-called hydrogen donor substrates to Compound I was discussed in Chapter 3. The electron reduces the porphyrin π-cation and the proton adds to the imidazole side chain of distal His42. The pK_a value for dissociation of this proton from Compound II is 8.6. Above pH 8.6 Compound II becomes

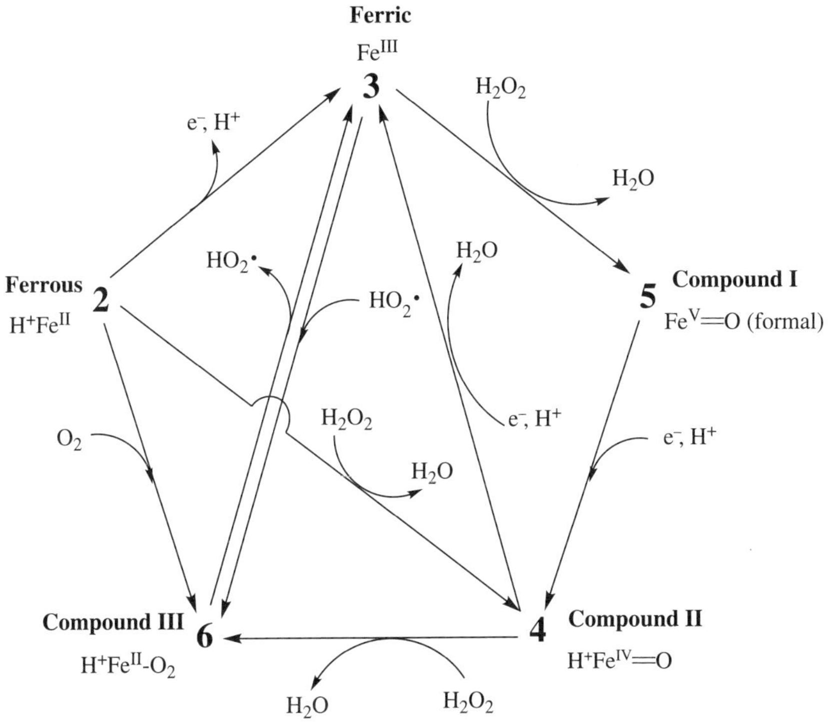

Figure 4.2. The five oxidation states of horseradish peroxidase, their interconversions, and proton and electron balances.

unreactive (Critchlow and Dunford, 1972; Hayashi and Yamazaki, 1979). By definition, when a hydrogen donor substrate reacts with Compound II another electron and proton are supplied. The sum from reductions of Compound I and II is two electrons and two protons donated by substrates. These reduce the ferryl oxygen to a water leaving group as Compound II is reduced to native enzyme.

$$O + 2H^+ + 2e^- \longrightarrow H_2O$$

Other proton balances were established in spectrophotometric and pH stat experiments (Yamada and Yamazaki, 1974). The pK_a value for proton dissociation from the ferrous enzyme is 7.2. Peroxidases are designed to have stable ferric states, whereas the myoglobins have stable ferrous states (Harbury, 1957).

Although the normal peroxidase cycle often appears irreversible in all three steps, each of the steps can be reversed using suitable reagents. Indeed, these reversible reactions provided the basis for redox measurements (Hayashi and Yamazaki, 1979). The redox potentials of the two compounds are similar in acid and neutral solutions. A variety of reagents can be used to oxidize native enzyme and Compound II. Thus oxidations of ferric HRP to Compound II, and Compound II to

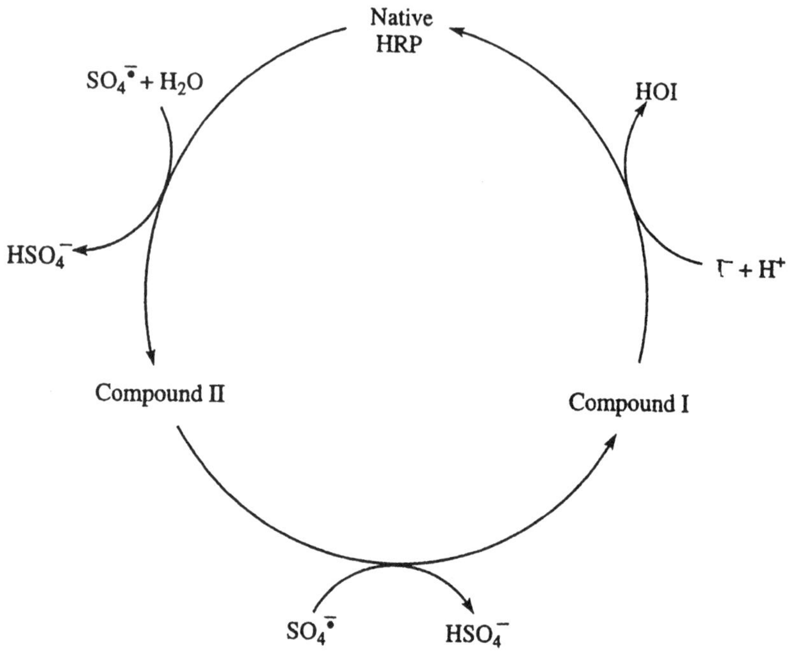

Figure 4.3. The horseradish peroxidase cycle running backwards.

Compound I, have been accomplished using chlorite (Hewson and Hager, 1979a, 1979b), periodate, and radical anions generated photochemically (Nadezhdin and Dunford, 1979, 1980), and the photochemically generated bipyridine complex of ruthenium(III) (Berglund et al., 1997).

Thus a completely reversed peroxidase cycle can be made to occur, one step at a time. An example is shown in Figure 4.3, in which oxidations by the sulfate radical anion are used to convert native enzyme to Compound II and Compound II to Compound I (Nadezhdin and Dunford, 1979, 1980). See also the discussion in Chapter 3. Iodide reduces Compound I directly to native enzyme (Roman and Dunford, 1972).

THE CATALATIC REACTION

Horseradish peroxidase has a small amount of catalase-like (catalatic) reactivity.

$$\underset{\text{Native HRP}}{Fe^{III}} \quad H_2O_2 \longrightarrow \underset{\text{Compound I}}{(+\bullet)Fe^{IV}{=}O} \quad H_2O$$

$$\underset{\text{Compound I}}{(+\bullet)Fe^{IV}{=}O} \quad H_2O_2 \longrightarrow \underset{\text{Native HRP}}{Fe^{III}} \quad H_2O + O_2$$

$$2\,H_2O_2 = 2\,H_2O + O_2$$

Evolved oxygen is readily detected with a Clarke electrode. Thus hydrogen peroxide reacts with Compound I, either in a one-electron reaction leading eventually to Compound III formation (described above), or in a two-electron reaction causing O_2 formation (the catalatic reaction). Neither of these reactions would be observed in the presence of a good hydrogen donor substrate such as a phenol, but when the only substrate present is hydrogen peroxide they are both important reactions. It was observed that when Compound III formation was optimal so was oxygen evolution, indicating essentially a constant ratio of one- and two-electron reactions of hydrogen peroxide (Nakajima and Yamazaki, 1987).

THE HRP CLOCK REACTION

A clock reaction can be observed in the HRP-catalyzed oxidation of iodide, in a pseudo-catalatic reaction. Hypoiodous acid will react with excess hydrogen peroxide to liberate dioxygen and regenerate iodide.

$$\begin{array}{rcl} H_2O_2 + H^+ + I^- & \longrightarrow & HOI + H_2O \\ HOI + H_2O_2 & \longrightarrow & H_2O + H^+ + I^- \\ \hline \end{array}$$

$$2\,H_2O_2 = O_2 + 2\,H_2O$$

When hydrogen peroxide is consumed then triiodide anion formation occurs:

$$\begin{array}{rcl} HOI + H^+ + I^- & \rightleftharpoons & I_2 + H_2O \\ I_2 + I^- & \rightleftharpoons & I_3^- \end{array}$$

Thus at pH 6.5 a solution, colorless except for a trace amount of HRP, instantly turns the deep purple color of triiodide in a reaction dependent upon the time required to consume all of the hydrogen peroxide (Kummer et al., 1996). The system HRP-hydrogen peroxide-iodide-EDTA has been studied intensively (Bhattacharyya et al., 1993, 1994; Adak et al., 1995).

ENZYME INACTIVATION

A third important reaction occurs when hydrogen peroxide is the only substrate present. Horseradish peroxidase is inactivated. The most intensive and accurate studies of peroxidase inactivation by hydrogen peroxide and other oxidants have been conducted by Manuel Acosta and coworkers at the University of Murcia. The scheme shown in Figure 4.4 summarizes their mechanism of inactivation using hydrogen peroxide (Arnao et al., 1990; Acosta et al., 1993).

All of the species shown in boldface in Figure 4.4 are known. The inactive species P-670, named after its strong absorption band at 670 nm (Bagger and Williams, 1971), is a verdohemoprotein (Nakajima and Yamazaki, 1980). One mole of CO is evolved

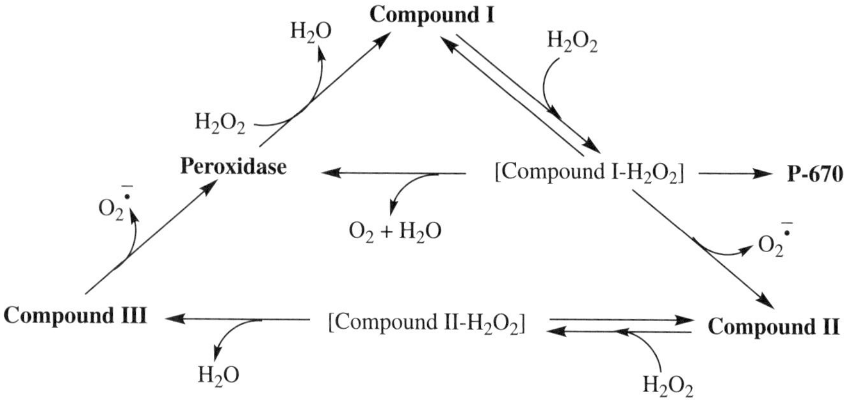

Figure 4.4. The reactions of horseradish peroxidase with hydrogen peroxide in the absence of reducing substrate. (Based on Acosta M, Arnao MB, Hernandez-Ruiz J, Garcia-Canovas F. 1993. In: Welinder KG, Rasmussen SK, Penel C, Greppin H, editors. *Plant Peroxidases. Biochemistry and Physiology.* University of Geneva, pp. 202–205.)

in the conversion of heme *b* to verdoheme. The transient complexes Compound I-H_2O_2 and Compound II- H_2O_2 must have a fleeting existence.

The catalatic pathway of horseradish peroxidase, leading to oxygen evolution, is well established (Arnao et al., 1990; Acosta et al., 1993; Baker et al., 2000; Hernández-Ruiz et al, 2001; Hiner et al., 2001b). It is the most important route for regeneration of native enzyme when oxidizing substrates are the only reactants. The similar reactions when hydrogen peroxide is replaced by *m*-chloroperbenzoic acid have been thoroughly quantitated (Rodriguez-Lopez et al., 1997). The ratio between turnover and inactivation can be quantitated (Hiner et al., 1995). It was shown that native HRP C is more resistant to denaturation than HRP C*, which in turn is more resistant than HRC* mutants. Inactivation of HRP isoenzyme A2 has also been examined (Hiner et al., 2001a).

REFERENCES

Acosta M, Arnao MB, Hernandez-Ruiz J, Garcia-Canovas F. 1993. Inactivation of peroxidases by hydroperoxides. In: Welinder KG, Rasmussen SK, Penel C, Greppin H, editors. *Plant Peroxidases: Biochemistry and Physiology*. University of Geneva, Geneva, pp. 202–205.

Adak S, Bhattacharyya DK, Mazumder A, Bandyopadhyay U, Banerjee RK. 1995. Concurrent reduction of iodine and oxidation of EDTA at the active site of horseradish peroxidase: probing the iodine binding site by optical difference spectroscopy and steady state kinetic analysis for the formation of active enzyme-I^+-EDTA ternary complex for iodine reductase activity. *Biochemistry* **34**: 12998–13006.

Adak S, Mazumdar A, Banerjee RK. 1997. Low catalytic turnover of horseradish peroxidase in thiocyanate oxidation. *J. Biol. Chem.* **272**: 11049–11056.

Adediran SA, Lambeir A-M. 1989. Kinetics of the reaction of Compound II of horseradish peroxidase with hydrogen peroxide to form Compound III. *Eur. J. Biochem.* **186**: 571–576.

Anzenbacher P, Niwa T, Tolbert LM, Sirimanne SR, Guengerich FP. 1996. Oxidation of 9-alkylanthracenes by cytochrome P450 2B1, horseradish peroxidase, and iron tetraphenylphorphine/iodosylbenzene systems: anaerobic and aerobic mechanisms. *Biochemistry* **35**: 2512–2520.

Araiso T, Miyoshi K, Yamazaki I. 1976. Mechanisms of electron transfer from sulfite to horseradish peroxidase-hydroperoxide compounds. *Biochemistry* **15**: 3059–3063.

Arnao MB, Acosta M, Del Rio JA, Varón R, Garcia-Canovas F. 1990. A kinetic study on the suicide inactivation of peroxidase by hydrogen peroxide. *Biochim. Biophys. Acta* **1041**: 43–47.

Baciocchi E, Lanzaluga O, Malandrucco S. 1996. Oxidation of sulfides by peroxidases: involvement of radical cations and the rate of the oxygen rebound step. *J. Am. Chem. Soc.* **118**: 8973–8974.

Bagger S, Williams RJP. 1971. Intermediates in the reaction between hydrogen peroxide and horseradish peroxidase. *Acta Chem. Scand.* **25**: 976–982.

Baker CJ, Deahl K, Domek J, Orlandi EW. 2000. Scavenging of H_2O_2 and production of oxygen by horseradish peroxidase. *Arch. Biochem. Biophys.* **382**: 232–237.

Bandyopadhyay D, Walda KN, Grogan TM, Madge D, Traylor TG, Sharma VS. 1996. Evidence for a slow tertiary relaxation in the reaction of *tert*-butyl isocyanide with horseradish peroxidase. *Biochemistry* **35**: 1500–1505.

Bhattacharyya DK, Adak S, Bandyopadhyay U, Banerjee RK. 1994. Mechanism of inhibition of horseradish peroxidase-catalysed iodide oxidation by EDTA. *Biochem. J.* **298**: 281–288.

Bhattacharyya DK, Bandyopadhyay U, Chatterjee R, Banerjee RK. 1993. Iodide modulation of the EDTA-induced iodine reductase activity of horseradish peroxidase by interaction at or near the EDTA-binding site. *Biochem. J.* **289**: 575–580.

Battistuzzi G, Borsari M, Ranieri A, Sola M. 2002. Redox thermodynamics of the Fe^{3+}/Fe^{2+} couple in horseradish peroxidase and its cyanide complex. *J. Am. Chem. Soc.* **124**: 26–27.

Berglund GI, Carlsson GH, Smith AT, Szöke H, Henriksen A, Hajdu J. 2002. The catalytic pathway of horseradish peroxidase at high resolution. *Nature* **417**: 463–468.

Berglund J, Pascher T, Winkler JR, Gray HB. 1997. Photoinduced oxidation of horseradish peroxidase. *J. Am. Chem. Soc.* **119**: 2464–2469.

Björkstén F. 1968. A kinetic study of the horse-radish peroxidase-catalyzed oxidation of iodide. *Eur. J. Biochem.* **5**: 133–142.

Chen JK, Nobe K. 1993. Oxidation of dimethylaniline by horseradish peroxidase and electrogenerated peroxide. *J. Electrochem. Soc.* **140**: 299–303.

Critchlow JE, Dunford HB. 1972. Studies on horseradish peroxidase. IX. Kinetics of the oxidation of *p*-cresol by Compound II. *J. Biol. Chem.* **247**: 3703–3713.

Doerge DR, Taurog A, Dorris ML. 1994. Evidence for a radical mechanism in peroxidase-catalyzed coupling. II. Single turnover experiments with horseradish peroxidasc. *Arch. Biochem. Biophys.* **315**: 90–99.

Dunford HB. 2005. Kinetics and mechanisms of mammalian heme peroxidase reactions. *Progr. Reaction Kinetics Mechanisms* **30**: 245–265.

Dunford HB, Ralston IM. 1983. On the mechanism of iodination of tyrosine. *Biochem. Biophys. Res. Commun.* **116**: 639–643.

Feis A, Angeloni L. 2001. Photodissociation of the CO complex of horseradish peroxidase studied by laser-induced optoacoustic spectroscopy. *J. Phys. Chem. B.* **105**: 2638–2643.

Goto Y, Matsui T, Ozaki S, Watanabe Y, Fukuzumi S. 1999. Mechanisms of sulfoxidation catalyzed by high-valent intermediates of heme enzymes: electron-transfer vs oxygen-transfer mechanism. *J. Am. Chem. Soc.* **121**: 9497–9502.

Griffin BW, Ting PL. 1978. Mechanism of N-demethylation of aminopyrine by hydrogen peroxide catalayzed by horseradish peroxidase, metmyoglobin, and protohemin. *Biochemistry* **17**: 2206–2211.

Harbury HA. 1957. Oxidation-reduction potentials of horseradish peroxidase. *J. Biol. Chem.* **225**: 1009–1024.

Harris RZ, Newmyer SL, Ortiz de Montellano PR. 1993. Horseradish peroxidase-catalyzed two-electron oxidations: oxidation of iodide, thioanisoles, and phenols at distinct sites. *J. Biol. Chem.* **268**: 1637–1645.

Hashimoto S, Tatsuno Y, Kitagawa T. 1986. Resonance Raman evidence for oxygen exchange between the $Fe^{IV}{=}O$ heme and bulk water during enzymic catalysis of horseradish peroxidase and its relation with the heme-linked ionization. *Proc. Natl. Acad. Sci. USA* **83**: 2417–2421.

Hayashi Y, Yamazaki I. 1979. The oxidation-reduction potentials of Compound I/Compound II and Compound II/ferric couples of horseradish peroxidases A_2 and C. *J. Biol. Chem.* **254**: 9101–9106.

Hewson WD, Hager LP. 1979a. Mechanism of the chlorination reaction catalyzed by horseradish peroxidase wih chlorite. *J. Biol. Chem.* **254**: 3175–3181.

Hewson WD, Hager LP. 1979b. Oxidation of horseradish peroxidase compound II to compound I. *J. Biol. Chem.* **254**: 3182–3186.

Hernández-Ruíz J, Arnao MB, Hiner ANP, García-Cánovas F, Acosta M. 2001. Catalase-like activity of horseradish peroxidase: relationship to enzyme inactivation by H_2O_2. *Biochem. J.* **354**: 107–114.

Hiner ANP, Hernández-Ruíz J, García-Cánovas F, Smith AT, Arnao MB, Acosta M. 1995. A comparative study of the inactivation of wild-type, recombinant and two mutant horseradish peroxidase isoenzymes C by hydrogen peroxide and *m*-chloroperbenzoic acid. *Eur. J. Biochem.* **234**: 506–512.

Hiner ANP, Hernández-Ruiz, Rodriguez-López JN, Arnao MB, Varón R, Garcia-Cánovas, Acosta M. 2001a. The inactivation of horseradish peroxidase isoenzyme A2 by hydrogen peroxide: an example of partial resistance due to the formation of a stable enzyme intermediate. *J. Biol. Inorg. Chem.* **6**: 504–516.

Hiner ANP, Hernández-Ruíz J, Williams GA, Arnao MB, García-Cánovas F, Acosta M. 2001b. Catalase-like oxygen production by horseradish peroxidase must predominantly be an enzyme-catalyzed reaction. *Arch. Biochem. Biophys.* **392**: 295–302.

Huang L, Wojciechowski G, Ortiz de Montellano PR. 2005. Prosthetic heme modification during halide ion oxidation: demonstration of chloride oxidation by horseradish peroxidase. *J. Am. Chem. Soc.* **127**: 5345–5353.

Huang L, Wojciechowski G, Ortiz de Montellano PR. 2006. Role of heme-protein covalent bonds in mammalian peroxidases: protection of the heme by a single engineered heme-protein link in horseradish peroxidase. *J. Biol. Chem.* **281**: 18983–18988.

Kedderis GL, Hollenberg PF. 1983. Characaterization of the *N*-demethylation reactions catalyzed by horseradish peroxidase. *J. Biol. Chem.* **258**: 8129–8138.

Keilin D, Hartree EF. 1951. Purification of horseradish peroxidase and comparison of its properties with those of catalase and methaemoglobin. *Biochem. J.* **49**: 88–106.

Keilin D, Mann T. 1937. On the haematin compound of peroxidase. *Proc. R. Soc. (London) B* **122**: 119–133.

Kelder PP, de Mol NJ, Fischer MJE, Janssen LHM. 1994. Kinetic evaluation of the oxidation of phenothiazine derivatives by methemoglobin and horseradish peroxidase in the presence of hydrogen peroxide. Implications for the reaction mechanisms. *Biochim. Biophys. Acta* **1205**: 230–238.

Kertesz D, Antonini E, Brunori M, Wyman J, Zito R. 1965. Studies on the equilibria and kinetics of the reactions of peroxidases with ligands. I. The reaction of ferroperoxidases with carbon monoxide. *Biochemistry* **4**: 2672–2676.

Kobayashi S, Nakano M, Kimura T, Schaap AP. 1987. On the mechanism of the peroxidase-catalyzed oxygen-transfer reaction. *Biochemistry* **26**: 5019–5022.

Kummer U, Wegmann K, Baier G. 1996. A peroxidase switch. In Obinger C, Burner U, Ebermann R, Penel C, Greppin H, editors. *Plant Peroxidases: Biochemistry and Physiology*. University of Geneva Press, Geneva, pp. 101–105.

Nadezhdin AD, Dunford HB. 1979. Horesradish peroxidase. XXXIV. Oxidation of compound II to I by periodate and inorganic anion radicals. *Can. J. Biochem.* **57**: 1080–1083.

Nadezhdin AD, Dunford HB. 1980. Horesradish peroxidase. XL. The oxidation of native enzyme to Compound II. *Can. J. Chem.* **58**: 2504–2507.

Nakajima R, Yamazaki I. 1980. The conversion of horseradish peroxidase C to a verdohemoprotein by a hydroperoxide derived enzymatically from indole-3-acetic acid and *m*-nitroperoxybenzoic acid. *J. Biol. Chem.* **255**: 2067–2071.

Nakajima R, Yamazaki I. 1987. The mechanism of oxyperoxidase formation from ferryl peroxidase and hydrogen peroxide. *J. Biol. Chem.* **262**: 2576–2581.

Noble RW, Gibson QH. 1970. The reaction of ferrous horseradish peroxidase with hydrogen peroxide. *J. Biol. Chem.* **245**: 2409–2413.

Ortiz de Montellano PR, Choe YS, DePillis G, Catalano CE. 1987. Structure-mechanism relationships in hemoproteins: oxygenations catalyzed by chloroperoxidase and horseradish peroxidase. *J. Biol. Chem.* **262**: 11641–11646.

Ortiz de Montellano PR, Ozaki S-I, Newmyer SL, Miller VP, Hartmann C. 1995. Structural determinants of the catalytic activites of peroxidases. *Biochem. Soc. Trans.* **23**: 223–227.

Ozaki S, Ortiz de Montellano PR. 1994. Molecular engineering of horseradish peroxidase: highly enantioselective sulfoxidation of aryl alkyl sulfides by the Phe-41 → Leu mutant. *J. Am. Chem. Soc.* **116**: 4487–4488.

Pérez U, Dunford HB. 1990a. Transient-state kinetics of the reactions of 1-methoxy-4-(methylthio)benzene with horseradish peroxidase Compounds I and II. *Biochemistry* **29**: 2757–2763.

Pérez U, Dunford HB. 1990b. Spectral studies on the oxidation of organic sulfides (thioanisoles) by horseradish peroxidase Compounds I and II. *Biochim. Biophys. Acta* **1038**: 98–104.

Phelps CF, Antonini E, Giacometti G, Brunori M. 1974. The kinetics of oxidation of ferroperoxidase by molecular oxygen. *Biochem. J.* **141**: 265–272.

Rodriguez-López JN, Hernández-Ruiz J, Garcia-Cánovas F, Thorneley RNF, Acosta M, Arnao MB. 1997. The inactivation and catalytic pathways of horseradish peroxidase with *m*-chloroperoxybenzoic acid: a spectrophotometric and transient kinetic study. *J. Biol. Chem.* **272**: 5469–5476.

Rodriguez-López JN, Smith AT, Thorneley RNF. 1997. Effect of distal cavity mutations on the binding and activation of oxygen by ferrous horseradish peroxidase. *J. Biol. Chem.* **272**: 389–395.

Roman R, Dunford HB. 1972. pH dependence of the oxidation of iodide by Compound I of horseradish peroxidase. *Biochemistry* **11**: 2076–2082.

Roman R, Dunford HB. 1973. Studies on horseradish peroxidase. XII. A kinetic study of the oxidation of sulfite and nitrite by Compounds I and II. *Can. J. Chem.* **51**: 588–596.

Roman R, Dunford HB, Evett M. 1971. Studies on horseradish peroxidase. VIII. A kinetic study of the oxidation of iodide by horseradish peroxidase Compound II. *Can. J. Chem.* **49**: 3059–3063.

Schonbaum GR, Lo S. 1972. Interaction of peroxidases with aromatic peracids and alkyl peroxides. *J. Biol. Chem.* **247**: 3353–3360.

Sun W, Dunford HB. 1993. Kinetics and mechanism of the peroxidase-catalyzed iodination of tyrosine. *Biochemistry* **32**: 1324–1331.

Van Wart HE, Zimmer J. 1985. Resonance Raman evidence for the activation of diogygen in horseradish oxyperoxidase. *J. Biol. Chem.* **260**: 8372–8377.

Wittenberg JB, Noble RW, Wittenberg BA, Antonini E, Brunori M, Wyman J. 1967. Studies on the equilibria and kinetics of the reactions of peroxidases with ligands. II. The reaction of ferroperoxidase with oxygen. *J. Biol. Chem.* **242**: 626–634.

Yamada H, Yamazaki I, 1974. Proton balance in conversions between five oxidation-reduction states of horseradish peroxidase. *Arch. Biochem. Biophys.* **165**: 728–738.

CHAPTER 5

HORSERADISH PEROXIDASE. III. OSCILLATIONS AND PEROXIDASE-OXIDASE REACTIONS WITH NADH, INDOLE-3-ACETIC ACID, AND ISOBUTYRALDEHYDE. LIGHT EMISSION

OSCILLATIONS AND THE NADH PEROXIDASE-OXIDASE REACTION

An oxidase is an enzyme that transfers electrons to molecular oxygen. The most famous example is cytochrome *c* oxidase, a heme- and copper-containing enzyme. It accepts four electrons from cytochrome *c* and transfers them, along with four protons, to oxygen, reducing each oxygen molecule to two molecules of water. Oxidase activity of horseradish peroxidase with dihydroxyfumarate as substrate was detected in the absence of added hydrogen peroxide (Swedin and Theorell, 1940), and the incorporation of molecular oxygen into the dihydroxyfumarate was proven using $^{18}O_2$ (Anan and Mason, 1963). Many other substrates were also shown to induce oxidase activity, for example indole-3-acetic acid (Kenten, 1955), hydroquinones (Klapper and Hackett, 1963a, 1963b), and triose reductone (Yamazaki and Piette, 1963). The most famous example of oxidase activity in a peroxidase is the HRP (horseradish peroxidase)-NADH-oxygen system, which has become known in the literature as the peroxidase-oxidase reaction, to the detriment of all the other known examples.

A note of caution about peroxidase oxidase activity: as discussed below, the peroxidase oxidase mechanisms are different from true oxidase reactions. Dioxygen is not a direct substrate of the enzyme; rather, oxygen is converted to hydroperoxyl radical, either by free radicals produced in a peroxidase reaction or by reaction with NADH. If carbon radicals are generated in a peroxidase reaction they are rapidly scavenged by dioxygen.

Peroxidases and Catalases: Biochemistry, Biophysics, Biotechnology, and Physiology, Second Edition. By H. Brian Dunford

The peroxidase-oxidase reaction of the HRP-NADH-O_2 system was first discovered in a search for a route to Compound III formation (Yamazaki and Yokota, 1965). Compound III was formed, then disappeared after complete consumption of oxygen, and a mixture of ferric and ferrous peroxidase was obtained. By addition of more oxygen Compound III was regenerated. When an open system was used (continuous supply of oxygen), damped oscillations were observed in the amounts of Compound III and oxygen in solution (Yamazaki et al., 1965). Damping is caused by P-670 formation, which effectively removes active peroxidase.

When a phenol and methylene blue were added, sustained oscillations were attained (Yamazaki and Yokota, 1973). The HRP-NADH-O_2 system provided the first known example of oscillations in a single enzyme system. It provides the simplest model to help understand oscillatory and chaotic behavior in more complex biological systems, and it well deserves the attention lavished upon it for a period now in excess of 40 years.

Ferrous peroxidase adds dioxygen to form Compound III, and if the ferrous enzyme is in excess then all oxygen is consumed. In an open system more oxygen diffuses into the solution. In the meantime, if ferrous enzyme is regenerated then the cycle will repeat. This naïve picture accounts for one in-phase cycle for O_2 consumption and Compound III formation. Since native peroxidase is stable and Compound III fairly stable, the oscillatory behavior can also be described as bistability (Degn, 1968; Olsen and Degn, 1978).

The reaction is started with ferric enzyme, so how important is ferrous peroxidase and how does peroxidase in the form of Compound III reenter the reaction cycle? An Urbanalator model was proposed by the University of Illinois group (Olson and Scheeline, 1995a, 1995b; Olson et al., 1995; Scheeline et al., 1997). There have been many theoretical contributions (Aguda and Clarke, 1987; Aguda and Larter, 1990; Alexandre and Dunford, 1991; Bronnikova et al., 1995). Examples of oscillatory behavior are shown in Figures 5.1 and 5.2.

Chaotic behavior has also been observed (Olsen and Degn, 1977). The peroxidase-oxidase reaction has also been observed with lacto-, soy bean, and *Coprinus cinerus* peroxidases (Valeur and Olsen, 1996).

A fairly complete model can be constructed using the following reactions.

$$\begin{array}{lcl} \mathrm{NADH} + \mathrm{O_2} & \longrightarrow & \mathrm{NAD^{\bullet}} + \mathrm{HO_2} \\ \mathrm{NADH} + \mathrm{HO_2^{\bullet}} & \longrightarrow & \mathrm{NAD^{\bullet}} + \mathrm{H_2O_2} \\ \hline \end{array}$$

$$2\mathrm{NADH} + \mathrm{O_2} = 2\mathrm{NAD^{\bullet}} + \mathrm{H_2O_2}$$

Thus in a slow induction period some hydrogen peroxide is generated to start a normal peroxidase cycle as well as some $NAD^{\bullet}$ radicals.

$$\begin{array}{rcl} \text{Peroxidase} + \mathrm{H_2O_2} & \longrightarrow & \text{Compound I} + \mathrm{H_2O} \\ \text{Compound I} + \mathrm{NADH} & \longrightarrow & \text{Compound II} + \mathrm{NAD^{\bullet}} \\ \text{Compound II} + \mathrm{NADH} & \longrightarrow & \text{Peroxidase} + \mathrm{H_2O} + \mathrm{NAD^{\bullet}} \\ \hline \end{array}$$

$$\mathrm{H_2O_2} + 2\mathrm{NADH} = 2\mathrm{H_2O} + 2\mathrm{NAD^{\bullet}}$$

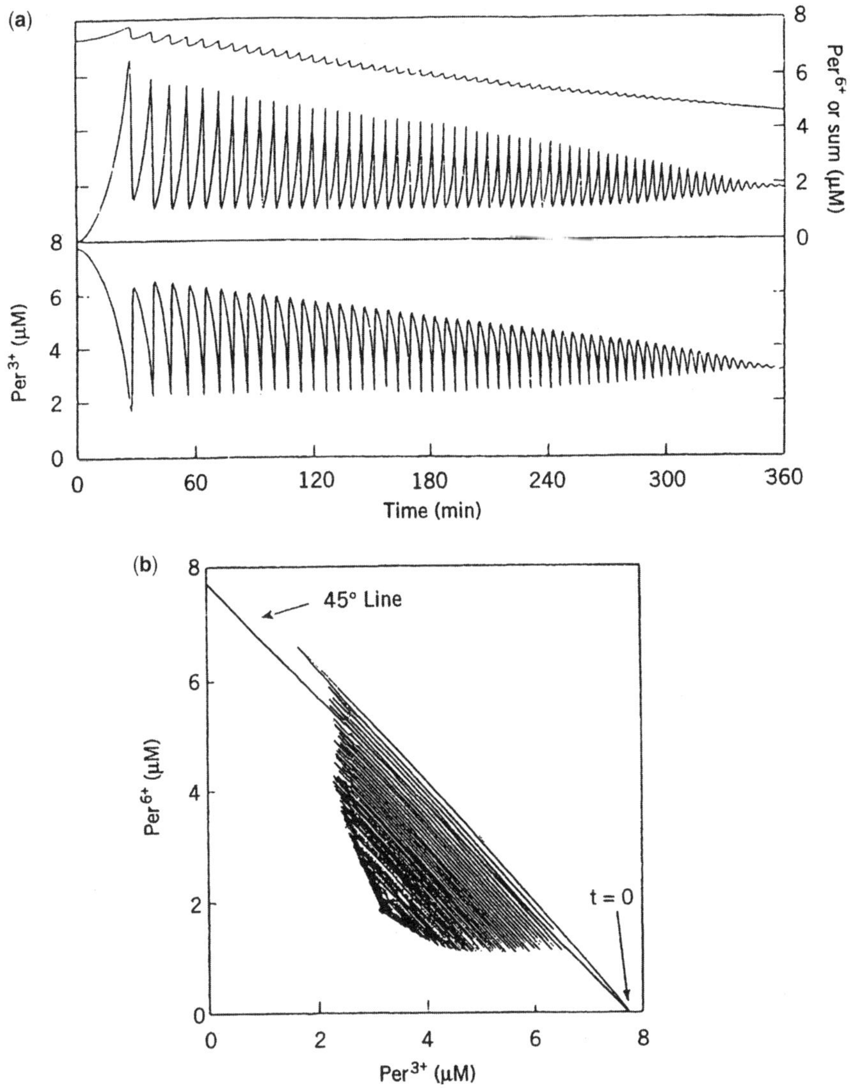

Figure 5.1. Oscillations in the peroxidase-oxidase reaction in an open stirred system. (a) Upper curve, sum of Compound III (Per^{6+}) and native enzyme (Per^{3+}); Middle curve, Compound III; and lower curve, native enzyme. (b) Plot of Compound III versus native enzyme concentrations. (Reproduced with permission from Olson DL, Williksen EP, Scheeline AJ. 1995. *J. Am. Chem. Soc.* **117**: 2–15.)

The rate constants for reactions of Compounds I and II with NADH have been measured as a function of pH (Kashem and Dunford, 1986).

The normal peroxidase cycle (above) produces a copious supply of $NAD^{\bullet}$ radicals that reduce dioxygen to superoxide, or its more active form, hydroperoxyl radicals. The pK_a value for the acid dissociation of hydroperoxyl radicals is 4.9. The

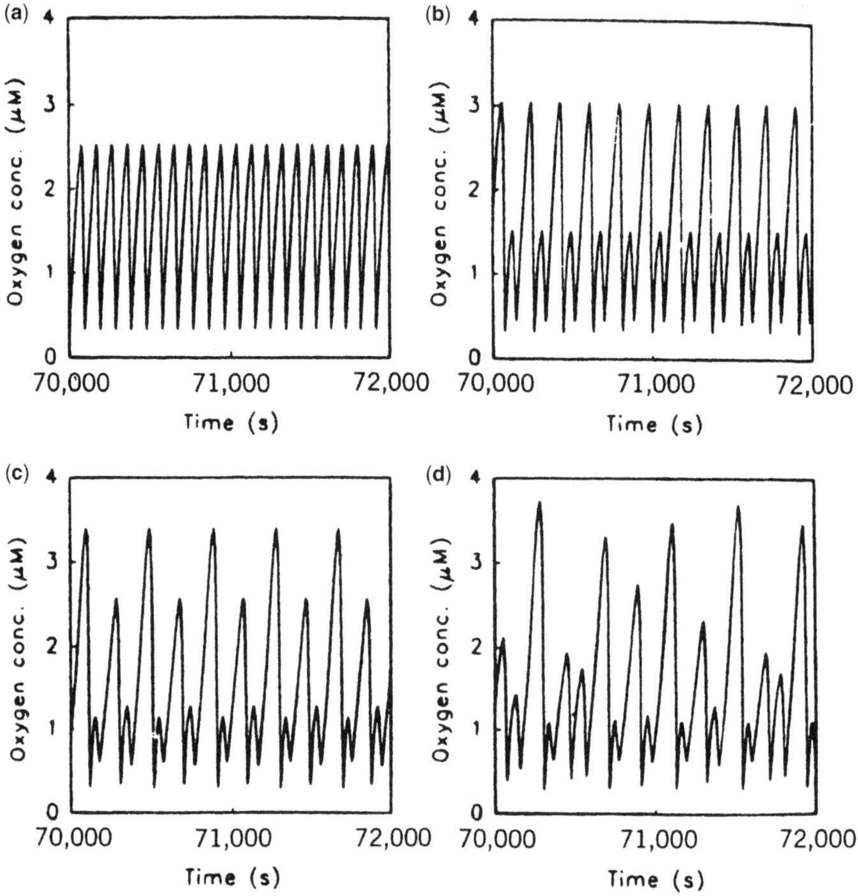

Figure 5.2. Modeled oscillations in oxygen concentration obtained by varying the rate constant ($M^{-1}\,s^{-1}$) for the reduction of Compound III to Compound I by the $NAD^{\bullet}$ radical: (a) 1.1×10^8, (b) 1.2×10^8, (c) 1.317×10^8, (d) 1.35×10^8. (Reproduced with permission from Bronnikova TV, Fed'kina VR, Schaffer WM, Olsen LP. 1995. *J. Phys. Chem.* **99**: 9309–9312.)

hydroperoxyl radicals reduce more NADH so more hydrogen peroxide is produced.

$$\begin{array}{rcl} NAD^{\bullet} + O_2 & \xrightarrow{H^+} & NAD^+ + HO_2^{\bullet} \\ HO_2^{\bullet} + NADH & \longrightarrow & NAD^{\bullet} + H_2O_2 \\ \hline \multicolumn{3}{c}{NADH + O_2 + H^+ = NAD^+ + H_2O_2} \end{array}$$

The hydrogen peroxide in turn generates more $NAD^{\bullet}$ radicals in the normal peroxidase cycle. One now has a full-fledged branching chain reaction so there is no steady state.

There are two routes to Compound III formation. In one, ferrous enzyme is formed which adds dioxygen directly.

$$\text{Peroxidase} + \text{NAD}^{\bullet} \xrightarrow{H^+} \text{Ferroperoxidase} + \text{NAD}^+$$
$$\text{Ferroperoxidase} + O_2 \longrightarrow \text{Compound III}$$
$$\text{Peroxidase} + O_2 + H^+ + \text{NAD}^{\bullet} = \text{Compound III} + \text{NAD}^+$$

In the other, native enzyme reacts with hydroperoxyl radical.

$$\text{Peroxidase} + HO_2^{\bullet} \rightleftarrows \text{Compound III}$$

The addition of hydroperoxyl radical to ferric peroxidase occurs in two detectable steps, which are completely inhibited by superoxide dismutase. The intermediate species appears to be a complex of superoxide/hydroperoxyl radical with native enzyme (Metodiewa and Dunford, 1989). The reverse reaction occurs with no detectable intermediates and is accelerated by superoxide dismutase.

Compound III may also reenter the reaction cycle via Compound I formation. The following reaction is the most likely route.

$$\text{Compound III} + \text{NAD}^{\bullet} \xrightarrow{H^+} \text{Compound I} + H_2O + \text{NAD}^+$$

Compound III conversion to Compound I also occurs via the following sequence.

$$\text{Compound III} \longrightarrow \text{Peroxidase} + HO_2^{\bullet}$$
$$\text{NADH} + HO_2^{\bullet} \longrightarrow H_2O_2 + \text{NAD}^{\bullet}$$
$$\text{Peroxidase} + H_2O_2 \longrightarrow \text{Compound I} + H_2O$$
$$\text{Compound III} + \text{NADH} = \text{Compound I} + H_2O + \text{NAD}^{\bullet}$$

Striking dependence of the Compound III concentration on the amount of NADH added was demonstrated (Kummer et al., 1997).

Various types of behavior were obtained in modeling studies: complex oscillations, period doubling, and transient chaos, by making small adjustments in the assumed rate constant for the above reaction (Bronnikova et al., 1995). The correctness of the model was shown experimentally (Hauser et al., 1997). Further modeling led to the fat torus scenario (Bronnikova et al., 1998).

The peroxidase-oxidase reaction requires a chain termination step which was thought to be

$$2\text{NAD}^{\bullet} \longrightarrow (\text{NAD})_2$$

Limitations of earlier models were examined and it was suggested that their failure to explain certain types of behavior was caused by reactions of the $(\text{NAD})_2$ dimer (Kirkor and Scheeline, 2000; Bronnikova et al., 2001; Schaffer et al., 2001). HRP adsorbs on silica surfaces, and surface effects were indicated to be important at low enzyme cocncentrations (Kirkor and Scheeline 2001; Lewis et al., 2006).

Addition of methylene blue and dichlorophenol enhances certain aspects of the overall reaction. Methylene blue accelerates initiation of oscillations and eliminates irreproducible induction periods. Two-electron reactions involving completely oxidized methylene blue (MB^+) and completely reduced (MBH) lead to enhanced hydrogen peroxide production (Sevcík and Dunford, 1991).

$$NADH + MB^+ \longrightarrow NAD^+ + MBH$$

$$MBH + O_2 + H^+ \longrightarrow MB^+ + H_2O_2$$

Methylene blue is sensitive to light, and this affects the oscillations (Carson and Walleczek, 2003).

Dichlorophenol is a competitor to NADH for reactions with Compounds I and II in the normal peroxidase cycle. This diminishes the amplitude of the oscillations, and lowers the concentration of Compound I, hence diminishing or eliminating P-670 formation. Thus sustained oscillations occur. Some other phenols and aromatic amines have the same effect as dichlorophenol (Kummer et al., 1997). The effect of a large number of naturally occurring phenols was examined and found to vary from causing inhibition or monotonic reaction, to damped, sustained, or complex oscillations (Hauser and Olsen, 1998). A magnetic field affected the amplitude of oscillations and was attributed to influence on the reductions of Compounds I and II (Møller and Olsen, 1999, 2000).

One might intuitively assume that oscillations would cause instability in biological systems. However, the opposite may be true. The ability of in vivo systems to control their environment by oscillatory versus chaotic behavior has been discussed (Møller et al., 1998). There are lower concentrations of active oxygen species in the oscillatory state, a potential advantage in in vivo systems (Olsen et al., 2003).

The unlikely production of hydroxyl radicals has been reported in the HRP-H_2O_2-NADH system (Chen and Schopfer, 1999), an apparent result that I have not seen duplicated.

A system analogous to HRP-NADH-O_2 is one in which NADH is replaced by an analog nifedipine. A reaction mechanism involving Compound III formation was proposed on the basis of kinetic magnetic field effects (Afanasyeva et al., 2006).

Another peroxidase-oxidase reaction occurs with a product of lipid peroxidation, malonaldehyde. In the presence of malonaldehyde, Mn^{2+}, and acetate buffer, horseradish peroxidase catalyzes a reaction in which oxygen is consumed. Some malonaldehyde is oxidized to a peracid, which initiates a peroxidase cycle. Carbon radicals are generated, which are scavenged by molecular oxygen. The same reaction occurs if Mn^{2+} is replaced with hydrogen peroxide (MacDonald and Dunford, 1989).

For background information on oscillatory and chaotic behavior (nonequilibrium thermodynamics), consult the publications of Irving R. Epstein and John Ross. Ilya Prigogine won the Nobel Prize in Chemistry in 1977 for his contributions to nonequilibrium thermodynamics.

PEROXIDASE OXIDASE REACTION WITH INDOLE-3-ACETIC ACID

Indole-3-acetic acid, like NADH, can react with horseradish peroxidase in the absence of added hydrogen peroxide, and oscillations in concentrations of oxygen and peroxidase species can be observed. In addition, the reaction emits light (Kenten, 1955). The study of light emission in peroxidase catalyzed reactions became the mainstay of the scientific career of Giuseppe Cilento at the University of São Paulo (Vidigal et al., 1975; Escobar et al., 1992).

Indole-3-acetic acid (IAA) is the major plant growth hormone (auxin; Bandurski et al., 1995). Partly for this reason, the HRP-IAA-O_2 system deserves at least equal billing with the HRP-NADH-O_2 system as a peroxidase-oxidase reaction. An entire agricultural industry has grown by finding suitable derivatives of IAA for control of broad-leaf weeds.

An IAA inactivating substance was first described (Larsen, 1936) and later named as the enzyme indoleacetic acid oxidase. Indoleacetic acid oxidase is now known to be a plant peroxidase (Kenten, 1955). Many studies of the reaction of horseradish peroxidase with IAA followed (Hinman and Lang, 1965; Fox et al., 1965; Ricard and Nari, 1967; Ricard et al., 1969; Gaspar, 1970; Yamazaki and Yamazaki, 1973; Ricard and Job, 1974). Labeling experiments using $^{18}O_2$ confirmed the oxidase nature of the HRP-IAA reaction (Kobayashi et al., 1984).

The horseradish peroxidase-indole-3-acetic acid-oxygen reaction generated more than the usual amount of controversy. The following major points were disputed: whether IAA directly reduces native HRP, the mechanism of initiation, the mechanism of oxygen uptake, the pH dependence of the reaction, and the role if any of ferrous peroxidase and Compound III.

IAA does not directly reduce HRP, as shown in very careful experiments (Nakajima and Yamazaki, 1979). In addition to HRP and IAA, either ascorbate and ascorbate oxidase, or glucose and glucose oxidase, plus catalase, was used to ensure strict anaerobic conditions. The solution was covered with either nitrogen or carbon monoxide; the latter as a trap for any ferrous enzyme that might be generated.

The initiator of the HRP-IAA reaction proved to be particularly elusive. Careful purification of IAA by sublimation and recrystallization had no effect on the uptake of oxygen, indicating that initiation was not caused by a contaminant (Metodiewa et al., 1992). The most likely candidate as initiator is the cation radical of IAA, formed by loss of an electron in an extremely slow reaction. From modeling studies the rate constant for IAA cation radical formation was estimated to be $3 \times 10^{-7}\ s^{-1}$ (Krylov and Dunford, 1996a).

The skatole radical was trapped and identified by ESR spectroscopy as the dominant free radical in the IAA reaction (Mottley and Mason, 1986). It was detected in a reaction initiated by hydrogen peroxide. However, it most probably is also formed in the absence of hydrogen peroxide, and is rapidly scavenged by molecular oxygen.

At pH 7 the mechanism of oxygen uptake is the following (Nakajima and Yamazaki, 1979). The IAA cation radical decarboxylates to form the skatole radical, which is the oxygen scavenger. The resultant peroxyl radical $ROO^{\bullet}$ abstracts a hydrogen atom, the most likely source for which is IAA, to form the peroxide ROOH. The stage is now set for a classic peroxidase cycle to occur. The sequence from decarboxylation to ROOH formation is extremely rapid, compensating for the slow initiation. The pathway from IAA to $ROO^{\bullet}$ is shown in Figure 5.3, and the overall mechanism at pH 7 is summarized in Figure 5.4 (Krylov and Dunford, 1996b). According to the mechanism in Figure 5.4, the principal product is indole-3-methanol ROH formed in the initial step of the peroxidase cycle (Nakajima and Yamazaki, 1979). The equilibrium between $IAA^{\bullet +}$ and $IAA^{\bullet}$ is shown in Figure 5.5. The pK_a for the acid dissociation of $IAA^{\bullet +}$ is 5.09 (Candeias et al., 1994). Rate constants for the reactions of HRP-I and HRP-II with IAA are $(2.2 \pm 0.1) \times 10^3\ M^{-1}\ s^{-1}$ and $(2.3 \pm 0.2) \times 10^2\ M^{-1}\ s^{-1}$, respectively, at pH 7.4 (Krylov and Dunford, 1996c).

P-670 is a verdohemoprotein and is an inactive form of HRP (Nakajima and Yamazaki, 1980). It is formed in the HRP-IAA-O_2 system by reaction of the organic hydroperoxide with Compound I (Acosta et al., 1988, 1989). From the accumulation

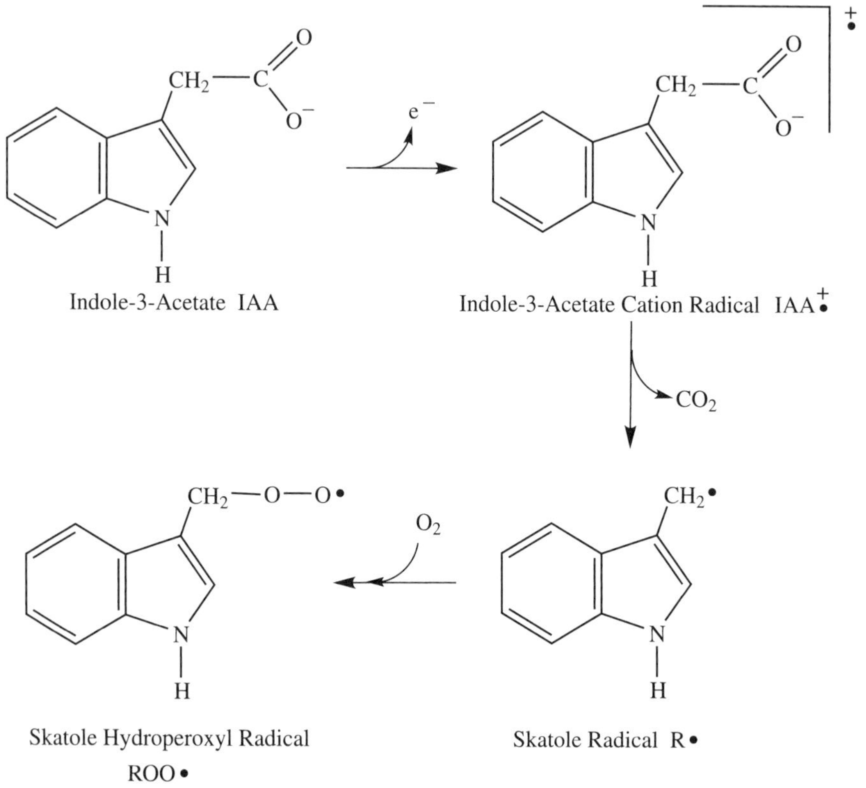

Figure 5.3. Initial steps in the reaction of indole-3-acetic acid with peroxidase.

$$\mathrm{IAA} \longrightarrow \mathrm{IAA}^{\overset{+}{\bullet}} + e^-$$

$$\mathrm{IAA}^{\overset{+}{\bullet}} \longrightarrow \mathrm{R}\bullet + \mathrm{CO_2}$$

$$\mathrm{R}\bullet + \mathrm{O_2} \longrightarrow \mathrm{ROO}\bullet$$

$$\mathrm{ROO}\bullet + \mathrm{IAA} \longrightarrow \mathrm{ROOH} + \mathrm{IAA}\bullet$$

$$\mathrm{IAA}\bullet + \mathrm{H^+} \rightleftharpoons \mathrm{IAA}^{\overset{+}{\bullet}}$$

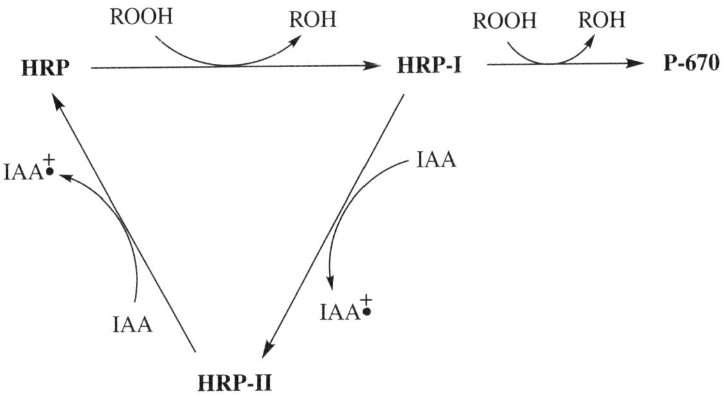

Figure 5.4. Complete mechanism for the reaction of indole-3-acetic acid with horseradish peroxidase at pH 7.

of P-670 its extinction coefficient was determined to be $(2.4 \pm 0.2) \times 10^4\ \mathrm{M^{-1}\,cm^{-1}}$ at 670 nm (Nakajima and Yamazaki, 1980; Krylov and Dunford, 1996c).

It is now established in the HRP-IAA-O_2 system that at neutral pH the dominant enzyme species observed in the steady state is Compound II and, as the

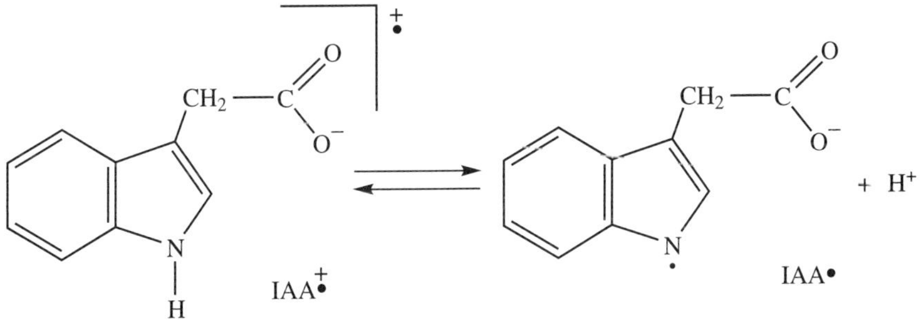

Figure 5.5. Equilibrium between the indole acetate cation radical and the indolyl free radical.

reaction progresses, increasing amounts of P-670 are formed. No Compound III is formed.

At acid pHs (4 to 6), different behaviors can be observed. It was claimed that the same mechanism as at neutral pH occurred (Nakajima and Yamazaki, 1979) but this contradicted observation of ferrous peroxidase and Compound III formation (Ricard and Job, 1974). An apparent resolution was obtained in a study where widely different ratios of HRP/IAA were used (Smith et al., 1982). At pHs in the range 4.0 to 5.8 they found that with high HRP/IAA ratios the mechanism was the same as observed at neutral pH. At low ratios the shuttle ferrous peroxidase-Compound III was observed. Thus a pathway analogous to that in the NADH peroxidase-oxidase is obtained.

The controversy boils down to a choice between two peroxides: hydrogen peroxide and the organic hydroperoxide ROOH derived from the skatole radical. Here I present a unique resolution to the problem. Contaminant hydrogen peroxide was present in the buffers (Smith et al., 1982). Under certain conditions this contaminant hydrogen peroxide wins the battle, and dominates the mechanism at the expense of ROOH, as demonstrated experimentally. How can this occur? There must be a chain reaction that generates hydroperoxyl radical so Compound III formation can occur which in turn generates ferrous peroxidase. The alternative mechanism, when contaminant hydrogen peroxide dominates, is shown in Figure 5.6.

As shown in Figure 5.6, the hydroperoxyl radical is generated by reduction of hydrogen peroxide by the IAA radical.

Indole-3-methanol is oxidized by HRP only when hydrogen peroxide is added (Arnao et al., 1993). Therefore, quantitation of indole-3-methanol formation appears to provide a method to distinguish between the two chain mechanisms, one involving

$$\text{Peroxidase} + H_2O_2 \longrightarrow \text{Compound I} + H_2O$$

$$\text{Compound I} + \text{IAA} \longrightarrow \text{Compound II} + \text{IAA}\bullet$$

$$\text{Compound II} + \text{IAA} \longrightarrow \text{Peroxidase} + \text{IAA}\bullet + H_2O$$

$$\boxed{H_2O_2 + \text{IAA}\bullet \longrightarrow \text{IAA} + HO_2\bullet}$$

$$\text{Peroxidase} + HO_2\bullet \longrightarrow \text{Compound III}$$

$$\text{Compound III} + \text{IAA} \longrightarrow \text{IAA}\bullet + HO_2\bullet + \text{Ferrous Peroxidase}$$

$$\text{Ferrous Peroxidase} + HO_2\bullet \rightleftharpoons \text{Compound III}$$

Figure 5.6. Reaction mechanism at low pH for indole-3-acetic acid and horseradish peroxidase when some hydrogen peroxide is present.

ROOH only, and the other hydrogen peroxide. To the best of my knowledge quantitation of indole-3-methanol formation has never been performed.

In conclusion, I agree with Isao Yamazaki that in an in vitro HRP-IAA-O_2 system using very pure reagents the normal pathway involves the ROOH radical, and not hydrogen peroxide. Compound III is not formed. IAA does not directly reduce ferric peroxidase.

The damped oscillations observed in the HRP-IAA-O_2 system have received scant attention since their original discovery (Degn, 1969). The effect of the enhancer umbelliferone (7-hydroxycoumarin) on the IAA reaction was studied experimentally (Krylov and Dunford, 1996d), and a modeling study of the same system under conditions suitable for damped oscillations has been presented (Krylov, 1998).

The reactions of indoleacetic acid and a series of derivatives with Compound I was investigated at pH 5, and correlated via the Hammett equation (Candeias et al., 1996). Reactions of IAA with HRP-I and HRP-II are much faster in acid, compared to neutral solutions.

Light Emission from Indole-3-Acetic Acid

Finally, I turn attention to light emitting reactions observed in peroxidase-IAA systems. These are caused by secondary reactions and, despite intensive study, they are not well understood. Weak emission at 1268 nm has been observed with singlet oxygen $^1\Delta_g$ as the source (Kanofsky, 1988). Disproportionation of hydroperoxyl radicals is a likely cause. The chemiluminescence of indole derivatives oxidized by HRP has been examined (Ximenes et al., 2001).

Involvement of different dioxetanes has been postulated (Vidigal et al., 1975; Pires de Melo et al., 1992). In studies involving added dyes the conclusion was reached that "photochemistry without light" was being observed; that is, electronic energy was being transferred from excited states of IAA products to the dyes, causing them to react. Reviews are found in Cilento (1980, 1984) and Cilento and Adam (1988).

In a study involving xanthene dyes added to the HRP-IAA system, emitted light centered on three wavelengths was observed, 420, 465, and 535 nm, with a fourth emission caused by dye fluorescence (Krylov et al., 1990; Krylov and Chebotareva, 1993). Although the species involved at each wavelength is unknown, the peroxide ROOH is likely to be involved. The important conclusion from the latter work is that the dyes are being co-oxidized along with IAA, and transfer of electronic energy is not occurring.

The physiological mechanism of action of indole-3-acetic acid (auxin) is now becoming understood. Auxin binds to a protein called transport inhibitor response 1 (TIR1) and the crystal structure of TIR1 and its auxin complex has recently been solved (Tan et al., 2007). The presence of auxin greatly enhances the association of TIR1 with a peptide that blocks gene transcription. Gene transcription is then liberated for the start of a myriad number of plant growth processes.

IAA has very low toxicity, but the radicals formed in its peroxidase reaction are potent anticancer agents (Folkes and Wardman, 2001). "Because of this, IAA and peroxidase conjugated to an extracellular tumor antigen are currently in phase II clinical trials" (Tafazoli and O'Brien, 2004).

Reactions of IAA with different plant peroxidases have been studied (Gazaryan et al., 1996; Savitsky et al., 1999). The rates of reactions of the indole derivatives tryptamine, *N*-acetyltryptamine, trptophan, melatonin, and serotonin, with Compounds I and II of horseradish peroxidase, lactoperoxidase, and myeloperoxidase have been measured and compared (Jantschko et al., 2002).

REACTION OF ISOBUTYRALDEHYDE WITH HORSERADISH PEROXIDASE

The most spectacular finding of the group of Giuseppe Cilento was the discovery that radiation from triplet state acetone is obtained from the horseradish peroxidase-isobutyraldehyde-oxygen system (Durán et al., 1977; Bechara et al., 1978; Faria Oliveira et al., 1978).

If the isobutyraldehyde is vacuum distilled and kept out of contact with air until the final mixing of reagents then no reaction occurs. The initiation of the reaction is caused

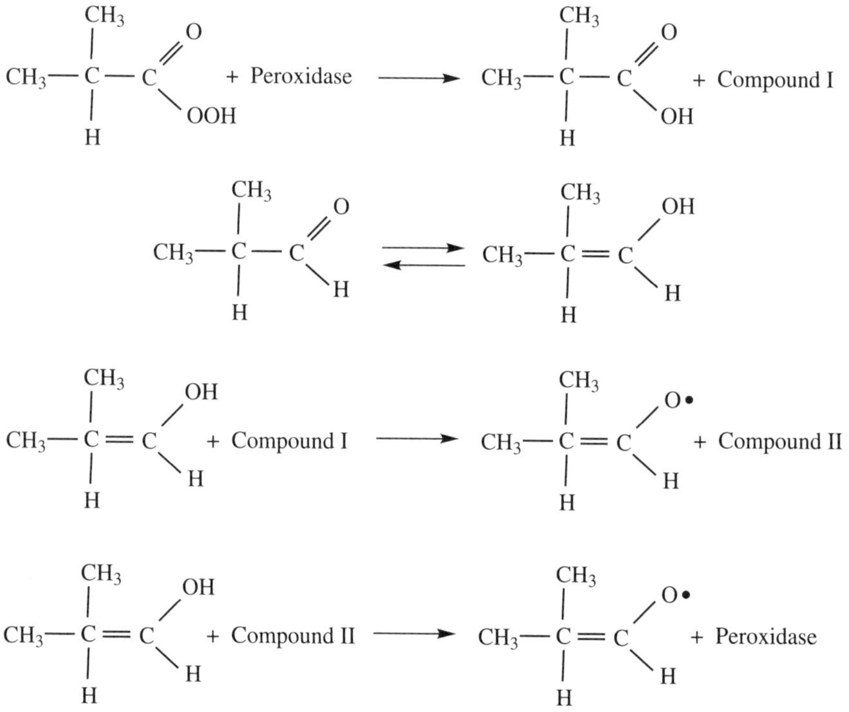

Figure 5.7. Mechanism for the reaction cycle of isobutyraldehyde with horseradish peroxidase. The reaction is initiated by autoxidation of isobutyraldehyde to the corresponding peracid. The enol form of isobutyraldehyde reacts in an analogous way to phenols.

by two factors. Trace autoxidation of isobutyraldehyde to the corresponding peroxy acid occurs by a chain reaction, the mechanism of which is well understood (Zaikov et al., 1973). Phosphate buffer is a catalyst of the keto-enol tautomerism of isobutyraldehyde. Thus, the two key starting reagents are generated: a peroxy acid and the enol form of isobutyraldehyde. The isobutyric peroxy acid ROOH reacts with native peroxidase to form Compound I, and the enol form of the aldehyde reacts in analogous fashion to phenols to reduce Compound I in two steps via Compound II back to native enzyme (Dunford et al., 1984; Baader et al., 1985; Bohne et al., 1987). The free radicals generated by oxidation by Compounds I and II are scavenged by molecular oxygen, leading to formation of a dioxetane, which fragments into triplet state acetone and formic acid. The peroxidase cycle with isobutyraldehyde is shown in Figure 5.7, and the subsequent formation of triplet state acetone is shown in Figure 5.8.

The light emission consists of an initial burst followed by a steady-state phase that can last up to 10 minutes. The initial burst occurs from rapid consumption of the equilibrium amount of enol in the isobutyraldehyde, which is followed by a steady-state replenishment of the enol. The only chain reaction occurs in the autoxidation of isobutyraldehyde. Thus a principal product of a conventional peroxidase cycle is light-generating triplet state acetone.

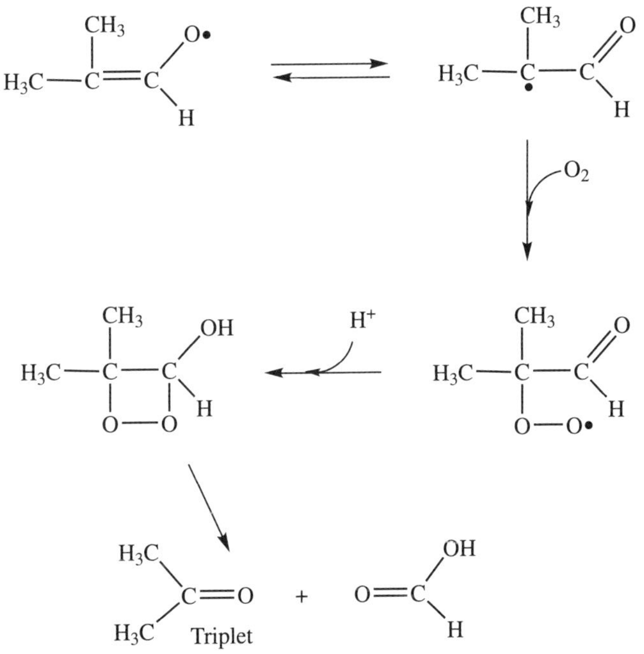

Figure 5.8. Rearrangement of the oxy radical of the enol form of isobutyraldehyde to a carbon-centered radical, followed by oxygen addition, dioxetane formation, and cleavage to formic acid and triplet state acetone. Light is emitted by the electronically excited acetone.

The rate of disappearance of horseradish peroxidase Compound I both with enol in excess and Compound I in excess was exploited to obtain the rate constants and equilibrium constants of aldehyde keto-enol tautomerism (Bohne et al., 1986).

Light Emission from Luminol

Cyclic hydrazides such as luminol are key reagents in many light-emitting reactions, utilized in interesting ways by many organisms. Both timing of light-emitting pulses and wavelength can be used to signal other members of the same species. Predators have learned to mimic these signals. The female *Photuris* species of firefly can generate the appropriate response light signal to that used by *Photinus* males as a sexual attractant, and lure the innocent male into the biggest and last surprise of his life (Agosta, 1996). The enzyme luciferase, not a peroxidase, oxidizes the light emitter luciferin in the North American firefly.

HRP will cause strong light emission from luciferin (Lind et al., 1983; Dunford, 1999). Aryl boronic acids are light enhancers in the HRP-luciferin reaction (Sun et al., 1994; Easton et al., 1996). Light enhancers have practical application in non-isotopic immunoassays (Whitehead et al., 1983) and nucleic acid sequencing (Kricka, 1995).

REFERENCES

Acosta M, Arnao MB, del Rio JA, Garcia-Cánovas F. 1989. Kinetic characterization of the inactivation process of two peroxide isoenzymes in the oxidation of indolyl-3-acetic acid. *Biochim. Biophys, Acta* **996**: 7–12.

Acosta M, del Rio JA, Arnao MB, Sánchez-Bravo J, Garcia-Carmona F, Garcia-Cánovas F. 1989. Oxygen consumption and enzyme inactivation in the indolyl-3-acetic acid oxidation catalyzed by peroxidase. *Biochim. Biophys, Acta* **955**: 194–202.

Afanasyeva MS, Taraban MB, Polyakov NE, Purtov PA, Leshina TV, Grissom CB. 2006. Elementary steps of enzymatic oxidation of nifedipine catalyzed by horseradish peroxidase. *J. Phys. Chem. B.* **110**: 21232–21237.

Agosta W. 1996. *Bombadier Beetles and Fever Trees*. Addison-Wesley, Reading MA.

Aguda BD, Clarke BL. 1987. Bistability in chemical reaction networks: theory and application to the peroxidase-oxidase reaction. *J. Chem. Phys.* **87**: 3461–3470.

Aguda BD, Larter R. 1990. Sustained oscillations and bistability in a detailed mechanism of the peroxidase-oxidase reaction. *J. Am. Chem. Soc.* **112**: 2167–2174.

Aguda BD, Larter R. 1991. Periodic-chaotic sequences in a detailed mechanism of the peroxidase-oxidase reaction. *J. Am. Chem. Soc.* **113**: 7913–7916.

Alexandre S, Dunford HB. 1991. A new model for oscillations in the peroxidase-oxidase reaction. *Biophys. Chem.* **40**: 189–195.

Anan FK, Mason HS. 1963. Peroxidase as oxidase: reaction between $^{18}O_2$ and dihydroxyfumarate. *Biochim. Biophys. Acta* **67**: 321–323.

Arnao MB, Sanchez-Bravo J, Acosta M. 1993. Indole-3-methanol as a substrate of horseradish peroxidase. In: Welinder KG, Rasmussen SK, Penel C, Greppin, editors. *Plant Peroxidases: Biochemistry and Physiology*. University of Geneva, 1993, pp. 181–184.

Baader WJ, Bohne C, Cilento G, Dunford HB. 1985. Peroxidase-catalyzed formation of triplet acetone and chemiluminescence from isobutyraldehyde and molecular oxygen. *J. Biol. Chem.* **260**: 10217–10225. Correction: 1986. *J. Biol. Chem.* **261**: 2994.

Bandurski RS, Cohen JD, Slovin JP, Reinecke DM. 1995. Auxin biosynthesis and metabolism. In: *Plant Hormones*, 2nd edition. Davies PJ, editor. Kluwer Academic, Dordrecht, pp. 39–65.

Bechara EJH, Faria Oliveira OMM, Durán N, Caseidei de Baptista R, Cilento G. 1978. *J. Biol. Chem.* **253**: 4707–4712.

Bohne C, MacDonald ID, Dunford HB. 1986. Measurement of rates and equilibria from keto-enol tautomerism of aldehydes using horseradish peroxidase Compound I. *J. Am. Chem. Soc.* **108**: 7867–7868.

Bohne C, MacDonald ID, Dunford HB. 1987. Transient state kinetics of the reactions of isobutyraldehyde with Compounds I and II of horseradish peroxidase. *J. Biol. Chem.* **262**: 3572–3578.

Bronnikova TV, Fed'kina VR, Schaffer WM, Olsen LF. 1995. Period-doubling bifurcations and chaos in a detailed model of the peroxidase-oxidase reaction. *J. Phys. Chem.* **99**: 9309–9312.

Bronnikova TV, Schaffer WM, Hauser MJB, Olsen LF. 1997. Routes to chaos in the peroxidase-oxidase reaction. II. The fat torus scenario. *J. Phys. Chem. B.* **102**: 632–640.

Bronnikova TV, Schaffer WM, Olsen LF. 2001. Nonlinear dynamics of the peroxidase-oxidase reaction. I. Bistability and bursting oscillations at low enzyme concentrations. *J. Phys. Chem. B.* **105**: 310–321.

Candeias LP, Folkes LK, Dennis MF, Patel KB, Everett SA, Stratford MRL, Wardman P. 1994. Free-radical intermediates and stable products in the oxidation of indole-3-acetic acid. *J. Phys. Chem.* **98**: 10131–10137.

Candeias LP, Folkes LK, Porssa M, Parrick J, Wardman P. 1996. Rates of reaction of indoleacetic acids with horseradish peroxidase Compound I and their dependence on the redox potentials. *Biochemistry* **35**: 102–108.

Carson JJL, Walleczek J. 2003. Response of the peroxidase-oxidase oscillator to light is controlled by MB^+-NADH photochemistry. *J. Phys. Chem. B.* **107**: 8637–8642.

Chen S, Schopfer P. 1999. Hydroxyl-radical production in physiological reactions: a novel function of peroxidase. *Eur. J. Biochem.* **260**: 726–735.

Cilento G. 1980. Photobiochemistry in the dark. *Photochem. Photobiol. Rev.* **5**: 109–228.

Cilento G. 1984. Generation of electronically excited triplet species in biochemical systems. *Pure Appl. Chem.* **56**: 1179–1190.

Cilento G, Adam W. 1988. Photochemistry and photobiology without light. *Photochem. Photobiol.* **48**: 109–228.

Degn H. 1968. Bistability caused by substrate inhibition of peroxidase in an open reaction system. *Nature* **217**: 1047–1050.

Degn H. 1969. Compound III kinetics and chemiluminescence in oscillatory oxidation reactions catalyzed by horseradish peroxidase. *Biochim. Biophys. Acta* **180**: 271–290.

Dunford HB. 1999. *Heme Peroxidases*. Wiley-VCH, New York.

Dunford HB, Baader WJ, Bohne C, Cilento G. 1984. On the mechanism of peroxidase-chemiluminescence from isobutyraldehyde. *Biochem. Biophys. Res. Commun.* **122**: 28–32.

Durán N, Faria Oliveira OMM, Haun M, Cilento G. 1977. Enzymically generated triplet acetone. *J. Chem. Soc. Chem. Commun.* 442–443.

Easton PM, Simmonds AC, Rakishev A, Egorov AM, Candeias LP. 1996. Quantitative model of the enhancement of peroxidase-induced luminol luminescence. *J. Am. Chem. Soc.* **118**: 6619–6624.

Escobar JA, Vásquez-Vivar, Cilento G. 1992. Free radicals and excited species in the metabolism of indole-3-acetic acid and its ethyl ester by horseradish peroxidase and by neutrophils. *Photochem. Photobiol.* **55**: 895–902.

Faria Oliveira OMM, Haun M, Durán N, O'Brien PJ, O'Brien CR, Bechara EJH, Cilento G. 1978. Enzyme-generated electronically excited carbonyl compounds. Acetone phosphorescence during the peroxidase-catalyzed aerobic oxidation of isobutanal. *J. Biol. Chem.* **253**: 4707–4712.

Folkes LK, Wardman P. 2001. Oxidative action of indole-3-acetic acids to cytotoxic species: a potential new role for plant auxins in cancer therapy. *Biochem. Pharmacol.* **61**: 129–136.

Fox LR, Purves WK, Nakada HI. 1965. The role of horseradish peroxidase in indole-3-acetic acid oxidation. *Biochemistry* **4**: 2754–2763.

Gaspar T. 1970. Dégradation de l'acide β-indolylacrylique par le système peroxydase-acide β-indolylacétique en l'absence de peroxide exogène. *C.R. Acad. Sci. Paris* **271**: 928–931.

Gazaryan IG, Lagramini LM, Ashby GA, Thorneley RNF. 1996. Mechanism of indole-3-acetic acid oxidation by plant peroxidases: anaerobic stopped-flow spectrophotometric studies on horseradish and tobacco peroxidases. *Biochem. J.* **313**: 841–847.

Hauser MJB, Olsen LF. 1998. The role of naturally occurring phenols in inducing oscillations in the peroxidase-oxidase reaction. *Biochemistry* **37**: 2458–2469.

Hauser MJB, Olsen LF, Bronnikova TV, Schaffer WM. 1997. Routes to chaos in the peroxidase-oxidase reaction: period-doubling and period-adding. *J. Phys. Chem. B.* **101**: 5075–5083.

Hinman RL, Lang J. 1965. Peroxidase-catalyzed oxidation of indole-3-acetic acid. *Biochemistry* **4**: 144–158.

Jantschko W, Furtmüller PG, Allegra M, Livrea MA, Jakopitsch C, Regelsberger G, Obinger C. 2002. Redox intermediates of plant and mammalian peroxidases: a comparative transient-kinetic study of their reactivity toward indole derivatives. *Arch. Biochem. Biophys.* **398**: 12–22.

Kanofsky JR. 1988. Singlet oxygen production from the peroxidase-catalyzed oxidation of indole-3-acetic acid. *J. Biol. Chem.* **263**: 14171–14175.

Kashem MA, Dunford HB. 1986. Kinetics of the oxidation of reduced nicotinamide adenine dinucleotide by horseradish peroxidase compounds I and II. *Biochem. Cell Biol.* **64**: 323–327.

Kenten RH. 1955. The oxidation of indolyl-3-acetic acid by waxpod bean root sap and peroxidase systems. *Biochem. J.* **59**: 110–121.

Kirkor ES, Scheeline A. 2000. Nicotinamide adenine dinucleotide species in the horseradish peroxidase-oxidase oscillator. *Eur. J. Biochem.* **267**: 5014–5022.

Kirkor ES, Scheeline A. 2001. Horseradish peroxidase adsorption on silica surfaces as an oscillatory dynamical behavior. *J. Phys. Chem. B.* **105**: 6278–6280.

Klapper MH, Hackett DP. 1963a. The oxidatic activity of horseradish peroxidase. I. Oxidation of hydro- and naphthohydroquinones. *J. Biol. Chem.* **238**: 3736–3742.

Klapper MH, Hackett DP. 1963b. The oxidatic activity of horseradish peroxidase. II. Participation of ferroperoxidase. *J. Biol. Chem.* **238**: 3743–3749.

Kobayashi S, Sugioka K, Nakano H, Nakano M, Tero-Kubato S. 1984. Analysis of the stable end products and intermediates of oxidative decarboxylation of indole-3-acetic acid by horseradish peroxidase. *Biochemistry* **23**: 4589–4597.

Kricka LJ, Editor. 1995. *Nonisotopic Probing, Blotting, and Sequencing*, 2nd edition. Academic Press, New York.

Krylov SN. 1998. Computer simulation of damped oscillations during peroxidase-catalyzed oxidation of indole-3-acetic acid. *Biophys. Chem.* **72**: 285–295.

Krylov SN, Chebotareva AB. 1993. Peroxidase-catalyzed co-oxidation of indole-3 acetic acid and xanthene dyes in the absence of hydrogen peroxide. *FEBS Lett* **324**: 6–8.

Krylov SN, Dunford HB. 1996a. Detailed mechanism of phenol-inhibited peroxidase-catalyzed oxidation of indole-3-acetic acid at neutral pH. *Photochem. Photobiol.* **63**: 735–741.

Krylov SN, Dunford. 1996b. Detailed model of the peroxidase-catalyzed oxidation of indole-3-acetic acid at neutral pH. *J. Phys. Chem.* **100**: 913–920.

Krylov SN, Dunford HB. 1996c. Evidence for a free radical mechanism in the reaction between peroxidase and indole-3-acetic acid at neutral pH. *Biophys. Chem.* **58**: 325–334.

Krylov SN, Dunford HB. 1996d. Accelerating effect of umbelliferone on peroxidase-catalyzed oxidation of indole-3-acetic acid at neutral pH. *J. Phys. Chem.* **100**: 19719–19727.

Krylov SN, Lazarev VV, Rubin LB. 1990. Kinetics associated with chemiluminescence spectra during oxidation of heterauxin by peroxidase. *Doklady Biophys.* **310**: 28–31.

Kummer U, Hauser MJB, Wegmann K, Olsen LF, Baier G. 1997. Oscillations and complex dynamics in the peroxidase-oxidase reaction induced by naturally occurring substrates. *J. Am. Chem. Soc.* **119**: 2084–2087.

Larsen P. 1936. Über einen wuchsstoff-inaktvierenden stoff aus phaseolus-keimplanzen. *Planta* **25**: 311–314.

Lewis DD, Ruane ML, Scheeline A. 2006. Biofilm effects on the peroxidase-oxidase reaction. *J. Phys. Chem. B.* **110**: 8100–8104.

Lind J, Merényi G, Eriksen TE. 1983. Chemiluminescence mechanism of cyclic hyrazides such as luminol in aqueous solutions. *J. Am. Chem. Soc.* **105**: 7655–7661.

MacDonald ID, Dunford HB. 1989. Mechanism of horseradish peroxidase-catalyzed oxidation of malonaldehyde. *Arch. Biochem. Biophys.* **272**: 185–193.

Metodiewa D, Dunford HB. 1989. The reactions of horseradish peroxidase, lactoperoxidase and myeloperoxidase with enzymatically generated superoxide. *Arch. Biochem. Biophys.* **272**: 245–253.

Metodiewa D, Pires de Melo M, Escobar JA, Cilento C, Dunford HB. 1992. Horseradish peroxidase-catalyzed aerobic oxidation and peroxidation of indole-3-acetic acid. I. Optical spectra. *Arch. Biochem. Biophys.* **296**: 27–33.

Møller AC, Hauser MJB, Olsen LF. 1998. Oscillations in peroxidase-catalyzed reactions and their potential function in vivo. *Biophys. Chem.* **72**: 63–72.

Møller AC, Olsen LF. 1999. Effect of magnetic fields on an oscillating enzyme reaction. *J. Am. Chem. Soc.* **121**: 6351–6354.

Møller AC, Olsen LF. 2000. Perturbations of simple oscillations and complex dynamics in the peroxidase-oxidase reaction using magnetic fields. *J. Phys. Chem. B.* **104**: 140–146.

Mottley C, Mason RP. 1986. An electron spin resonance study of free radical intermediates in the oxidation of indole acetic acid by horseradish peroxidase. *J. Biol. Chem.* **261**: 16860–16864.

Nakajima R, Yamazaki I. 1979. The mechanism of indole-3-acetic acid oxidation by horseradish peroxidase. *J. Biol. Chem.* **254**: 872–878.

Nakajima R, Yamazaki I. 1980. The conversion of horseradish peroxidase C to a verdohemoprotein by a hydroperoxide derived enzymaticalaly from indole-3-acetic acid and by *m*-nitroperoxybenzoic acid. *J. Biol. Chem.* **255**: 2067–2071.

Olsen LF, Degn H. 1977. Chaos in an enzyme reaction. *Nature* **267**: 177–178.

Olsen LF, Degn H. 1978. Oscillatory kinetics of the peroxidase-oxidase reaction in an open system. Experimental and theoretical studies. *Biochim. Biophys. Acta* **523**: 321–334.

Olsen LF, Hauser MJB, Kummer U. 2003. Mechanism of protection of peroxidase activity by oscillatory dynamics. *Eur. J. Biochem.* **270**: 2796–2804 (2003).

Olson DL, Scheeline A. 1995a. The peroxidase-NADH biochemical oscillator. I. Examination of oxygen mass transport, the effect of light, and the role of methylene blue. *J. Phys. Chem.* **99**: 1204–1211.

Olson DL, Scheeline A. 1995b. The peroxidase-NADH biochemical oscillator. II. Examination of the role of hydrogen peroxide and superoxide. *J. Phys. Chem.* **99**: 1212–1217.

Olson DL, Williksen EP, Scheeline A. 1995. An experimentally based model of the peroxidase-NADH biochemical oscillator: an enzyme-mediated chemical switch. *J. Am. Chem. Soc.* **117**: 2–15.

Pires de Melo M, Escobar JA, Metodiewa D, Dunford HB, Cilento G. 1992. Horseradish peroxidase-catalyzed aerobic oxidation and peroxidation of indole-3-acetic acid. II. Oxygen uptake and chemiexcitation. *Arch. Biochem. Biophys.* **296**: 27–33.

Ricard J, Job D. 1974. Reaction mechanisms of indole-3-acetate degredation by peroxidases. *Eur. J. Biochem.* **44**: 359–374.

Ricard J, Job D, Santimone M, Cambe H. 1969. Cinétique de formation et de decomposition de la ferroperoxydase, de la ferriperoxydase et du compose III, en présence d'acid β-indodlylacétique. *C.R. Acad. Sci. Paris* **269**: 1445–1448.

Ricard J, Nari J. 1967. The formation and reactivity of peroxidase Compound III. *Biochim. Biophys. Acta* **132**: 321–329.

Savitsky PA, Gazaryan IG, Tishkov VI, Lagrimini LM, Ruzgas T, Gorton L. 1999. Oxidation of indole-3-acetic acid by dioxygen catalyzed by plant peroxidases: specificity for the enzyme structure. *Biochem. J.* **340**: 579–583.

Schaffer WM, Bronnikova TV, Olsen LF. 2001. Nonlinear dynamics of the peroxidase-oxidase reaction. II. Compatibility of an extended model with previously reported model-data correspondences. *J. Phys. Chem. B.* **105**: 5331–5340.

Scheeline A, Olson DL, Williksen EP, Horras GA, Klein ML, Larter R. 1997. The peroxidase-oxidase oscillator and its constituent chemistries. *Chem. Rev.* **97**: 739–756.

Sevcík P, Dunford HB. 1991. Kinetics of the oxidation of NADH by methylene blue in a closed system. *J. Phys. Chem.* **95**: 2411–2415.

Smith AM, Morrison WL, Milham PJ. 1982. Oxidation of indole-3-acetic acid by peroxidase: involvement of reduced peroxidase and Compound III with superoxide as product. *Biochemistry* **21**: 4414–4419.

Sun W, Ji X, Kricka LJ, Dunford HB. 1994. Rate constants for reactions of horseradish peroxidase compounds I and II with 4-substituted arylboronic acids. *Can. J. Chem.* **72**: 2159–2162.

Swedin B, Theorell H. 1940. Dioximaleic acid oxidase activity of peroxidases. *Nature* **145**: 71–72.

Tafazoli S, O'Brien PJ. 2004. Prooxidant activity and cytotoxic effects of indole-3-acetic acid derivative radicals. *Chem. Res. Toxicol.* **17**: 1350–1355.

Tan X, Calderon-Villalobos LIA, Sharon M, Zheng C, Robinson CV, Estelle M, Zheng N. 2007. Mechanism of auxin perception by the TIR1 ubiquitin ligase. *Nature* **446**: 640–645.

Valeur KR, Olsen LF. 1996. Kinetic studies of the oscillatory dynamics in the peroxidase-oxidase reaction catalyzed by four different peroxidases. *Biochim. Biophys. Acta* **1289**: 377–384.

Vidigal CCC, Zinner K, Durán N, Bechara EJH, Cilento G. 1975. Generation of electronic energy in the peroxidase catalyzed oxidation of indole-3-acetic acid. *Biochem. Biophys. Res. Commun.* **65**: 138–145.

Whitehead TP, Thorpe GHG, Carter TJN, Groucutt C, Kricka LJ. 1983. Enhanced luminescence procedure for sensitive determination of peroxidase-labelled conjugates in immunoassay. *Nature* **305**: 158–159.

Ximenes VF, Campa A, Catalani LH. 2001. The oxidation of indole derivatives catalyzed by horseradish peroxidase is highly chemiluminescent. *Arch. Biochem. Biophys.* **387**: 173–179.

Yamazaki I, Piette LH. 1963. The mechanism of aerobic oxidase reaction catalyzed by peroxidase. *Biochim. Biophys. Acta* **77**: 47–64.

Yamazaki H, Yamazaki I. 1973. The reaction between indole-3-acetic acid and horseradish peroxidase. *Arch. Biochem. Biophys.* **154**: 147–159.

Yamazaki I, Yokota K. 1965. Conversion of ferrous peroxidase into Compound III in the presence of NADH. *Biochem. Biophys. Res. Commun.* **19**: 249–254.

Yamazaki I, Yokota K. 1973. Oxidation states of peroxidase. *Mol. Cell. Biochem.* **2**: 39–52.

Yamazaki I, Yokota K, Nakajima R. 1965. Oscillatory oxidations or reduced pyridine nucleotide by peroxidase. *Biochem. Biophys. Res. Commun.* **21**: 582–586.

Yokota K, Yamazaki I. 1977. Analysis and computer simulation of aerobic oxidation of reduced nicotinamide adenine dinucleotide catalyzed by horseradish peroxidase. *Biochemistry* **16**: 1913–1920.

CHAPTER 6

YEAST CYTOCHROME *c* PEROXIDASE: REACTIONS WITH SMALL SUBSTRATES

INTRODUCTION

Yeast cytochrome *c* peroxidase (C*c*P) is unique for structural, mechanistic, and historic reasons. It is a peroxidase without carbohydrate attachments, the site of its free radical in Compound I (formerly misnamed Complex ES, then Compound ES) is on the protein, not the porphyrin, and its natural reducing substrate is a macromolecule, cytochrome *c*. Its crystal structure was the first obtained for a peroxidase. This led to a proposed mechanism for Compound I formation by the oxidizing substrate hydrogen peroxide, the salient features of which have stood the test of time, and are applicable to other heme peroxidases. The classification of heme peroxidase families and superfamilies was based on the C*c*P structure as the reference point. Yeast cytochrome *c* peroxidase was one of the first enzymes to have been cloned and modified by site-directed mutagenesis.

Discovered in 1940 (Altschul et al., 1940; Abrams et al., 1942), extensive further work did not start until 1965, mainly in the laboratory of Takashi Yonetani, who introduced chromatographic purification procedures. This work has been well reviewed (Yonetani, 1970, 1976). The detailed story follows.

PROPERTIES OF YEAST CYTOCHROME *c* PEROXIDASE

Cytochrome *c* peroxidase (C*c*P) is found in bakers' or brewers' yeast. It consists of a single polypeptide chain of 294 residues with a heme *b* prosthetic group. Its

Peroxidases and Catalases: Biochemistry, Biophysics, Biotechnology, and Physiology, Second Edition.
By H. Brian Dunford

molecular weight is 35,350. The protein charge varies from +32 at pH 2 to −48 at pH 12 (Conroy and Erman, 1978). The isoelectric point of the apoenzyme is 5.45. The final correct amino acid sequence has been published (Bosshard et al., 1991).

Two valuable papers describe many of the enzyme's properties (Yonetani and Anni, 1987; Vitello et al., 1990b). The Soret maximum absorbance at 408 nm is $9.8 \times 10^4\ M^{-1}\ cm^{-1}$, and that of Compound I at 420 nm is $\sim 1.05 \times 10^5\ M^{-1}\ cm^{-1}$. The latter was obtained by adding a stoichiometric amount of hydrogen peroxide to native enzyme. Accurate spectra of native enzyme and Compound I have been published (Bosshard et al., 1991). Because the free radical site of Compound I is on the protein not the porphyrin, it has little or no effect on the heme chromophore. Thus the Compound I optical spectrum of C*c*P is very similar to that of typical Compound II spectra of other heme peroxidases. The small differences between the Compound I and Compound II spectra of yeast cytochrome *c* peroxidase have been measured by two different methods (Coulson et al., 1971; Ho et al., 1983). The differences may be attributed largely to the free radical site but there is not much agreement between the two determinations.

Freshly prepared C*c*P has five-coordinate high spin iron(III) (Sievers, 1978; Yonetani and Anni, 1987). Freezing or aging, particularly at acid pH, converts the iron to low spin, six coordinate. Different kinetic activities within a single sample were shown (Balny et al., 1987). Acetate buffers should be avoided (Mathews and Wittenberg, 1979) as should nitrate, used as a counterion (Vitello et al., 1990b). In aged samples it appears that water bound to distal His52, or distal His itself may coordinate to the iron. The Reinheitzahl (RZ) value (*A* 408 nm/*A* 280 nm) of pure freshly prepared C*c*P is 1.25 to 1.30 at pH 7 (Yonetani and Anni, 1987). Preparations contaminated with six-coordinate iron have higher RZ values, a unique occurrence. Normally higher RZ values indicate higher purity.

Spectral properties of different peroxidases have been compared (Smulevich, 1996). Native C*c*P contains more purely high spin iron(III) than horseradish peroxidase and its Soret maximum absorbance occurs at longer wavelength: 408 compared to 403 nm. Differences in reactivity result.

CRYSTAL STRUCTURES OF YEAST CYTOCHROME *c* PEROXIDASE, ITS COMPOUNDS AND COMPLEXES

Although crystals of yeast cytochrome *c* peroxidase were reported in various laboratories (Ellfolk, 1967; Yonetani, 1967; Larsson et al., 1970), a detailed 2.5 Å resolution x-ray determination of the crystal structure was not reported until several years later (Poulos et al., 1980). It was the first crystal structure for a heme-containing enzyme. It was followed by a refined structure at 1.7 Å resolution (Finzel et al., 1984), and a 2.5 Å resolution of Compound I (Edwards et al., 1987). One view of the structure of yeast cytochrome *c* peroxidase is shown in Figure 6.1.

Proximal and distal histidines were identified, as was a distal arginine. The proximal His forms a hydrogen bond with a buried Asp side chain. Several water molecules occupy sites on the distal side. Thus the heme environment is more polar than in the

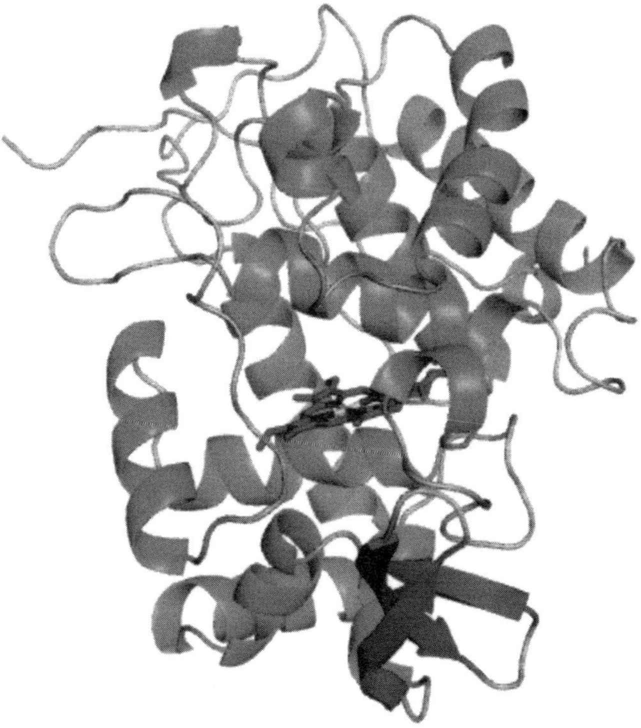

Figure 6.1. Structure of yeast cytochrome *c* peroxidase. (Kindly supplied by Dr. Tom Poulos.)

globins and cytochromes *c*, so that higher oxidation states can be stabilized. In this and a subsequent crystal structure determination (Finzel et al., 1984; Bonagura et al., 2003), water was identified as occupying the sixth coordination position on the distal side of the heme. This form of C*c*P reacts with hydrogen peroxide, but at a slower rate than the five-coordinate form where the sixth position is vacant. In the crystal structure of a plasmid-encoded form of C*c*P expressed in *E. coli*, the sixth coordination position is vacant and remains vacant in the Trp191Phe mutant (Wang et al., 1990). Replacement of proximal His175 with Gln or Glu had little effect on Compound I formation (Choudhury et al., 1994).

Crystals of Compound I were obtained by soaking native C*c*P crystals in a solution containing hydrogen peroxide (Edwards et al., 1987). Comparison of Compound I and native C*c*P crystal structures, both obtained at 2.5 Å resolution, showed the most pronounced differences occurred near the heme. The iron moved 0.2 Å toward the distal side of the heme. The second most pronounced feature involved the side chain of Thr180 which moved 0.15 Å toward Met230. This, combined with mutagenesis experiments that eliminated Trp51 and Met172 as Compound I free radical sites, led to the proposal that the Compound I free radical site occurs in the Met230, Met231, Trp191 cluster, a prescient prediction that was slow to be accepted. In another crystallographic study, no significant structural changes were observed at the Trp191 site (Fülöp et al., 1994). Crystal structures of Trp191Gly and Trp191Gln mutants revealed that the site occupied by Trp191 is a cation binding site in other enzymes. Thus the Trp191 cation radical in Compound I is stabilized compared to the same cation radical in solution (Miller et al., 1994a).

The His52Leu mutant loses most of its ability to form Compound I, and the loss cannot be attributed to important structural changes (Erman et al., 1993). Compound 0, the Fe—OOH precursor of Compound I, was modeled by determining the structure of oxyperoxidase, also known as Compound III: Fe^{II}—O_2 (Miller et al., 1994b). Crystal structures were obtained for cyanide, nitric oxide, carbon monoxide, and fluoride complexes (Edwards and Poulos, 1990).

In the latest crystal structures for native C*c*P and Compound I, obtained at 1.2 and 1.3 Å resolutions, it was shown that Fe moves into the porphyrin plane upon Compound I formation, and the Fe^{IV}—O bond distance is 1.87 Å, claimed to be too long for a double bond (Bonagura et al., 2003). The implication is that Compound I contains Fe^{IV}—OH, a claim that is disputed (see Chapter 8).

MECHANISM OF COMPOUND I FORMATION

The crystal structure of yeast cytochrome *c* peroxidase led to a mechanism for Compound I formation, which has largely stood the test of time and, aside from location of the final free radical site, is applicable to all heme peroxidases of Classes I–III (Poulos and Kraut, 1980). The essential features are first formation of Compound 0: a proton from the oxidant ROOH is transferred to distal His and ROO^- is bound to the iron, changing its coordination number from five to six. Starting with Compound 0, features of Compound I formation are shown in Figure 6.2.

Formation of the transition state is stabilized by distal Arg, which is followed by departure of the ROH leaving group, and formation of the ferryl group Fe═O.

An additional role of gatekeeper for distal Arg was proposed on the basis of results obtained in a directed molecular evolution study (Iffland et al., 2000). "After three

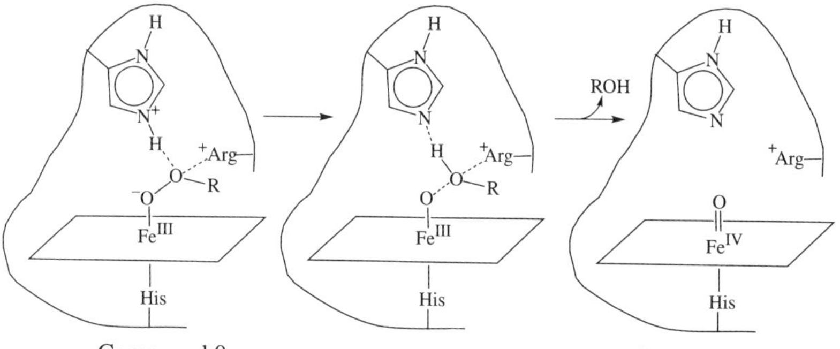

Figure 6.2. Conversion of Compound 0 to Compound I.

rounds of DNA shuffling and screening, mutants were isolated which possessed a 300-fold increased activity against guaiacol" (Iffland et al., 2000, p. 10790). In all cases the distal Arg was mutated to His. Thus Arg plays a role in controlling access to the ferryl oxygen and distal His. The complicating effects of replacement of the proximal His by free imidazole have been studied in detail (Hirst et al., 2001a, 2001b).

The transfer of an oxygen atom from ROOH to iron brings two oxidizing equivalents to the enzyme through iron(III). In all heme peroxidase Compound I, the final state is iron(IV) plus a free radical site. For nearly all peroxidases that free radical site is on the porphyrin. For yeast cytochrome *c* peroxidase it is different, and its exact location involved a ~30-year quest.

The free radical site gives a strong and unusual ESR signal with $g_{\parallel} = 2.04$ and $g_{\perp} = 2.01$ (Yonetani and Schleyer, 1967; Wittenberg et al., 1968). In the original Poulos–Kraut mechanism (1980), the following postulates were made, later shown to be wrong. It was assumed that water was in the sixth coordination position of the iron and Trp 51, which lies about 3.6 Å above and is parallel to the heme, was the free radical site. In one electron nuclear double resonance (ENDOR) and ESR study a methione residue, perhaps Met 172, was implicated (Hoffman et al., 1979, 1981), whereas a later ESR study indicated a Trp residue, either Trp 51 or 191 (Hori and Yonetani, 1985). Met172 was eliminated as the free radical site by site-directed mutagenesis (Goodin et al., 1986). Comparison of the 2.5 Å resolution structures of native and Compound I enzymes showed structural differences in the region of the cluster Met230, Met231, and Trp191 (Edwards et al., 1987). Replacement of Trp191 by Phe eliminated the signal and greatly decreased enzyme activity (Mauro et al., 1988). The ENDOR spectrum of the enzyme containing selectively deuterated Trp191 also indicated that Trp191 is the site of the free radical (Sivaraja et al., 1989). The ESR spectrum is not typical of a planar π radical. It was accounted for by weak exchange interaction between the $S = 1$ oxyferryl $(Fe{=}O)^{2+}$ moiety and the radical on Trp191 (Houseman et al., 1993). A further ENDOR study in which Trp191 was enriched with ^{13}C, ^{15}N, and ^{2}H provided proof that a π-cation radical form of Trp191 is the dominant free radical site (Huyett et al., 1995).

Yeast C*c*P Compound I was prepared electrolytically in the absence of hydrogen peroxide (Mondal et al., 1996). Tyr39 and Tyr153 have been identified by spin trapping as potential second free radical sites (Zhang et al., 2002). Yeast C*c*P contains a large number of oxidizable amino acid residues. Double replacements of Tyr and Trp residues by less oxidizable residues led to increased stability of Compound I (Pfister et al., 2001). One of Tyr36, Tyr39, or Tyr42 was shown to be involved in protein cross-links, formed by long exposure to excess hydrogen peroxide.

Sulfur centers provide extra electron density. Thus, iron sites are commonly associated with oxidation reactions, whereas iron-sulfur clusters are reductants. Trp191 is located in a cluster with Met230 and Met231. The latter sites may be responsible for the extra electron density that enables Trp191 to surpass the porphyrin ring as the preferred electron donor to the transient iron(V) formed by oxygen atom transfer from hydrogen peroxide. Compound I can also be formed by substituting peroxy acids for hydrogen peroxide, but at a slower rate (Palamakumbura et al., 1999a, 1999b).

THE REACTION CYCLE FOR YEAST CYTOCHROME *c* PEROXIDASE

A great deal of valuable information was obtained in an elegant kinetic, ESR, and stoichiometric study (Coulson et al., 1971). It was confirmed that the enzyme and hydrogen peroxide react in a 1 : 1 ratio. It was shown that two equivalents of reducing substrate are required to reduce Compound I to the native enzyme, and that reductions occur in two sequential steps. Ferrocyanide was used as the reducing agent since it is known to be oxidized in a one-electron reaction to ferricyanide. The first reduction step is faster than the second by a factor of about two. A further intensive study was made of the ferrocyanide reaction (Jordi and Erman, 1974a, 1974b).

The total cycle is complicated by the existence of two forms of Compound II. One of them is the conventional ferryl $Fe^{IV}{=}O$ form with no free radical. The other is an Fe^{III} form with the free radical remaining on the protein. The latter form would have an optical spectrum very similar to that of native enzyme. Thus, judging by the optical spectra, this complicated enzyme has a Compound I that looks like a conventional Compound II, and a Compound II that partly looks like a conventional Compound II but also looks like native enzyme.

Three mechanisms were examined for the reduction of Compound I to native enzyme: (1) a conventional peroxidase cycle, (2) an independent sites mechanism, and (3) an equilibrium mechanism. The conventional cycle was eliminated on the basis of kinetic, spectral, and ESR data. Nevertheless, it provides a well known reference point for the discussion that follows.

In the independent sites mechanism, reduction of $Fe^{IV}{=}O$ and of the Trp191 radical proceed independently. In the equilibrium mechanism for reduction, the faster phase would correspond to formation of the two forms of Compound II, in equilibrium with each other, followed by the slower phase in which Compound II is reduced to native enzyme (Coulson et al., 1971). The interconversion is not attained simply but must involve a complex series of steps. Not only is there an intramolecular electron transfer, but two protons and a water molecule are also involved.

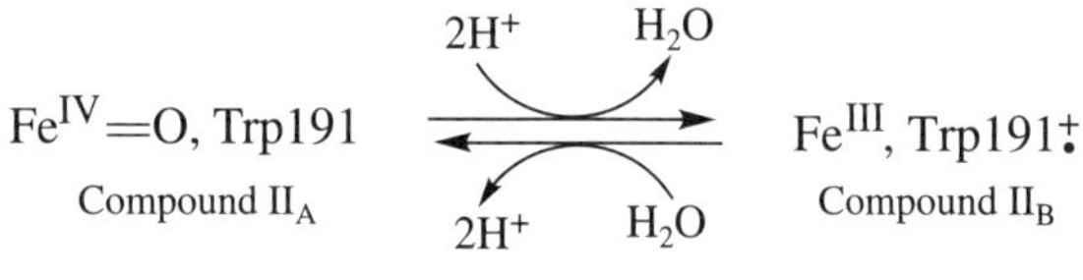

Evidence for this equilibrium was obtained by trapping Compound II_B with fluoride, which can bind to iron(III) but not to the ferryl $Fe^{IV}{=}O$ form with its blocked sixth site.

With the equilibrium shifted by added fluoride, a more intense ESR signal of the $Trp191^{\bullet+}$ radical was obtained. No one has ever observed pure Compound II_B since it cannot be formed in the absence of Compound II_A. However, by use of fluoride or other ligands the shift to the Compound II_B-bound ligand form can be attained (Ho et al., 1984).

A stable iron(IV) form of C*c*P, without the free radical, was prepared below pH 8 by two methods: acetophenone/isopropyl alcohol-mediated photochemical reduction of ferric enzyme, or reduction with dithionite and trapping with CO (Ho et al., 1983).

The photochemical method is vastly superior because dithionite reduction is incomplete and has complications. The resultant ferrous enzyme was oxidized by hydrogen peroxide at pH 7 to form a pure stable Compound II_A containing Fe^{IV}=O and no protein free radical. Difference spectra between the pure Fe^{IV}=O form and Compound I were obtained. Compound I has a broad weak absorption band centered at ~570 nm, not present in the iron(IV) enzyme, and hence is attributable to the free radical site. In the discussion of the nature of the free radical site, some possibilities were eliminated on the assumption that large shifts from "normal" pK_a values will not occur (Ho et al., 1983), an assumption with which I disagree. (See discussion of electron density circuits in Chapter 10.) The Fe^{IV}=O species formed by hydrogen peroxide oxidation of the reduced enzyme is stable and does not contain a protein free radical site. It reacts with ferrocyanide with a rate that corresponds to neither of the rates obtained by the biphasic reduction of Compound I by ferrocyanide (Jordi and Erman, 1974a, 1974b).

Two Different Free Radicals?

The existence of an Fe^{IV}=O form without any equilibrium formation of a free radical led to the postulate that there are two forms of free radical in oxidized C*c*P, one of which is more stable than the other (Ho et al., 1983). It was proposed that there is a switch that is open only when oxidants or reductants are reacting. The closed switch bars transfers of oxidizing equivalents within the enzyme. This intriguing possibility may be applicable to prostaglandin H synthase, which also has an oxidizable site on the proximal side of the heme.

The switch mechanism provides an explanation for the fact that the rate of reduction of pure Compound II_A by ferrocyanide (Ho et al., 1983) does not coincide with either the steady-state or transient-state kinetic results (Jordi and Erman, 1974a, 1974b). The ferrous peroxidase-derived, stable form of Fe^{IV}=O can be reacted with fluoride and the result is formation of the Fe^{III}-fluoride complex and transfer of the oxidizing equivalent to the free radical site! The transfer process is first order with a rate constant of $0.11_5\ s^{-1}$ which is independent of either ligand or protein concentration. It was assumed that this rate constant was applicable to the electron transfer process (Ho et al., 1984).

Contradictory evidence to the switch mechanism has been obtained (Goodin and McRee, 1993). Ferrous enzyme was prepared by electrolytic reduction of C*c*P using methyl viologen as mediator. Addition of a stoichiometric amount of hydrogen peroxide produced the pure Fe^{IV}=O form of Compound II with no free radical. This form of Compound II reacted with mesidine at the same rate as the slower phase of Compound I reduction to native enzyme by mesidine (Roe and Goodin, 1993).

STEADY-STATE KINETICS

Because of the difficulty of distinguishing spectrally the various oxidized forms of yeast cytochrome *c* peroxidase, perhaps the most revealing mechanistic evidence has come from steady-state kinetics. Below the three steady-state mechanisms analyzed by Coulson et al. (1971) plus a further study by Yandell and Yonetani (1983) are examined in detail.

Conventional Mechanism

For simplicity the reducing substrate is indicated as e^- and proton balances are neglected.

$$\underset{\text{C}c\text{P}}{\text{Fe}^{\text{III}}, \text{Trp}} + \text{H}_2\text{O}_2 \xrightarrow{k_1} \underset{\text{Compound I}}{\text{Fe}^{\text{IV}}{=}\text{O}, \text{Trp}^{+\bullet}} + \text{H}_2\text{O}$$

$$\underset{\text{Compound I}}{\text{Fe}^{\text{IV}}{=}\text{O}, \text{Trp}^{+\bullet}} + e^- \xrightarrow{k_2} \underset{\text{Compound II}_\text{A}}{\text{Fe}^{\text{IV}}{=}\text{O}, \text{Trp}}$$

$$\underset{\text{Compound II}_\text{A}}{\text{Fe}^{\text{IV}}{=}\text{O}, \text{Trp}} + e^- \xrightarrow{k_3} \underset{\text{C}c\text{P}}{\text{Fe}^{\text{III}}, \text{Trp}}$$

The steady-state velocity v is given by:

$$\frac{v}{[\text{E}]_0} = \frac{1}{\dfrac{1}{k_1[\text{H}_2\text{O}_2]} + \dfrac{k_2 + k_3}{k_2 k_3[e^-]}} = \frac{1}{\dfrac{1}{k_1[\text{H}_2\text{O}_2]} + \dfrac{1}{k'[e^-]}}$$

where $[\text{E}]_0$ is the total enzyme concentration and $k' = k_2k_3/(k_2 + k_3)$. For the condition that a sufficiently large concentration of hydrogen peroxide is used, the velocity is reduced to

$$\frac{v}{[\text{E}]_0} = k'[e^-]$$

Equivalent Sites Mechanism

In this mechanism either oxidized site can be reduced independently of the other. Showing only the various enzyme species, the equivalent sites mechanism can be represented as

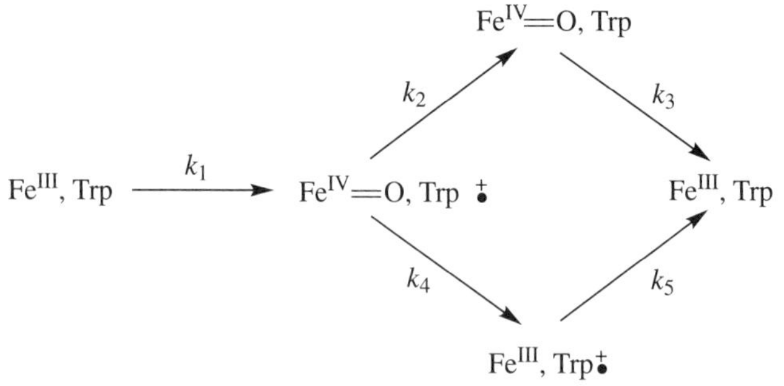

With H_2O_2 as oxidant and reductant designated simply as e^-, the rate equation is

$$\frac{v}{[E]_0} = \frac{1}{\dfrac{1}{k_1[H_2O_2]} + \dfrac{k_2k_5 + k_3k_5 + k_3k_4}{k_3k_5(k_2 + k_4)[e^-]}} = \frac{1}{\dfrac{1}{k_1[H_2O_2]} + \dfrac{1}{k''[e^-]}}$$

where $k'' = k_3k_5(k_2 + k_4)/(k_2k_5 + k_3k_5 + k_3k_4)$. For large $[H_2O_2]$

$$\frac{v}{[E]_0} = k''[e^-]$$

Thus the equivalent sites mechanism is kinetically indistinguishable from the conventional peroxidase mechanism.

Compound II Equilibrium Mechanism

If there is rapid equilibrium between Compounds II_A and II_B then it is not possible to tell whether one or both forms are reacting. One substitutes [Compound $II]_{total}$ for [Compound II] in the conventional mechanism and the form of the rate equations is identical. Thus the three mechanisms: conventional, equivalent sites, and rapid equilibrium between the two forms of Compound II are kinetically indistinguishable.

$Fe^{IV}{=}O$ Reduction Mechanism

This mechanism takes into account the slow intramolecular conversion of $Fe^{IV}{=}O$ and $Trp^{+\bullet}$ oxidizing equivalents. It is assumed here that the $Fe^{IV}{=}O$ site of Compound I is reduced first. The resultant Fe^{III} is reoxidized by the Trp free radical site to $Fe^{IV}{=}O$ which is again reduced, this time by the second molecule of reducing substrate. For simplicity protons and water molecules are not shown.

$$\underset{CcP}{Fe^{III}, Trp} + H_2O_2 \xrightarrow{k_1} \underset{\text{Compound I}}{Fe^{IV}{=}O, Trp^{+\bullet}}$$

$$\underset{\text{Compound I}}{Fe^{IV}{=}O, Trp^{+\bullet}} + e^- \xrightarrow{k_2} \underset{\text{Compound } II_B}{Fe^{III}, Trp^{+\bullet}}$$

$$\underset{\text{Compound } II_B}{Fe^{III}, Trp^{+\bullet}} \xrightarrow{k_A} \underset{\text{Compound } II_A}{Fe^{IV}{=}O, Trp}$$

$$\underset{\text{Compound } II_A}{Fe^{IV}{=}O, Trp} \xrightarrow{k_B} \underset{\text{Compound } II_B}{Fe^{III}, Trp^{+\bullet}}$$

$$\underset{\text{Compound } II_A}{Fe^{IV}{=}O, Trp} + e^- \xrightarrow{k_3} \underset{CcP}{Fe^{III}, Trp}$$

For the $Fe^{IV}{=}O$ reduction mechanism the rate is given by

$$\frac{v}{[E]_0} = \frac{1}{\dfrac{1}{k_1[H_2O_2]} + \dfrac{k_Ak_2 + k_Ak_3 + k_Bk_2}{k_Ak_2k_3} \bullet \dfrac{1}{[e^-]} + \dfrac{1}{k_A}}$$

$$= \frac{1}{\dfrac{1}{k_1[H_2O_2]} + \dfrac{1}{k'''[e^-]} + \dfrac{1}{k_A}}$$

where $k''' = k_Ak_2k_3/(k_Ak_2 + k_Ak_3 + k_Bk_2)$. For large (H_2O_2) this reduces to

$$\frac{v}{[E]_0} = \frac{1}{\dfrac{1}{k'''[e^-]} + \dfrac{1}{k_A}}$$

and the reciprocal is

$$\frac{[E]_0}{v} = \frac{1}{k'''[e^-]} + \frac{1}{k_A}$$

There are several important conclusions.

1. According to this mechanism the same oxidized species is reduced, reoxidized, and reduced the second time. Compound II_A and Compound II_B can be interchanged so the final equation contains k_B rather than k_A. Thus, the rate equations do not distinguish whether it is $Fe^{IV}{=}O$ or Fe^{III}, $Trp^{\dot{+}}$ that is reduced twice and reoxidized once.
2. The term $1/k_A$ is unique to the $Fe^{IV}{=}O$ reduction mechanism. Data that fit the above mechanism exclude three other mechanisms: conventional, equivalent sites, and rapid equilibrium.
3. The data of Yandell and Yonetani (1983) provide a beautiful example of the power of the kinetic method to exclude mechanisms. These authors studied the steady-state reactions of several inorganic complexes of different charges with hydrogen peroxide catalyzed by yeast C*c*P. At a given pH the plots of $[E]_0/v$ versus $1/[e^-]$ (where $[e^-]$ is reducing substrate concentration), had different slopes, to be expected for reactants of different charges, but had a common and finite intercept, $1/k_A$, to be expected for the intramolecular conversion of two forms of Compound II. The values of k_A are $10^3\ s^{-1}$ at pH 5.0, $300\ s^{-1}$ at pH 6.0, $50\ s^{-1}$ at pH 7.0, and $6.4\ s^{-1}$ at pH 8.0. The pH 7 results are reproduced in Figure 6.3. It was concluded that "the rate data require the assignment of the rate-limiting step . . . to a slow electron-transfer within the enzyme, and not, as previously suggested, to saturation of substrate binding to the enzyme" (Yandell and Yonetani, 1983). This conclusion is exactly correct. I do not agree with some other details of their mechanism. Compound 0, the precursor of Compound I, is postulated to react directly with reducing substrate.

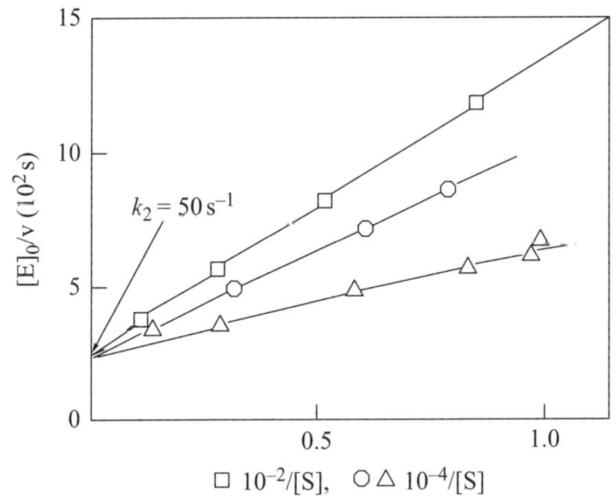

Figure 6.3. Double reciprocal plot of steady-state kinetic data for reactions catalyzed yeast cytochrome *c* peroxidase at pH 7. Oxidation by hydrogen peroxide of: □ ferrocyanide, ○ $[Co(terpyridine)_2]^{2+}$, Δ $[Ru(NH_3)_5pyridine]^{2+}$. (Reproduced with permission from Yandell JK, Yonetani T. 1983. *Biochim. Biophys. Acta* **748**: 263–270.)

4. Since the three simpler mechanisms have been excluded, the $Fe^{IV}{=}O$ reduction mechanism remains as the simplest alternative. The choice of $Fe^{IV}{=}O$ over $Trp^{\bullet +}$ as the doubly reduced site is not proven. If small reducing substrates do not bind at the cytochrome *c* docking site, then the $Trp^{\bullet +}$ site on the proximal side of the heme is less accessible than $Fe^{IV}{=}O$. Also the Trp site is the preferred site over the porphyrin to be oxidized in Compound I formation of C*c*P, and therefore thermodynamically it is more difficult to reduce. The important papers of Yandell and Yonetani (1983) and Ho et al. (1983) were not cross-referenced. Now, 26 years later, it would appear about time that total agreement on the yeast C*c*P mechanism of reaction with small substrates was reached.

Ten years ago I came to the conclusion that the same site on cytochrome *c* peroxidase was reduced twice in a normal cycle with small reducing substrates (Dunford, 1999), and I come to the same conclusion today. Yeast C*c*P operates by a different mechanism than other peroxidases, not only in its Compound I structure, but in the sequence of events in the reaction cycle involving small reducing substrates.

Supporting evidence for the $Fe^{IV}{=}O$ reduction mechanism is found in the results of chemical modification of the peroxidase heme. Only the edge of pyrrole ring IV is accessible to solvent. Phenylhydrazine attacks the δ-meso position and the 8-methyl position to form δ-meso-phenyl and 8-hydroxymethyl derivatives, which inactivates the ability of the peroxidase to oxidize guaiacol, but does not prevent reaction with hydrogen peroxide. Similarly, the azidyl radical binds to the δ-meso position blocking reaction of guaiacol. Thus, small reducing substrates react at the edge of pyrrole ring IV (DePillis et al., 1991). Since the $Fe^{IV}{=}O$ group is covalently bound to the pyrrole nitrogen, it is much closer to the substrate binding site than the proximal Trp191 cation radical site.

Other Ligand-Binding Reactions and Compound I Formation

The most comprehensive stopped-flow investigations of the kinetics of yeast cytochrome *c* peroxidase reactions have been conducted in the laboratory of James Erman beginning in 1974 and continuing to the present day. Reactions of ferrocyanide and a host of substitution-inert iron(II) complexes as reducing substrates have been studied (Erman, 1974a, 1974b; Purcell and Erman, 1976).

The iron(III) of native yeast C*c*P is more purely high spin than in horseradish peroxidase, and its reaction with fluoride is faster. For fluoride the binding rate constant to C*c*P is $5.1 \times 10^7\ M^{-1}\ s^{-1}$ compared to $4.2 \times 10^6\ M^{-1}\ s^{-1}$ for horseradish peroxidase. Crystal structures show that the distal heme environment is greatly perturbed upon small ligand binding (Edwards and Poulos, 1990). Electronic absorption and resonance Raman spectra showed that bound fluoride is strongly hydrogen bonded to distal Arg48 and less strongly bonded to distal His51 presumably through an intervening water molecule. In metmyoglobin only the distal His is hydrogen bonded to fluoride (Neri et al., 1997).

For cyanide the maximum rate of binding is $1.1 \times 10^5\ M^{-1}\ s^{-1}$ to C*c*P (Erman, 1974b). The acid form of azide HN_3 has a binding constant of $(1.05 \pm 0.06) \times 10^5\ M^{-1}$ to C*c*P (Jacobson et al., 2004).

In all cases it is the acid forms of the ligands that bind, with the proton binding to distal His and the anion to Fe^{III}. This mimics the first half of Compound I formation in which Compound 0 is formed. Compound I formation of C*c*P using hydrogen peroxide has a rate constant of $4.5 \times 10^7\ M^{-1}\ s^{-1}$, again a faster reaction than that for HRP (Loo and Erman, 1975). Several different oxidants were shown to produce identical Compounds I, indicating a true covalent compound, not an ES complex (Coulson and Yonetani, 1975). With no nitrate or acetate present Compound I formation is pH independent over the range 4.5 to 8.0 (Balny et al., 1987; Vitello et al., 1990b). Studies using viscous fluids showed that Compound I formation is not diffusion controlled (Loo and Erman, 1977). When a dimethyl ester of ferriprotoporphyrin IX was substituted for the natural heme, little change in the kinetics of Compound I formation was observed, indicating that the propionate side chains are not directly involved in the mechanism (Dowe and Erman, 1982).

Site-Directed Mutagenesis

Cloning may provide a better way to isolate an enzyme in high yield. One of the more difficult parts is to prove that there is no folding problem and a pure intact enzyme, identical to its native form, is obtained. Careful characterization using a battery of techniques is required. Techniques for site-directed mutagenesis have been reviewed (Smith, 1985). Because yeast cytochrome *c* peroxidase contains no carbohydrate and because for many years it was the only peroxidase with a known crystal structure, it was especially attractive for site-directed mutagenesis experiments. The nuclear gene was cloned in 1982 (Goltz et al., 1982; Kaput et al., 1982), and site-directed mutagenesis experiments began shortly after. Expression of the gene contained in a plasmid of *Saccharomyces cerevisiae* was accomplished. A 68-residue leader sequence was necessary. The first units expressed in yeast were obtained in the laboratory of Michael Smith (Pielak et al., 1985; Goodin et al., 1986). It was concluded that Met172 is not directly involved in Compound I formation.

A better expression system was developed by inserting the plasmid into *Escherichia coli*. The isolated enzyme was largely in its *apo* form but was readily converted to *holo* enzyme by addition of ferriprotoporphyrin IX. The leader sequence for the yeast system was replaced by the much simpler two-residue sequence Met-Ile for the *E. coli* expression system. Thus the cloned enzyme was labeled C*c*P(MI). It appeared to be identical in all its enzyme properties to wild-type C*c*P. The first mutant produced in *E. coli* was Trp51Phe. It was concluded that Trp51 contributed to the stability of Compound I but was not essential for its formation (Fishel et al., 1987).

A different clone of C*c*P was produced in *E. coli* in which the N-terminal end was Met-Lys-Thr and was thus designated as C*c*P(MKT) (Goodin et al., 1991). Some mutants at Trp51 enhanced the rate of reaction with hydrogen peroxide, others decreased it. Enhanced rate of oxidation of aniline derivatives was observed in the mutants where Trp51 was replaced by phenylalanine and alanine (Roe and Goodin, 1993). This was attributed to easier removal of oxygen from $Fe^{IV}{=}O$ as water. Rate constants for the aniline derivatives fit a Brown–Okamoto σ^+ plot slightly better than a Hammett σ plot. The expression system was developed to the point where enzyme yields of up to 1 g/L were obtained (D.B. Goodin, Scripps Research Institute, personal communication, December 30, 1997).

The Trp191Phe mutant greatly affects the ability of the enzyme to oxidize cytochrome *c* (Mauro et al., 1988). It reacted readily with hydrogen peroxide to produce Compound I containing a porphyrin π-cation radical (Vitello et al., 1990a).

The Compound I radical was probed in site-directed mutants by both ESR and ENDOR (Fishel et al., 1991). Met230 and 231 have sulfur atoms within 4 Å of the Trp191 indole ring. Mutations of the methionine residues perturbed the ESR line shapes but did not eliminate them. Both the Trp191Phe and Asp235Asn mutants eliminated the broad ESR signal. The mutants were prepared using *E. coli* (Fishel et al., 1987).

Distal His52 is crucial for Compound I formation. The His52Leu mutant has a 10^5-fold decrease in rate with hydrogen peroxide compared to the cloned enzyme (Erman et al., 1992, 1993). Replacement of His52 with lysine leads to several different enzyme forms. The side chain ε-amino group of the lysine is able to act like the distal imidazole of His, but not as effectively (Foshay et al., 2004).

The effects of Asp235Asn replacement on the C*c*P reaction with hydrogen peroxide were examined (Vitello et al., 1992, 1993). Asp235 forms a hydrogen bond with the proximal imidazole of His175. The Asp235Asn mutant changes the coordination number of the iron(III) from five to six but only decreases the rate of reaction with hydrogen peroxide by a factor of 4 to 5. The oxyferryl form of the enzyme is quite stable, indicating that Asp235 is not necessary to stabilize the higher oxidation states (Vitello et al., 1990a, 1992).

The His175 proximal ligand was replaced by Gln, Glu, or Cys. The Gln and Glu mutants retained a high level of enzyme activity. A double mutant with His175 replaced by Gln and Trp191 by Phe also forms a Compound I rapidly, most likely a long-lived porphyrin π-cation radical that slowly decays to another species (Choudhury et al., 1994).

The Asp235Ala mutant was characterized by electronic, ESR, NMR (Ferrer et al., 1994), and X-ray Absorption Spectroscopy (XAS) (Sinclair et al., 1996). Rupture of the hydrogen bond between Asp235 and proximal His175 causes an elongation of the iron-proximal imidazole nitrogen bond from ~1.9 to ~2.1 Å and greatly reduces the rate of reaction with cytochrome *c*. The most important function of the proximal Asp-His-Fe linkage "may be to correctly orientate Trp191 for efficient coupling of the free radical to the heme and to maintain a high-spin 5-coordinate heme center" (Goodin and McRee, 1993, p. 3313).

Arg48 was replaced by both leucine and lysine. Crystallographic structures, electronic absorption spectra, and hydrogen peroxide reactivity were determined (Vitello et al., 1993). Both crystal structure and electronic absorption spectra of Arg48Leu are consistent with pentacoordinate iron(III). The Arg48Lys mutant forms an alkaline bis-imidazole complex. Arg48 is necessary to stabilize Compound I. The His175Glu mutation results in a more planar heme (Smulevich et al., 1995).

Yeast cytochrome *c* peroxidase has been engineered to become a manganese peroxidase mimic (Wang and Lu, 1999; Gengenbach et al., 1999). Conversion of tryptophan to phenylalanine residues played an important role in the conversion. The yeast enzyme has also been modified to accommodate an ascorbate binding site (Meharenna et al., 2008; Murphy et al., 2008).

Ferrous Cytochrome *c* Peroxidase

The reduction potential for the ferric enzyme has been measured over the pH range 4.5 to 8.0. At pH 7 and 25°C the value is −194 mV relative to the standard hydrogen electrode (Conroy et al., 1978). This value can be compared to −270 mV for ferric horseradish peroxidase (Harbury, 1957), and −109 mV for turnip peroxidase isoenzyme P_7 (Ricard et al., 1972).

Biphasic geminate recombination of CO and C*c*P has been detected in a flash photolysis study (English et al., 1984). The CO adducts of *E. coli* generated C*c*P and many mutants have been examined in great detail by both resonance Raman and infrared spectroscopy. The Fe^{II}-proximal His bond, which has two components, occurs at ~240 cm^{-1}. It is not affected by distal mutations. The two components, at 233 and 246 cm^{-1}, were suggested to form a double well potential caused by protonated and deprotonated imidazole. In the Asp235Asn mutant the double well is missing and is replaced by a new band at 205 cm^{-1}, attributed to Fe—His stretch associated with non-H-bonded imidazole (Smulevich et al., 1988a). A correlation between υ_{Fec} and υ_{CO} was used to distinguish between back-bonding and axial ligand donor effects (Smulevich et al., 1988b). Laser excitation of the CO adduct at pH 8.5 appears to photodissociate the distal His52 residue (Smulevich et al., 1989). The important roles of both distal and proximal histidines were shown in a resonance Raman study of Leu52 and Gly181 mutants. At alkaline pH mutants were shown to perturb the important hydrogen bonding network that extends from the distal to the proximal side of the heme (Smulevich et al., 1991). Replacement of Arg48 increases CO dissociation rates of the acid form of the enzyme, indicating Arg—CO hydrogen bonding (Miller et al., 1990).

A unique route to a Compound I-like species was obtained by photolysis of CO-bound C*c*P in the presence of dioxygen. The $Fe^{II}—O_2$ complex was transiently formed with a rate constant of $(3.8 \pm 0.3) \times 10^4\ M^{-1}\ s^{-1}$ at pH 6.0 and 23°C. This was followed by a slow conversion ($k = 4 \pm 1\ s^{-1}$) to a species with electronic absorption spectrum and ESR signal identical to those of Compound I. Thus the $Fe^{II}—O_2$ complex (oxidation state +6) appeared to rearrange to a species similar to Compound I but with an extra oxidizing equivalent. Mutations of Trp191 and Asp235 that prevented normal Compound I formation also prevented formation of the form with an apparent +6 oxidation state (Miller et al., 1992). Spontaneous conversion of Compound III to Compound I has been observed with other peroxidases (see for example the discussion of lignin peroxidase). Rather than Compound I with an extra oxidizing equivalent, it is likely that the old peroxidase bugbear of exogenous/endogenous electron donor is operating to change the +6 oxidation state of Compound III to the +5 oxidation state of Compound I.

REFERENCES

Abrams R, Altschul AM, Hogness TR. 1942. Cytochrome *c* peroxidase. II. The peroxidase-hydrogen peroxide complex. *J. Biol. Chem.* **142**: 303–316.

Altschul AM, Abrams R, Hogness TR. 1940. Cytochrome *c* peroxidase. *J. Biol. Chem.* **136**: 777–794.

Balny C, Anni H, Yonetani T. 1987. A stopped-flow study of the reaction of cytochrome *c* peroxidase with hydroperoxides. *FEBS Lett.* **221**: 349–354.

Bonagura CA, Bhaskar B, Shimizu H, Li H, Sundaramoorthy M, McRee DE, Goodin DB, Poulos TL. 2003. High-resolution crystal structures and spectroscopy of native and Compound I cytochrome *c* peroxidase. *Biochemistry* **42**: 5600–5608.

Bosshard HR, Anni H, Yonetani T. 1991. Yeast cytochrome *c* peroxidase. In: *Peroxidases in Chemistry and Biology*, Vol. II. Everse J, Everse KE, Grisham MB, editors. CRC Press, Boca Raton, FL, pp. 51–84.

Choudhury K, Sundaramoorthy M, Hickman A, Yonetani T, Woehl E, Dunn MF, Poulos TL. 1994. Role of the proximal ligand in peroxidase catalysis: crystallographic, kinetic and spectral studies of cytochrome *c* peroxidase proximal ligand mutants. *J. Biol. Chem.* **269**: 20239–20249.

Conroy CW, Erman JE. 1978. pH titration study of cytochrome *c* peroxidase and apocytochrome *c* peroxidase. *Biochim. Biophys. Acta* **537**: 396–405.

Conroy CW, Tyma P, Daum PH, Erman JE. 1978. Oxidation-reduction potential measurements of cytochrome *c* peroxidase and pH dependent spectral transitions in the ferrous enzyme. *Biochim. Biophys. Acta* **537**: 62–69.

Coulson AFW, Erman JE, Yonetani T. 1971. Studies on cytochrome *c* peroxidase. XVII. Stoichiometry and mechanism of the reaction of Compound ES with donors. *J. Biol. Chem.* **246**: 917–924.

Coulson AFW, Yonetani T. 1975. Mechanism of cytochrome *c* peroxidase: *O*-benzoylhydroxylamine as an analog of hydrogen peroxide. *Biochemistry* **14**: 2389–2396.

DePillis GD, Sishta BP, Mauk AG, Ortiz de Montellano PR. 1991. Small substrates and cytochrome *c* are oxidized at different sites of cytochrome *c* peroxidase. *J. Biol. Chem.* **266**: 19334–19341.

Dowe RJ, Erman JE. 1982. The reaction of hydrogen peroxide with the dimethyl ester heme derivative of cytochrome *c* peroxidase. *J. Biol. Chem.* **257**: 11186–11190.

Dunford HB. 1999. *Heme Peroxidases*. Wiley-VCH, New York.

Edwards SL, Poulos TL. 1990. Ligand binding and structural perturbations in cytochrome *c* peroxidase: a crystallographic study. *J. Biol. Chem.* **265**: 2588–2595.

Edwards SL, Xuong N, Hamlin RC, Kraut J. 1987. Crystal structure of cytochrome *c* peroxidase Compound I. *Biochemistry* **26**: 1503–1511.

Ellfolk N. 1967. Cytochrome *c* peroxidase. I. Preparation of the crystalline enzyme from Baker's yeast. *Acta Chem. Scand.* **21**: 175–181.

English AM, McLendon G, Taylor K. 1984. Kinetics of CO binding to cytochrome *c* peroxidase: pH and CO concentration effects. *J. Am. Chem. Soc.* **106**: 6448–6449.

Erman JE. 1974a. Kinetic studies of fluoride binding by cytochrome *c* peroxidase. *Biochemistry* **13**: 34–39.

Erman JE. 1974b. Kinetic and equilibrium studies of cyanide binding by cytochrome *c* peroxidase. *Biochemistry* **13**: 39–44.

Erman JE, Vitello LB, Miller MA, Kraut J. 1992. Active site mutations in cytochrome *c* peroxidase: a critical role for histidine-52 in the rate of formation of Compound I. *J. Am. Chem. Soc.* **114**: 6592–6593.

Erman JE, Vitello LB, Miller MA, Shaw A, Brown KA, Kraut J. 1993. Histidine 52 is a critical residue for rapid formation of cytochrome *c* peroxidase Compound I. *Biochemistry* **32**: 9798–9806.

Ferrer JC, Turano P, Banci L, Bertini I, Morris IK, Smith KM, Smith M, Mauk AG. 1994. Active site coordination chemistry of the cytochrome *c* peroxidase Asp235Ala variant: spectroscopic and functional characterization. *Biochemistry* **33**: 7819–7829.

Finzel BC, Poulos TL, Kraut J. 1984. Crystal structure of yeast cytochrome *c* peroxidase refined at 1.7-Å resolution. *J. Biol. Chem.* **259**: 13027–13036.

Fishel LA, Farnum MF, Mauro JM, Miller MA, Kraut J, Liu Y, Tan X, Scholes CP. 1991. Compound I radical in site-directed mutants of cytochrome *c* peroxidase as probed by electron paramagnetic resonance and electron-nuclear double resonance. *Biochemistry* **30**: 1986–1996.

Fishel LA, Villafranca JE, Mauro JM, Kraut J. 1987. Yeast cytochrome *c* peroxidase: mutagenesis and expression in *Escherichia coli* show tryptophan-51 is not the radical site in Compound I. *Biochemistry* **26**: 351–360.

Foshay MC, Vitello LB, Erman JE. 2004. pH dependence of heme iron coordination, hydrogen peroxide reactivity, and cyanide binding in cytochrome *c* peroxidase (H52K). *Biochemistry* **43**: 5065–5072.

Fülöp V, Phizackerley RP, Soltis SM, Clifton IJ, Wakatsuki S, Erman J, Hajdu J, Edwards SL. 1994. Laue diffraction study on the structure of cytochrome *c* peroxidase compound I. *Structure* **2**: 201–208.

Gengenbach A, Syn S, Wang X, Lu Y. 1999. Redesign of cytochrome *c* peroxidase into a manganese peroxidase: role of tryptophans in peroxidase activity. *Biochemistry* **38**: 11425–11432.

Goltz S, Kaput J, Blobel G. 1982. Isolation of the yeast nuclear gene encoding the mitochondrial protein, cytochrome *c* peroxidase. *J. Biol. Chem.* **257**: 11186–11190.

Goodin DB, Davidson MG, Roe JA, Mauk AG, Smith M. 1991. Amino acid substitutions at tryptophan-51 of cytochrome *c* peroxidase: effects on coordination, species preference for cytochrome *c*, and electron transfer. *Biochemistry* **30**: 4953–4962.

Goodin DB, Mauk AG, Smith M. 1986. Studies of the radical species in compound ES of cytochrome *c* peroxidase altered by site-directed mutagenesis. *Proc. Natl. Acad. Sci. USA* **83**: 1295–1299.

Goodin DB, McRee DE. 1993. The Asp-His-Fe triad of cytochrome *c* peroxidase controls the reduction potential, electronic structure and coupling of the tryptophan free radical to the heme. *Biochemistry* **32**: 3313–3324.

Harbury HA. 1957. Oxidation-reduction potentials of horseradish peroxidase. *J. Biol. Chem.* **225**: 1009–1023.

Hirst J, Wilcox SK, Ai J, Moënne-Loccoz P, Loehr TM, Goodin DB. 2001a. Replacement of the axial histidine ligand with imidazole in cytochrome *c* peroxidase. II. Effects on heme coordination and function. *Biochemistry* **40**: 1265–1273.

Hirst J, Wilcox SK, Williams PA, Blankenship, McRee DE, Goodin DB. 2001b. Replacement of the axial histidine ligand with imidazole in cytochrome *c* peroxidase. I. Effects on structure. *Biochemistry* **40**: 1265–1273.

Ho PS, Hoffman BM, Kang CH, Margoliash E. 1983. Control of the transfer of oxidizing equivalents between heme iron and free radical site in yeast cytochrome *c* peroxidase. *J. Biol. Chem.* **258**: 4356–4363.

Ho PS, Hoffman BM, Solomon N, Kang CH, Margoliash E. 1984. Kinetics and energetics of intramolecular electron transfer in yeast cytochrome *c* peroxidase. *Biochemistry* **23**: 4122–4128.

Hoffman BM, Roberts JE, Brown TG, Kang CH, Margoliash E. 1979. Electron-nuclear double resonance of the hydrogen peroxide compound of cytochrome *c* peroxidase: identification of the free radical site with a methionyl cluster. *Proc. Natl. Acad. Sci. USA* **76**: 6132–6136.

Hoffman BM, Roberts JE, Brown TG, Kang CH, Margoliash E. 1981. Electron paramagnetic and electron nuclear double resonance of the hydrogen peroxide compound of cytochrome *c* peroxidase. *J. Biol. Chem.* **256**: 6556–6564.

Hori H, Yonetani T. 1985. Powder and single-crystal electron paramagnetic resonance studies of yeast cytochrome *c* peroxidase and its peroxide compound, Compound ES. *J. Biol. Chem.* **260**: 349–355.

Houseman ALP, Doan PE, Goodin DB, Hoffman BM. 1993. Comprehensive explanation of the anomalous EPR spectra of wild-type and mutant cytochrome *c* peroxidase Compound ES. *Biochemistry* **32**: 4430–4443.

Huyett JE, Doan PE, Gurbiel R, Houseman ALP, Sivaraja M, Goodin DB, Hoffman BM. 1995. Compound ES of cytochrome *c* peroxidase contains a Trp π-cation radical: characterization by CW and pulsed Q-band ENDOR spectroscopy. *J. Am. Chem. Soc.* **117**: 9033–9041.

Iffland A, Tafelmeyer P, Saudan C, Johnsson K. 2000. Directed molecular evolution of cytochrome *c* peroxidase. *Biochemistry* **39**: 10790–10798.

Jacobson T, Williamson J, Wasilewski A, Felesik J, Vitello LB, Erman JE. 2004. Azide binding to yeast cytochrome *c* peroxidase and horse metmyoglobin: comparative thermodynamic investigation using isothermal titration calorimetry. *Arch. Biochem. Biophys.* **422**: 125–136.

Jordi HC, Erman JE. 1974a. Cytochrome *c* peroxidase catalyzed oxidation of ferrocyanide by hydrogen peroxide: transient state kinetics. *Biochemistry* **13**: 3734–3741.

Jordi HC, Erman JE. 1974b. Cytochrome *c* peroxidase catalyzed oxidation of ferrocyanide by hydrogen peroxide: steady-state kinetics. *Biochemistry* **13**: 3741–3745.

Kaput J, Goltz S, Blobel G. 1982. Isolation of the yeast nuclear gene encoding the mitochondrial protein, cytochrome *c* peroxidase. *J. Biol. Chem.* **257**: 15054–15058.

Larsson LO, Hagman L-O, Kierkegaard P, Yonetani T. 1970. Preliminary crystallographic data for cytochrome *c* peroxidase from bakers' yeast. *J. Biol. Chem.* **245**: 902–903.

Loo S, Erman JE. 1975. A kinetic study of the reaction between cytochrome *c* peroxidase and hydrogen peroxide: dependence on pH and ionic strength. *Biochemistry* **14**: 3467–3470.

Loo S, Erman JE. 1977. The rate of reaction between cytochrome *c* peroxidase and hydrogen peroxide is not diffusion limited. *Biochim. Biophys. Acta* **481**: 279–282.

Mathews RA, Wittenberg JB. 1979. Cytochrome *c* peroxidase. Interconversion of chemically and enzymatically reactive and unreactive forms of the ferric protein. *J. Biol. Chem.* **254**: 5991–5996.

Mauro JM, Fishel LA, Hazzard JT, Meyer TE, Tollin G, Cusanovich MA, Kraut J. 1988. Tryptophan-191 → phenylalanine, a proximal-side mutation in yeast cytochrome *c* peroxidase that strongly affects the kinetics of ferrocytochrome *c* oxidation. *Biochemistry* **27**: 6243–6256.

Meharenna YT, Oertel P, Bhaskar B, Poulos TL. 2008. Engineering ascorbate peroxidase activity into cytochrome *c* peroxidase. *Biochemistry* **47**: 10324–10332.

Miller MA, Bandyopadhyay D, Mauro JM, Traylor TG, Kraut J. 1992. Reaction of ferrous cytochrome *c* peroxidase with dioxygen: site-directed mutagenesis provides evidence for rapid reduction of dioxygen by intramolecular electron transfer from the Compound I radical site. *Biochemistry* **31**: 2789–2797.

Miller MA, Han GW, Kraut J. 1994a. A cation binding motif stabilizes the compound I radical of cytochrome *c* peroxidase. *Proc. Natl. Acad. Sci. USA* **91**: 11118–11122.

Miller MA, Mauro JM, Smulevich G, Coletta M, Kraut J, Traylor TG. 1990. CO dissociation in cytochrome *c* peroxidase: site-directed mutagenesis shows that distal Arg 48 influences CO dissociation rates. *Biochemistry* **29**: 9978–9988.

Miller MA, Shaw A, Kraut J. 1994b. 2.2 Å structure of oxy-peroxidase as a model for the transient enzyme:peroxidase complex. *Struct. Biol.* **1**: 524–531.

Mondal MS, Fuller HA, Armstrong FA. 1996. Direct measurement of the reduction potential of catalytically active cytochrome *c* peroxidase Compound I: voltammetric detection of a reversible, cooperative two-electron transfer reaction. *J. Am. Chem. Soc.* **118**: 263–264.

Murphy EJ, Metcalfe CL, Basran J, Moody PCE, Raven EL. 2008. Engineering the substrate specificity and reactivity of a heme protein: creation of an ascorbate binding site in cytochrome c peroxidase. *Biochemistry* **47**: 13933–13941.

Neri F, Kok D, Miller MA, Smulevich G. 1997. Fluoride binding in hemoproteins: the importance of distal cavity structure. *Biochemistry* **36**: 8947–8953.

Palamakumbura AH, Foshay MC, Vitello LB, Erman, JE. 1999a. Oxidation of cytochrome *c* peroxidase to Compound I by peroxyacids: evidence for rate-limiting diffusion through the protein matrix. *Biochemistry* **38**: 15647–15652.

Palamakumbura AH, Vitello LB, Erman, JE. 1999b. Oxidation of the His-52 $\rightarrow$ Leu mutant of cytochrome *c* peroxidase by *p*-nitroperoxybenzoic acid: role of the distal histidine in hydroperoxide activation. *Biochemistry* **38**: 15653–15658.

Pfister TD, Gengenbach AJ, Syn S, Lu Y. 2001. The role of redox-active amino acids on Compound I stability, substrate oxidation, and protein cross-linking in yeast cytochrome *c* peroxidase. *Biochemistry* **40**: 14942–14951.

Pielak GJ, Mauk AG, Smith M. 1985. Site-directed mutagenesis of cytochrome *c* shows that an invariant Phe is not essential for function. *Nature* **313**: 152–154.

Poulos TL, Freer ST, Alden RA, Edwards SL, Skogland U, Takio K, Eriksson B, Xuong N, Yonetani T, Kraut J. 1980. The crystal structure of cytochrome *c* peroxidase. *J. Biol. Chem.* **255**: 575–580.

Poulos TL, Kraut J. 1980. The stereochemistry of peroxidase catalysis. *J. Biol. Chem.* **255**: 8199–8205.

Purcell WL, Erman JE. 1976. Cytochrome *c* peroxidase catalyzed oxidations of substitution inert iron(II) complexes. *J. Am. Chem. Soc.* **98**: 7033–7037.

Ricard J, Mazza G, Williams RPJ. 1972. Oxidation-reduction potentials and ionization states of two turnip peroxidases. *Eur. J. Biochem.* **28**: 566–578.

Roe JA, Goodin DB. 1993. Enhanced oxidation of aniline derivatives by two mutants of cytochrome *c* peroxidase at tryptophan 51. *J. Biol. Chem.* **268**: 20037–20045.

Sievers G. 1978. Circular dichroism studies on cytochrome *c* peroxidase from Baker's yeast (*Saccharomyces cerevisiae*). *Biochim. Biophys. Acta* **536**: 212–225.

Sinclair R, Hallam S, Chen M, Chance B, Powers L. 1996. Active site structure in cytochrome *c* peroxidase and myoglobin mutants: effects of altered hydrogen bonding to the proximal histidine. *Biochemistry* **35**: 15120–15128.

Sivaraja M, Goodin DB, Smith M, Hoffman BM. 1989. Identification by ENDOR of Trp^{191} as the free-radical site in cytochrome *c* peroxidase Compound ES. *Science* **245**: 738–740.

Smith M. 1985. In vitro mutagenesis. *Annu. Rev. Genet.* **19**: 423–462.

Smulevich G. 1996. A comparative analysis of the structures of various peroxidases as defined by resonance Raman and electronic absorption spectroscopies. In: Obinger C, Burner U, Ebermann R, Penel C, Greppin H, editors. *Plant Peroxidases. Biochemistry and Physiology. IV International Symposium*. University of Geneva, pp. 13–19.

Smulevich G, Mauro JM, Fishel LA, English AM, Kraut J, Spiro TG. 1988a. Heme pocket interactions in cytochrome c peroxidase studied by site-directed mutagenesis and resonance Raman spectroscopy. *Biochemistry* **27**: 5477–5485.

Smulevich G, Mauro JM, Fishel LA, English AM, Kraut J, Spiro TG. 1988b. Cytochrome *c* peroxidase mutant active site structures probed by resonance Raman and infrared signatures of the CO adducts. *Biochemistry* **27**: 5486–5492.

Smulevich G, Miller MA, Gosztola D, Spiro TG. 1989. Photodissociable endogenous ligand in alkaline-reduced cytochrome *c* peroxidase implicates distal protein tension. *Biochemistry* **28**: 9905–9908.

Smulevich G, Miller MA, Kraut J, Spiro TG. 1991. Conformational change and histidine control of heme chemistry in cytochrome *c* peroxidase: resonance Raman evidence from Leu-52 and Gly-181 mutants of cytochrome *c* peroxidase. *Biochemistry* **30**: 9546–9558.

Smulevich G, Neri F, Willemsen O, Choudhury K, Marzocchi MP, Poulos TL. 1995. Effect of the His 175 → Glu mutation on the heme pocket architecture of cytochrome *c* peroxidase. *Biochemistry* **34**: 13485–13490.

Vitello LB, Erman JE, Mauro JM, Kraut J. 1990a. Characterization of the hydrogen peroxide-enzyme reaction for two cytochrome *c* peroxidase mutants. *Biochim. Biophys. Acta* **1038**: 90–97.

Vitello LB, Erman JE, Miller MA, Mauro JM, Kraut J. 1992. Effect of Asp-235 → Asn substitution on the absorption spectrum and hydrogen peroxide reactivity of cytochrome *c* peroxidase. *Biochemistry* **31**: 11524–11535.

Vitello LB, Erman JE, Miller MA, Wang J, Kraut J. 1993. Effect of Arginine-48 replacement on the reaction between cytochrome *c* peroxidase and hydrogen peroxide. *Biochemistry* **32**: 9807–9818.

Vitello LB, Huang M, Erman JE. 1990b. pH-dependent spectral and kinetic properties of cytochrome *c* peroxidase: comparison of freshly isolated and stored enzyme. *Biochemistry* **29**: 4283–4288.

Wang J, Mauro JM, Edwards SL, Oatley SJ, Fishel LA, Ashford VA, Xuong N-H, Kraut J. 1990. X-ray structures of recombinant yeast cytochrome *c* peroxidase and three heme-cleft mutants prepared by site-directed mutagenesis. *Biochemistry* **29**: 7160–7173.

Wang X, Lu Y. 1999. Proton NMR investigation of the heme active site structure of an engineered cytochrome c peroxidase that mimics manganese peroxidase. *Biochemistry* **38**: 9146–9157.

Wittenberg BA, Kampa L, Wittenberg JB, Blumberg WE, Peisach J. 1968. The electronic structure of protoheme proteins. II. An electron paramagnetic resonance and optical study of cytochrome *c* peroxidase and its derivatives. *J. Biol. Chem.* **243**: 1863–1870.

Yandell JK, Yonetani T. 1983. Steady-state kinetics of yeast cytochrome *c* peroxidase catalyzed oxidation of inorganic reductants by hydrogen peroxide. *Biochim. Biophys. Acta* **748**: 263–270.

Yonetani T. 1967. Studies on cytochrome *c* peroxidase. X. Crystalline apo- and reconstituted holoenzymes. *J. Biol. Chem.* **242**: 5008–5013.

Yonetani T. 1970. Cytochrome *c* peroxidase. In: *Advances in Enzymology*. Nord FF, editor. Wiley, New York, 33: 309–335.

Yonetani T. 1976. Cytochrome *c* peroxidase. In: *The Enzymes*, Vol. 13, 3rd edition. Boyer PD, editor. Academic Press, New York, pp. 345–361.

Yonetani T, Anni H. 1987. Yeast cytochrome *c* peroxidase: coordination and spin states of heme prosthetic group. *J. Biol. Chem.* **262**: 9547–9554.

Yonetani T, Schleyer H. 1967. Electromagnetic properties of hemoproteins. I. Electron paramagnetic resonance absorptions of single crystals of ferrimyoglobin and cytochrome *c* peroxidase. *J. Biol. Chem.* **242**: 3919–3925.

Zhang H, He S, Mauk AG. 2002. Radical formation at Tyr 39 and Tyr 153 following reaction of yeast cytochrome *c* peroxidase with hydrogen peroxide. *Biochemistry* **41**: 13507–13513.

CHAPTER 7

YEAST CYTOCHROME *c* PEROXIDASE: REACTION WITH CYTOCHROME *c*

INTRODUCTION

The reaction of oxidized yeast cytochrome *c* peroxidase with reduced cytochrome *c* provides a model to examine the mechanism of electron transfer between two proteins. The crystal structures of both the enzyme and many cytochromes *c* are known, as well as for a cytochrome *c* peroxidase-cytochrome *c* complex. Marcus theory, developed and proven for electron transfer between small molecules, can be tested for long-range electron transfer. Furthermore, whether electron transfer occurs along well-defined bonding pathways, including σ, π, and hydrogen bonds, or through "free" space can, in principle, be tested. The natural reductants for oxidized yeast cytochrome *c* peroxidase are the cytochromes found in yeast, labeled iso-1 and iso-2. In many studies other sources of cytochrome *c* have been used.

Complications include the presence of two oxidized sites in Compound I, the utilization of either or both sites in Compound II, and the presence of at least two binding sites on the enzyme for cytochrome *c*. Rates, and perhaps mechanism, are dependent upon ionic strength. Following are some pertinent questions: Which redox center in Compound I is reduced first, the oxyferryl heme or the Trp191 cation radical? Is the same cytochrome *c* binding site used for sequential reduction of Compound I to native enzyme, or are different sites used?

EXPERIMENTAL RESULTS

The discovery of the yeast enzyme included a description of its specificity for cytochrome *c* as reducing substrate (Altschul et al., 1940). After a hiatus of 25 years

Peroxidases and Catalases: Biochemistry, Biophysics, Biotechnology, and Physiology, Second Edition.
By H. Brian Dunford

intensive studies were undertaken, mainly in the laboratory of Takashi Yonetani (Yonetani, 1965; Yonetani and Ray, 1965, 1966; Bosshard et al., 1991). There are two reviews of extensive earlier work (Yonetani, 1970, 1976).

Binding of Cytochrome *c* to Cytochrome *c* Peroxidase in a 1:1 Ratio

Cytochrome *c* peroxidase is negatively charged at pH 7, whereas cytochrome *c* is positively charged. Thus electrostatic interactions play a role in binding, an effect diminished by increasing ionic strength. A binding ratio of 1:1 was established between enzyme and several modifications of horse cytochrome *c*, and the role of polyelectrolytes in inhibiting binding was examined (Mochan, 1970). Kinetics and equilibria of C*c*P-C*c* complex formation were described (Mochan and Nicholls, 1971; Nicholls and Mochan, 1971a, 1971b), and the Brownian dynamics of the association of cytochrome *c* and cytochrome *c* peroxidase have been examined (Northrup et al., 1988).

A model based on the crystal structures of the two components was constructed based largely on hydrogen bonding between oppositely charged groups on the two proteins. The two heme groups are parallel in the model (Poulos and Kraut, 1980).

Chemical modifications of the heme showed that whereas the δ-meso edge is the reactive site for small reducing substrates, it is modifications to the γ-meso edge that affect cytochrome *c* oxidation (DePillis et al., 1991). Thus small reducing substrates approach the δ-meso edge of the heme plane, whereas, based on the Poulos–Kraut model, cytochrome *c* approaches as closely as possible to the proximal underside of the enzyme's heme group.

The crystal structure of the 1:1 complex was obtained at 2.3 Å resolution and is shown in Figure 7.1. The "front" (heme solvent exposed) side of cytochrome *c*, whether from yeast (iso-1) or horse heart, binds to the enzyme. "The shortest straight line between the two hemes closely follows the enzyme backbone chain of Ala194, Ala193, Gly192 and finally Trp191, the indole ring of which is perpendicular to, and in van der Waals contact with, the peroxidase heme" (Pelletier and Kraut, 1992, p. 1748). The proposed electron transfer pathway is shown in Figure 7.2.

Thus although significantly different from the electrostatic model, the crystal structure of the complex confirms the binding of cytochrome *c* to a site closest to the proximal side of the peroxidase and closest to Trp191. The two heme groups are not parallel but make an angle of about 60° with each other. At the higher ionic strength at which the crystals were grown, hydrophobic and van der Waals interactions are more important than electrostatic interactions. At the conclusion of diffraction studies it was shown that the cytochrome *c* was fully oxidized, showing that the structure is that of a true electron transfer complex (Pelletier and Kraut, 1992). The Trp191Phe mutant reduces the rate of oxidation of ferrocytochrome *c* by ~3000-fold, whereas ferrocyanide oxidation is reduced only about 5-fold (Mauro et al., 1988), further evidence of different binding sites for small reducing substrates and cytochrome *c*. A further study of the Trp191Phe mutant showed that it reacted with hydrogen peroxide to form $Fe^{IV}{=}O$ and a transient porphyrin π-cation radical. The transient rate of reduction of the $Fe^{IV}{=}O$ group was reduced about 10,000-fold in the mutant

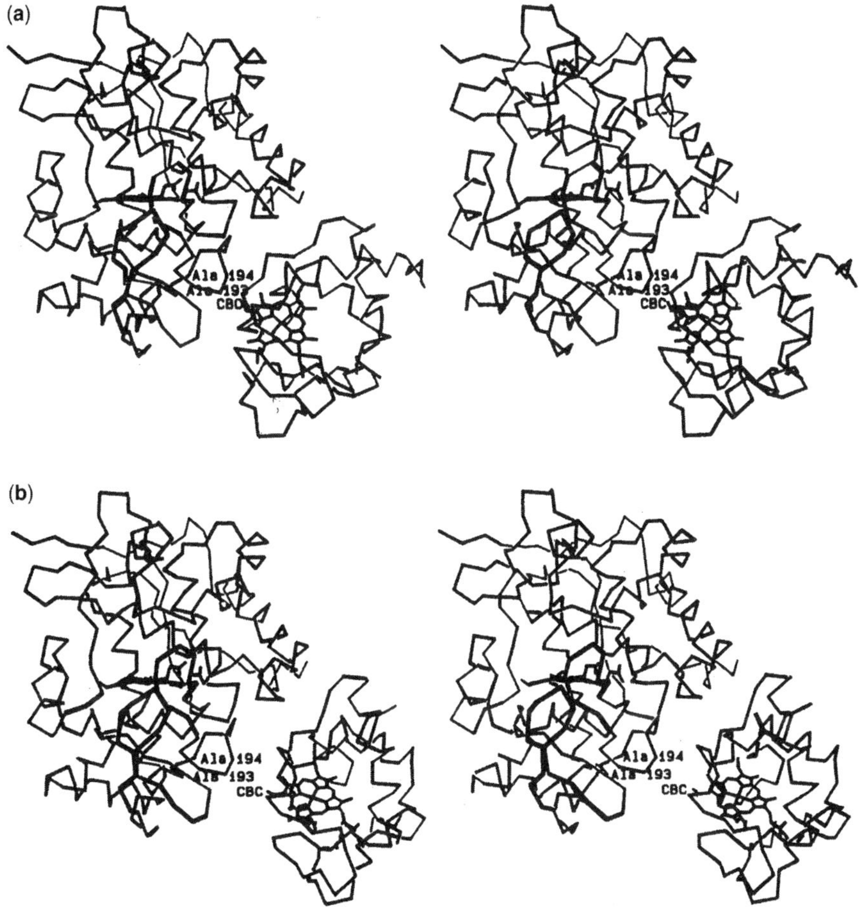

Figure 7.1. α-Carbon stereo representations of: (a) the yeast cytochrome *c* complex with yeast iso-1-cytochrome *c*; (b) the yeast cytochrome *c* complex with horse heart cytochrome *c*. (Reproduced with permission from Pelletier H, Kraut J. 1992. *Science* **258**: 1748–1755.)

enzyme, and was not observed in the steady state (Miller et al., 1995). Part of the electron transfer pathway was excised from the enzyme and replaced with surrogate peptide. Electron transfer no longer occurred, indicating the importance of the Trp191 radical (Putnam et al., 2009).

The transient-state (Summers and Erman, 1988) and steady-state kinetics (Kim et al., 1990) of reaction of cytochrome *c* peroxidase with horse heart cytochrome *c* indicate that 1:1 binding will explain the observed results; 2:1 binding will not. Despite the fact that the $Fe^{IV}{=}O$ form of Compound II is favored over the Trp191 cation radical form in an equilibrium mixture, it is the $Fe^{IV}{=}O$ site that is first reduced in Compound I. This is followed by formation of some of the $Trp^{\bullet +}$ species. Reduction of Compound II is a very complicated process.

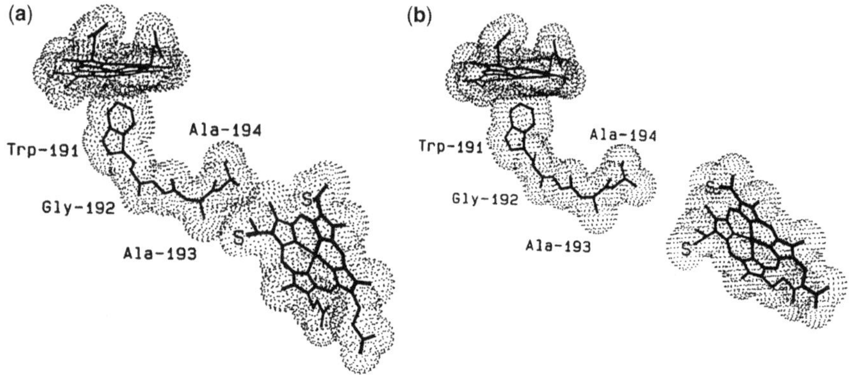

Figure 7.2. Proposed electron pathway from the cytochrome *c* heme (shown on the right in each diagram) to the yeast cytochrome *c* heme. (a) yeast iso-1-cytochrome *c*; (b) horse heart cytochrome *c*. (Reproduced with permission from Pelletier H, Kraut J. 1992. *Science* **258**: 1748–1755.)

The influence of surface amino acid replacements in the enzyme on complex formation and reactivity with cytochrome *c* was examined (Corin et al., 1991, 1993). An NMR investigation of yeast iso-1 cytochrome *c*, the yeast enzyme, and their 1 : 1 complex showed that pyrrole ring B of C*c* is most affected by complex formation (Sukits et al., 1997), in agreement with the crystal structure (Pelletier and Kraut, 1992). Low spin ($S = 1/2$) ferric C*c* and high spin ferric C*c*P ($S = 5/2$) were utilized in the NMR study, so the C*c* resonances were not obliterated upon complex formation. The magnetic axis orientation of C*c* shifts significantly upon complex formation. The crystal structure of the C*c*P-horse C*c* complex was favored over the theoretical binding model in results obtained in isothermal titration calorimetry experiments conducted using native C*c*P and six binding site mutants (Erman et al., 1997).

Results on a much faster time scale can be obtained using laser photoexcitation of ruthenium(II) complexes covalently bound to cytochrome *c*. Excitation of Ru^{II} leads to electron transfer to the Fe^{III} of cytochrome *c* which in turn is transferred to the Trp191 cation radical of the peroxidase Compound I (Mei et al., 1999; Millett and Durham, 2002). Yeast iso-1 C*c* reacts so rapidly in the $Trp^{\bullet +}$ reduction in Compound I that no configurational gating occurs; gating is required for the slower reacting horse C*c* (Mei et al., 1999).

The electron transfer rate constant from yeast iso-1 C*c* to $Trp^{\bullet +}$ is $2 \times 10^6\ s^{-1}$. This step is too fast to be seen on a stopped-flow (ms) time scale, where the subsequent step, transfer to the $Fe^{IV}{=}O$ group, can be measured (of the order of $10^3\ s^{-1}$ depending upon experimental conditions). Therefore, a consistent picture of reduction of Compound I by C*c* emerges; rapid electron transfer to $Trp^{\bullet +}$ (leading to a $Fe^{IV}{=}O$, Trp structure for Compound II) followed by slower transfer from Trp to the ferryl heme (leading to a Fe^{III}, $Trp^{\bullet +}$ structure for Compound II). Reduction of $Trp^{\bullet +}$ thus appears to occur by the same mechanism in both Compounds I and II.

Interactions between 19 different complexes involving combinations of yeast iso-1-cytochrome *c*, its peroxidase, six C*c*P variants, and three C*c* variants were studied. The results confirmed the crystallographic binding site for the 1:1 complex (Pielak and Wang, 2001).

Enzyme reactions inside a cell are very crowded in contrast to the dilute solutions studied in vitro. Crowding was modeled using concentrated saccharide solutions (Morar et al., 2001; Morar and Pielak, 2001). The sugars cause a small increase in the stability of both the 1:1 and 2:1 complexes.

Binding of Cytochrome *c* to Cytochrome *c* Peroxidase in a 2:1 Ratio

With higher concentrations of yeast (iso-1 and iso-2), horse, tuna, and cicada cytochromes *c*, 2:1 binding was proven in a steady-state kinetic and equilibrium binding study (Kang et al., 1977). Equilibrium dissociation constants and kinetic K_M values correlated remarkably well. At low ionic strength binding is too tight, and at high ionic strength binding is too loose for optimal rate of reaction. Either biphasic or monophasic kinetics were observed, dependent largely upon cytochrome *c* concentration and ionic strength. Large differences were observed in the behaviors of horse and yeast cytochromes *c*, and these differences were exploited to demonstrate an allosteric effect between the two binding sites. A value of k_{cat}/K_M of $5.0 \times 10^8\,M^{-1}\,s^{-1}$ for iso-1 yeast cytochrome *c* was obtained at pH 6.0. This maximum (saturation) rate is about 1000-fold faster than for the reaction of ferrocyanide with yeast C*c*P. A fluorescence study using porphyrin cytochrome *c* confirmed the 2:1 binding to C*c*P (Kornblatt and English, 1986).

A series of studies on Zn-substituted cytochrome *c* peroxidase, an analog of the ferrous enzyme, has been completed in the laboratory of Brian Hoffman (Ho et al., 1985; Stemp and Hoffman, 1993). A ratio of 2:1 for ferric cytochrome *c* binding to the zinc enzyme was obtained. The zinc enzyme can photochemically reduce ferric cytochrome *c* resulting in formation of a porphyrin π-cation radical. This is followed by a "hopping" electron transfer in which the cation radical is transferred to Trp191 before decay back to the starting Zn-porphyrin. These studies have been extended to single crystals of both horse and yeast C*c* complexes with Zn-substituted C*c*P in the laboratory of Brian Crane (Kang et al., 2004).

Binding of iron-free cytochrome *c* is similar to but considerably weaker than the iron-containing protein. Cyanide does not inhibit cytochrome *c* binding (Kornblatt and English, 1986). The rate constant for electron transfer within the complex of ferrous cytochrome *c* peroxidase-ferric cytochrome *c* is $0.23 \pm 0.02\,s^{-1}$ (Cheung et al., 1986).

From an NMR study of H/D exchange rates on bound and free yeast iso-1 cytochrome *c*, it was concluded that both the front side and the opposite back side were involved with binding. The front side coincides with the strong binding observed in the crystal structure of the 1:1 complex (Yi et al., 1994). Properties of the high- and low-affinity sites were examined by potentiometric titration. Proton release occurs upon binding to the high affinity site, where a proton uptake occurs at the low affinity site (Mauk et al., 1994).

By photoinduced electron transfer from horse Zn-substituted cytochrome *c* combined with site-directed mutagenesis and isothermal titration calorimetry, the first evidence for the second binding site of C*c* was obtained, involving residues 146 to 150 on C*c*P (Leesch et al., 2000).

Stopped-flow kinetic studies have led to a detailed mechanism of the reaction of hydrogen peroxide with cytochrome *c* catalyzed by cytochrome *c* peroxidase (Miller, 1996; Mei et al., 2002). Strong evidence was obtained that turnover is strongly affected by the rate of dissociation of oxidized cytochrome *c* from the C*c*P-C*c* complex, as originally proposed by Kang et al. (1977). The following equations provide a basis for discussion.

$$\underset{\text{Compound I}}{\mathrm{Fe^{IV}{=}O,\ Trp^{\bullet +}}} + \mathrm{C}c^{\mathrm{II}} \rightleftharpoons \mathrm{Fe^{IV}{=}O,\ Trp^{\bullet +},\ C}c^{\mathrm{II}} \longrightarrow \underset{\text{Compound II}_{\text{A}}}{\mathrm{Fe^{IV}{=}O,\ Trp}} + \mathrm{C}c^{\mathrm{II}}$$

$$\underset{\text{Compound II}_{\text{A}}}{\mathrm{Fe^{IV}{=}O,\ Trp}} + \mathrm{H^+} \rightleftharpoons \underset{\text{Compound II}_{\text{B}}}{\mathrm{Fe^{III},\ Trp^{\bullet +}}} + \mathrm{H_2O}$$

$$\underset{\text{Compound II}_{\text{B}}}{\mathrm{Fe^{III},\ Trp^{\bullet +}}} + \mathrm{C}c^{\mathrm{II}} \rightleftharpoons \mathrm{Fe^{III},\ Trp^{\bullet +},\ C}c^{\mathrm{II}} \longrightarrow \underset{\mathrm{C}c\mathrm{P}}{\mathrm{Fe^{III},\ Trp}} + \mathrm{C}c^{\mathrm{III}}$$

Compound $\mathrm{II_B}$ is shown as it was proposed in the paper describing its discovery (Coulson et al., 1971). It is shown as $\mathrm{Fe^{III}}$-OH, $\mathrm{Trp^{\bullet +}}$ by Miller (1996), formed by a simple proton addition to the ferryl oxygen. Proof is lacking for the structure of the intermediate species when an electron is transferred through one oxidized site on the enzyme to the other.

The above equations, coupled with data for the 2:1 complex of C*c*-C*c*P, account for all the salient features of the total reduction of Compound I to native enzyme (Mei et al., 2002). Under conditions for 2:1 complex formation the rate of reduction of $\mathrm{Trp^{\bullet +}}$ by yeast iso-1 C*c* is greater than $3000\ \mathrm{s^{-1}}$, with subsequent reduction of the $\mathrm{Fe^{IV}{=}O}$ site of $10\ \mathrm{s^{-1}}$. If however the tighter binding site on Compound I is pre-occupied by ferric C*c* then upon addition of ferrous C*c* the rate of reduction of both sites is $10\ \mathrm{s^{-1}}$. The dissociation rate constant of oxidized product is $<5\ \mathrm{s^{-1}}$ for the 1:1 complex and $>2000\ \mathrm{s^{-1}}$ for the 2:1 complex.

Conclusions are: (1) occupancy of the second (weaker binding) site by C*c* greatly accelerates the rate-controlling dissociation of oxidized C*c* from the first site; (2) both redox reactions occur with ferrous C*c* in the first (stronger binding) site, and both reduce the $\mathrm{Trp^{\bullet +}}$ radical. These conclusions are in accord with an earlier study that showed Met230 plays an important role in electron transfer from $\mathrm{Fe^{IV}{=}O}$ to $\mathrm{Trp^{\bullet +}}$, regenerating the site for the second redox step (Liu et al., 1994).

Both steady-state and transient-state kinetics were also studied in an earlier investigation (Matthis and Erman, 1995; Matthis et al., 1995). The requirement of a 2:1 complex was shown, but it was concluded that ferryl oxygen was the initial site of reduction on the enzyme, in disagreement with the above conclusions (Mei et al., 2002).

To further complicate matters, the possibility of a tyrosyl free radical site on the enzyme must be considered (Zhang et al., 2002).

REFERENCES

Altschul AM, Abrams R, Hogness TR. 1940. Cytochrome *c* peroxidase. *J. Biol. Chem.* **136**: 777–794.

Bosshard HR, Anni H, Yonetani T. 1991. Yeast cytochrome *c* peroxidase. In: *Peroxidases in Chemistry and Biology*. Vol. II. Everse J, Everse KE, Grisham MB, editors. CRC Press, Boca Raton, FL, pp. 51–84.

Cheung E, Taylor K, Kornblatt JA, English AM, McLendon G, Miller JR. 1986. Direct measurements of intramolecular electron transfer rates between cytochrome *c* and cytochrome *c* peroxidase: effects of exothermicity and primary sequence on rate. *Proc. Natl. Acad. Sci. USA* **83**: 1330–1333.

Corin AF, Hake RA, McLendon G, Hazzard JT, Tollin G. 1993. Effects of surface amino acid replacements in cytochrome *c* peroxidase on intracomplex electron transfer from cytochrome *c*. *Biochemistry* **32**: 2756–2762.

Corin AF, McLendon G, Zhang Q, Hake RA, Falvo J, Lu KS, Ciccarelli RB, Holzschu D. 1991. Effects of surface amino acid replacements in cytochrome *c* peroxidase on complex formation with cytochrome *c*. *Biochemistry* **30**: 11585–11595.

Coulson AFW, Erman JE, Yonetani T. 1971. Studies on cytochrome *c* peroxidase. XVII. Stoichiometry and mechanism of the reaction of Compound ES with donors. *J. Biol. Chem.* **246**: 917–924.

DePillis GD, Sishta BP, Mauk AG, Ortiz de Montellano PR. 1991. Small substrates and cytochrome *c* are oxidized at different sites of cytochrome *c* peroxidase. *J. Biol. Chem.* **266**: 19334–19341.

Erman JE, Kresheck GC, Vitello LB, Miller MA. 1997. Cytochrome *c*/cytochrome *c* peroxidase complex: effect of binding site mutations on the thermodynamics of complex formation. *Biochemistry* **36**: 4054–4060.

Ho PS, Sutoris C, Liang N, Margoliash E, Hoffman BM. 1985. Species specificity of long-range electron transfer within the complex between zinc-substituted cytochrome *c* peroxidase and cytochrome *c*. *J. Am. Chem. Soc.* **107**: 1070–1071.

Kang CH, Ferguson-Miller S, Margoliash E. 1977. Steady state kinetics and binding of eukaryotic cytochromes *c* with yeast cytochrome *c* peroxidase. *J. Biol. Chem.* **252**: 919–926.

Kang SA, Marjavaara PJ, Crane BR. 2004. Electron transfer between cytochrome *c* and cytochrome *c* peroxidase in single crystals. *J. Am. Chem. Soc.* **126**: 10836–10837.

Kim KL, Kang DS, Vitello LB, Erman JE. 1990. Cytochrome *c* peroxidase catalyzed oxidation of ferrocytochrome *c* by hydrogen peroxide: ionic strength dependence of the steady state parameters. *Biochemistry* **29**: 9150–9159.

Kornblatt JA, English AM. 1986. The binding of porphyrin cytochrome *c* to yeast cytochrome *c* peroxidase: a fluorescence study of the number of sites and their sensitivity to salt. *Eur. J. Biochem.* **155**: 505–511.

Leesch VW, Bujons J, Mauk AG, Hoffman BM. 2000. Cytochrome *c* peroxidase-cytochrome *c* complex: locating the second binding domain on cytochrome *c* peroxidase with site-directed mutagenesis. *Biochemistry* **39**: 10132–10139.

Liu R-Q, Miller MA, Han GW, Hahm S, Geren L, Hibdon S, Kraut J, Durham B, Millett F. 1994. Role of methionine 230 in intramolecular electron transfer between the oxyferryl heme and tryptophan 191 in cytochrome *c* peroxidase Compound II. *Biochemistry* **33**: 8678–8685.

Matthis AL, Erman JE. 1995. Cytochrome *c* peroxidase-catalyzed oxidation of yeast iso-1 ferrocytochrome *c* by hydrogen peroxide: ionic strength dependence of the steady state parameters. *Biochemistry* **34**: 9985–9990.

Matthis AL, Vitello LB, Erman JE. 1995. Oxidation of yeast iso-1 ferrocytochrome *c* by cytochrome *c* peroxidase Compounds I and II: dependence upon ionic strength. *Biochemistry* **34**: 9991–9999.

Mauk MR, Ferrer JC, Mauk AG. 1994. Proton linkage in formation of the cytochrome *c*-cytochrome *c* peroxidase complex: electrostatic properties of the high- and low-affinity cytochrome binding sites on the peroxidase. *Biochemistry* **33**: 12609–12614.

Mauro JM, Fishel LA, Hazzard JT, Meyer TE, Tollin G, Cusanovich MA, Kraut J. 1988. Tryptophan-191 → phenylalanine, a proximal-side mutation in yeast cytochrome *c* peroxidase that strongly affects the kinetics of ferrocytochrome *c* oxidation. *Biochemistry* **27**: 6243–6256.

Mei H, Geren L, Miller MA, Durham B, Millett F. 2002. Role of the low-affinity binding site in electron transfer from cytochrome *c* to cytochrome *c* peroxidase. *Biochemistry* **41**: 3968–39376.

Mei H, Wang K, Peffer N, Weatherly G, Cohen DS, Miller M, Pielak G, Durham B, Millett F. 1999. Role of configurational gating in intracompex electron transfer from cytochrome *c* to the radical cation in cytochrome *c* peroxidase. *Biochemistry* **38**: 6846–6854.

Miller MA. 1996. A complete mechanism for steady-state oxidation of yeast cytochrome *c* by yeast cytochrome *c* peroxidase. *Biochemistry* **35**: 15791–15799.

Miller MA, Vitello L, Erman JE. 1995. Regulation of interprotein electron transfer by Trp 191 of cytochrome *c* peroxidase. *Biochemistry* **34**: 12048–12058.

Millett F, Durham B. 2002. Design of photoactive ruthenium complexes to study interprotein electron transfer. *Biochemistry* **41**: 11315–11324.

Mochan E. 1970. The nature of complex formation between cytochrome *c* and cytochrome *c* peroxidase. *Biochim. Biophys. Acta* **216**: 80–95.

Mochan E, Nicholls P. 1971. Complex-formation between cytochrome *c* and cytochrome *c* peroxidase: equilibrium and titration studies. *Biochem. J.* **121**: 69–82.

Morar AS, Pielak GJ. 2001. Crowding by trisaccharides and the 2:1 cytochrome *c*-cytochrome *c* peroxidase complex. *Biochemistry* **41**: 547–551.

Morar AS, Wang X, Pielak GJ. 2001. Effects of crowding by mono-, di-, and tetrasaccharides on cytochrome *c*-cytochrome *c* peroxidase binding: comparing experiment to theory. *Biochemistry* **40**: 281–285.

Nicholls P, Mochan E. 1971a. Complex-formation between cytochrome *c* and cytochrome *c* peroxidase: kinetic studies. *Biochem. J.* **121**: 55–67.

Nicholls P, Mochan E. 1971b. Formation of a stable "active" complex between cytochrome *c* and yeast peroxidase. *Nature New Biol.* **230**: 276–277.

Northrup SH, Boles JO, Reynolds JCL. 1988. Brownian dynamics of cytochrome *c* and cytochrome *c* peroxidase association. *Science* **241**: 67–70.

Pelletier H, Kraut J. 1992. Crystal structure of a complex between electron transfer partners cytochrome *c* peroxidase and cytochrome *c*. *Science* **258**: 1748–1755.

Pielak GJ, Wang X. 2001. Interactions between iso-1 cytochrome *c* and its peroxidase. *Biochemistry* **40**: 422–428.

Poulos TL, Kraut J. 1980. A hypothetical model of the cytochrome *c* peroxidase · cytochrome *c* electron transfer complex. *J. Biol. Chem.* **255**: 10322–10330.

Putnam A-MAH, Lee Y-T, Goodin DB. 2009. Replacement of an electron transfer pathway in cytochrome *c* peroxidase with a surrogate peptide. *Biochemistry* **48**: 1–3.

Stemp EDA, Hoffman BM. 1993. Cytochrome *c* peroxidase binds two molecules of cytochrome *c*: evidence for a low-affinity, electron-transfer-active site on cytochrome *c* peroxidase. *Biochemistry* **32**: 10848–10865.

Sukits SF, Erman JE, Satterlee JD. 1997. Proton NMR assignments and magnetic axes orientations for wild-type yeast iso-1-ferricytochrome *c* free in solution and bound to cytochrome *c* peroxidase. *Biochemistry* **36**: 5251–5259.

Summers FE, Erman JE. 1988. Reduction of cytochrome *c* peroxidase Compounds I and II by ferrocytochrome *c*: a stopped-flow kinetic investigation. *J. Biol. Chem.* **263**: 14267–14275.

Yi Q, Erman JE, Satterlee JD. 1994. Studies of protein-protein association between yeast cytochrome *c* peroxidase and yeast iso-1 ferricytochrome *c* by hydrogen-deuterium exchange labeling and proton NMR spectroscopy. *Biochemistry* **33**: 12032–12041.

Yonetani T. 1965. Studies on cytochrome *c* peroxidase. II. Stoichiometry between enzyme, H_2O_2, and ferrocytochrome *c* and enzymic determination of extinction coefficients of cytochrome *c*. *J. Biol. Chem.* **240**: 4509–4514.

Yonetani T. 1970. Cytochrome *c* peroxidase. *Adv. Enzymol.* **33**: 309–335.

Yonetani T. 1976. Cytochrome *c* peroxidase. In: *The Enzymes*, Vol. 13, 3rd edition, Boyer PD, editor. Academic Press, New York, pp. 345–361.

Yonetani T, Ray GS. 1965. Studies on cytochrome *c* peroxidase. I. Purification and some properties. *J. Biol. Chem.* **240**: 4503–4508.

Yonetani T, Ray GS. 1966. Studies on cytochrome *c* peroxidase. III. Kinetics of the peroxidatic oxidation of ferrocytochrome *c* catalyzed by cytochrome *c* peroxidase. *J. Biol. Chem.* **241**: 700–706.

Zhang H, He S, Mauk AG. 2002. Radical formation at Tyr39 and Tyr153 following reaction of yeast cytochrome *c* peroxidase with hydrogen peroxide. *Biochemistry* **41**: 13507–13513.

CHAPTER 8

SPECTROSCOPY. I. OPTICAL, RESONANCE RAMAN, AND X-RAY ABSORPTION

A wealth of detail on peroxidase structure can be obtained from the application of spectroscopic techniques. In this chapter an introduction will be given to optical, resonance Raman, and x-ray absorption spectroscopies. The latter include XAS, EXAFS, XANES, and XAFS, acronyms that will be defined later. The common theme of the chapter is the interaction with peroxidases of photons from all regions, except microwave and radio frequency, of the electromagnetic spectrum.

In the following chapter magnetic spectroscopic techniques are described: electron spin resonance (ESR) and nuclear magnetic resonance (NMR) spectroscopies, as well as Mössbauer spectroscopy. After the introduction of each technique, examples of results in the literature are cited. These are dominated by the two enzymes most ubiquitous in the literature, horseradish peroxidase HRP, and yeast cytochrome *c* peroxidase C*c*P.

OPTICAL ABSORPTION SPECTRA

Optical absorption spectroscopy of heme peroxidases is controlled by the *d* electrons of the iron, the π electrons of the porphyrin, and the interaction between the two. The properties of the five *d* electrons of iron(III) with six ligands provide a good starting point for discussion. In the absence of external influence, the five *d* electrons are identical in energy, and in accord with Hund's rule they each occupy a separate *d* orbital (Fig. 8.1a).

Peroxidases and Catalases: Biochemistry, Biophysics, Biotechnology, and Physiology, Second Edition.
By H. Brian Dunford

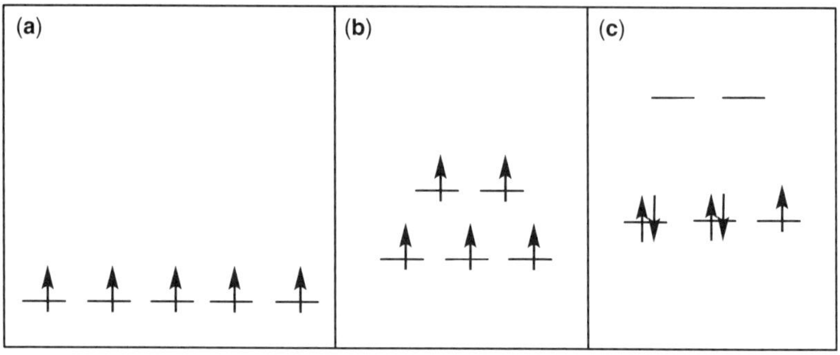

Figure 8.1. The five *d* electrons of iron(III) under the influence of ligand fields (a) of zero strength, (b) weak, (c) strong.

Simple crystal field theory is adequate to describe the influence on the *d* electron energies of six ligands located on the x-, y-, and z-axes. Each ligand has a lone pair of electrons to form a chemical bond with the ferric ion. The repulsion between the incoming electron pair on the ligand and a *d* electron is dependent on the *d* orbital geometry. The orbitals $d_{x^2-y^2}$ and d_{z^2} are located along the axes of ligand approach, whereas the d_{xy}, d_{xz}, and d_{yz} are not. Repulsion is greater for electrons in the $d_{x^2-y^2}$ and d_{z^2} orbitals. Thus two orbitals are raised in energy more than the other three. For a weak ligand field the energy splitting is small, so Hund's rule still applies and each *d* electron remains in a separate orbital. A high-spin complex is obtained (Fig. 8.1b), with a spin-only magnetic moment of 5.92 Bohr magnetons. For a strong ligand field, large splitting occurs and occupancy of the lowest possible energy level possible has priority over Hund's rule (Fig. 8.1c). Pairing occurs so only a single unpaired electron remains, with a spin-orbital magnetic moment in the range 1.7 to 2.8. If the six ligands are identical then the geometry is described as octahedral since the complex can be enclosed by a symmetric eight-sided figure.

If the iron has four identical ligands in a plane with two other ligands at right angles, then a tetragonal rather than octahedral field exists. The tetragonal field provides an appropriate description for the iron of peroxidases, whether five- or six-coordinate. The four pyrrole nitrogens of the heme coordinated to the iron are all equivalent, but the influence of the fifth and sixth coordination sites is different. For discussion of whether the iron of a peroxidase complex is low or high spin, the simple crystal field concepts used to construct Figure 8.1 are satisfactory. However for a discussion of spectral properties, one must switch to a more realistic energy diagram.

Heme electronic absorption spectra originate from porphyrin π-electrons. The most populated porphyrin π electron energy levels are an almost degenerate pair (meaning of equal energy) of orbitals designated as a_{1u} and a_{2u}. Absorbed light will cause electrons to jump into higher energy orbitals, either a π* porphyrin orbital (where the star indicates an antibonding orbital) or a vacant *d* orbital of the iron.

Absorption bands caused by transitions within the porphyrin, in order of increasing energy, are labeled Q_{00}, Q_{0v}, and B. Corresponding older nomenclature is α, β, and Soret bands (Fig. 8.2). The subscript pair 00 indicates a transition from the lowest

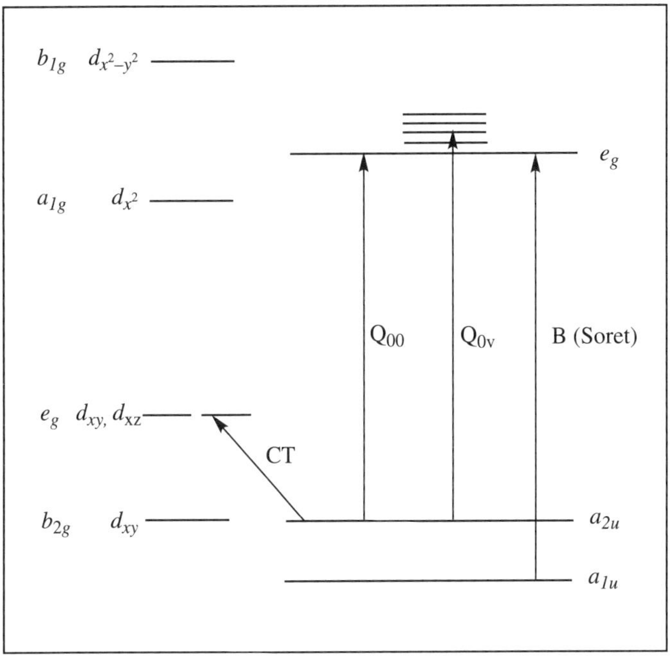

Figure 8.2. Absorption bands typically observed for a peroxidase: CT charge transfer, Q_{00} transition from the zeroth vibrational level of the a_{2u} electronic state to the e_g state, Q_{0v} transitions to higher vibrational levels. Iron energy levels on the left, porphyrin on the right. (Based on Smulevich G. 1996. In: Obinger C, et al., editors. *Plant Peroxidases: Biochemistry and Physiology.* University of Geneva Press, pp. 13–19.)

vibrational level of the lower electronic state to the lowest vibrational level of the excited state, whereas the v in 0v indicates a transition to an envelope of higher vibrational levels in the excited state. If no vacant *d* orbitals are available in the central metal ion then one obtains a regular porphyrin spectrum: Q_{00}, Q_{0v}, and B (α, β, and Soret) bands only. Transitions from porphyrin to iron sometimes result in charge transfer (CT) bands that can mask Q_{00} and Q_{0v}, making it difficult to assign transitions.

Examples of Class I and Class II native peroxidases, yeast cytochrome *c* and *Coprinus cinereus*, are purely high-spin, five-coordinate. Plant peroxidases such as horseradish peroxidase exhibit some six-coordinate high-spin properties indicating partial occupancy of the sixth coordination position by water (Smulevich, 1996). The optical spectrum of native *Coprinus cinerus* peroxidase is shown in Figure 8.3, illustrating the B (Soret), Q_{00} (α), Q_{0v} (β), and charge transfer bands.

Optical Spectral Evidence for Coordination Number

High spin complexes have B (Soret) bands in the 398 to 408 nm region. Using HRP as an example, its Soret band is centered at 403 nm. When fluoride is bound, a pure six-coordinate high-spin species exists with the Soret maximum essentially

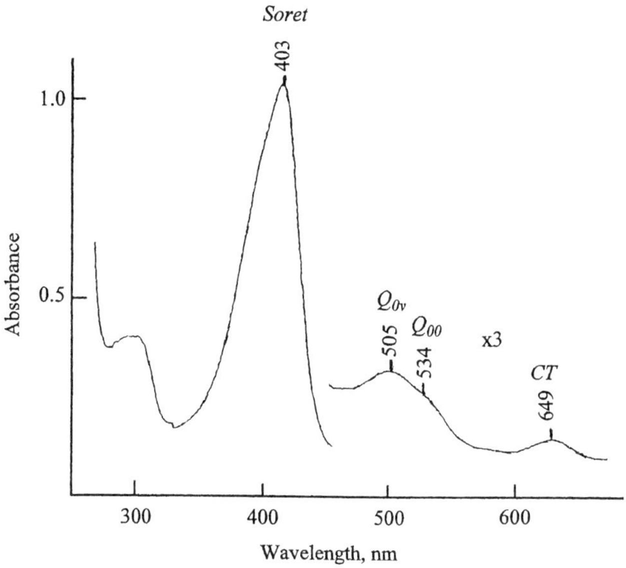

Figure 8.3. Optical spectrum of native *Coprinus cinereus* peroxidase, illustrating the α (Q_{00}), β (Q_{0v}), Soret (B) and charge transfer (CT) bands. (Based on Smulevich G. 1996. In: Obinger C, et al., editors. *Plant Peroxidases: Biochemistry and Physiology.* University of Geneva Press, pp. 13–19.)

unchanged at 404 nm. However, the molar absorptivity is increased. The average molar absorptivity in the Soret region of a series of native peroxidases is $(105 \pm 6)\ mM^{-1}\ cm^{-1}$. For a series of globins containing ferric ion, where it is known that water fully occupies the sixth coordination position, it is $(177 \pm 9)\ mM^{-1}\ cm^{-1}$. The molar absorptivities of the fluoride complexes of both peroxidases and globins is $(138 \pm 7)\ mM^{-1}\ cm^{-1}$. The simplest explanation is that the native peroxidase iron is always essentially five-coordinate, whereas the native ferric globins and all fluoride complexes are six-coordinate species (Dunford, 1982).

Whereas fluoride provides a weak ligand field, cyanide presents a very strong one. The low-spin six-coordinate cyanide complex of HRP has its Soret band maximum shifted to 420 nm. A summary of wavelengths and molar absorptivities of optical absorption bands of HRP, its compounds, and its complexes has been published (Dunford and Stillman, 1976; Dunford, 1999).

Optical Detection of Ferrous, Compound II, Compound III Conversions

The Soret region spectra of Compounds II and III are virtually identical so reliable distinction between these two species is not possible by measuring B band intensities. In a classic paper, accurate visible spectra were exploited to follow the hydrogen peroxide-initiated conversions of horseradish peroxidase from its ferrous form to

Compound II to Compound III (Noble and Gibson, 1970). The visible spectral data were supplied by Jonathan Wittenberg and these provided the basis for stopped-flow measurements. Whereas the maxima in the Q_{00} and Q_{0v} bands of Compound II occur at 555 and 527 nm, the corresponding bands in Compound III are at 578 ± 2 nm and 544 ± 1 nm (Wittenberg et al., 1967; Adediran and Lambeir, 1989).

RESONANCE RAMAN SPECTRA

Introduction

To a first approximation, a chemical bond may be regarded as a simple harmonic oscillator. The two atoms at each end of the bond oscillate back and forth, even at 0 K. The lowest energy oscillation is defined by the vibrational quantum number $v = 0$. At room temperature most bonds occupy the vibrational level defined by $v = 0$. Irradiation by infrared light of exactly the right energy (the resonance frequency) will cause transition to the next highest energy vibrational level defined by $v = 1$. The stronger a chemical bond, the higher the energy of the vibrational frequencies. Thus, infrared spectroscopy provides a means of identifying the structure of a molecule. For example, a carbon–carbon single bond will have a lower stretching frequency than a stronger carbon–carbon double bond.

For a diatomic molecule only a stretching mode is observed. In a polyatomic molecule there are also other types of vibrational motion. The oxygen and hydrogen atoms of the OH group of a phenol molecule will undergo stretching and contraction, just as in a diatomic molecule. In addition the OH group can wag back and forth, characterized by a bending frequency. A symmetric polyatomic molecule can undergo symmetric stretches akin to a breathing motion. The same molecule can vibrate asymmetrically, in which one bond is contracting while its symmetric counterpart is stretching.

Isotopic substitution changes a vibrational frequency. Thus, if a heavier deuterium atom (^2H) replaces a hydrogen atom (^1H), the bond will have a substantially lower vibrational energy. For macrocycles such as hemes such substitutions may affect the motions of all the atoms, and hence the entire resonance Raman spectrum. However, some individual vibrational frequencies are sufficiently uncoupled that they can be detected directly. Examples are those of Fe=O and Fe-proximal His. Similarly, stretching frequencies of C=C bonds in vinyl substituents were identified by vinyl deuteration. Their frequencies are indicative of the degree of conjugation of the vinyl double bond with the porphyrin ring (Choi et al., 1982).

One way to obtain a vibrational spectrum is to scan the sample by continuously changing the frequency of the incident radiation. Modern infrared spectrometers use continuous radiation, and the bands are detected using Fourier transforms, as in NMR spectroscopy. In resonance Raman spectroscopy a light source of higher energy than the vibrational resonance frequencies is used.

The effect of water is to largely mask other vibrational spectra in the infrared region. The development of resonance Raman spectroscopy essentially eliminated the problem of water interference. High energy laser light is used corresponding in energy to an electronic transition giving rise to what is known as the resonance Raman effect. Using laser light in the Soret region of the spectrum of a heme excites the B electronic band and greatly enhances heme vibrational transitions. Analysis of heme vibrations using resonance Raman spectroscopy has been built upon data using simpler models. Nickel(II) octaethyl porphyrin is a symmetric molecule with 35 Raman active bands (Abe et al., 1978; Li et al., 1990). Studies were extended to nickel(II) protoporphyrin IX and iron protoporphyrin complexes so that new bands caused by reduced symmetry could be assigned. Finally, myoglobin was used as an excellent model for assignment of all protoheme Raman resonances (Hu et al., 1996). The motions of all of the atoms can be accounted for by a technique called normal coordinate analysis.

The resonance Raman spectrum of a heme protein is characterized by fairly strong porphyrin skeletal bands in the 1300 to 1650 cm^{-1} region. These bands are highly sensitive to the size of the metal binding site (the core size) and the oxidation state, coordination number, and spin state of the metal ion. One particular vibrational motion of the porphyrin, the breathing mode v_4, shown in Figure 8.4, is responsible for the strongest resonance Raman band under Soret excitation, which provides a marker for the oxidation state of the metal ion and other effects. Spin states, determined from resonance Raman spectra, are in agreement with direct magnetic measurements.

Horseradish Peroxidase Resonance Raman Spectra

The first resonance Raman spectral studies involving a peroxidase were comparisons of ferrous HRP, cytochrome *c*, and microperoxidase (Loehr and Loehr, 1973). In a more comprehensive investigation, spectra for the native, fluoro, and cyano derivatives

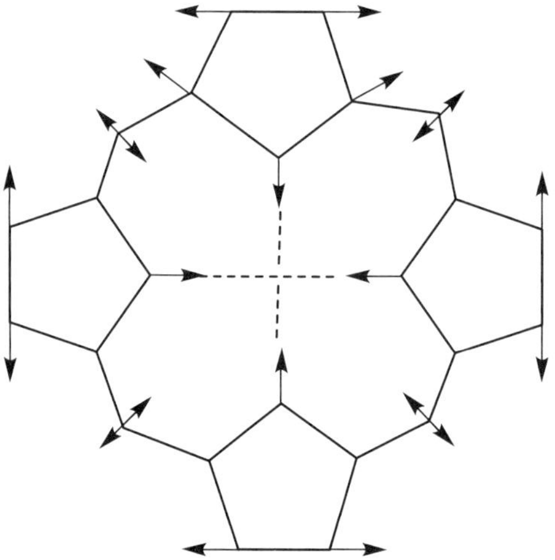

Figure 8.4. Breathing vibration of a porphyrin.

of ferric HRP were obtained (Rakshit and Spiro, 1974). The oxidation state of the central metal ion can be determined by the frequency of the core size marker bands since these are affected by vibrations of the pyrolle nitrogens.

Native HRP bands at 1500, 1575, and 1630 cm^{-1} are close to those for five-coordinate model compounds (Spiro et al., 1979). By means of isotopic iron substitution in HRP, the iron-proximal histidine stretching frequency was detected (Teraoka and Kitagawa, 1981). For HRP isoenzyme C, v(Fe—His) is 243 cm^{-1} (Smulevich et al., 1994). Low temperature spectra of HRP A2 have been analyzed (Howes et al., 2000).

Binding of benzhyroxamic acid, but not hydroquinone, changed the iron(III) from five- to six-coordination, with likely binding of water in the sixth position.

The controversy over the nature of the alkaline form of Class III peroxidases (see Sitter et al., 1988, for references) was resolved by resonance Raman spectroscopy (Sitter et al., 1988; Feis et al., 1994; Fig. 8.5). The transition to the alkaline form of HRP C occurs at pH 11.0. It consists of a change from five-coordinate high-spin iron(III) to six-coordinate low-spin iron(III) with a hydroxyl group occupying the sixth coordination position. The assignment was made iron-clad by the use of ^{18}O, ^{2}H, ^{54}Fe, and ^{57}Fe substitutions on HRP C (Fig. 8.6). The Fe^{III}—OH stretching frequency at 503 cm^{-1} shifts to 484 cm^{-1} upon ^{18}O substitution.

At pH 12 the alkaline form Fe—OH is oxidized to Fe=O by ferricyanide (Sitter et al., 1988). From studies using sight-directed mutants, it appears that Arg38 is essential for stabilization of the OH^- ligand, while His42 acts as a hydrogen bond acceptor (Howes et al., 1997). Similar results were found for the fluoride complex (Neri et al., 1997).

Further details on native HRP as well as its aromatic donor complexes have been obtained by resonance Raman spectroscopy (Smulevich et al., 1991a). At room temperature two sets of skeletal porphyrin vibrations are observed that correspond to mixture of five- and six-coordinate high-spin forms, with the five-coordinate form being dominant. As the temperature is lowered to 20 K, intermediate- and low-spin states are obtained. Two different C=C stretching modes are observed for both the enzyme and its complexes, indicating that the vinyl groups are not equivalent. Hydroquinone has little effect on the observed spectra, but benhydroxamic acid shifts the predominantly five-coordinate spectra of the native enzyme to six-coordinate, and one of the vinyl modes is enhanced.

Dramatic changes to native HRP when the pH is lowered to 3.1 have been detected by both resonance Raman and electronic absorption spectra (Smulevich et al., 1997). Native HRP is first converted to a six-coordinate species by bound water, then back to a five-coordinate species as the proximal His—Fe bond is cleaved.

Recombinant HRP isoenzyme C, and three mutants: Phe41Trp, Phe41Val, and Arg38Lys were examined by resonance Raman spectroscopy (Smulevich et al., 1994). The active enzymes were obtained by in vitro refolding after expression in *E. coli*. Spectra of the iron(II) and iron(III) forms and complexes with benzhydroxamic acid were recorded at neutral pH. The mutations, all on the distal side, dramatically affect the ability to bind benzhydroxamic acid. The substitution at position 38 and of Trp at position 41 eliminate binding altogether, whereas that of Val at position 441 leads to a twofold increase in affinity. These studies were extended to

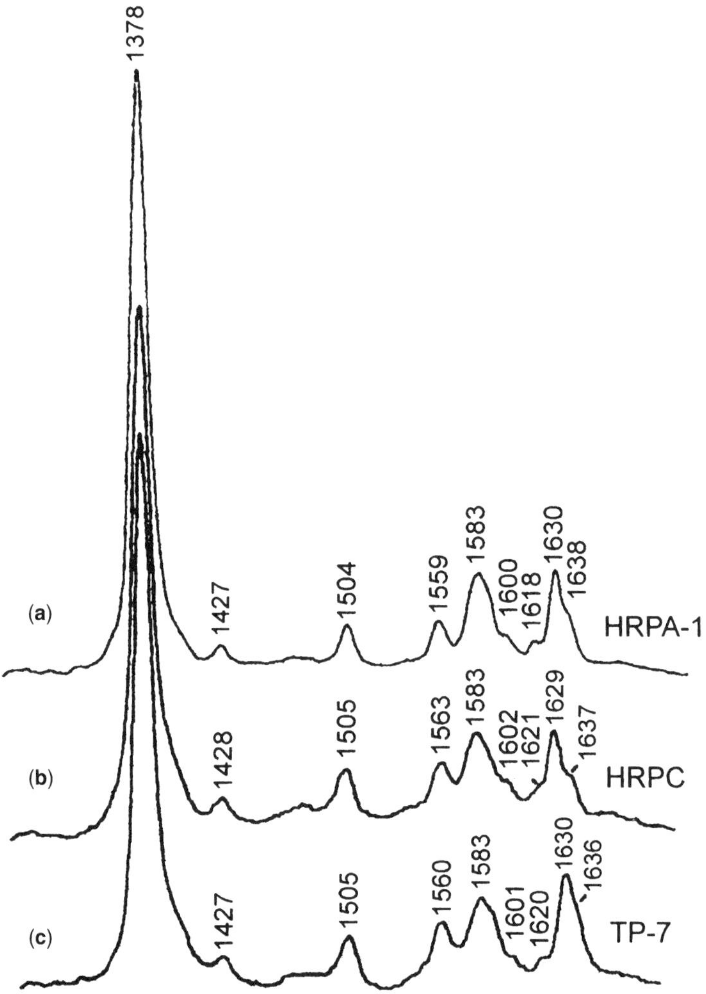

Figure 8.5. Resonance Raman spectra of peroxidase alkaline forms, with patterns of six-coordinate low-spin hemes. (a) HRP A1 at pH 10.2, (b) HRP C at pH 12.0, (c) turnip peroxidase 7 at pH 9.6. (Figure kindly supplied by Dr. James Terner.)

His42Leu, His42Arg, Arg38Gly, and Arg38Leu mutants (Howes et al., 1997). The results support the conclusion that the iron in wild-type HRP is predominantly five-coordinate high-spin. It also contains two forms of six-coordinate high-spin: one with water coordinated to distal Arg and the other water coordinated to distal His.

Further resonance Raman spectral studies explored selective effects on the reduced state by a position 171 mutant (Howes et al., 2005), and a proximal side position 221 mutant (Howes et al., 2001b). The Class III peroxidases such as HRP display a distinctive five-coordinate species that consists of an admixture of $S = 5/2$ and $3/2$

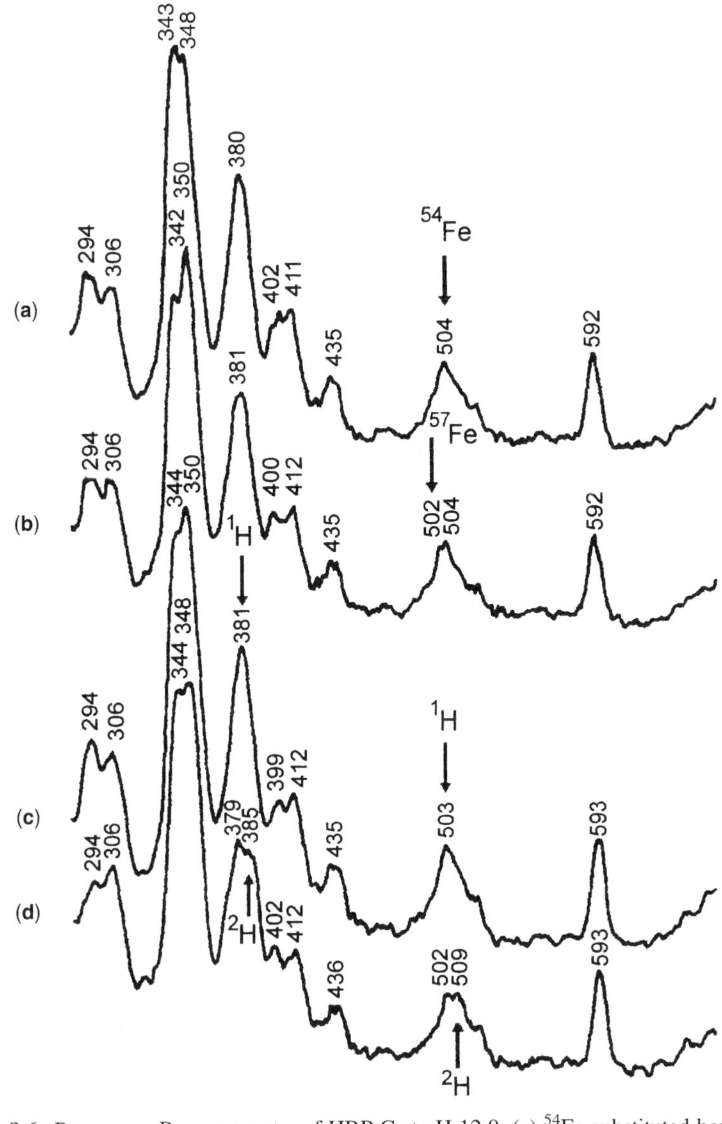

Figure 8.6. Resonance Raman spectra of HRP C at pH 12.0. (a) ^{54}Fe-substituted heme compared to (b) ^{57}Fe-substituted heme. The shift of (c) H_2O to (d) ^2H_2O changes $\upsilon_{Fe(III)-OH}$ from 503 cm^{-1} to 509 cm^{-1}. An out-of-plane mode also shifts from (c) 381 cm^{-1} to (d) 385 cm^{-1}. (Figure kindly supplied by Dr. James Terner.)

states. Benhydroxamic acid binding leads to a six-coordinate species, also a quantum-mechanically mixed state (Indiani et al., 2000).

Differences in resonance Raman spectra of HRP compared to cytochrome *c* and *Coprinus cinereus* peroxidases have been described (Smulevich, 1995). Calcium ion plays a critical role in the structural integrity of HRP (Howes et al., 2001a).

Resonance Raman Spectra of Horseradish Peroxidase Compound II

Raman spectral results for oxidized forms of peroxidase have been complicated by photoreduction, particularly of Compound I, the spectrum of which is discussed later (Van Wart and Zimmer, 1985a). Because of the inherent instability and photosensitivity of heme proteins, flow systems have been designed (Woodruff and Spiro, 1974; Paeng and Kincaid, 1988; Ogura and Kitagawa, 1987; Oertling and Babcock, 1988). One system was designed so that both resonance Raman and optical spectra could be monitored on the same flowing sample (Ogura and Kitagawa, 1987).

The Fe=O vibrational frequency in a heme was first identified for iron tetraphenylporphyrin in a frozen oxygen matrix. The Fe=O frequency at 852 cm^{-1} was shifted to 818 cm^{-1} upon ^{18}O substitution (Bajdor and Nakamoto, 1984). In this system there are no hydrogen bonds and the iron is five-coordinate. In solutions of six-coordinate model compounds containing axial ligands such frequencies are shifted to lower frequencies and, in proteins such as peroxidases, to even lower frequencies. Thus, in Compound II of HRP isoenzyme C the Fe^{IV}=O frequency occurs at 775 cm^{-1}, which is shifted to 745 cm^{-1} upon ^{18}O substitution (Fig. 8.7; Terner et al., 1985; Sitter et al., 1985; Makino et al., 1986).

For Compound II of HRP isoenzyme A1 the Fe=O frequency is 779 cm^{-1}. Alkaline pH values are observed to shift the Compound II Fe^{IV}=O stretching frequency of the A1 and A2 isoenzymes to 790 cm^{-1} and to 755 cm^{-1} upon ^{18}O substitution. The pK_a values for these transitions were found to be 6.9 for isoenzyme A2, and 8.5 for isoenzymes B and C (Fig. 8.8) (Sitter et al., 1985; Hashimoto et al., 1986).

The changes have been attributed to the ionization of the imidazole side chain of distal His42 which, it was originally thought, broke the hydrogen bond to the ferryl oxygen atom. Recent crystal structure results indicate there is an intervening water molecule between Fe^{IV}=O and distal His (Berglund et al., 2002), but the importance of the linkage between iron and distal imidazole below pH 8.6 is established. By use of

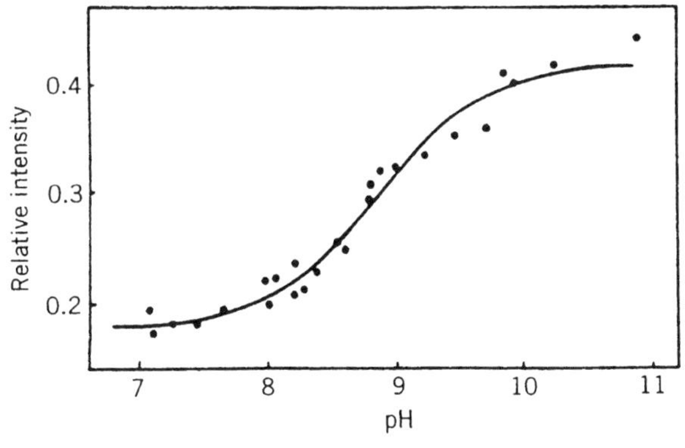

Figure 8.7. pH-dependent intensity change of the Fe^{IV}=O stretching frequency of horseradish peroxidase C Compound II. At pH values below the inflection point Compound II is reactive, and at higher pH values is unreactive. (Reproduced with permission from Hashimoto S, Tatsuno Y, and Kitagawa T. 1986. *Proc. Natl. Acad. Sci. USA* **83**: 2417–2421.)

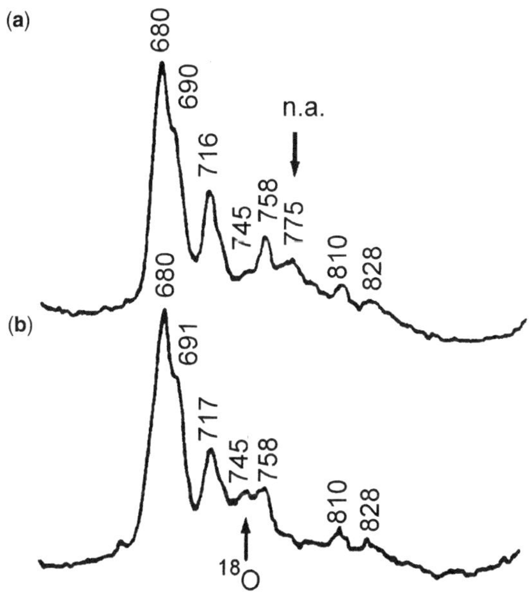

Figure 8.8. Resonance Raman spectra of HRP Compound II (a) with normal isotope abundance (n.a.) and (b) ^{18}O introduced using $H_2^{18}O_2$. (Figure kindly supplied by Dr. James Terner.)

$H_2{}^{18}O$ it was shown that the ferryl oxygen exchanges with water below pH 8.6 (Hashimoto et al., 1986).

The alkaline forms of the Compounds II are unreactive (Critchlow and Dunford, 1972). Compound III (oxyperoxidase) resonance Raman spectra have been obtained (Van Wart and Zimmer, 1985b).

Is the Ferryl Oxygen in Compound II Protonated? A controversy has arisen over whether the ferryl $Fe^{IV}{=}O$ form of peroxidases really exists or whether it is really $Fe^{IV}{-}OH$ as supported by a long 1.9 Å bond observed in crystallographic measurements (Rovira and Fita, 2003; Bonagura et al., 2003; Silaghi-Dumitresco, 2004; Conradie et al., 2004; Green, 2006).

The assignment of $Fe^{IV}{=}O$, not $Fe^{IV}{-}OH$, appears justified by resonance Raman frequencies that are appreciably higher than those observed for $Fe^{III}{-}OH$ and $Fe^{II}{-}O_2$ hemes (Terner et al., 2006). The $Fe^{IV}{=}O$ structure is confirmed in an EPR and ENDOR investigation (Davydov et al., 2008). The exception is Compound II of chloroperoxidase, where an $Fe^{IV}{-}OH$ frequency is observed at 565 cm^{-1} (Stone et al., 2006).

Resonance Raman Spectra of Horseradish Peroxidase Compound I A band at 791 cm^{-1} was correctly assigned to HRP-I (Oertling and Babcock, 1988). HRP-I bands are weaker than those of HRP-II so photoreduction of HRP-I has to be minimized. This can be accomplished by using ultraviolet rather than deep violet (Soret) excitation (Palaniappan and Terner, 1989; Kincaid et al., 1996). The latest results place this HRP-I band at 794 cm^{-1}, with an additional band at 786 cm^{-1} observed at lower pH (Terner et al., 2006). For chloroperoxidase Compound I the corresponding band occurs at 792 cm^{-1} (Egawa et al., 1992, 2001; Hosten et al., 1994). Interference also occurs from fluorescence. To date only samples of HRP isoenzyme C and of chloroperoxidase have exhibited a sufficiently low fluorescent background to detect the weak porphyrin π-cation skeletal bands of Compound I species (Terner et al., 2006). Resonance Raman spectra of horseradish peroxidase Compound I are shown in Figure 8.9.

Yeast Cytochrome *c* Peroxidase Resonance Raman Spectra

The yeast C*c*P iron(III) coordination of five was determined in two resonance Raman studies (Sievers et al., 1979; Smulevich et al., 1990b). Isotopically labeled hemes were used to assign bands of the recombinant enzyme, and the Asp235Asn mutant (Smulevich et al., 1996).

In a study of Fe^{II} and Fe^{III} forms of mutants and the enzyme cloned in *E. coli*, relations among hydrogen-bonded networks were established. The Fe^{III} coordination number is changed from five to six in the Trp51Phe mutant, showing that the hydrogen bond from Trp51 to the distal water inhibits its coordination to iron. In the Asp235Asn mutant, the iron coordination number is also changed to six, showing that the Asp235-proximal His175 interaction helps prevent the iron from entering the heme plane, which would facilitate water coordination in the sixth position (Smulevich et al., 1988).

The effects of glycerol and temperature at pH 7 on the resonance Raman spectra of native enzyme and the mutants Trp51Phe, Arg48Leu, Arg48Lys, Asp235Asn, and Trp191Phe were investigated (Smulevich et al., 1989). Glycerol retards conversion to a six-coordination state at low temperature. As the temperature is lowered, the iron gradually moves into the heme plane, facilitating water binding in the sixth coordination position (Smulevich et al., 1990a).

The effect of mutations on both the distal and proximal His residues of Fe(II), Fe(III), and Fe(II)CO has been examined by resonance Raman spectroscopy (Smulevich et al., 1991b, 1995). Mutations disrupt a hydrogen bonding network that extends from the distal to the proximal side (Fig. 8.10). In addition to C*c*P, this proximal-distal H—bonded network has been established for horseradish and *Coprinus cinereus* peroxidases (Smulevich et al., 2005). Reviews on the resonance Raman spectra of yeast cytochrome *c* peroxidase have been published (Spiro et al., 1990; Smulevich, 1993; Smulevich et al., 2005).

Shape of Peroxidase Hemes

Metalloporphyrins and other tetrapyrolle macrocycles can have many different conformations. The nonplanar distortions can be described as saddling, ruffling, doming, waving, and propelloring (Fig. 8.11). The types and extents of these distortions have been analyzed in the research program of John Shelnutt and coworkers. The

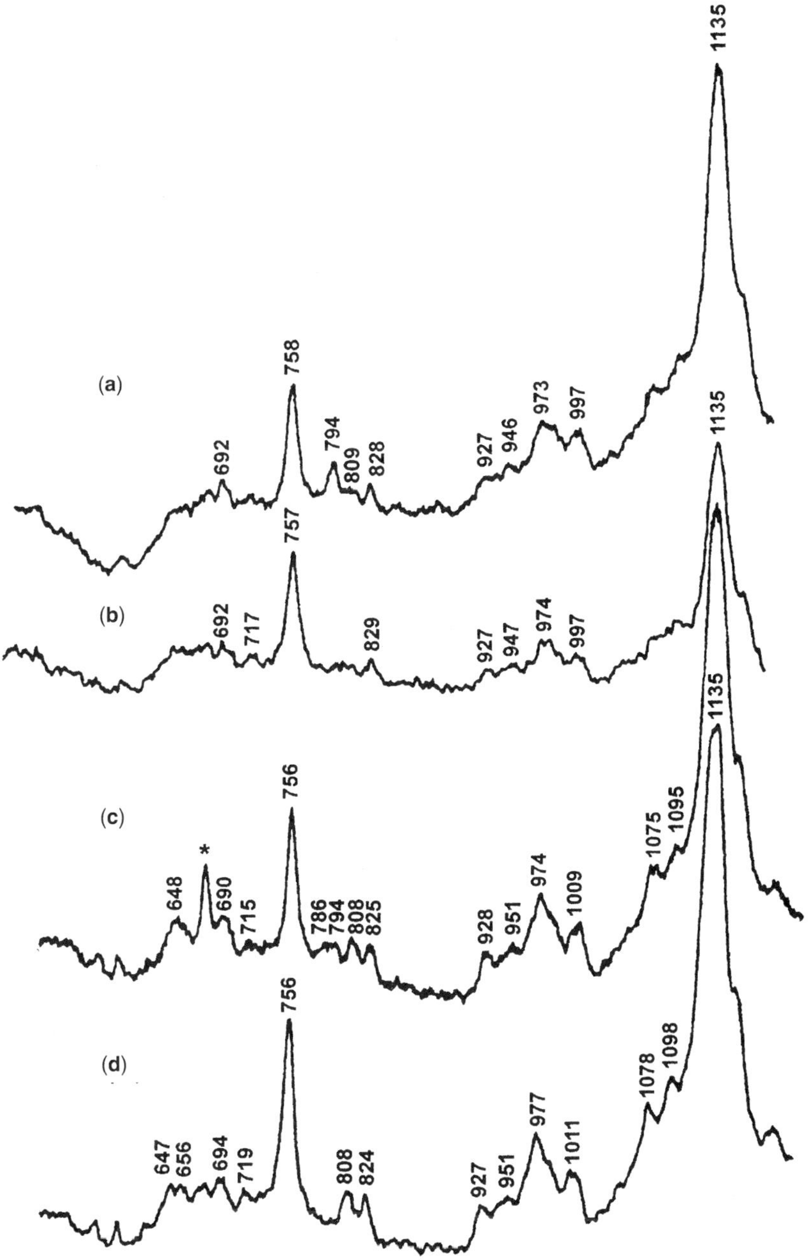

Figure 8.9. pH changes in resonance Raman spectra of horseradish peroxidase Compound I. Figures (a) and (b) pH 6.8 with oxidants H_2O_2 and $H_2^{18}O_2$, respectively. Figures (c) and (d) pH 4.4 with H_2O_2 and $H_2^{18}O_2$. The (*) in (c) marks a laser plasma line that could not be eliminated. (Figure kindly supplied by Dr. James Terner.)

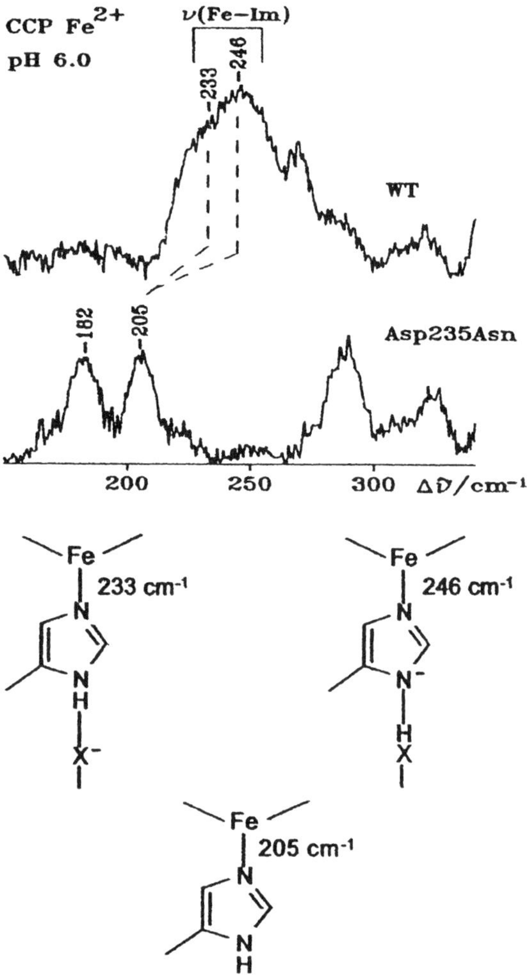

Figure 8.10. Effect of Asp235Asn mutant on H—bonds of ferrous cytochrome *c* peroxidase. (Reproduced with permission from Smulevich G, Feis A, Howes BD. 2005. *Acc. Chem. Res.* **38**: 433–440.)

principal tools are resonance Raman spectroscopy and normal coordinate analysis. X-ray structures, molecular modeling, and optical spectra are also utilized. The analysis utilizes normal coordinate structural decomposition (NSD) (Shelnutt et al., 1998). By use of NSD an accurate static crystal structure of the heme in a protein can be deconvoluted into the relative contributions of fundamental vibrational frequencies. This leads to a tabulation of +/− distortions from planarity. Distortions of the heme of HRP C rule against D_{4h} symmetry (Huang et al., 2003).

The heme *b* of cytochrome *c* peroxidase was shown to have a saddle shape, markedly different from that of a wide variety of other heme proteins (Jentzen et al., 1997).

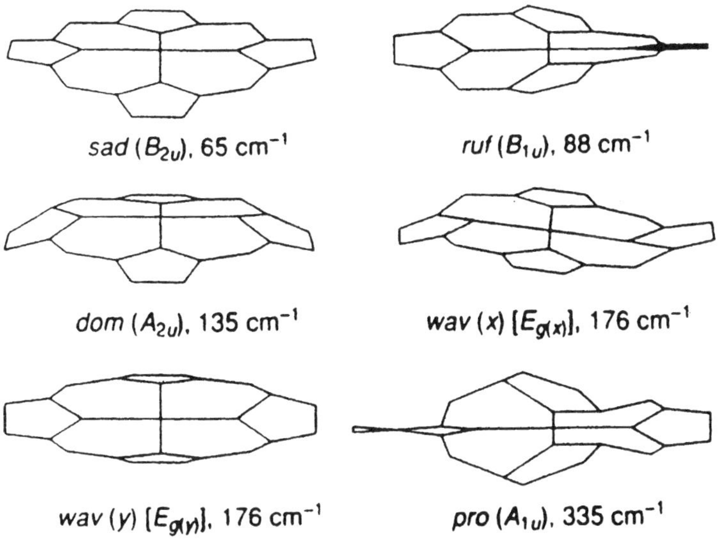

Figure 8.11. Symmetric deformations of hemes for saddling (sad), ruffling (ruf), doming (dom), waving (wav), and propelloring (por). (Reproduced with permission from Shelnutt JA, et al., 1998. *Chem. Soc. Rev.* **27**: 31–41.)

Other peroxidase heme *b*s have similar saddle shapes. The saddle shape of peroxidase hemes appears to be an intrinsic property present when the heme iron is either five- or six-coordinate. Even the structurally unrelated heme of myeloperoxidase has a saddle shape. An exception is the heme of some prostaglandin synthases.

X-RAY ABSORPTION SPECTROSCOPY

X-ray absorption spectroscopy (XAS) refers to an experimental technique of obtaining chemical bond lengths, applicable to a compound which may be in solution, frozen, amorphous, crystalline, or even in the gas phase. The discussion here concentrates on data involving a central metal ion. XAS data can be divided into two parts, the edge component and the absorption that occurs at higher energies (Fig. 8.12).

K edge absorption is determined by the transition energy of innermost 1*s* electrons to unoccupied higher energy orbitals of the same ion, such as 3*d*. *L* edge absorption originates from 2*s* and 2*p* electrons. The edge absorption extends over a range of ~20 electron volts (eV). At somewhat higher energies these electrons are liberated from the metal ion and are called photoelectrons. Like all electrons they exhibit wave as well as particle properties; the waves interact with nearest neighbor atoms in such a way that backscattering occurs. If the backscattering is in phase, then wave reinforcement occurs; out-of-phase scattering causes troughs. Thus oscillations are superimposed on the high-energy absorption. The backscattering can be measured up to 1000 eV in excess of the energy at the absorption edge and is called extended

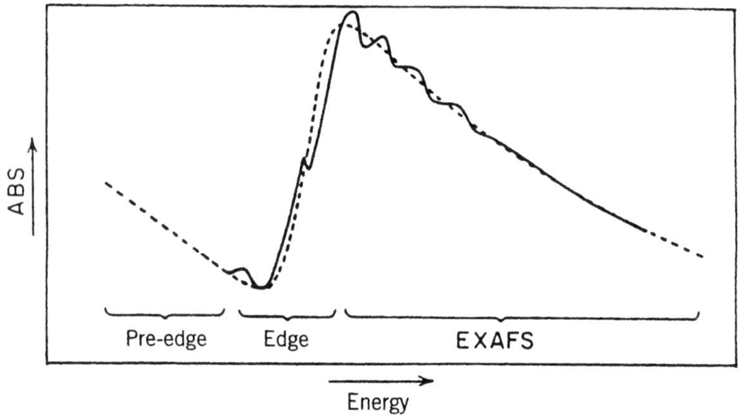

Figure 8.12. Schematic representation of data from x-ray absorption spectroscopy. (Reproduced with permission from Cramer SP, Hodgson KO. 1979. *Prog. Inorg. Chem.* **25**: 1–39.)

x-ray absorption spectroscopy (EXAFS). It yields information about fine structure, such as bond lengths to ± 0.02 Å, and coordination numbers to $\pm 20\%$. Detailed analysis near the edge region is called x-ray absorption near edge structure (XANES). The location of the near edge gives information about the electron density on the metal ion and hence about its oxidation state. Comprehensive analysis of both the edge and higher energy regions is combined in the name x-ray absorption fine structure (XAFS).

The phenomenon of EXAFS has been known since 1931, but little progress was made until the availability of synchrotron radiation, capable of delivering x-ray beams up to a million times more intense than obtained from a conventional x-ray tube. The quantity known as χ is the difference between the total oscillating signal and the smoothed curve. This χ wave is defined in terms of its amplitude and phase shift. The amplitude varies with the inverse of the square of the distance between absorber and scatterer; the phase shift is a function of both atoms. Analysis is simplest if all the backscattering atoms are identical. A possible complication is multiple backscattering, especially important for collinear atoms. Data analysis is most reliable when based on data from a suitable model compound. Reliable structural data for an unknown compound is unlikely. The strength of the technique is determination of bond lengths with great precision for compounds of known structure.

XAS of Horseradish Peroxidase

The first report of XAS of horseradish peroxidase analyzed edge data and contained a preliminary analysis of the EXAFS region (Penner-Hahn et al., 1983). The edge data were consistent with the presence of iron(IV) in both Compounds I and II. Oxygen and nitrogen ligands could not be distinguished, but in accord with other data a short bond length of ~1.6 Å was assigned to Fe=O and ~2.0 Å to the Fe—N bonds connecting the metal ion to the porphyrin. Later the same group obtained a bond length of 1.62 ± 0.02 Å for Fe=O in both Compounds I and II. K_2FeO_4 was used as a model for Fe=O EXAFS (Penner-Hahn et al., 1986). Very concentrated solutions of HRP were used in both studies, >5 mM.

Agreement was obtained on the Fe^{IV}=O bond length in HRP-I (1.64 Å) (Chance et al., 1984) and C*c*P-I (1.67 Å) (Chance et al., 1986). A much longer bond length was observed for HRP-II: 1.93 Å (Chance et al., 1984). A longer bond length for HRP-II was confirmed in a later study in which the pH dependence was examined (Chang et al., 1993). At pH 7 it is 1.90 ± 0.02 Å, but at pH 10 it decreases to 1.72 ± 0.02 Å. The change in bond length coincides with a change in reactivity of HRP-II, which is fully active at pH 7 and has negligible reactivity at pH 10 (Critchlow and Dunford, 1972). Crystallographic results for HRP Fe^{IV}=O distances are in qualitative agreement: 1.7 Å for HRP-I and 1.8 Å for HRP-II at pH 6.5 (Berglund et al., 2002).

Nearly all available evidence indicates longer Fe^{IV}=O bond lengths in enzyme systems compared to model compounds (Wolter et al., 2000). The controversy is whether the ferryl species really exists or whether it is better represented as Fe—OH. As described in the discussion of resonance Raman spectra, the available evidence, with the exception of chloroperoxidase Compound II, would appear to favor Fe^{IV}=O modified by the stable hydrogen bonding networks that exist in enzymes but not in model systems.

REFERENCES

Abe M, Kitagawa T, Kyogoku Y. 1978. Resonance Raman spectra of octaethylporphinato-Ni(II) and meso-deuterated ^{15}N substituted derivatives. II. A normal coordinate analysis. *J. Chem. Phys.* **69**: 4526–4534.

Adediran SA, Lambeir A-M. 1989. Kinetics of the reaction of Compound II of horseradish peroxidase with hydrogen peroxide to form Compound III. *Eur. J. Biochem.* **186**: 571–576.

Bajdor K, Nakamoto K. 1984. Formation of ferryltetraphenylporphyrin by laser irradiation. *J. Am. Chem. Soc.* **106**: 3045–3046.

Berglund GL, Carlsson GH, Smith AT, Szöke H, Henriksen A, Hajdu J. 2002. The catalytic pathway of horseradish peroxidase at high resolution. *Nature* **417**: 463–468.

Bonagura CA, Bhaskar B, Shimizu H, Li H, Sundaramoorthy M, McRee DE, Goodin DB, Poulos TL. 2003. High-resolution crystal structures and spectroscopy of native and Compound I cytochrome *c* peroxidase. *Biochemistry* **42**: 5600–5608.

Chance B, Powers L, Ching Y, Poulos T, Schonbaum GR, Yamazaki I, Paul KG. 1984. X-ray absorption studies of intermediates in peroxidase activity. *Arch. Biochem. Biophys.* **235**: 596–611.

Chance M, Powers L, Poulos T, Chance B. 1986. Cytochrome *c* peroxidase Compound ES is identical with horseradish peroxidase Compound I in iron-ligand distances. *Biochemistry* **25**: 1266–1270.

Chang CS, Yamazaki I, Sinclair R, Khalid S, Powers L. 1993. pH dependence of the active site of horseradish peroxidase Compound II. *Biochemistry* **32**: 923–928.

Choi S, Spiro TG, Langry KC, Smith KM. 1982. Structural correlations and vinyl influences in resonance Raman spectra of protoheme complexes and proteins. *J. Am. Chem. Soc.* **104**: 4337–4344.

Conradie J, Swarts JC, Ghosh A. 2004. Models of high-valent heme protein intermediates: a quantum chemical study of iron(IV) porphyrins with two univalent axial π-bonding ligands. *J. Phys. Chem. B* **108**: 452–456.

Critchlow JE, Dunford HB. 1972. Studies on horseradish peroxidase. IX. Kinetics of the oxidation of *p*-cresol by Compound II. *J. Biol. Chem.* **247**: 3703–3713.

Davydov R, Osborne RL, Kim SH, Dawson JH, Hoffman BM. 2008. EPR and ENDOR studies of cryoreduced Compounds II of peroxidases and myoglobin: proton-coupled electron transfer and protonation status of ferryl hemes. *Biochemistry* **47**: 5147–5155.

Dunford HB. 1982. Peroxidases. *Adv. Inorg. Biochem.* **4**: 41–68.

Dunford HB. 1999. *Heme Peroxidases*. Wiley-VCH, New York.

Dunford HB, Stillman JS. 1976. On the function and mechanism of action of peroxidases. *Coord. Chem. Rev.* **19**: 187–251.

Egawa T, Miki H, Ogura T, Makino R, Ishimura Y, Kitagawa T. 1992. Observation of the $Fe^{IV}{=}O$ stretching Raman band for thiolate-ligated heme protein Compound I of chloroperoxidase. *FEBS Lett.* **305**: 206–208.

Egawa T, Proshlyakov DA, Miki H, Makino R, Ishimura Y, Kitagawa T. 2001. Effects of a thiolate axial ligand on the $\pi \rightarrow \pi^*$ electronic states of oxoferryl porphyrins: a study of the optical and resonance Raman spectra of compounds I and II of chloroperoxidase. *J. Biol. Inorg. Chem.* **6**: 46–54.

Feis A, Marzocchi MP, Paoli M, Smulevich G. 1994. Spin state and axial ligand bonding in the hydroxide complexes of metmyoglobin, methemoglobin, and horseradish peroxidase at room and low temperatures. *Biochemistry* **33**: 4577–4583.

Green MT. 2006. Application of Badger's rule to heme and non-heme iron-oxygen bonds: an examination of ferryl protonation states. *J. Am. Chem. Soc.* **128**: 1902–1906.

Hashimoto S, Tatsuno Y, Kitagawa T. 1986. Resonance Raman evidence for oxygen exchange between $Fe^{IV}{=}O$ heme and bulk water during enzymic catalysis of hprseradish peroxidase and its relation with the heme-linked ionization. *Proc. Natl. Acad. Sci. USA* **83**: 2417–2421.

Hosten CM, Sullivan AM, Palaniappapan V, Fitzgerald MM, Terner J. 1994. Resonance Raman spectroscopy of the catalytic intermediates and derivatives of chloroperoxidase from *Caldariomyces fumago*. *J. Biol. Chem.* **269**: 13966–13978.

Howes BD, Brissett NG, Doyle WA, Smith AT, Smulevich G. 2005. Spectroscopic and kinetic properties of the horseradish peroxidase mutant T171S: evidence for selective effects on the reduced state of the enzyme. *FEBS J.* **272**: 5514–5521.

Howes BD, Feis A, Indiani C, Marzocchi MP, Smulevich G. 2000. Formation of two types of low-spin heme in horseradish peroxidase isoenzyme A2 at low temperature. *J. Biol. Inorg. Chem.* **5**: 227–235.

Howes BD, Feis A, Raimondi L, Indiani C, Smulevich G. 2001a. The critical role of the proximal calcium ion in the structural properties of horseradish peroxidase. *J. Biol. Chem.* **276**: 40704–40711.

Howes BD, Rodriguez-Lopez JN, Smith AT, Smulevich G. 1997. Mutation of distal residues of horseradish peroxidase: influence on substrate binding and cavity properties. *Biochemistry* **36**: 1532–1543.

Howes BD, Veitch NC, Smith AT, White CG, Smulevich G. 2001b. Haem-linked interactions in horseradish peroxidase revealed by spectroscopic analysis of the Phe-221 → Met mutant. *Biochem. J.* **353**: 181–191.

Hu S, Smith KM, Spiro TG. 1996. Assignment of protoheme resonance Raman spectrum by heme labeling in myoglobin. *J. Am. Chem. Soc.* **118**: 12638–12646.

Huang Q, Szigeti K, Fidy J, Schweitzer-Stenner R. 2003. Structural disorder of native horseradish peroxidase C probed by resonance Raman and low-temperature optical absorption spectroscopy. *J. Phys. Chem. B.* **107**: 2822 2830.

Indiani C, Feis A, Howes BD, Marzocchi MP, Smulevich G. 2000. Benhydroxamic acid-peroxidase complexes: spectroscopic characterization of a novel heme spin species. *J. Am. Chem. Soc.* **122**: 7368–7376.

Jentzen W, Song X-Z, Shelnutt JA. 1997. Structural characterization of synthetic and protein-bound porphyrins in terms of the lowest-frequency normal coordinates of the macrocycle. *J. Phys. Chem. B* **101**: 1684–1699.

Kincaid JR, Zheng Y, Al-Mustafa J, Czarnecki K. 1996. Resonance Raman spectra of native and mesoheme-reconstituted horseradish peroxidase and their catalytic intermediates. *J. Biol. Chem.* **271**: 28805–28811.

Li X-Y, Czernuszewicz RS, Kincaid, JR, Stein P, Spiro TG. 1990. Consistent porphyrin force field. II. Nickel octaethylporphyrin skeletal and substituent mode assignments from ^{15}N, meso-d_4, and methylene-d_{16} Raman and infrared isotope shifts. *J. Phys. Chem.* **94**: 47–61.

Loehr TM, Loehr JS. 1973. Determination of oxidation and spin states of heme iron: resonance Raman spectroscopy of cytochrome *c*, microperoxidase, and horseradish peroxidase. *Biochem. Biophys. Res. Commun.* **55**: 218–223.

Makino R, Uno T, Nishimura Y, Iizuka T, Tsuboi M, Ishimura Y. 1986. Coordination structures and reactivities of Compound II in iron and manganese horseradish peroxidase. *J. Biol. Chem.* **261**: 8376–8382.

Neri F, Kok D, Miller MA, Smulevich G. 1997. Fluoride binding in hemoproteins: the importance of distal cavity structure. *Biochemistry* **36**: 8947–8953.

Noble RW, Gibson QH. 1970. The reaction of ferrous horseradish peroxidase with hydrogen peroxide. *J. Biol. Chem.* **245**: 2409–2413.

Oertling WA, Babcock GT. 1988. Time-resolved and static resonance Raman spectroscopy of horseradish peroxidase intermediates. *Biochemistry* **27**: 3331–3338.

Ogura T, Kitagawa T. 1987. Device for simultaneous measurements of transient Raman and absorption spectra of enzymic reactions: application to Compound I of horseradish peroxidase. *J. Am. Chem. Soc.* **109**: 2177–2179.

Paeng K-J, Kincaid JR. 1988. The resonance Raman spectrum of horseradish peroxidase Compound I. *J. Am. Chem. Soc.* **110**: 7913–7915.

Palaniappan V, Fitzgerald MM, Sullivan AM, Shifflett JR, Terner J. 1992. Rapid mixing resonance Raman spectroscopy of peroxidase intermediates. In: *Time Resolved Vibrational Spectroscopy*, Vol. 5, Takahashi H, editor. Springer, Berlin, pp. 32–35.

Palaniappan V, Terner J. 1989. Resonance Raman spectroscopy of horseradish peroxidase derivatives and intermediates with excitation in the near ultraviolet. *J. Biol. Chem.* **264**: 16046–16053.

Penner-Hahn JE, Eble KM, McMurry TJ, Renner M, Balch AL, Groves JT, Dawson JH, Hodgson KO. 1986. Structural characterization of horseradish peroxidase using EXAFS spectroscopy: evidence for Fe=O ligation in Compounds I and II. *J. Am. Chem. Soc.* **108**: 7819–7825.

Penner-Hahn JE, McMurry TJ, Renner M, Latos-Grazynsky L, Eble KM, Davis IM, Balch AL, Groves JT, Dawson JH, Hodgson KO. 1983. X-ray absorption spectroscopic studies of high valent iron porphyrins. *J. Biol. Chem.* **258**: 12761–12764.

Rakshit G, Spiro TG. 1974. Resonance Raman spectra of horseradish peroxidase: evidence for anomalous heme structure. *Biochemistry* **36**: 5317–5323.

Rovira C, Fita I. 2003. The proximal hydrogen-bonded residue controls the stability of the Compound II intermediate of peroxidases and catalases. *J. Phys. Chem. B* **107**: 5300–5305.

Shelnutt JA, Song X-Z, Ma J-G, Jia S-L, Jentzen W, Medforth CJ. 1998. Non-planar porphyrins and their significance in proteins. *Chem. Soc. Rev.* **27**: 31–41.

Sievers G, Österland K, Ellfolk N. 1979. Resonance Raman study on yeast cytochrome *c* peroxidase: effect of coordination and axial ligands. *Biochim. Biophys. Acta* **581**: 1–14.

Silaghi-Dumitrescu R. 2004. The nature of the high-valent complexes in the catalytic cycles of hemoproteins. *J. Biol. Inorg. Chem.* **9**: 471–476.

Sitter AJ, Reczek CM, Terner J. 1985. Heme-linked ionization of horseradish peroxidase Compound II monitored by the resonance Raman Fe(IV)=O stretching vibration. *J. Biol. Chem.* **260**: 7515–7522.

Sitter AJ, Shifflett JR, Terner J. 1988. Resonance Raman spectroscopic evidence for heme iron-hydroxide ligation in peroxidase alkaline forms. *J. Biol. Chem.* **263**: 13032–13038.

Smulevich G. 1993. Structure-function relationships in peroxidases via resonance Raman spectroscopy and site-directed mutagenesis: cytochrome *c* peroxidase. In: *Biomolecular Spectroscopy, Part A*. Clark RJH, Hester RE, editors. Wiley, New York, pp. 163–193.

Smulevich G. 1995. The functional role of key residues in the active site of peroxidases. *Biochem. Soc. Trans.* **23**: 240–243.

Smulevich G. 1996. A comparative analysis of the structures of various peroxidases as defined by resonance Raman and electronic absorption spectroscopies. In: *Plant Peroxidases: Biochemistry and Physiology*. Obinger C, Burner U, Ebermann R, Penel C, Greppin H, editors. University of Geneva, Switzerland, pp. 13–19.

Smulevich G, English AM, Mantini AR, Marzocchi MP. 1991a. Resonance Raman investigation of ferric iron in horseradish peroxidase and its aromatic donor complexes at room and low temperatures. *Biochemistry* **30**: 772–779.

Smulevich G, Feis A, Howes BD. 2005. Fifteen years of Raman spectroscopy of engineered heme containing peroxidases: what have we learned? *Acc. Chem. Res.* **38**: 433–440.

Smulevich G, Hu S, Rodgers KR, Goodin DB, Smith KM, Spiro TG. 1996. Heme-protein interactions in cytochrome *c* peroxidase revealed by site-directed mutagenesis and resonance Raman spectroscopy of isotopically labeled hemes. *Biospectroscopy* **2**: 365–376.

Smulevich G, Mantini AR, English AM, Mauro JM. 1989. Effects of temperature and glycerol on the resonance Raman spectra of cytochrome *c* peroxidase and selected mutants. *Biochemistry* **28**: 5058–5064.

Smulevich G, Mauro JM, Fishel LA, English AM, Kraut J, Spiro TG. 1988. Heme pocket interactions in cytochrome *c* peroxidase studied by site-directed mutagenesis and resonance Raman spectroscopy. *Biochemistry* **27**: 5477–5485.

Smulevich G, Miller MA, Kraut J, Spiro TG. 1991b. Conformational change and histidine control of heme chemistry in cytochrome *c* peroxidase: resonance Raaman evidence from Leu-52 and Gly-181 mutants of cytochrome *c* peroxidase. *Biochemistry* **30**: 9546–9558.

Smulevich G, Neri F, Willemsen O, Choudhury K, Marzcocchi MP, Poulos TL. 1995. Effect of the His 175 → Glu mutation on the heme pocket architecture of cytochrome *c* peroxidase. *Biochemistry* **34**: 13485–13490.

Smulevich G, Paoli M, Burke JF, Sanders SA, Thorneley RNF, Smith AT. 1994. Characterization of recombinant horseradish peroxidase C and three site directed mutants, F41V, F41W, and R38K, by resonance Raman spectroscopy. *Biochemistry* **33**: 7398–7407.

Smulevich G, Paoli M, De Sanctis G, Mantini AR, Ascoli F, Coletta M. 1997. Spectroscopic evidence for a conformational transition in horseradish peroxidase at very low pH. *Biochemistry* **36**: 640–649.

Smulevich G, Wang Y, Edwards SL, Poulos TL, English AM, Spiro TG. 1990a. Resonance Raman spectroscopy of cytochrome *c* peroxidase single crystals on a variable-temperature microscope stage. *Biochemistry* **29**: 2586–2592.

Smulevich G, Wang Y, Mauro JM, Fishel LA, Kraut J, Spiro TG. 1990b. Single-crystal resonance Raman spectroscopy of site-directed mutants of cytochrome *c* peroxidase. *Biochemistry* **29**: 7174–7180.

Spiro TG, Smulevich G, Su C. 1990. Probing structure and dynamics with resonance Raman spectroscopy: cytochrome *c* peroxidase and hemoglobin. *Biochemistry* **29**: 4497–4508.

Spiro TG, Stong JD, Stein P. 1979. Porphyrin core expansion and doming in heme proteins: new evidence from resonance Raman spectra of six-coordinate high-spin iron(III) hemes. *J. Am. Chem. Soc.* **101**: 2648–2655.

Stone KL, Behan RK, Green MT. 2006. Resonance Raman spectroscopy of chloroperoxidase compound II provides direct evidence for the existence of an iron(IV)-hydroxide. *Proc. Natl. Acad. Sci. USA* **103**: 12307–12310.

Teraoka J, Kitagawa T. 1981. Structural implications of the heme-linked ionization of horseradish peroxides probed by the Fe-histidine stretching Raman line. *J. Biol. Chem.* **256**: 3969–3977.

Terner J, Palaniappan V, Gold A, Weiss R, Fitzgerald MM, Sullivan AM, Hosten CM. 2006. Resonance Raman spectroscopy of oxoiron(IV) porphyrin π-cation and oxoiron(IV) hemes in peroxidase intermediates. *J. Inorg. Biochem.* **100**: 480–501.

Terner J, Sitter AJ, Reczek CM. 1985. Resonance Raman spectroscopic characterizations of horseradish peroxidase. Observation of the $Fe^{IV}{=}O$ stretching vibration of Compound II. *Biochim. Biophys. Acta* **828**: 73–80.

Van Wart HE, Zimmer J. 1985a. Resonance Raman studies of the photoreduction of horseradish peroxidase Compounds I and II. *J. Am. Chem. Soc.* **107**: 3379–3381.

Van Wart HE, Zimmer J. 1985b. Resonance Raman evidence for the activation of dioxygen in horseradish oxyperoxidase. *J. Biol. Chem.* **260**: 8372–8377.

Wittenberg JB, Noble RW, Wittenberg BA, Antonini E, Brunori M, Wyman J. 1967. Studies on the equilibria and kinetics of the reactions of peroxidases with ligands. II. The reaction of ferroperoxidase with oxygen. *J. Biol. Chem.* **242**: 626–634.

Wolter T, Meyer-Klauke W, Müther M, Mandon D, Winkler H, Trautwein AX, Weiss R. 2000. Generation of oxoiron(IV) tetramesitylporphyrin π-cation radical complexes by *m*-CPBA oxidation of ferric tetramesitylporphyrin derivatives in butyronitrile at $-78^{\circ}C$: evidence of formation of six-coordinate oxoiron(IV) tetramesitylporphyrrin π-cation radical complexes FeIV=O(tmp·)X (X = Cl^-, Br^-), by Mössbauer and X-ray absorption spectroscopy. *J. Inorg. Biochem.* **78**: 117–122.

Woodruff WH, Spiro TG. 1974. Resonance Raman spectroscopy of reaction intermediates by a rapid mixing, continuous flow technique. *Appl. Spectroscopy* **28**: 576–578.

CHAPTER 9

SPECTROSCOPY. II. NUCLEAR MAGNETIC RESONANCE, ELECTRON SPIN, AND MÖSSBAUER

NUCLEAR MAGNETIC RESONANCE (NMR) SPECTROSCOPY

Introduction

By far the most common type of NMR experiment is measurement of proton spectra. ^1H nuclei can be thought of as tiny magnets with spin quantum number I of $\frac{1}{2}$. Since these tiny magnets may align themselves with or opposed to a strong external magnetic field, indicated by magnetic quantum numbers m_I of $\pm\frac{1}{2}$, their energy levels are split in two, α and β. The spins that align themselves with the magnetic field β are of lower energy. The populations of the two levels follow a Boltzmann distribution so there is a slight excess of protons in the lower level. Thus the sample possesses a net magnetization. The magnetic component of electromagnetic radiation of exactly the right energy, the resonance frequency ν, will cause protons to flip from the lower to the higher energy level; ν occurs in the radio frequency range of the electromagnetic spectrum. Thus, the basic requirements for an NMR machine are a strong magnet, a radio frequency transmitter and receiver, and a data acquisition system.

The stationary field provided by the strong magnet is designated B_0, and its direction is defined as along the z axis. The net magnetization of the sample, M_0, aligns itself with B_0. In a basic NMR experiment a 90° pulse is applied from the radio frequency transmitter, creating an additional magnetic field, B_1, at right angles to B_0. The result is a transfer of the M_0 magnetization into the xy plane, now called M_1, which oscillates (precesses) at the resonance frequency ν. For example, in a magnetic field of 11.75 Tesla, the resonance frequency of protons is 500 MHz. The oscillating magnetic vector M_1 creates a time-dependent signal that is transformed into

Peroxidases and Catalases: Biochemistry, Biophysics, Biotechnology, and Physiology, Second Edition.
By H. Brian Dunford

recorded frequencies (NMR peaks). Protons are shielded in different ways from the magnetic field by the electron clouds in their immediate environment. Thus, nonequivalent protons have different resonance frequencies. From an arbitrary standard, chemical shifts δ are recorded for each unique proton for which unambiguous identification may be possible. Typically the single sharp peak of tetramethyl silane (TMS) or the water-soluble 2,2-dimethyl-2-silapentane-5-sulfonate (DDS) is used as reference. Chemical shifts are recorded in parts per million (ppm) according to the equation

$$\delta = 10^6(\nu_{\text{obs}} - \nu_{\text{ref}})/\nu_{\text{ref}}$$

where ν_{obs} and ν_{ref} are the resonance frequencies of protons on the compound of interest and of the reference compound.

In a poorly resolved spectrum of ethanol the three equivalent methyl protons have a unique NMR peak, as do the two methylene protons and the single hydroxyl proton. The peak areas are in the ratio 3 : 2 : 1, in accord with the number of protons. In a high-resolution NMR spectrum of ethanol, the influence of protons attached to one atom on protons attached to an adjacent atom can be detected. The interactions are called *spin-spin coupling*, the strength of which is defined by a coupling constant *J*. The result in ethanol is splitting of the methylene proton peak into a quartet, and the methyl proton peak into a triplet (Fig. 9.1), a result obtained using the simple statistics of Pascal's triangle (Wüthrich, 1976). The splittings give rise to *multiplet structure*. The OH proton of ethanol exchanges rapidly with other solvent molecules and so exhibits a single broad resonance. The electronegative oxygen atom tends to remove the electron cloud from the OH proton, which has the least screening so its resonance is shifted farthest downfield. The methyl group is an electron donor, so its proton resonances are most shielded and occur farthest upfield. The total peak areas for the methyl triplet, the methylene quartet, and the hydroxyl proton singlet maintain the ratio of 3 : 2 : 1.

NMR experiments are difficult to carry out in H_2O because the solvent protons obscure signals from solute molecules. The problem can be circumvented by using deuterium oxide D_2O as solvent. Alternatively, there are pulse sequences that can be used to eliminate the water proton signal.

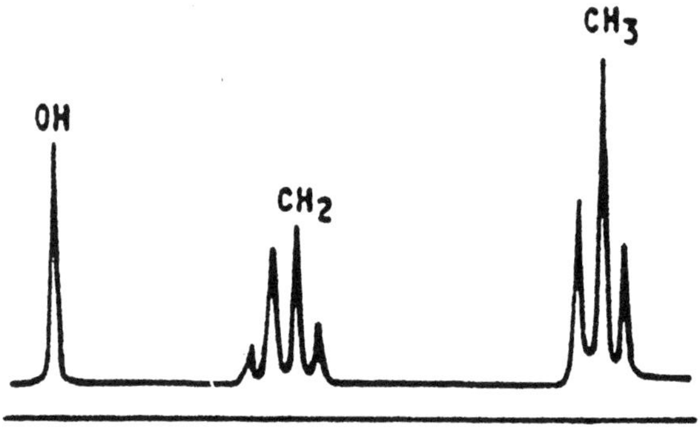

Figure 9.1. ^{1}H NMR spectrum of ethanol.

Aliphatic groups are electron donating so they provide strong shielding and small downfield shifts relative to the reference compound. Electronegative groups, on the other hand, withdraw electrons so protons in their immediate environment are less well shielded. These protons will have large downfield shifts. In the NMR spectrum of the diamagnetic compound Zn^{2+-}porphin, shown in the top part of Figure 9.2, all of the proton resonances are shifted downfield. In paramagnetic Fe^{3+}-porphin-$(CN)_2$ two of the proton resonances from the Zn^{2+}-porphin are shifted dramatically upfield (the lower part of Fig. 9.2).

The appearance of NMR spectra is dominated by two mechanisms: longitudinal, characterized by a relaxation time T_1, and transverse, characterized by T_2. First, a discussion of T_2. When the M_1 magnetization is created by the 90° pulse, magnetization vectors of the same type of protons on different molecules are aligned in the *xy* plane. As they precess, coherence is gradually lost and the magnetization spreads over the entire 360° resulting in a loss of signal. The speed of this process is characterized by T_2. Signals with short T_2 are broad and in extreme cases so broad they are undetectable. On the other hand, signals with long values of T_2 are sharp. Assuming a homogeneous magnetic field (i.e., a well-shimmed magnet), T_2 can be estimated from the inverse of the peak width at half-height.

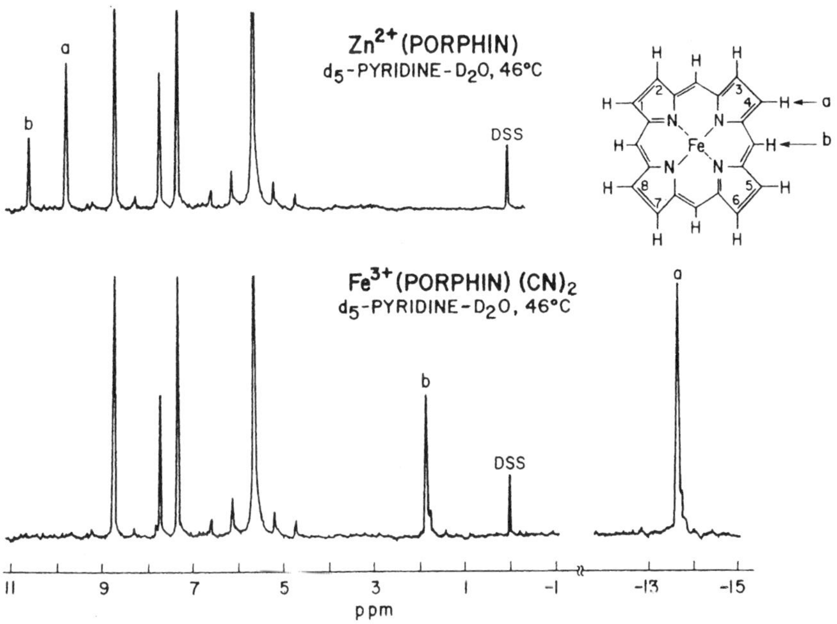

Figure 9.2. ^{1}H NMR spectra of Zn^{2+}-porphin and Fe^{3+}-porphin-$(CN)_2$ in D_2O at 220 MHz. The resonances *a* and *b* were identified by their relative intensities. The low-spin paramagnetic ferric ion causes resonances *a* and *b* to be shifted dramatically upfield. (Reproduced with permission from Wütrich K. 1976. *NMR in Biological Research: Peptides and Proteins.* North-Holland, Amsterdam, p. 228.)

T_1 is the time taken for M_1 magnetization to return to its original alignment with the z axis. Relaxation of T_1 follows an exponential time course, so at least five T_1s must elapse for the complete return of M_1 to the z axis. In order to increase the intensity of an NMR spectrum sequential 90° pulses may be applied.

Nuclear Overhauser Enhancement (NOE)

^{13}C, like ^{1}H, has a nuclear spin of $\frac{1}{2}$. Because of its small natural abundance and lower NMR sensitivity, it has an NMR signal roughly 10^{-4} that of a proton. However, it exhibits chemical shifts of up to 200 ppm, compared to 0–10 ppm for protons. Natural abundance ^{13}C can be detected, and enrichment enhances the signal. Cells can be fed a ^{13}C diet so ^{13}C-enriched proteins can be isolated.

The example of a C—H bond will be used to illustrate an NOE experiment. The ^{1}H and ^{13}C will interact strongly because they are in close proximity. Multiplet splitting will result from application of the radio frequency field B_1, which decreases the intensity of the ^{13}C resonance. The multiplet splitting can be destroyed by application of a second fixed radio frequency field B_2. This also shifts the two ^{13}C energy levels. Thus a single ^{13}C peak is observed with up to three times its original intensity. Since two fields B_1 and B_2 have been applied, a double resonance experiment has been performed.

Despite the example of atoms attached by a bond, an NOE experiment gives "through-space" not "through-bond" information. Its magnitude is dependent on $1/r^6$ where r is the internuclear distance. An NOE experiment can also be performed on nuclei of the same type. One can determine which of a pair of protons is nearest neighbor to a third proton, which is irradiated with its resonance frequency. The one with the greatest peak intensity enhancement is the nearer neighbor. Thus structural information can be gained.

Two-Dimensional Nuclear Magnetic Resonance

The use of two-dimensional NMR is dependent on improved instrumentation and computers, and software packages capable of handling large amounts of data. The basic technique is to use a series of pulse sequences, and to collect data as a function of incremented delay times between pulses. By proper choices of pulses and delay times extraneous signals are eliminated and only data of interest is collected.

In a two-dimensional NMR experiment, results are displayed on a grid with two frequency axes. The conventional one-dimensional NMR spectrum is displayed along a single-line diagonal. The peak for proton H_1 is defined by coordinates ν_1, ν_1, and for proton H_2 by coordinates ν_2, ν_2 along the diagonal line. The existence of cross-peaks, or off-diagonal peaks, defined by coordinates ν_1, ν_2, and ν_2, ν_1 provides valuable structural information. An example is shown in Figure 9.3.

In a correlated spectroscopy (COSY) experiment, through-bond or scalar interactions are measured. Through-space or dipolar interactions can be measured in a NOESY experiment, which is essentially NOE in two dimensions. NMR spectroscopy

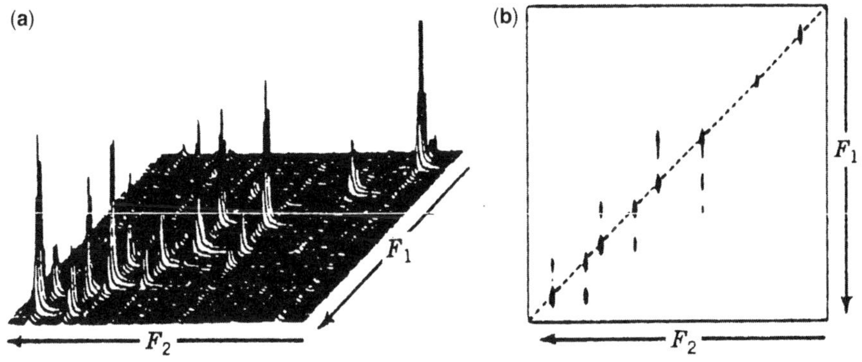

Figure 9.3. Different representations of a 2D NMR spectrum: (a) as a stacked plot, and (b) as a contour plot. F_1 and F_2 are the two frequency axes. The cross (off-diagonal) peaks are readily apparent in the contour plot. (Reproduced with permission from Günther H. 1995. *NMR Spectroscopy*, 2nd edition. Wiley, New York, p. 277.)

has advanced to the point where entire protein structures can be determined, so a viable alternative and complement exists to crystal structure determinations.

Influence of a Paramagnetic Center

The magnetic moment of an unpaired electron is about 660 times as great as that of a proton. Therefore, if a molecule has an unpaired electron (paramagnetic center) there will be profound effects on the NMR spectrum. Whereas an unknown paramagnetic contaminant is the bane of NMR spectroscopists, a great deal of valuable information can be gained from a naturally occurring unpaired electron or one that is added to a known location.

The *hyperfine shifts* caused by a paramagnetic center can remove some resonances from a massive envelope of overlapping peaks. When combined with other information such as amino acid sequence data, the shifted peaks may be identified and their distance from the paramagnetic center determined. The effect of changing from a diamagnetic to paramagnetic center is shown in Figure 9.2.

Too many unpaired electrons may hopelessly complicate interpretation of the hyperfine shifts. Thus cyanide complexes of a peroxidase are commonly used, so the iron is changed to low spin with only one unpaired electron.

Nuclear Magnetic Resonance Spectra of Horseradish Peroxidase

In a study ahead of its time, it was shown that water is not present in the sixth coordination position of the heme ferric ion (Lanir and Schejter, 1975). The values of the relaxivity $1/T_1$ for the water protons in a native horseradish peroxidase solution were about the same as for the cyanide complex of the enzyme. In contrast, enhancement of $1/T_1$ was observed for cytochrome *c* and metmyoglobin where it is known that water is coordinated to iron. A later NMR study confirmed the five-coordinate structure of native HRP (La Mar et al., 1988).

A series of studies on the NMR spectra of HRP was undertaken in the laboratory of Isao Morishima. The alkaline transition of HRP at p^2H 11 could be detected in which the transition from high spin to low spin iron occurs (Iizuka et al., 1976). Aromatic donor binding (Morishima and Ogawa, 1979) and influence of calcium binding were studied (Shiro et al., 1986; Morishima et al., 1986). The four heme methyl resonances at 77.6, 70.1, 67.0, and 50.1 ppm are shifted to 70.9, 63.3, 61.1, and 46.1 ppm upon calcium removal. The results indicate one calcium is bound near the heme and is necessary to maintain the structural integrity of the enzyme. Titration with calcium ion restores the original NMR spectrum of the native enzyme.

Comprehensive NMR investigations of horseradish peroxidase have come from the laboratory of Gerd La Mar (Fig. 9.4). By means of deuterium labels in key positions of ferriprotoporphyrin IX and deuteroheme, many of the hyperfine shifted resonances of reconstituted holoenzyme were assigned (La Mar et al., 1980). NMR spectra of deuteroheme-reconstituted HRP-I were consistent with a porphyrin π-cation structure (La Mar and de Ropp, 1980; La Mar et al., 1981).

The nuclear Overhauser effect was utilized to assign all of the heme resonances (Thanabal et al., 1987a). Resonances were also correctly assigned to amino acid residues: Arg38, His42, and His170 (Thanabal et al., 1987b), Phe152 (Veitch and Williams, 1990), and Ile244 (Veitch et al., 1992). It was shown that despite the presence of a strong paramagnetic species, two-dimensional NMR spectra of HRP could be obtained (de Ropp and La Mar, 1991; Chen et al., 1994).

The NMR spectra of the cyanide complexes of isoenzymes A1 and A2 of HRP were compared to isoenzyme C (de Ropp et al., 1995). The heme environment is very similar for all the isoenzymes examined. A review of all NMR results on cyanide-bound HRP has been published (La Mar et al., 1995). Work has been extended to complete identification of nearest heme neighbor residues and to extend identification to second-sphere residues (de Ropp et al., 1999). Examination of the His42Ala mutant of HRP-CN by NMR led to the conclusion that proximal Asp247 is more important than distal His42 in imparting imidazolate character to the proximal His 170 (de Ropp et al., 2002).

The binding of aromatic molecules to HRP was classified according to whether the native enzyme optical spectrum was red- or blue-shifted (Paul and Ohlsson, 1978). It was concluded from an NMR study using one example of each type of optical spectral shift, resorcinol and 2-methoxy-4-methylphenol, that both bind in similar fashion close to the 8-methyl group of the heme (Sakurada et al., 1986). *p*-Cresol was also shown to bind near the 8-methyl group, and its binding to HRP was modeled (Banci et al., 1993, 1994). HRP mutants were prepared in an attempt to improve luminol binding and were examined by NMR (Tanaka et al., 1999).

Unusual interactions occur between benzhydroxamic acid and derivatives and HRP. The magnetic moment of native enzyme is shifted from 5.27 to 5.97 Bohr magnetons upon benzhydroxamic binding. Both ESR (Gupta et al., 1979) and resonance Raman spectra (Smulevich et al., 1991) indicate that upon binding the iron(III) is converted from five-coordinate to six-coordinate with water occupying the sixth

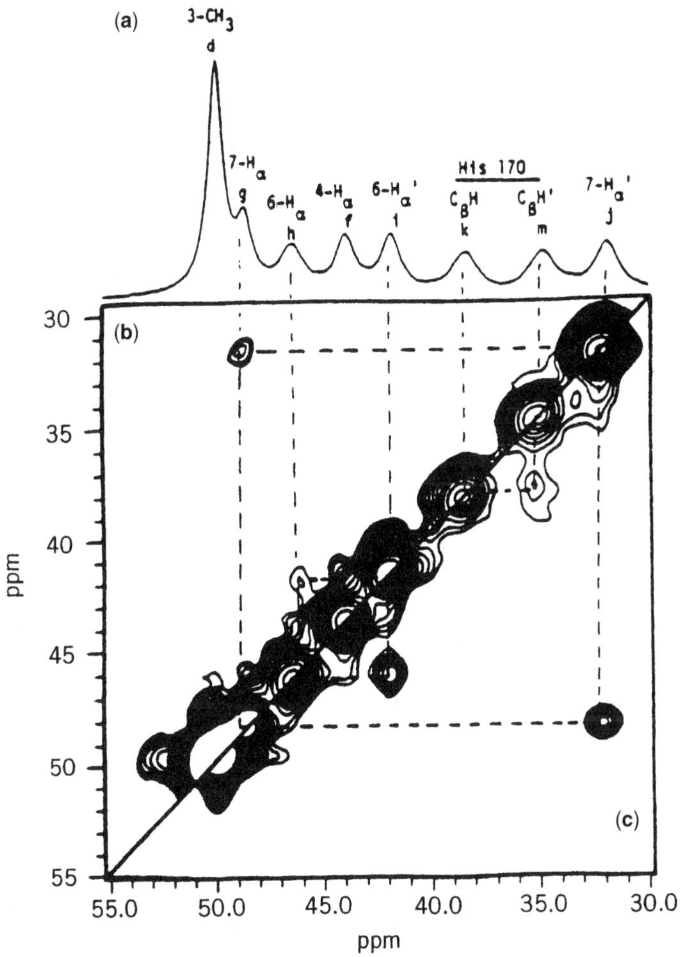

Figure 9.4. NMR spectrum of native horseradish peroxidase. (a) Reference one-dimensional spectrum. (b) COSY map. (c) NOESY map showing the $C_\beta H$ cross peak. (Reproduced with permission from de Ropp JS, La Mar GN. 1991. *J. Am. Chem. Soc.* **113**: 4348–4350.)

coordination position. In the ternary system HRP-benzhydroxamic acid-cyanide, benzyhroxamic acid slows the rate of cyanide binding and the exchange rate of the His42 imidazole proton with water (La Mar et al., 1992). The NMR spectra of cyanide-ligated HRP isoenzymes A2 and C, with and without bound benzhydroxamic acid, is shown in Figure 9.5 (Veitch, 1995).

Two-dimensional NMR spectra of methyl-substituted benzhydroxamic acids were obtained (Veitch and Williams, 1995). The results show that the substrates bind towards the distal side of the heme and that a Phe residue is at the entrance to the binding site. All substrates interact strongly with the heme methyl $C18H_3$. The Phe residue was later identified as Phe179 (Veitch et al., 1997).

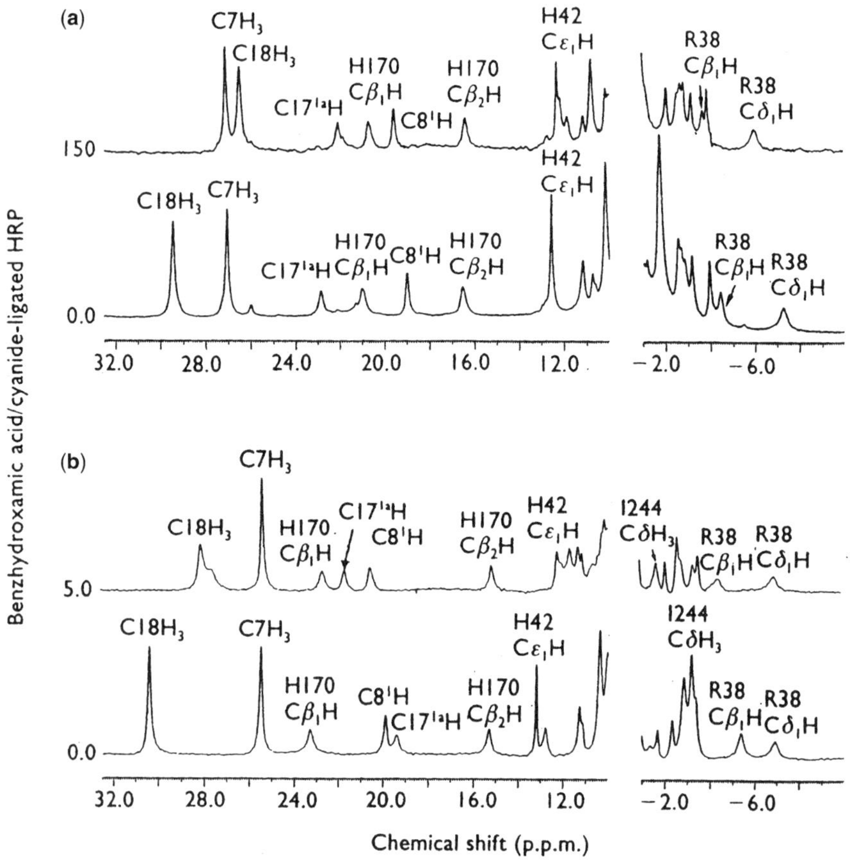

Figure 9.5. Comparison of the NMR spectra of cyanide-ligated horseradish peroxidase isoenzymes A2 (a) and C (b) in the presence (upper spectra) and absence (lower spectra) of benzhydroxamic acid. (Reproduced with permission from Veitch NC. 1995. *Biochem. Soc. Trans.* **23**: 232–240.)

Perturbations to the heme pocket of horseradish peroxidase were observed in the Phe221Met mutant. These were attributed to increased steric freedom and solvent accessibility of the His170 ligand, as indicated by the ^{1}H-NMR data, and loss of the π-π interaction between His170 and Phe221. The NMR results are shown in Figure 9.6.

^{15}N chemically induced dynamic nuclear polarization (CIDNP) was measured in a model nitrating system. The substrate was *N*-acetyl-L-tyrosine and two nitrating systems were used: nitrite/hydrogen peroxide/horseradish peroxidase, and nitrite/hypochlorous acid (Lehnig, 2001). In the CIDNP technique either emission or enhanced absorption resulting from radical pair formation ($NO_2^{\bullet}$ and phenyl ring radical) can be detected. Nitration occurs by recombination of caged radicals.

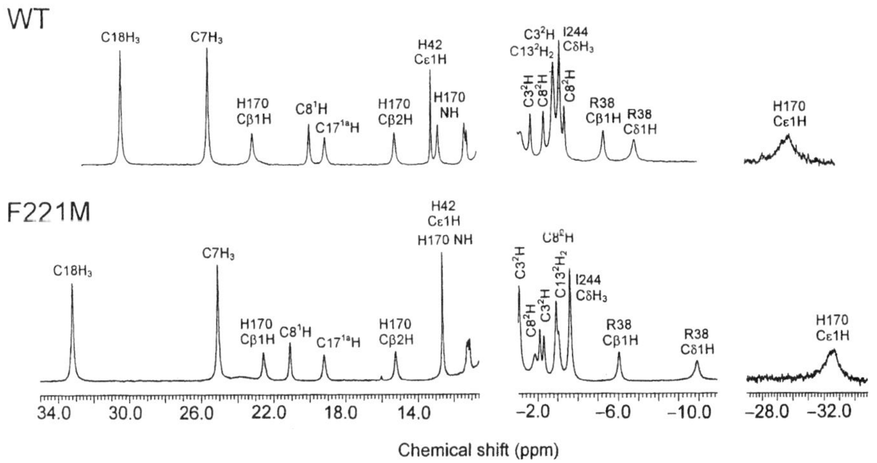

Figure 9.6. Hyperfine-shifted regions from the ^{1}H-NMR spectra of cyanide-ligated wild type (WT) and Phe221Met mutant. (Reproduced with permission from Howes BD, et al., 2001. *Biochem. J.* **353**: 181–191.)

Nuclear Magnetic Resonance Spectra of Yeast Cytochrome *c* Peroxidase

Extensive NMR studies on yeast C*c*P and C*c*P–cytochrome *c* complexes have been conducted, mainly in the laboratory of James Satterlee. First, the studies on C*c*P and complexes with small molecules.

In an early NMR investigation of yeast C*c*P, conducted with KNO_3 present (Satterlee and Erman, 1980), pD dependences of hyperfine shifts were detected, indicating pK_a values of 4.9 and 5.8. Heme asymmetry was detected in the deuterohemin-reconstituted enzyme, evidence for more than one orientation of the bound heme (Satterlee and Erman, 1981a). Unlike the NMR spectrum of horseradish peroxidase Compound I, the C*c*P Compound I spectrum exhibits only broad resonances near 17 and 30 ppm (Satterlee and Erman, 1981b). By reconstitution with selectively deuterated hemins, hyperfine shifts were assigned in both low- and high-spin forms of C*c*P (Satterlee et al., 1983a, 1983b). A major cause of rhombic distortion in low-spin complexes (cyanide and azide) is the orientation of the proximal histidine's imidazole plane. In high-spin forms (native and fluoride complex) essential correspondence was obtained between solution and crystal structural data on the heme crevice. An NMR study of deuterium exchangeable protons appeared to rule out an imidazolate form on the proximal imidazole (Satterlee and Erman, 1983a). Later evidence did indicate imidazolate character (Satterlee et al., 1990). Evidence was obtained for hindered vinyl group rotation in the C*c*P–cyanide complex (Satterlee and Erman, 1983b). Proton assignments are possible in the presence of a paramagnetic ion as demonstrated in the NMR spectrum of the CcP—CN complex, shown in Figure 9.7 (Satterlee et al., 1991).

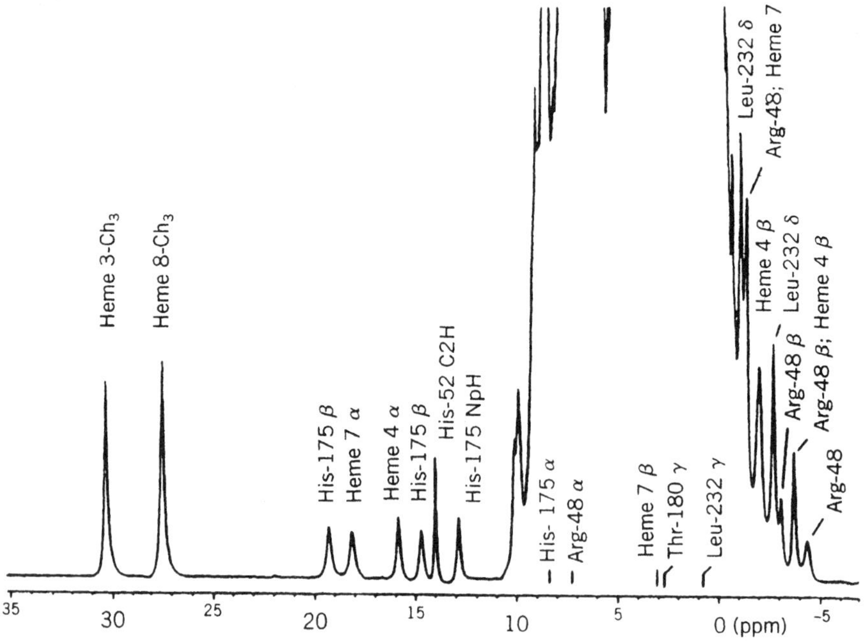

Figure 9.7. Assignments of proton resonances at 500 MHz in the cyanide complex of yeast cytochrome *c* peroxidase. (Reproduced with permission from Satterlee JD, Russell DJ, Erman JE. 1991. *Biochemistry* **30**: 9072–9077.)

By use of the one-dimensional nuclear Overhauser effect, all of the downfield proton hyperfine resonances were assigned. In addition resonances of His175, Thr180, and His52 were identified (Satterlee et al., 1987a). In a later two-dimensional NOE study it was shown that the cyanide complex provides the best NMR model for the low-spin intermediates Compounds I and II. Resonances for both distal and proximal histidines were completed, and partial assignments were made for Trp51 and Arg48 (Satterlee and Erman, 1991). The utility of both COSY and NOESY NMR experiments on paramagnetic proteins was demonstrated using yeast C*c*P as an example (Satterlee et al., 1991). Cyanide binding to H52L cytochrome *c* peroxidase was studied by a variety of techniques (Bidwai et al., 2003; Satterlee et al., 2003). Cyanide binds more strongly to the site-specific mutated enzyme, which exists in multiple forms.

NMR studies on the yeast cytochrome *c* peroxidase-cytochrome *c* complex present a double challenge because there are two separate molecules, each with a paramagnetic center. By use of a carbodiimide the two molecules were cross-linked (Waldmeyer and Bosshard, 1985). At least two major cross-links occurred which essentially shut down the peroxidase activity of the enzyme. Further studies showed that a 1 : 1 complex could be obtained by using ferrocytochrome *c* rather than the ferric form (Moench et al., 1987). Studies were extended to noncovalent complexes. In both covalently and noncovalently bound complexes large changes occurred in the hyperfine shifts of cytochrome *c* (Satterlee et al., 1987b). Results indicated the principal site of interaction is the exposed heme edge of cytochrome *c* in proximity to lysine 13 and heme pyrrole II. The minimum off rate for exchanging cytochrome *c* in the noncovalent complex is $(1.1 \pm 0.1) \times 10^3\ s^{-1}$ (Satterlee et al., 1987c). Studies were extended to a comparison of horse, tuna, and yeast cytochrome *c* complexes with yeast cytochrome *c* peroxidase (Moench et al., 1992). Complexes of the physiological partners yeast cyctochromes *c* iso-1 and iso-2 had distinctly different NMR spectra than those for the horse and tuna cytochromes *c*. An intimate relationship exists in the magnetic environments around the 3-methyl group and Phe82 of the cytochromes.

Changes in the NMR spectrum of the C*c*P—CN complex were detected upon cytochrome *c* binding. Using the C*c*P—CN complex as a model for Compounds I and II, this provided the first evidence for changes in the oxidized enzyme spectrum upon cytochrome *c* binding (Yi et al., 1992). Two-dimensional NMR was utilized to unscramble the overlap between hyperfine shifts of the enzyme and cytochrome *c* (Yi et al., 1993). Proton NMR studies of the yeast peroxidase-cytochrome *c* complexes were reviewed (Yi et al., 1993). The lifetime for exchange between bound and free forms of cytochrome *c* is in the range of 1 to 4 ms. Evidence supporting a second binding site was obtained (Yi et al., 1994a). Exchange rates of backbone amide protons of yeast iso-1 cytochrome *c* were slowed upon complex formation. The slowing occurred near the exposed heme edge, the so-called "front" side (Yi et al., 1994b).

Hyperfine-shifted resonances for free yeast iso-1-ferricytochrome *c* were completely assigned, and 70% assigned in the complex with C*c*P (Sukits et al., 1997). Isotopically enriched ^{15}N cytochrome *c* was used in a combined ^{15}N, ^{1}H NMR study of binding to C*c*P. For ferrous cytochrome *c* 34 amides at the front face are affected by binding. In the ferric protein residues at both the front and back are affected. There is no change in the geometry of the magnetic axes of cytochrome *c* upon binding (Worrall et al., 2001).

ELECTRON SPIN RESONANCE (ESR) SPECTROSCOPY

Introduction

Electron spin resonance, also called electron paramagnetic resonance, spectroscopy provides a means for detecting unpaired electrons and the interactions between the electron and the molecule in which it is contained. The technique is based on the interaction of the magnetic properties of the unpaired electron with an external magnetic field. It is similar to NMR spectroscopy, but the magnitudes of the resonance energy transitions are very different. In NMR the oscillating external field is in the radio frequency region of the electromagnetic spectrum, whereas in ESR it is in the microwave region. The sample is placed in a cavity in a static magnetic field of flux density B_0, and an oscillating magnetic field supplied by the microwave generator is used to find the resonance frequency. The resonance condition is defined by

$$h\nu_0 = g\beta B$$

where h is Planck's constant, ν_0 is the resonance frequency, β is the Bohr magneton, and B is the magnetic flux density. The factor g is determined by the location and the environment of the unpaired electron. For a free electron $g = 2.0023$, and this is well approximated by a free radical electron in a highly aromatic molecule.

Standard cavities are designed to resonate at a frequency of about 9 GHz (X band) or about 35 GHz (Q band). First-derivative spectra are recorded so that instead of an absorption peak, a curve with positive and negative components is obtained. The reason a derivative curve is obtained is because the slow sweep to detect the resonance frequency is subject to large noise levels. In order to minimize the noise, the sweep is modulated by a high frequency signal that samples the curve in very small increments, thereby providing tangents to, and hence the derivative of, the absorption curve (Fig. 9.8).

There are four factors that determine the nature of an ESR signal: the intensity, the linewidth, the g value, and multiplet splitting. The linewidth is influenced by many factors, but simplistically it can be described in terms of the reciprocals of relaxation times T_1 and T_2, as in NMR spectroscopy. In transition metal ions T_1 may be so short that the linewidth is too large for the signal to be detected. The linewidth may be dramatically decreased at temperatures approaching 0 K.

The g values may range from about 1.4 to 10. The method of reporting the positions of resonance frequencies differs in ESR compared to NMR spectroscopy. For NMR the g values of the nuclei are assumed to be fixed, and different resonance frequencies

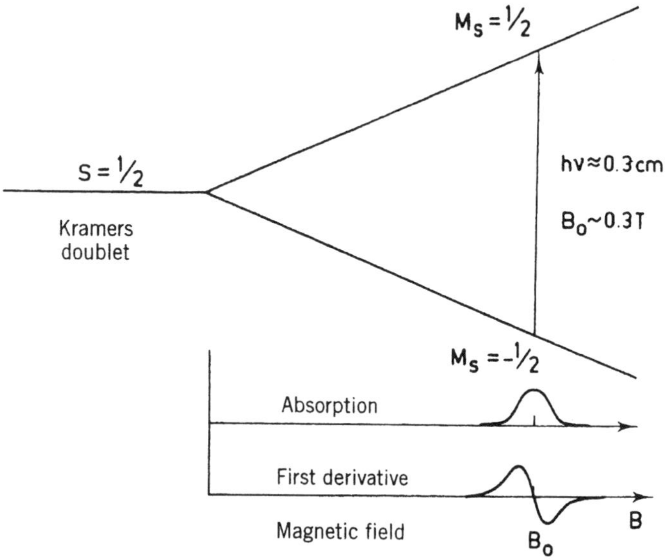

Figure 9.8. Magnetic splitting of the energy level of one unpaired electron, $S = \frac{1}{2}$, into two. Thus a Kramers doublet is obtained in which degeneracy is removed by a magnetic field. The absorption of energy occurs for the resonance condition. Spectra are recorded as the first derivative of absorption.

are attributed to different shielding effects caused by variations in the electron densities in the vicinity of the nuclei. For ESR the differences in resonance frequencies are attributed entirely to differences in g values, which therefore are empirical parameters.

Multiplet splitting can arise from two sources. Zero field splitting is the name given to splitting caused by the interactions of two or more unpaired electrons within the same molecule or ion, as in a transition metal ion. It does not require the external magnetic field. Hyperfine interactions, which occur between electron and nuclear spins, also lead to multiplet structure. The general rule for the splitting of the ESR spectrum of an electron that interacts with a single nucleus is

$$\text{Number of lines} = 2I + 1$$

where I is the spin quantum number of the nucleus. For a hydrogen atom, I for the nucleus is $\frac{1}{2}$, so the ESR spectrum will be a doublet. For a nitrogen atom $I = 1$, so an attached electron will give a triplet ESR signal.

In practice, a free radical electron feels the influence of several nuclei because it occupies a molecular orbital that is not localized. An illustration of hyperfine splitting is shown in Figure 9.9, the ESR spectrum of the $^{\bullet}CH_2OH$ radical. The ^{12}C nucleus

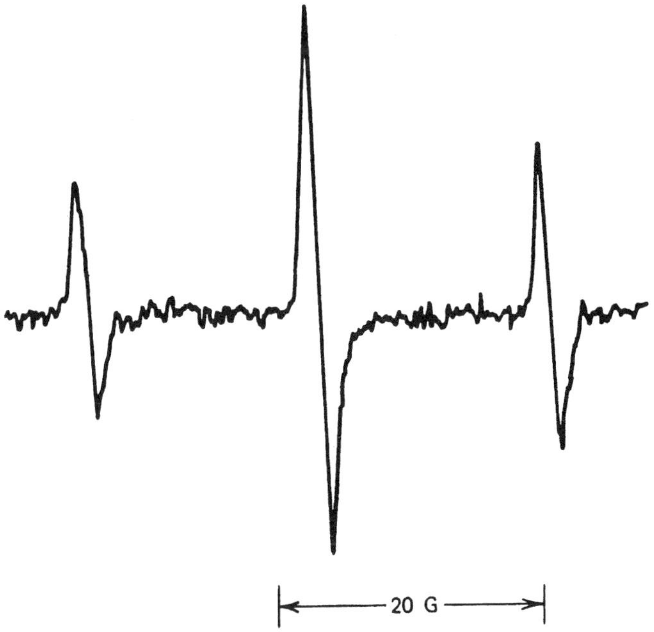

Figure 9.9. First-derivative ESR spectrum of the $^{\bullet}CH_2OH$ radical at pH 1. Only the protons attached to carbon influence the spectrum. (Reproduced with permission from Bolton JR. 1972. In: Bolton JR, Borg DC, editors. *Biological Applications of Electron Spin Resonance.* Wiley-Interscience, New York, pp. 11–60.)

with six protons and six neutrons has no net nuclear spin so it does not cause hyperfine splitting. However, the unpaired electron interacts with the two equivalent protons, which split the ESR signal into a triplet with 1 : 2 : 1 intensities.

A π electron may feel the influence of many nuclei. If all the nuclei are identical, then the extent of hyperfine splitting can be predicted using Pascal's triangle. A different splitting constant exists for each different nucleus. The oxidation product of 3,3′,5,5′-tetramethylbenzidine reacting with horseradish peroxidase and hydrogen peroxide is the tetramethylbenzidine cation radical. The cation radical possesses two amine N nuclei, four amine protons, and 12 methyl protons. Of a theoretical number of 1625 lines, about 31 were resolved (Josephy et al., 1982). The obvious advantage of ESR free radical detection is direct identification of the radical. In favorable cases free radical ESR detection is possible in living cells and tissues.

Spin Trapping and Spin Labels

A free radical localized on a single atom may be so short-lived that it cannot be detected by ESR spectroscopy. The possibility of detection may be greatly increased by using a spin trap, a molecule S that reacts with the highly reactive or unstable radical •R to form a stable radical •RS, which can be studied at leisure (Fig. 9.10).

A spin label already possesses a stable free radical. It may be attached at a specific point on a protein and used as a probe of its environment. However, without an exact knowledge of its orientation, spurious results may be obtained.

ESR Spectral Studies on Horseradish Peroxidase Reaction Products

A landmark study was published in 1960, the first ESR detection of free radicals in a biological system. It coincided with the commercial development of ESR machines. Semiquinone, ascorbate, and dihydroxyfumarate radicals were detected by ESR in reactions catalyzed by Japanese radish peroxidase (Yamazaki et al., 1960). Studies were extended to free radical detection in products of horseradish peroxidase reactions (Yamazaki and Piette, 1961, 1963; Griffen, 1978; Griffen and Ting, 1978; Kalyanaraman et al., 1982; Josephy et al., 1982, 1983a, 1983b; Harman et al., 1984; West et al., 1984; Fischer and Mason, 1984, 1986; Fischer et al., 1986; Morehouse et al., 1987; Curtis et al., 1988; Kolodziejczyk et al., 1988; Valoti et al., 1989; Gadelha et al., 1992). In all cases the free radical product was detected directly by ESR; no spin traps were used. The first ESR spectrum obtained from an HRP reaction was that of the oxidation product of reductic acid (Yamazaki and Piette, 1961). Most of

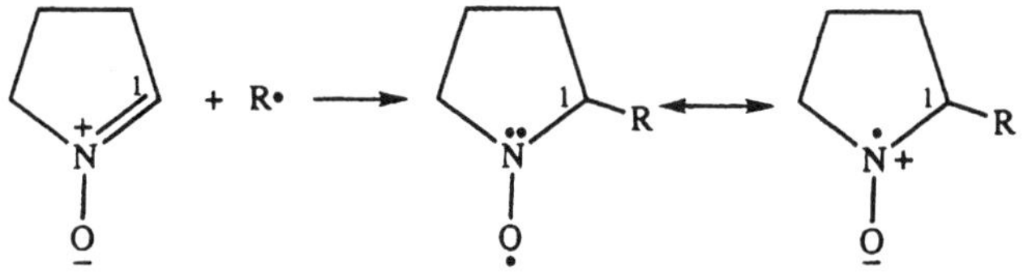

Figure 9.10. Use of pyrrolidine nitroxide as a spin trap.

the ESR work on HRP, cited above, was conducted in the laboratory of Ronald Mason. A few examples of the excellent work conducted on diverse substrates in his laboratory are provided here: sulfate anion free radical formation (Mottley and Mason, 1988), proof that porphyrins can be oxidized to π-cation radicals by peroxidase (Morehouse et al., 1989), and a study of reduction of the acetaminophen radical by glutathione and ascorbate (Rao et al., 1990).

Work in Mason's laboratory was extended by use of spin traps (Kalyanaraman et al., 1985; Harman et al., 1986; Stock et al., 1986; Stolze et al., 1989; Van der Zee et al., 1989; Mottley et al., 1991). The cyanyl radical was shown to be produced by the HRP catalyzed oxidation of cyanide (Moreno et al., 1988), and to attack the heme (Chen et al., 2000). The ESR spectrum of the skatole radical formed in the reaction of the plant growth hormone indoleacetic acid with horseradish peroxidase was obtained by spin trapping (Mottley and Mason, 1986). See Chapter 5 for the mechanism of this important reaction.

ESR Studies on Native Horseradish Peroxidase and its Compounds

In two ESR studies the coordination number of the iron(III) in horseradish peroxidase was addressed. In the first study, substitution of $H_2{}^{17}O$ for $H_2{}^{16}O$ caused line broadening for metmyoglobin, known to have water in the sixth coordination position of its iron, but no broadening for the peroxidase (Vuk-Pavlovic and Siderer, 1977). The conclusion: the iron is five-coordinate in the peroxidase. In the second study, the low temperature properties of metmyoglobin and horseradish peroxidase, their nitric oxide complexes, and their photolyzed products were investigated (Kobayashi et al., 1980). At low temperature the photolyzed NO is unable to return to its binding site. The spectrum of the photolyzed metmyoglobin was different from the original protein, whereas photolyzed HRP had the same ESR spectrum as the native enzyme. Again the difference is attributable to metmyoglobin having water in its sixth coordination position, whereas this position is vacant in native HRP.

The low-spin nature of the alkaline form of HRP was shown by ESR (Tamura and Hori, 1972). The fluoride complex of HRP is more high spin than the native enzyme (Morita and Mason, 1965). Benhydroxamic acid changes the iron of HRP to a six-coordinate species (Schulz et al., 1984), unlike other aromatic substrates (Maltempo et al., 1979).

The first ESR signal of horseradish peroxidase Compound I that correlated with the amount of HRP-I present was obtained in 1975, and is shown in Figure 9.11 (Aasa et al., 1975). This was confirmed in a later study which was extended to Mössbauer spectral measurements (Schulz et al., 1979, 1984). The weak ESR signal of HRP Compound I is caused by interaction of iron with the porphyrin π-cation radical. By means of electron nuclear double resonance (ENDOR), it was confirmed that HRP Compound I contains a single spin-coupled center with the free radical portion on the porphyrin (Roberts et al., 1981a, 1981b). In an ESR and ENDOR study of Compound II it was shown that the oxygen of the ferryl group is not protonated (Davydov et al., 2008). [For an introductory discussion of ENDOR spectroscopy see Dunford (1999).]

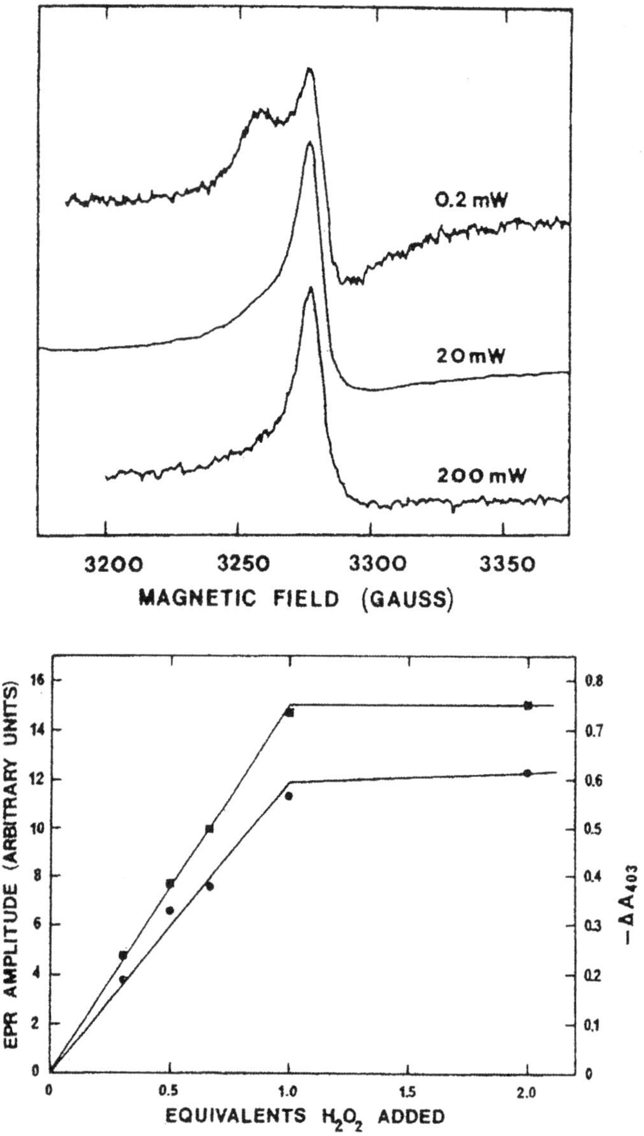

Figure 9.11. ESR spectrum of horseradish peroxidase Compound I (top) and titration of HRP with H_2O_2 (bottom) showing 1:1 correlation with Compound I formation: by change in optical absorption at 403 nm (●), and by amplitude of the $g = 1.995$ ESR signal (■). (Reproduced with permission from Aasa R, Vanngård T, Dunford HB. 1975. *Biochim. Biophys. Acta* **391**: 259–264.)

When a frozen solution of HRP Compound I is photolyzed, a doublet ESR signal centered on $g = 2$ is generated (Chu et al., 1977; McIntosh and Stillman, 1977). An electron appears to be liberated from an aromatic protein residue, which results in the π-cation radical being transferred from porphyrin to protein.

MÖSSBAUER SPECTROSCOPY

For a brief introduction to Mössbauer spectroscopy, see Dunford (1999), and for a more comprehensive review, Lang (1970). The technique utilizes absorption of γ radiation for which ^{57}Fe has a particularly large cross section. Thus heme is ideally suited for Mössbauer studies, particularly if it is further enriched in ^{57}Fe above the 2% natural abundance. A Mössbauer spectrum is characterized by its chemical shift δE, a function of the spin and oxidation states of the iron, and its quadrupole splitting ΔE_Q, a function of the nuclear spins of ground and excited states. The number of lines and shape of a spectrum is dependent on whether an external magnetic field is present and whether it is parallel or at right angles to the incident γ beam. The Mössbauer spectrum of HRP Compound I (Schulz et al., 1979) confirms the presence of iron(IV) with an electron spin S of 1. HRP Compound I has essentially the same Mössbauer parameters as Compound II (Schulz et al., 1984), but the spin S weakly couples to a spin S' of $\frac{1}{2}$ for the free radical. A model based on this weak coupling satisfactorily accounts for the Compound I Mössbauer spectra at all temperatures and magnetic field strengths. This work was extended to Compound I obtained from horseradish peroxidase reconstituted with mesoheme, with the same conclusions (Rutter et al., 1983). The latter Compound I Mössbauer spectrum is shown in Figure 9.12.

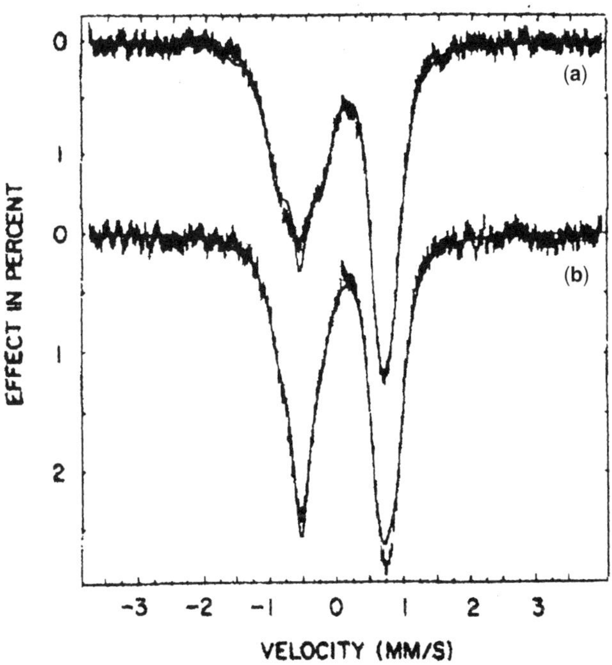

Figure 9.12. Mössbauer spectra of mesoheme horseradish peroxidase Compound I at 4.2 K. (Reproduced with permission from Rutter R, et al., 1983. *Biochemistry* **22**: 4769–4774.)

REFERENCES

Aasa R, Vänngård T, Dunford HB. 1975. EPR studies on Compound I of horseradish peroxidase. *Biochim. Biophys. Acta* **391**: 259–264.

Banci L, Bertini I, Bini T, Tien M, Turano P. 1993. Binding of horseradish, lignin, and manganese peroxidases to their respective substrates. *Biochemistry* **32**: 5825–5831.

Banci L, Carloni P, Savellini GG. 1994. Molecular dynamics studies on peroxidases: a structural model for horseradish peroxidase and a substrate adduct. *Biochemistry* **33**: 12356–12366.

Bidwai A, Witt M, Foshay M, Vitello LB, Satterlee JD, Erman JE. 2003. Cyanide binding to cytochrome *c* peroxidase (H52L). *Biochemistry* **42**: 10764–10771.

Bolton JR. 1972. In: Bolton JR, Borg DC, editors. *Biological Applications of Electron Spin Resonance*. Wiley-Interscience, New York, pp. 11–60.

Chen Y-R, Deterding LJ, Tomer KB, Mason RP. 2000. Nature of the inhibition of horseradish peroxidase and mitochondrial cytochrome *c* oxidase by cyanyl radical. *Biochemistry* **39**: 4415–4422.

Chen Z, de Ropp JS, Hernández G, La Mar GN. 1994. 2D NMR approaches to characterizing the molecular structure and dynamic stability of the active site for cyanide-inhibited horseradish peroxidase. *J. Am. Chem. Soc.* **116**: 8772–8783.

Chu M, Dunford HB, Job D. 1977. Horseradish peroxidase. XXV. An electron spin resonance study of the low temperature photochemical reaction of Compound I. *Biochem. Biophys. Res. Commun.* **74**: 159–164.

Curtis JF, Hughes MF, Mason RP, Eling TE. 1988. Peroxidase-catalyzed oxidation of (bi)sulfite: reaction of free radical metabolites with ($\pm$)-7,8-dihydroxy-7,8-di-hydro-benzo[*a*]pyrene. *Carcinogenesis* **9**: 2015–2021.

Davydov R, Osborne RL, Kim SH, Dawson JH, Hoffman BM. 2008. EPR and ENDOR studies of cryoreduced Compounds II of peroxidases and myoglobin: proton-coupled electron transfer and protonation status of ferryl hemes. *Biochemistry* **47**: 5147–5155.

de Ropp JS, Chen Z, La Mar GN. 1995. Identification of residues in the aromatic substrate binding site of horseradish peroxidase by ^{1}H NMR studies of isozymes. *Biochemistry* **34**: 13477–13484.

de Ropp JS, La Mar GN. 1991. 2D NMR assignment of hyperfine-shifted resonances in strongly paramagnetic metalloproteins: resting-state horseradish peroxidase. *J. Am. Chem. Soc.* **113**: 4348–4350.

de Ropp JS, Mandal PK, La Mar GN. 1999. Solution ^{1}H NMR investigation of the heme cavity and substrate binding site in cyanide-inhibited horseradish peroxidase. *Biochemistry* **38**: 1077–1086.

de Ropp JS, Sham S, Asokan A, Newmyer S, Ortiz de Montellano PR, La Mar GN. 2002. Influence of the distal His in imparting imidazolate character to the proximal His in heme peroxidase: ^{1}H NMR spectroscopic study of cyanide-inhibited His42 $\rightarrow$ Ala horseradish peroxidase. *J. Am. Chem. Soc.* **124**: 11029–11037.

Dunford HB. 1999. *Heme Peroxidases*. Wiley-VCH, New York.

Fischer V, Harman LS, West PR, Mason RP. 1986. Direct electron spin resonance detection of free radical intermediates during the peroxidase catalyzed oxidation of phenacetin metabolites. *Chem.-Biol. Interactions* **60**: 115–127.

Fischer V, Mason RP. 1984. Stable free radical and benzoquinone imine metabolites of an acetaminophen analogue. *J. Biol. Chem.* **259**: 10284–10288.

Fischer V, Mason RP. 1986. Formation of iminoxyl and nitroxide free radicals from nitrosonaphthols: an electron spin resonance study. *Chem.-Biol. Interactions* **57**: 129–142.

Gadelha FR, Hanna PM, Mason RP, Docampo R. 1992. Evidence for free radical formation during horseradish peroxidase-catalyzed *N*-demethylation of crystal violet. *Chem.-Biol. Interactions* **85**: 35–48.

Griffen BW. 1978. Evidence for a free radical mechanism of *N*-demethylation of *N,N*-dimethylaniline and an analog by hemeprotein-H_2O_2 systems. *Arch. Biochem. Biophys.* **190**: 850–853.

Griffen BW, Ting PL. 1978. Mechanism of *N*-demethylation of aminopyrine by hydrogen peroxide catalyzed by horseradish peroxidase, metmyoglobin, and protohemin. *Biochemistry* **17**: 2206–2211.

Günther H. 1995. *NMR Spectroscopy*, 2nd edition. Wiley, New York.

Gupta RK, Mildvan AS, Schonbaum GR. 1979. Heme environment of horseradish peroxidase as observed by oxygen-17 superfine interaction in EPR. *Biochem. Biophys. Res. Commun.* **89**: 1334–1340.

Harman LS, Carver DK, Schreiber J, Mason RP. 1986. One- and two-electron oxidation of reduced glutathione by peroxidases. *J. Biol. Chem.* **261**: 1642–1648.

Harman LS, Mottley C, Mason RP. 1984. Free radical metabolites of L-cysteine oxidation. *J. Biol. Chem.* **259**: 5606–5611.

Howes BD, Veitch NC, Smith AT, White CG, Smulevich G. 2001. Haem-linked interactions in horseradish peroxidase revealed by spectroscopic analysis of the Phe-221 → Met mutant. *Biochem. J.* **353**: 181–191.

Iizuka T, Ogawa S, Inubushi T, Yonezawa T, Morishima I. 1976. NMR studies of hemoproteins: pH dependence of ferric horseradish peroxidase and horse heart myoglobin. *FEBS Lett.* **64**: 156–158.

Josephy PD, Eling TE, Mason RP. 1982. The horseradish peroxidase-catalyzed oxidation of 3,5,3′,5′-tetramethyl benzidine: free radical and charge-transfer complex intermediates. *J. Biol. Chem.* **257**: 3669–3675.

Josephy PD, Eling TE, Mason RP. 1983a. Oxidation of *p*-aminophenol catalyzed by horseradish peroxidase and prostaglandin synthase. *Mol. Pharmacol.* **23**: 461–466.

Josephy PD, Eling TE, Mason RP. 1983b. An electron spin resonance study of the activation of benzidine by peroxidases. *Mol. Pharmacol.* **23**: 766–770.

Kalyanaraman B, Janzen EG, Mason RP. 1985. Spin trapping of the azidyl radical in azide/catalase/H_2O_2 and various azide/peroxidase/H_2O_2 peroxidizing systems. *J. Biol. Chem.* **260**: 4003–4006.

Kalyanaraman B, Mason RP, Sivarajah K. 1982. An electron spin resonance study of a novel radical cation produced during the horseradish peroxidase-catalyzed oxidation of tetramethylhydrazine. *Biochem. Biophys. Res. Commun.* **105**: 217–224.

Kobayashi K, Tamura M, Hayashi K, Hori H, Morimoto H. 1980. Electron paramagnetic resonance and optical absorption spectra of the pentacoordinated ferrihemoproteins. *J. Biol. Chem.* **255**: 2239–2242.

Kolodziejczyk P, Reszka K, Lown JW. 1988. Enzymatic activation and transformation of the antitumor agent mitoxantrone. *Free Radical Biol. Med.* **5**: 13–25.

La Mar GN, Chatfield MJ, Peyton DH, de Ropp JS, Smith WS, Krishnamoorthi R, Satterlee JD, Erman JE. 1988. Solvent isotope effects on NMR spectral parameters in high-spin ferric hemoproteins: an indirect probe for distal hydrogen bonding. *Biochim. Biophys. Acta* **956**: 267–276.

La Mar GN, Chen Z, de Ropp JS. 1995. Assignment strategies and structure determination in cyanide-inhibited heme peroxidases. In: *Nuclear Magnetic Resonance of Paramagnetic Molecules*. La Mar GN, editor. Kluwer Academic, Amsterdam.

La Mar GN, de Ropp JS. 1980. Proton nuclear magnetic resonance characterization of the electronic structure of horseradish peroxidase Compound I. *J. Am. Chem. Soc.* **102**: 395–397.

La Mar GN, de Ropp JS, Smith KM, Langry KC. 1980. Proton nuclear magnetic resonance study of the electronic and molecular structure of the heme crevice in horseradish peroxidase. *J. Biol. Chem.* **255**: 6646–6652.

La Mar GN, de Ropp JS, Smith KM, Langry KC. 1981. Proton nuclear magnetic resonance investigation of the electronic structure of Compound I of horseradish peroxidase. *J. Biol. Chem.* **256**: 237–243.

La Mar GN, Hernández G, de Ropp JS. 1992. ^{1}H NMR investigation of the influence of interacting sites on the dynamics and thermodynamics of subtrate and ligand binding to horseradish peroxidase. *Biochemistry* **31**: 9158–9168.

Lanir A, Schejter A. 1975. Nuclear magnetic resonance evidence for the absence of iron coordinated water in horseradish peroxidase. *Biochem. Biophys. Res. Commun.* **62**: 199–203.

Lang G. 1970. Mössbauer spectroscopy. *Quart. Rev. Biophys.* **3**: 1–60.

Lehnig M. 2001. ^{15}N chemically induced dynamic nuclear polarization during reaction of *N*-acetyl-L-tyrosine with nitrating systems nitrite/hydrogen peroxide/horseradish peroxidase and nitrite/hypochlorous acid. *Arch. Biochem. Biophys.* **393**: 245–254.

Maltempo MM, Ohlsson P-I, Paul K-G, Petersson L, Ehrenberg A. 1979. Electron paramagnetic resonance analyses of horseradish peroxidase in situ and after purification. *Biochemistry* **12**: 2935–2941.

McIntosh AR, Stillman MJ. 1977. Electron-paramagnetic resonance studies on a photochemically produced species of horseradish peroxidase Compound I. *Biochem. J.* **167**: 31–37.

Moench SJ, Chroni S, Lou B-S, Erman JE, Satterlee JD. 1992. Proton NMR comparison of noncovalently and covalently cross-linked complexes of cytochrome *c* peroxidase with horse, tuna and yeast ferricytochromes *c*. *Biochemistry* **31**: 3661–3670.

Moench SJ, Satterlee JD, Erman JE. 1987. Proton NMR and electrophoretic studies of the covalent complex formed by cross-linking yeast cytochrome *c* peroxidase and horse cytochrome *c* with a water-soluble carbodiimide. *Biochemistry* **26**: 3821–3826.

Morehouse KM, Flitter WD, Mason RP. 1987. The enzymatic oxidation of Desferal to a nitroxide free radical. *FEBS Lett.* **222**: 246–250.

Morehouse KM, Sipe HJ Jr, Mason RP. 1989. The one-electron oxidation of porphyrins to porphyrin-pi-cation radicals by peroxidases: an electron spin resonance investigation. *Arch. Biochem. Biophys.* **273**: 158–164.

Moreno SNJ, Stolze K, Janzen EG, Mason RP. 1988. Oxidation of cyanide to the cyanyl radical by peroxidase/H_2O_2 systems as determined by spin trapping. *Arch. Biochem. Biophys.* **265**: 267–271.

Morishima I, Kurono M, Shiro Y. 1986. Presence of endogenous calcium ion in horseradish peroxidase: elucidation of metal-binding site by substitutions of divalent and lanthanide ions for calcium and use of metal-induced NMR (^{1}H and ^{113}Cd) resonances. *J. Biol. Chem.* **261**: 9391–9399.

Morishima I, Ogawa S. 1979. Nuclear magnetic resonance studies of hemoproteins: binding of aromatic donor molecules to horseradish peroxidase. *J. Biol. Chem.* **254**: 2814–2820.

Morita Y, Mason HS. 1965. An electron spin study of some hemoproteins. *J. Biol. Chem.* **240**: 2654–2659.

Mottley C, Mason RP. 1986. An electron spin resonance study of free radical intermediates in the oxidation of indole acetic acid by horseradish peroxidase. *J. Biol. Chem.* **261**: 16860–16864.

Mottley C, Mason RP. 1988. Sulfate anion free radical formation by the peroxidation of (bi)sulfite and its reaction with hydroxyl radical scavengers. *Arch. Biochem. Biophys.* **267**: 681–689.

Mottley C, Robinson RE, Mason RP. 1991. Free radical formation in the oxidation of malonaldehyde and acetylacetone by peroxidase enzymes. *Arch. Biochem. Biophys.* **289**: 153–160.

Paul K-G, Ohlsson P-I. 1978. Equilibria between horseradish peroxidase and aromatic donors. *Acta Chem. Scand.* **B32**: 395–404.

Rao DNR, Fischer V, Mason RP. 1990. Glutathione and ascorbate reduction of the acetaminophen radical formed by peroxidase: detection of the glutathione disulfide radical anion and the ascorbyl radical. *J. Biol. Chem.* **265**: 844–847.

Roberts JE, Hoffman BM, Rutter R, Hager LP. 1981a. ^{17}O ENDOR of horseradish peroxidase Compound I. *J. Am. Chem. Soc.* **103**: 7654–7656.

Roberts JE, Hoffman BM, Rutter R, Hager LP. 1981b. Electron-nuclear double resonance of horseradish peroxidase Compound I: detection of the porphyrin π-cation radical. *J. Biol. Chem.* **256**: 2118–2121.

Rutter R, Valentine M, Hendrich MP, Hager LP, Debrunner PG. 1983. Chemical nature of the porphyrin π cation radical in horseradish peroxidase Compound I. *Biochemistry* **22**: 4769–4774.

Sakurada J, Takahashi S, Hosoya T. 1986. Nuclear magnetic resonance studies on the spatial relationship of aromatic donor molecules to the heme iron of horseradish peroxidase. *J. Biol. Chem.* **261**: 9657–9662.

Satterlee JD, Erman JE. 1980. Proton magnetic resonance of cytochrome *c* peroxidase: pD dependence of the isotropically shifted resonances. *Arch. Biochem. Biophys.* **202**: 608–616.

Satterlee JD, Erman JE. 1981a. Heme asymmetry in deuterohemin-reconstituted cytochrome *c* peroxidase. *J. Am. Chem. Soc.* **103**: 199–200.

Satterlee JD, Erman JE. 1981b. Proton nuclear magnetic resonance characterization of the oxidized intermediates of cytochrome *c* peroxidase. *J. Biol. Chem.* **256**: 1091–1093.

Satterlee JD, Erman JE. 1983a. Deuterium exchangeable proton hyperfine resonances of low-spin cytochrome *c* peroxidase and the mechanism of peroxidase catalysis. *Biochim. Biophys. Acta* **743**, 149–154.

Satterlee JD, Erman JE. 1983b. Temperature and pH dependence of the proton hyperfine resonances of cytochrome *c* peroxidase-cyanide. *J. Biol. Chem.* **258**: 1050–1056.

Satterlee JD, Erman JE. 1991. Proton NMR assignments of heme contacts and catalytically implicated amino acids in cyanide-ligated cytochrome *c* peroxidase determined from one- and two-dimensional nuclear Overhauser effects. *Biochemistry.* **30**: 4398–4405.

Satterlee JD, Erman JE, de Ropp JS. 1987a. Proton hyperfine resonance assignments in cyanide-ligated cytochrome *c* peroxidase using the nuclear Overhauser effect. *J. Biol. Chem.* **262**: 11578–11583.

Satterlee JD, Erman JE, La Mar GN, Smith KM, Langry KC. 1983a. Assignment of hyperfine-shifted resonances in low-spin forms of cytochrome *c* peroxidase by reconstitutions with deuterated hemins. *J. Am. Chem. Soc.* **105**: 2099–2104.

Satterlee JD, Erman JE, La Mar GN, Smith KM, Langry KC. 1983b. Assignment of hyperfine-shifted resonances in high-spin forms of cytochrome *c* peroxidase by reconstitutions with deuterated hemins. *Biochim. Biophys. Acta* **743**: 246–255.

Satterlee JD, Erman JE, Mauro JM, Kraut J. 1990. Comparative proton NMR analysis of wild-type cytochrome c peroxidase from yeast, the recombinant enzyme from *Escherichia coli*, and an Asp → Asn-235 mutant. *Biochemistry* **29**: 8797–8804.

Satterlee JD, Moench SJ, Erman JE. 1987b. The cytochrome *c* peroxidase-cytochrome *c* molecular redox complex: electron transfer kinetics, chemical crosslinking and docking effects. *Rec. Trav. Chim. Pays-Bas.* **1106**: 286.

Satterlee JD, Moench SJ, Erman JE. 1987c. A proton NMR study of the non-covalent complex of horse cytochrome *c* and yeast cytochrome *c* peroxidase and its comparison with other interacting protein complexes. *Biochim. Biophys. Acta* **912**: 87–97.

Satterlee JD, Russell DJ, Erman JE. 1991. Proton homonuclear correlated spectroscopy as an assignment tool for hyperfine-shifted resonances in medium-sized paramagnetic proteins: cyanide-ligated yeast cytochrome *c* peroxidase as an example. *Biochemistry.* **30**: 9072–9077.

Satterlee JD, Savenkova MI, Fosshay M, Erman JE. 2003. Temperature, pH, and solvent isotope dependent properties of the active sites of resting-state and cyanide-ligated recombinant cytochrome *c* peroxidase (H52L) revealed by proton hyperfine resonance spectroscopy. *Biochemistry* **42**: 10772–10782.

Schulz CE, Devaney PW, Winkler H, Debrunner PG, Doan N, Rutter R, Hager LP. 1979. Horseradish peroxidase Compound I: evidence for spin coupling between the heme iron and a "free" radical. *FEBS Lett.* **103**: 102–105.

Schulz CE, Rutter R, Sage JT, Debrunner PG, Hager LP. 1984. Mössbauer and electron paramagnetic resonance studies of horseradish peroxidase and its catalytic intermediates. *Biochemistry* **23**: 4743–4754.

Shiro Y, Kurono M, Morishima I. 1986. Presence of endogenous calcium ion and its functional and structural regulation in horseradish peroxidase. *J. Biol. Chem.* **261**: 9382–9390.

Smulevich G, English AM, Mantini AR, Marzocchi MP. 1991. Resonance Raman investigation of ferric iron in horseradish peroxidase and its aromatic donor complexes at room and low temperatures. *Biochemistry* **30**: 772–779.

Stock BH, Schreiber J, Guenat C, Mason RP, Bend JR, Eling TE. 1986. Evidence for a free radical mechanism of styrene-glutathione conjugate formation catalyzed by prostaglandin H synthase and horseradish peroxidase. *J. Biol. Chem.* **261**: 15915–15922.

Stolze K, Moreno SNJ, Mason RP. 1989. Free radical intermediates formed during the oxidation of cyanide by horseradish peroxidase/H_2O_2 as detected with nitroso spin traps. *J. Inorg. Biochem.* **37**: 45–53.

Sukits SF, Erman JE, Satterlee JD. 1997. Proton NMR assignments and magnetic axes orientations for wild-type yeast iso-1ferricytochrome *c* free and in solution and bound to cytochrome *c* peroxidase. *Biochemistry* **36**: 5251–5259.

Tamura M, Hori H. 1972. Optical and magnetic measurements of horseradish peroxidase. III. Electron paramagnetic resonance studies at liquid hydrogen and helium temperatures. *Biochim. Biophys. Acta* **284**: 20–29.

Tanaka M, Ishimori K, Morishima I. 1999. Luminol activity of horseradish peroxidase mutants mimicking a proposed binding site for luminol in *Arthromyces ramosus* peroxidase. *Biochemistry* **38**: 10463–10473.

Thanabal V, de Ropp JS, La Mar G. 1987a. ^{1}H NMR study of the electronic and molecular structure of the heme cavity in horseradish peroxidase: complete heme resonance assignments based on saturation transfer and nuclear Overhauser effects. *J. Am. Chem. Soc.* **109**: 265–272.

Thanabal V, de Ropp JS, La Mar G. 1987b. Identification of the catalytically important amino acid residue resonances in ferric low-spin horseradish peroxidase with nuclear Overhauser measurements. *J. Am. Chem. Soc.* **109**: 7516–7525.

Valoti M, Sipe HJ Jr, Sgaragli G, Mason RP. 1989. Free radical intermediates during peroxidase oxidation of 2-*t*-butyl-4-methoxyphenol, 2,6-di-*t*-butyl-4-methylphenol, and related phenol compounds. *Arch. Biochem. Biophys.* **269**: 423–432.

Van der Zee, Duling DR, Mason RP, Eling TE. 1989. The oxidation of N-substituted aromatic amines by horseradish peroxidase. *J. Biol. Chem.* **264**: 19828–19836.

Veitch NC. 1995. Aromatic donor molecule binding sites of haem peroxidases. *Biochem. Soc. Trans.* **23**: 232–240.

Veitch NC, Gao Y, Smith AT, White CG. 1997. Identification of a critical phenylalanine residue in horseradish peroxidase, Phe179, by site-directed mutagenesis and ^{1}H-NMR: implications for complex formation with aromatic donor molecules. *Biochemistry* **36**: 14751–14761.

Veitch NC, Williams RJP. 1990. Two-dimensional ^{1}H-NMR studies of horseradish peroxidase C and its interaction with indole-3-propionic acid. *Eur. J. Biochem.* **189**: 351–362.

Veitch NC, Williams RJP. 1995. The use of methyl-substituted benzhydroxamic acids as structural probes of peroxidase substrate binding. *Eur. J. Biochem.* **229**: 629–640.

Veitch NC, Williams RJP, Bray RC, Burke JF, Sanders SA, Thorneley RNF, Smith AT. 1992. Structural studies by proton-NMR spectroscopy of plant horseradish peroxidase C, the wild-type recombinant protein from *Escherichia coli* and two protein variants, Phe41 → Val and Arg38 → Lys. *Eur. J. Biochem.* **207**: 521–531.

Vuk-Pavlović S, Siderer Y. 1977. Probing axial ligands in ferric haemoproteins: an ESR study of myoglobin and horseradish peroxidase in $H_2{}^{17}O$. *Biochem. Biophys. Res. Commun.* **79**: 885–889.

Waldmeyer B, Bosshard HR. 1985. Structure of an electron transfer complex. I. Covalent cross-linking of cytochrome *c* peroxidase and cytochrome *c*. *J. Biol. Chem.* **260**: 5184–5190.

West PR, Harman LS, Josephy PD, Mason RP. 1984. Acetominophen: enzymatic formation of a transient phenoxyl radical. *Biochem. Pharmacol.* **33**: 2933–2936.

Worrall JAR, Kolczak U, Canters GW, Ubbink M. 2001. Interaction of yeast iso-1-cytochrome *c* with cytochrome *c* peroxidase investigated by (^{15}N, ^{1}H) heteronuclear NMR spectroscopy. *Biochemistry* **40**: 7069–7076.

Wüthrich K. 1976. *NMR in Biological Research: Peptides and Proteins*, North-Holland, Amsterdam.

Yamazaki I, Mason HS, Piette L. 1960. Identification, by electron paramagnetic resonance spectroscopy, of free radicals generated from substrates by peroxidase. *J. Biol. Chem.* **235**: 2444–2449.

Yamazaki I, Piette L. 1961. Mechanism of free radical formation and disappearance during ascorbate oxidase and peroxidase reactions. *Biochim. Biophys. Acta* **50**: 62–69.

Yamazaki I, Piette L. 1963. The mechanism of aerobic oxidase reaction catalyzed by peroxidase. *Biochim. Biophys. Acta* **77**: 47–64.

Yi Q, Alam S, Ge Y, Satterlee JD, Erman JE. 1993. Proton nuclear magnetic resonance studies of non-covalent complexes of yeast cytochrome *c* peroxidase with cytochromes *c*. *Techniques in Protein Chemistry IV*, pp. 605–613.

Yi Q, Erman JE, Satterlee JD. 1992. Ferricytochrome *c* binding induces detectable proton NMR shift changes in cytochrome *c* peroxidase-CN. *J. Am. Chem. Soc.* **114**: 7907–7909.

Yi Q, Erman JE, Satterlee JD. 1994a. 1H NMR evaluation of yeast isozyme-1 ferricytochrome *c* equilibrium exchange dynamics in noncovalent complexes with two forms of yeast cytochrome *c* peroxidase. *J. Am. Chem. Soc.* **116**: 1981–1987.

Yi Q, Erman JE, Satterlee JD. 1994b. Studies of protein-protein association between yeast cytochrome *c* peroxidase and yeast iso-1 ferricytochrome *c* by hydrogen-deuterium exchange labeling and proton NMR spectroscopy. *Biochemistry* **33**: 12032–12041.

CHAPTER 10

THEORETICAL

This chapter is devoted to theoretical calculations on peroxidase systems, as well as to the theoretical underpinnings to a large amount of experimental work. As an important example of the latter, a discussion of peroxidase kinetics follows.

PEROXIDASE KINETICS

Valuable information can be obtained from straight lines and simple curves. Most steady state kinetic measurements are made on a conventional spectrophotometer with a dead time of 1–5 s after reaction initiation before measurements can begin. Reaction rates are slowed by using enzyme concentrations of the order 10^{-9} M (nM). For many reactions a linear plot of absorbance of product versus time is obtained, an example of zero order kinetics. Each linear plot provides one point on a rectangular hyperbolic plot of velocity versus substrate concentration. The rate constant at infinite substrate concentration is k_{cat}.

The simplest equation used to fit the rectangular hyperbolic curve is

$$E + S \underset{}{\overset{K_M}{\rightleftharpoons}} ES \xrightarrow{k} E + P$$

This "mechanism" is inadequate for the simplest enzyme reaction but it has the virtue of simple interpretation. K_M is a binding equilibrium constant and k is a first order rate contant (units s^{-1}). A mechanism with multiple intermediates ES, ES′, ES″, will also fit the rectangular hyperbola, and although the meaning of K_M is more complicated the concept of reversible binding equilibria is maintained.

Enter horseradish peroxidase and an entirely different interpretation is required. The steady state rate data for a typical peroxidase reaction still fit a rectangular

Peroxidases and Catalases: Biochemistry, Biophysics, Biotechnology, and Physiology, Second Edition.
By H. Brian Dunford

hyperbola. However, for most substrates the mechanism contains no detectable reversibility. Now the value of K_M is typically $k_1(H_2O_2)/k_3$, where k_1 is the rate constant for Compound I formation and k_3 the rate constant for Compound II reaction (see Appendix). K_M now has nothing to do with reversible binding constants.

In steady state kinetics one tries to deduce what the enzyme is doing from observing rate of product formation from substrate. It is more informative if one can follow directly what is happening to the enzyme during the course of a reaction, which is possible in transient state kinetic studies. The highly colored peroxidases provide ideal systems since each intermediate has its own unique optical spectrum. Enzyme concentrations of $\sim 10^{-6}$ M (μM) provide optical absorbances ideal for accurate measurement, so reactions occur rapidly and require the use of a stopped flow apparatus. Now the dead times are of the order of ~ 1 ms.

As an example, it is possible to prepare horseradish peroxidase Compound I by reaction of native enzyme with hydrogen peroxide in a multimixing stopped flow apparatus and then react it with reducing substrate. Depending on the wavelength of the kinetic measurement, a curve of increasing absorbance may be obtained. The relevant equations are

$$
\begin{aligned}
-\frac{d[\text{Compound I}]}{dt} &= k'[\text{substrate}][\text{Compound I}] \\
&= k[\text{Compound I}]
\end{aligned}
$$

The substrate concentration, in excess of (Compound I), remains essentially constant during the course of the reaction. The exponential curve can be fit by a single constant k, a (pseudo) first order rate constant (units s^{-1}).

The unique virtue of first order kinetics is that the half life is independent of the starting concentration. The practical value of this virtue is that an accurate rate constant can be obtained even though the relevant molar absorptivities are unknown. From a series of first order measurements the variation of the first order rate constants with substrate concentration typically is linear, from which there are two important conclusions. The reaction is proven to be first order with respect to substrate, and the slope yields the second order rate constant (units $M^{-1}s^{-1}$) with an accuracy greater than for a single measurement.

Because the optical spectra are known, the enzyme product formed from Compound I can be determined; usually it is Compound II. This provides proof that a one-electron reduction of Compound I has occurred. The subsequent reaction of newly formed Compound II with the same substrate may occur to a small extent. Its influence can be eliminated by operating at the isosbestic point between Compound II and native enzyme, 411 nm. In the (so far) unique cases of iodide and sulfite the product of Compound I reduction is native horseradish peroxidase—proof that a two-electron reduction has occurred; thus the rate constants of each elementary reaction in a peroxidase cycle can be measured using transient state kinetics.

The ultimate aim of enzyme kinetics is to establish the most plausible mechanism. Enzyme kinetics does not provide proof of the correct mechanism, but when properly used, it can disprove mechanisms that otherwise appear plausible. If there appear to be two most plausible mechanisms and kinetic results disprove one of them, the remaining mechanism stands alone. In fact, the elimination of alternative explanations is the procedure in interpretation of experimental results from all techniques.

When the kinetic method is used properly, a theoretical fit to the experimental data is limited to the minimum number of required variables, and the effect of changing a variable is understood. When extra unnecessary parameters are introduced, the scientific method is abused, and the contribution to the scientific literature has negative value. For very complicated systems enough variables can always be introduced to fit the experimental data but there is no way to establish their significance (Fig. 10.1).

Of course, speculative interpretation of further significance of a given mechanism is the author's prerogative if restricted to the Discussion section of a research paper. However, addition of extra unnecessary variables to a postulated mechanism *ist verboten.* I cite three examples of abuse of the kinetic method, one following, and two in the discussion of prostaglandin H synthase in Chapter 19 (one of them my own). It is important to point out these errors, since they otherwise remain in the literature, decoys off the path to the truth.

The first example of forbidden addition of an unnecessary variable is found in a study of the reaction of myeloperoxidase with hydrogen peroxide (Kettle and Winterbourn, 2001). A major contribution by endogenous donor was postulated to occur, and a rate constant was assigned to it. In the resulting analysis it was concluded that Compound I underwent the catalatic reaction with hydrogen peroxide to form native enzyme with a rate constant of $2 \times 10^6\ M^{-1}s^{-1}$. When the spectra are observed,

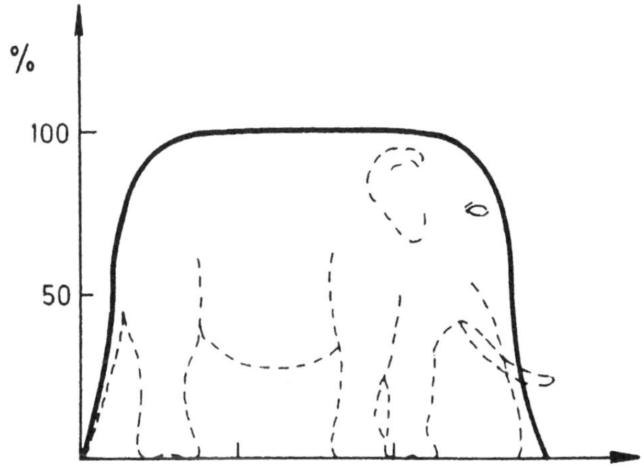

Figure 10.1. No comment. Reproduced with permission from Eigen M. Kinetics of reaction control and information transfer in enzymes and nucleic acids. In: Claesson S, editor. *Nobel Symposium 5. Fast Reactions and Primary Processes in Chemical Kinetics.* Almqvist and Wiksell, Stockholm. pp. 334–367.

it is Compound II that is formed from Compound I and hydrogen peroxide, and the rate constant is $8 \times 10^4\ M^{-1}s^{-1}$ (Marquez et al., 1994). I tried without success to dissuade my New Zealand friends, outstanding contributors to an overall understanding of myeloperoxidase reactions, from their postulate.

An example where, initially, too few variables were introduced is a study of the reaction of prostaglandin H synthase with 5-phenyl-4-pentenyl hydroperoxide (PPHP) and a series of phenols (Hsuanyu and Dunford, 1992). The missing variable was the arachidonic acid binding site, shown in the later crystal structure to be on the proximal side of the heme (Picot et al., 1994). The results were interpreted as all reaction occurring on the distal side of the heme. The experimental results are valid, and were readily accommodated to an expanded mechanism in which Compound I can be reduced by a phenol directly to Compound II or it can be transformed in an intramolecular electron transfer whereby a tyrosyl radical is formed in the arachidonic acid binding site.

Two mechanisms for prostaglandin H synthase have received the most attention. One is the branching chain mechanism in which, once Compound I is formed, the cyclooxygenase reaction proceeds independently. The other is the tightly coupled mechanism, in which the peroxidase and cyclooxygenase cycles occur together. Phenol at lower concentrations stimulates prostaglandin G_2 formation because it reduces Compound II back to native enzyme that can reenter the cycle, an effect that cannot be explained by the branching chain mechanism. Phenol at higher concentrations may divert Compound I directly to Compound II, preventing tyrosyl radical and PGG_2 formation. It also may block the arachidonic acid bindiung site, thus a small reducing substrate plays a continuous role. The branching chain mechanism requires an improbably complicated proton and electron transfer process (see Chapter 19). The peroxidase and cyclooxygenase cycles are coupled, and the branching chain mechanism is disproven.

MARCUS THEORY FOR ELECTRON TRANSFER REACTIONS

The usual representation of a simple chemical reaction is by means of a three-dimensional potential energy diagram as shown in Figure 10.2. So, for a reaction of the type

$$A + B - C \longrightarrow [A\cdot\cdot B\cdot\cdot C] \longrightarrow A - B + C$$

potential wells for A–B and B–C are at right angles to each other with a pass, called a saddle point, connecting one to the other.

The diagram shows the relative energies of A, B, and C as the reaction progresses. The pathway of the reaction, called the *reaction coordinate*, goes from the minimum in the B – C potential energy well, over the maxima at the saddle point, into the minimum of the A – B well. The intermediate species (A·····B·····C) occurs at the saddle point where the three atoms are in the *transition state*. Figure 10.2 provides a good representation of reaction in a three-atom system in the gas phase.

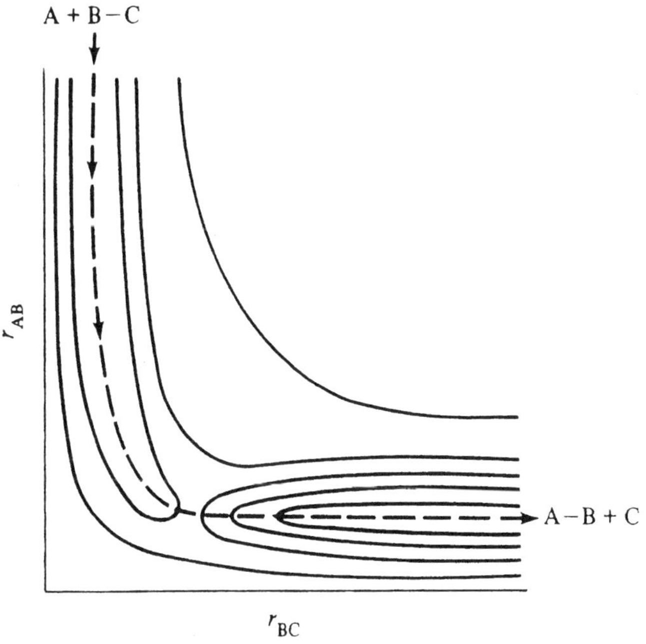

Figure 10.2. Potential energy diagram with energy at right angles to the plane of the paper. The reaction is shown to occur along the minimum energy pathway.

There appears to be nothing simpler than an electron transfer reaction. No chemical bonds are broken or are formed in a reaction that can be followed using isotopes:

$$Fe^{2+} + {}^{*}Fe^{3+} \longrightarrow Fe^{3+} + {}^{*}Fe^{2+}$$

The rate of the above reaction is much slower than that between the bigger ferro- and ferri-cyanide ions. Why? What are the complications? The different charges on the ferrous and ferric ions induce different but fluctuating arrangements of solvating water molecules, and electron transfer can only occur when the water molecules on each ion attain similar configurations; thus a rearrangement energy plays an important part in the overall energetics. In the bigger ferro- and ferri-cyanide ions the central iron ions are shielded and the required rearrangement energy is much smaller.

Because of the complicated role of solvating molecules, different from the behavior of bulk solvent, multi-dimensional N coordinate potential energy diagrams are required for accuracy. A great simplification was made possible by replacing potential energy diagrams with those depicting free energy changes (Marcus, 1993). Figure 10.3 shows free energy G wells for reactants plus environment R and products plus environment P. The reaction coordinate q occurs from left to right and the transition state is found where the two parabolae intersect.

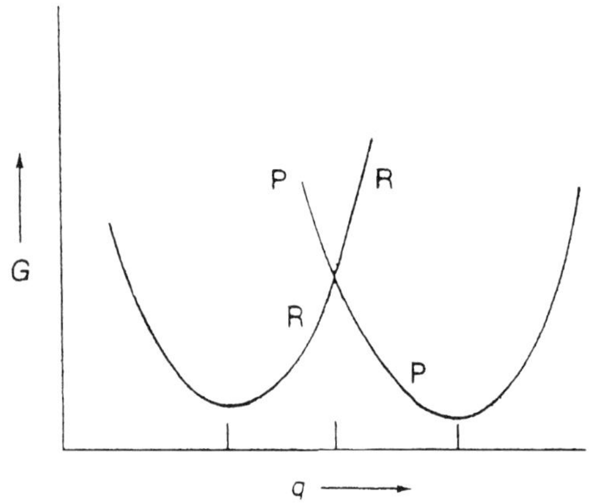

Figure 10.3. Plot of free energy of reactants R versus the reaction coordinate q. The three vertical lines on the abscissa indicate the free energies of reactants, transition state, and products P. Reproduced with permission from Marcus RA. 1993. *Angewandte Chemie* **32**: 1111–1121.

The rate constant k for reaction is given by

$$k = A\exp\left(-\frac{\Delta G^*}{RT}\right)$$

where ΔG^*, the free energy of activation, is the difference in free energies between the minimum in the G parabola for reactant R and the crossing point of the two parabolae.

The constant A depends on the type of electron transfer reaction: intramolecular, bimolecular, etc. The free energy of activation G^* is given by

$$\Delta G^* = \frac{\lambda}{4}\left(1 + \frac{\Delta G^{\circ}}{\lambda}\right)^2$$

where λ is the reorganization energy and G° is the standard free energy change, zero for a self-exchange reaction. Marcus derived specific expressions for the reorganization energy term λ in which interactions with dipoles of solvating molecules are added to vibrational energies.

One result was the prediction of an inverted region. Normally the smaller the free energy change for reaction G°, the smaller G^* becomes and hence the larger the rate constant k. When G° is decreased to $-\lambda$, the value of G^* reaches its minimum value of zero. As G° decreases further, because the $(1 + G^{\circ}/\lambda)$ is squared, G^* increases causing a decrease in k.

Another result was a prediction of the rate constant of a cross reaction k_{12} in terms of two self-exchange reactions $k_{11,}$ and k_{22}.

$$k_{12} \cong (k_{11}k_{22}K_{12}f_{12})^{\frac{1}{2}}$$

As an example, k_{11} could be for the Fe^{2+}/Fe^{3+} exchange, k_{22} for Ce^{4+}/Ce^{3+}, so k_{12} would be for Fe^{2+}/Ce^{4+}. K_{12} is the equilibrium constant for the cross-reaction and f is a known function of k_{11}, k_{22}, and K_{12}, usually close to unity.

Both predictions of an inverted region and of rate constants for cross reactions have been borne out by experiment, particularly in the work of Norman Sutin (Marcus, 2007; Sutin, 2007). For further details and predictions on electron transfer reactions see the review that summarizes fifty years of work (Marcus, 1993). Marcus theory is applicable to outer sphere electron transfer reactions, in which no formal covalent bond is formed between oxidant and reductant.

Marcus theory for cross reactions was applied to yeast cytochrome *c* peroxidase Compound II reactions with substitution-inert iron(II) complexes (Purcell and Erman, 1976). However, their interpretation of the experimental results has been questioned (Yandell and Yonetani, 1983).

Difficult electrochemical measurements have been made on horseradish peroxidase reactions (Andreu et al., 2007). The three processes that contribute to the electrochemical current—substrate mass transport, substrate reaction, and electron exchange with the electrode—were differentiated experimentally. The Marcus–DOS reorganization energy and maximum electron transfer rate for Compound II were determined.

Marcus theory was applied to kinetic results on phenol reactions with HRP Compounds I and II. Reorganization energies λ for the electron transfer reactions were estimated (Khopde and Priyadarsini, 2000).

ELECTRON TUNNELING

We intuitively can predict how certain systems will behave in a world governed by classical Newtonian mechanics, thus we know that an orange in a box will stay in the box. According to quantum mechanics there is a finite (although very small) probability that the orange can be found outside the box, because it can tunnel through the wall (Fig. 10.4).

Quantum mechanics is perhaps better described as wave mechanics for a non-expert, since wave properties of particles are responsible for their non-Newtonian behavior. The De Broglie wavelength λ of a particle of mass m and velocity v is given by

$$\lambda = h/mv$$

where h is Planck's constant. The smaller the mass of the particle the larger the wavelength associated with it, and the higher the probability that the particle can tunnel through a barrier. Electrons are then more likely to tunnel compared to protons which in turn are more likely to tunnel than deuterons. In terms of free energy diagrams it means that electron transfer can occur at lower energy than in the classical case.

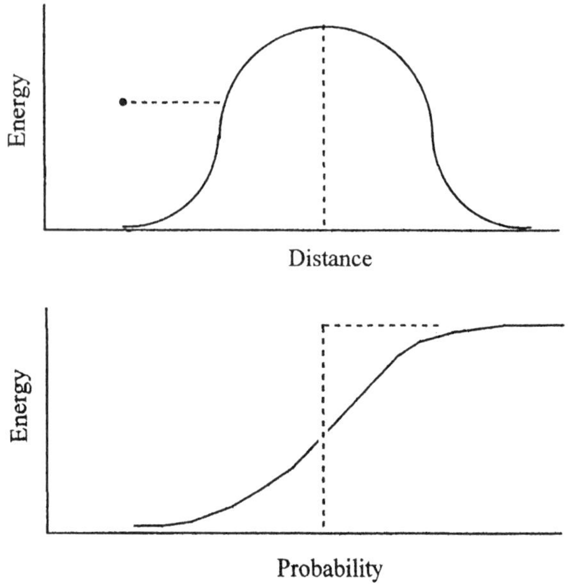

Figure 10.4. Representation of the pathway of a particle toward a potential energy barrier. In classical mechanics the particle cannot surpass the barrier unless its energy exceeds that of the barrier. In quantum mechanics there is a finite probability of tunneling through the barrier, which increases with increasing energy.

ELECTRON TRANSFER REACTIONS IN PROTEINS

A great deal of experimental work has been accumulated on the rates of intra- and inter-electron transfer in peroxidase systems by Francis Millett and coworkers. This section is concerned with both theoretical and experimental studies leading to the present day interpretation of experimental results.

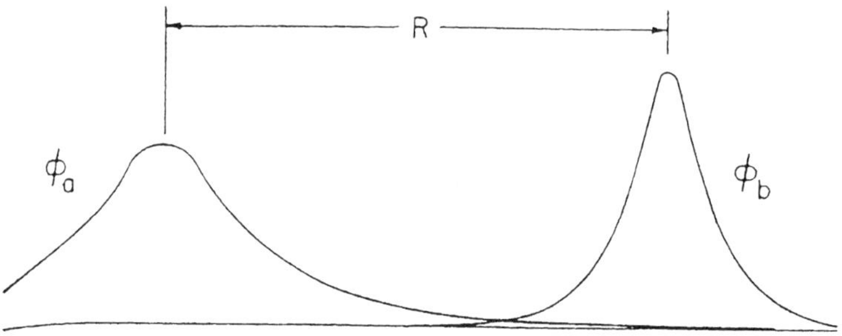

Figure 10.5. The wave functions ψ_a and ψ_b separated by a distance R, with exponentially decaying tails overlapping weakly. Reproduced with permission from Hopfield JJ. 1974. *Proc. Natl. Acad. Sci. USA.* **71**: 3640–3644.

There is general agreement that electrons do tunnel through protein matrixes. Wave functions for two sites for an electron separated by a distance R are shown in Figure 10.5.

Overlap between the two wave functions produces a tunneling matrix element. (Hopfield, 1974). As a first approximation, this matrix element decreases exponentially with distance, as shown by calculations on crystal structure with known sites for transfer (Page et al., 1999). Vibronic coupling facilitates the electron transfer at higher temperatures. Detailed analyses shows that electron transfer is faster along covalent bonds and non-bonded transfer can be over short distances only (Beretan et al., 1987, 1991, 1992; Gray and Winkler, 1996; Onuchic et al., 1992). Distances for electron transfer in proteins are less than 14 Å units.

ELECTRON DENSITY CIRCUITS

The influence of distal and proximal hydrogen bonding networks to delocalize charges and thus stabilize both resting state and intermediates was proposed before many plant peroxidase crystal structures were known. Calculations were based on a static model of resting peroxidases (Welinder et al., 1995). Electron density circuits provide a dynamic model (Dunford, 2001). Shifts in electron density are used to explain pK_a shifts, essential for enzyme catalysis. When an electron pair is held equally by two atoms, a covalent bond results. If the electron pair is held more closely by one atom then partial charges develop. The stronger the partial positive charge $\delta+$ on a proton the more likely it is to dissociate, that is, the stronger the acid.

$$\overset{\delta+}{\mathbf{H}}:\overset{\delta-}{\mathbf{A}} \longrightarrow \mathbf{H}^{+} + \mathbf{A}^{-}$$

Electron withdrawal by substituents on **A** results in it pulling the electron pair closer to **A**, thus increasing $\delta-$. Correspondingly, $\delta+$ is increased on **H** so **HA** has become a stronger acid. Electron density flow in the opposite direction decreases the acidity.

pH-dependent enzyme kinetics can yield accurate data on specific pK_a values in the active site of an enzyme (Alberty, 2007). Kinetic data has established that the distal imidazole side chain in horseradish peroxidase has a very low pK_a value, $\sim$2.5 in the resting enzyme. This is the site that accepts the proton from hydrogen peroxide as Compound 0 is formed, and the rate of Compound 0, followed by Compound I formation, is pH independent at least to pH 11 where the alkaline transition occurs. Thus a pK_a increase of at least 8.5 units has occurred! How is this possible? Distal and proximal sides of peroxidase hemes are known to be connected by hydrogen bond networks. Electron flow through one or more circuits starting with $^{-}$OOH, through the ferric ion, and through proximal-to-distal pathway(s) can increase the electron density at the crucial imidazole nitrogen, converting it from a weak to a very strong base (Fig. 10.6).

The $^{+}$N–H group can be formed at very high pH. After Compound 0 is formed, a reverse electron density flow through the same circuit(s) enables the imidazole proton

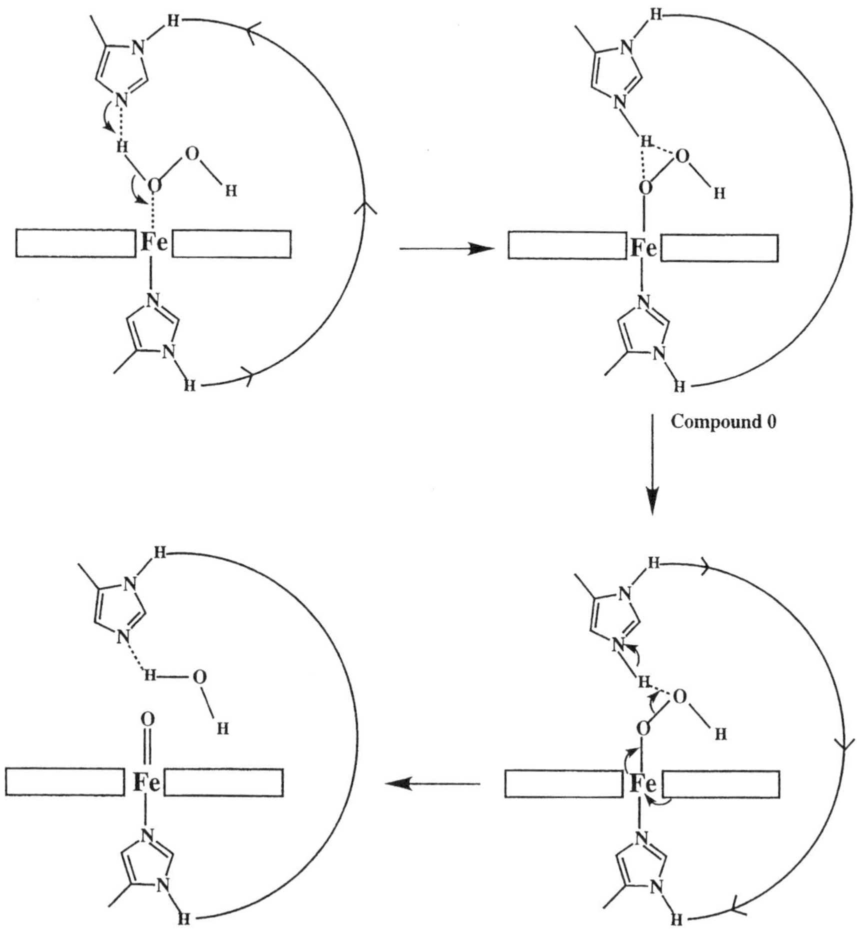

Figure 10.6. Electron flow (proximal to distal side) initiated by the binding of hydrogen peroxide in the peroxidase active site, with simultaneous start of proton transfer from hydrogen peroxide to distal imidazole. With proton transfer complete, Compound 0 is formed. This is followed by reverse electron flow (distal to proximal side) initiated by the π-back bonding from iron to its bound oxygen with a simultaneous start of proton transfer from distal imidazole. Final products are formed: Compound I and the water leaving group.

to help form the water leaving group (Dunford, 2001). The reverse flow is initiated by π-back bonding from iron as the Fe=O group is formed. Recently a large effect of magnetic particles on the activity of horseradish peroxidase has been observed, which is an effect expected of a magnetic field on an electron circuit (Chalkias et al., 2008). Similar circuits have been constructed that can help to explain catalysis in carbonic anhydrase, chymotrypsin, and fumarase (Dunford, 2003).

DIFFUSION CONTROL

In a diffusion controlled reaction of enzyme E and substrate S, an encounter complex must first be formed

$$E + S \underset{k_{-1}}{\overset{k_1}{\rightleftarrows}} E\text{---}S \longrightarrow \overset{k_2}{} \longrightarrow E + P$$

Assuming a steady state the overall rate constant k_a is given by

$$k_a = \frac{k_1 k_2}{k_{-1} + k_2}$$

For a true diffusion controlled reaction k_2 must greatly exceed k_{-1}, so that $k_a = k_1$. The second-order rate constant k_1 for the diffusion-controlled combination to form the encounter complex is given by

$$k_1 = 4\pi(D_E + D_S)(R_E + R_S)N/1000\ M^{-1}s^{-1}$$

where D is a diffusion coefficient, R radius, and N Avogadro's number (Smoluchowski, 1917). The first-order rate constant k_{-1} for diffusion-controlled separation of an ES encounter complex is

$$k_{-1} = 3(D_E + D_S)/(R_{ES})^3\ s^{-1}$$

where R_{ES} is the combined radius of the encounter complex (Eigen, 1974). Electrostatic interactions either enhance or slow diffusion controlled rates. The value of k_{-1} is of the order of $10^{10}\ s^{-1}$ for small molecules. The fastest known diffusion controlled reaction is the recombination rate of a proton and hydroxide ion: $1.4 \times 10^{11}\ M^{-1}s^{-1}$. The diffusion controlled rate constant for an enzyme reaction is estimated to be about $10^9\ M^{-1}s^{-1}$.

An experimental test of diffusion control is to measure reaction rate in viscous solvents such as water-glycerol or water-sucrose, since according to Stokes' law diffusion coefficients are proportional to the reciprocal of solvent viscosity η. For forward rate constants k_1 and reverse rate constants k_{-1} in two different solvent media

$$k_1^0\ \eta^0 = k_1\ \eta \qquad k_{-1}^0\ \eta^0 = k_{-1}\ \eta$$

The superscript 0 refers to aqueous solution as reference. It follows that

$$\frac{1}{k_a} = \frac{1}{k_1^0}\frac{\eta}{\eta^0} + \frac{k_{-1}^0}{k_1^0}\frac{1}{k_2}$$

From a plot of $1/k_a$ versus η/η^0, the value of k_1^0 is obtained from the slope and k_{-1}^0/k_2 from the intercept. The values obtained for the reaction of horseradish peroxidase with p-nitroperbenzoic acid are $k_1^0 = 1.3 \times 10^8\ M^{-1}\ s^{-1}$, and $k_{-1}^0/k_2 = 2.2$

(Nakatani and Dunford, 1979). So, of the encounter complexes formed, roughly one in three reacts, and two of three dissociate. There is partial diffusion control.

Viscosity tests for diffusion control using aqueous glycerol were made for the horseradish peroxidase reaction with hydrogen peroxide: no diffusion control, and for *m*-chloroperbenzoic acid: partial diffusion control (Dunford and Hewson, 1977). Similarly, no diffusion control was found for Compound I formation in yeast cytochrome *c* peroxidase (Loo and Erman, 1977).

QUANTUM MECHANICAL CALCULATIONS

There has been an explosion of quantum mechanical calculations on biological systems in the past ten years. Various theoretical approaches have been developed for the design of new drugs (Nogrady and Weaver, 2005). Applications of quantum mechanics to properties and reactions of peroxidases are outlined below.

HOMOs and LUMOs

Energies of Highest Occupied Molecular Orbitals (HOMOs) and Lowest Unoccupied Molecular Orbitals (LUMOs) have been calculated and correlated with Hammett $\rho\sigma$ values for substituted phenol and aniline reactions with HRP Compound II (Sakurada et al., 1990). Initial steady state rates of 15 estrogenic phenols reacted with HRP were correlated with calculated HOMO energies (Colosi et al., 2006).

Extended Hückel Calculations

The electronic heart of a porphyrin is the inner 16-member ring with 18 π electrons, which dominates electronic spectra of porphyrins (Gouterman, 1978). The inner ring includes the four pyrrole nitrogens. For a metalloporphyrin electrons in *d* orbitals of central metal ions such as Fe^{2+} and Fe^{3+} tend to interact weakly with the π electrons of the porphyrin ring as shown in the optical spectra. The 16-member ring can be treated by π-electron theory. The ring is like a copper wire in its ability to conduct electrons. However, π-electron theory by itself cannot treat the unpaired *d* electrons present in a central metal ion. Extended Hückel calculations provided the first theoretical treatment that included all metal ion valence electrons (Zerner and Gouterman, 1966). The calculations ordered the energies of the *d* orbitals with respect to the π, thus accounting for ground state magnetic properties. They confirmed the weak d-π electronic coupling.

Extended Hückel calculations were conducted on Compounds I and II of peroxidases and catalase (Hanson et al., 1981). Compound II was described as having a ferryl configuration ($Fe^{IV}{=}O$) and Compound I as a ferrylporphyrin π-cation radical. It was concluded that peripheral substituents and axial ligands determined whether the heme has an a_{1u} or a_{2u} ground state. The axial ligand of catalase was predicted to differ from that of peroxidases. Ferryl porphyrin π-cation models in the a_{1u} state have been prepared (Fujii and Ichikawa, 1992; Czarnecki et al., 1996).

Ab Initio Methods

Ab initio calculations take into account all electrons, not just valence electrons. They confirmed that a stable peroxidase-hydrogen peroxide complex is formed (Compound 0), that the peroxide is bound in an asymmetric end-on mode, and that the energy difference between either oxygen of the hydrogen peroxide being the iron ligand is small (Collins et al., 1992; Loew and Dupuis, 1996). In a subsequent study relative energies of sextet, quartet, and doublet spin states were examined for a peroxidase resting state. In the model it was assumed that water is a coordinating ligand, which is not the case (Loew and Dupuis, 1997). Peroxidase intermediates and benzhydroxamic acid binding to HRP were examined (Harris and Loew, 1996; Chang et al., 1998).

Density Functional Theory (DFT)

DFT applied to a ferrous deoxyheme model showed that for Fe-O distances greater than 2.5 Å several electronic states have similar energies. Therefore, spin crossing can explain how the O_2 in a triplet state can bind to a quintet deoxyheme to form a singlet oxyheme. This gives a large advantage to biological systems where oxygen binding occurs. A similar near degeneracy between spin states is observed in a model ferric heme peroxidase model (Jensen and Ryde, 2004).

A density functional study of model peroxidase Compounds I and II indicated that the a_{2u} π-cation radical state ($^4A_{2u}$) is the ground state of Compound I, not the a_{1u} π-cation radical state ($^4A_{1u}$) (Kuramochi et al., 1997). Very weak magnetic interaction occurred between the $S = 1$ Fe=O center and the porphyrin $S = \frac{1}{2}$ free radical. However, the difference in energy between the two states is small and could be changed by structural alteration of the axial ligand. It was proposed that Compound I may be an admixture of the A_{2u} and A_{1u} states.

A subsequent use of DFT showed essential agreement for the Poulos-Kraut mechanism for Compound 0 formation (Wirstam et al., 1999). Lowest energy occurred for proton transfer from hydrogen peroxide to the imidazole side chain of the distal His residue. Calculations were performed on a simplified model heme with only distal and proximal imidazole rings present. The O—O bond cleavage was studied next. It was found that the O—O bond distance and the distance from the distal oxygen of OOH^- and the proton on the histidine were the crucial structural parameters. A protonated proximal Trp residue was indicated for yeast cytochrome *c* peroxidase Compound I using the model shown in Figure 10.7.

For other calculations the proximal side was modeled with an imidazole ring only. The merits of DFT calculations compared to the larger systems used in quantum mechanical/molecular mechanical studies have been listed (Siegbahn and Borowski, 2006).

Potential energy surfaces for spin-forbidden recombination of CO, NO and H_2O with ferrous deoxyheme were calculated using DFT theory (Strickland and Harvey, 2007). A simplified model heme was used. CO binding had a significant barrier because of the change from a quintet ground state for the deoxyheme to a bound singlet state. NO binding involves a low barrier. Bond energies were calculated.

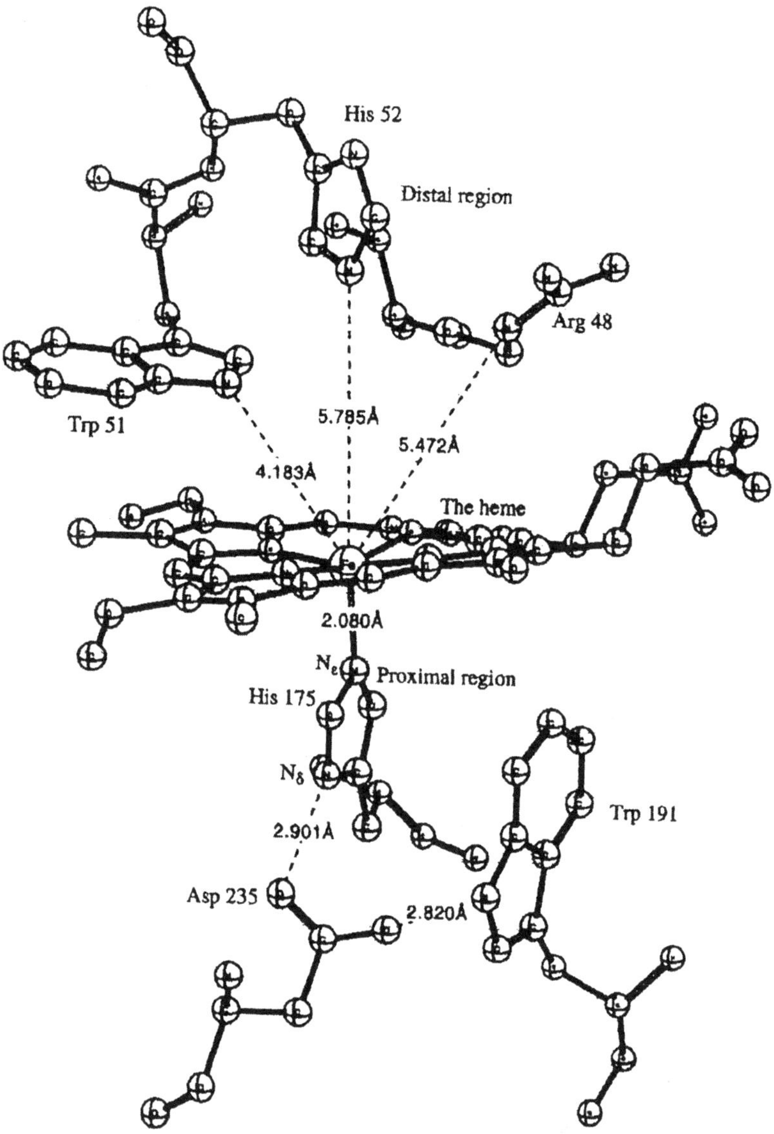

Figure 10.7. The active site of yeast cytochrome *c* peroxidase based on the crystal structure which was used in density functional theory (DFT) calculations. Reproduced with permission from Wirstam M, Blomberg MRA, Siegbahn PEM. 1999. *J. Am. Chem. Soc.* **121**: 10178–10185.

DFT calculations were used to examine why catalases, unlike peroxidases, rarely form Compound II (Rovira and Fita, 2003). Species that are hydrogen bonded to the proximal His were found to play an important role. It was erroneously suggested that the oxygen of the $Fe^{IV}{=}O$ group may be protonated. A horseradish peroxidase theoretical mutant in which axial nitrogen has been replaced by phosphorus was examined using DFT calculations (de Visser, 2007).

Jane Vanderkooi and coworkers have combined experimental and theoretical approaches. Carbonmonoxy horseradish peroxidase has been examined as a function of pH and in the presence of other substrates. Solvent dependent and independent motions of the HRP-CO complex were examined by infrared spectroscopy and molecular dynamic calculations (Kaposi et al., 2001, 2003). A DFT study was made of Fe-C-O conformers (Dalosto et al., 2003). Properties of the heme pocket of HRP were examined by studying the quenching of fluorescence of free base and Pd porphyrin derivatives by benzhydroxamic acid (Khajehpour et al., 2003). The influence of benzhydroxamic acid on the charge transfer absorption band of native HRP was examined experimentally and by use of DFT calculations (Zelent et al., 2004).

The influence of substrate binding on fluorescence of a Mg-mesoporphryin substituted horseradish peroxidase was examined both experimentally and with modeling studies. The results showed that the electrostatic potential of the prosthetic group is modified by the binding event (Laberge et al., 2001).

Quantum Mechanical/Molecular Mechanical (QM/MM) Studies

The following (abbreviated) quote describes QM/MM studies. "The QM/MM method dissects the total enzyme into two subsystems: the active center and the rest. The active center is described by a quantum mechanical method. The rest of the system is treated by molecular mechanics using a force field calibrated for proteins. The two subsystems are allowed to interact by electrostatic and van der Waals terms, such that the QM subsystem adapts its electronic structure and charge distribution to the field exerted by the protein environment" (Derat et al., 2005). Potentially it offers large advantages over theoretical methods applied to a restricted active site only, where possible influences of the rest of the protein are neglected. On the other hand, "In QM/MM studies, the only possibility applied so far for transition metal containing enzymes (to detect errors) is to go backward and forward along the reaction pathway until there are no more structural changes, . . . a procedure that is tedious and still not free from artifacts" (Siegbahn and Borowski, 2006). Detailed QM/MM studies have been conducted on the formation and structure of horseradish peroxidase Compound I (Derat et al., 2005, 2007; Derat and Shaik, 2006a), and formation of Compound II (Derat and Shaik, 2006b).

QM/MM Studies on Compound I The QM/MM method provides a tremendous amount of data on the models that are studied: spin states, bond lengths, relative energies. Formation of Compound I via the Poulos-Kraut mechanism was modeled starting with the crystal structure of Compound I (Derat and Shaik, 2006a). This means the distal imidazole is initially only singly protonated, i.e. Compound I is in a basic form. It was calculated that the energy for proton transfer from the hydrogen peroxide-HRP complex to the imidazole ring of distal His42 was about 20 kcal/mol (Derat et al., 2007), far higher than the experimental value, and in disagreement with the earlier DFT study (Wirstam et al., 1999). The model for direct proton transfer

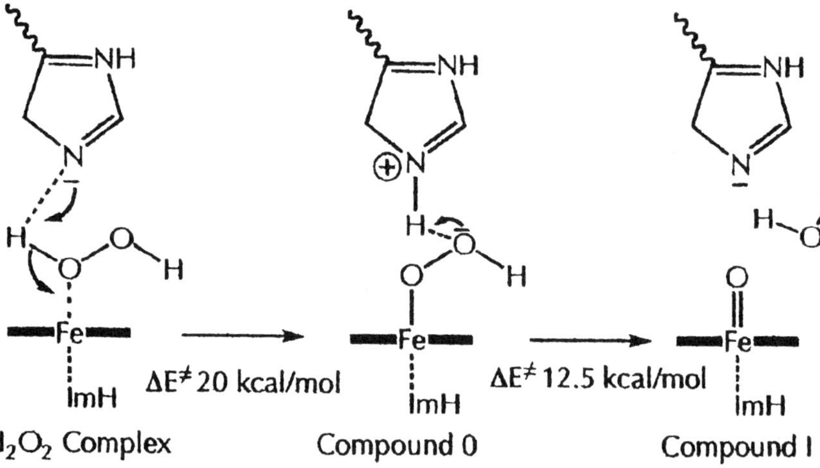

Figure 10.8. Scheme for Compound I formation in which water is absent in the initial step. Reproduced with permission from Derat E, Shaik S, Rovira C, Vidossich P, Alfonso-Prieto M. 2007. *J. Am. Chem. Soc.* **129**: 6346–6347.

from the hydrogen peroxide complex to distal imidazole can be labeled "dry" (Fig. 10.8).

In order to obtain harmony between the QM/MM result and experiment, a stealth water molecule was introduced in a later study, a "wet" model (Derat et al., 2007). Water2407 was assumed to enter the active site, assist the proton transfer, and then depart.

The high energy barrier for the proton transfer in the dry model rules against the conclusions that the influence of the Fe^{III} might just be strong enough to facilitate the hydrogen peroxide-imidazole proton transfer (Jones and Dunford, 2005). Experimental evidence favors the dry model for horseradish peroxidase, whereas the wet model is appropriate for catalase, where the water released from Compound I formation has been postulated to remain in the active site, blocking the approach of reducing substrates such as phenols (Jones, 2001). The latter process was dubbed "Watergate" by R.J.P. Williams. The exact count of directly participating water molecules in the active site for each step of the peroxidase cycle remains unknown. To solve this problem it is necessary first to have an agreed-upon definition of what constitutes a participating water molecule. It appears reasonable to exclude water molecules that are not immediately adjacent to those participating in proton transfer to or from protein. However, this condition may not exclude water molecules outside the region examined by the QM method.

The model that remains in accord with the experimental facts for HRP Compound I formation is that of electron density circuits, described earlier, in which flows of electron density between proximal and distal sides of the heme, along known proximal-distal connectivities, facilitate Compound 0 and Compound I formation (Dunford, 2001). The latter model cannot be examined by the QM/MM method, in which the protein is regarded as a fixed matrix with a role limited to its effect on the charge distribution and geometry of the active site.

Energy barriers to reaction can be calculated by DFT and QM/MM methods. These barriers can be lowered according to the electron density shift approach. Barriers to reaction (activation energies) can be measured experimentally for overall reactions, but at present data are sadly lacking. Experimental activation energies would provide a check and a restriction on theoretical calculations.

Ten different subsystems were used to test different aspects of Compound I structure (Derat et al., 2005). The model has the distal imidazole ring doubly protonated so the model is appropriate for an acid form of Compound I. Conclusions were: (a) The lowest doublet and quartet states of Compound I are virtually degenerate, with two unpaired electrons on Fe=O and one localized on the porphyrin; (b) The radical cation on the porphyrin has dominant A_{2u} character with electron density on the pyrrole nitrogens (Gouterman, 1978); (c) R(Fe=O = 1.70 Å and R Fe—N_{ImH} = 2.06 Å; (d) The proximal ligand is imidazole, not imidazolate; (e) Calculated Mössbauer parameters agree with experiment if models of distal Arg and His are included in the calculation; (f) Theory agrees with experiment except for NMR studies indicating proximal imidazolate; and (g) the protein shapes most of the subtle features of Compound I structure.

QM/MM Study on Compound I Reduction to Compound II Whereas the theoretical study of the structure of Compound I (not its formation) is essentially in agreement with experiment, the study on horseradish peroxidase Compound II formation can be labeled disagreement (Derat and Shaik, 2006b). Starting with the Compound I structure, two structures of Compound II are shown, Fe^{IV}=O and a Fe^{Z}—OH form (where Z = 3 or 4). The Fe^{Z}—OH form is almost certainly a very minor structure, despite recent claims that Compound II is basic. The strongest evidence that there is an Fe^{Z}—OH form of Compound II of horseradish peroxidase is that the oxygen of Fe^{IV}=O will exchange with bulk water, so there must be a transient state existence of Fe^{Z}—OH and Fe^{Z}—OH_2. Evidence for the existence of an appreciable amount of an Fe^{Z}—OH form of Compound II is controversial. The consensus is that the ferryl oxygen of Compound II is not protonated (Terner et al., 2006). The authors then attempt, unsuccessfully, to find a pathway from the Fe^{IV}=O to the Fe^{Z}—OH structure (Derat and Shaik, 2006b). They then attempted, again unsuccessfully, to find a reverse pathway from Fe^{Z}—OH to Fe^{IV}=O. The result was formation of two new states from Fe^{Z}—OH: singlet and triplet states of Fe^{III}—OH_2. Of the four states Fe^{IV}=O, Fe^{Z}—OH and singlet and triplet Fe^{III}—OH_2, the Fe^{IV}=O state was found to be of highest energy. Mössbauer parameters were calculated for the four states, and only those for Fe^{IV}=O agreed with experiment. The conclusion: "it is clear that the Mössbauer study has missed the various isomers." An alternative conclusion is that Compound II isomers other than Fe^{IV}=O exist in negligible quantities, which does not exclude them from being kinetically important. However the inability to find a pathway in either direction between Fe^{IV}=O and a Fe^{Z}—OH raises the question of the validity of the conclusions about the Compound II structure.

QM/MM Study on Ferulic Acid Reductions of Compounds I and II

Starting with the crystal structure of the native HRP-ferulic acid structure (Henriksen et al., 1999), QM calculations produced a model of the Compound I-ferulic acid complex in which $H_2O(402)$ undergoes a configuration change. With the combined QM/MM method, pathways were obtained for reduction of Compound I to II, and Compound II to native enzyme (Derat and Shaik, 2006c). It was concluded that $Fe^{IV}{=}O$ and $Fe^{IV}{-}OH$ intermediates precede Compound II formation with the structure of the latter being a (porphyrin $+\bullet$)$Fe^{III}{-}OH$. This Compound II structure is then reduced by ferulic acid through a $Fe^{III}{-}OH$ intermediate to Fe^{III} resting enzyme. The distal imidazole ring is alternately protonated and deprotonated in both reduction reactions. Possible mechanisms for ferulic acid oxidation were examined: (a) stepwise electron transfer followed by proton transfer; (b) proton coupled electron transfer (PCET); and (c) hydrogen atom transfer. Their conclusion is that PCET is the preferred mechanism (Derat and Shaik, 2006c). In Scheme 2 of their paper they show a "through-space" electron transfer from ferulic acid to the porphyrin ring, whether this is implied. An electron transfer along the hydrogen bonded network from ferulic acid to heme looks feasible, and is more in accord with the weight of opinion of the "through space" versus bonded electron transfer pathways.

All of the QM/MM studies on HRP are providing fresh light and raising provocative questions about an enzyme that has been rather thoroughly worked over experimentally. Compound II of HRP has a pK_a value of 8.6 for an active site acid group. When this group is deprotonated at high pH Compound II is totally unreactive (Dunford and Stillman, 1976). The conventional wisdom was that this acid group was the protonated imidazole side chain of distal His42. According to the Derat/Shaik mechanism, above, it would be the proton on the hydroxyl group of the (porphyrin $+\bullet$)$Fe^{III}{-}OH$ form of Compound II. A principle-of-least-motion, if applicable, would exclude the Derat/Shaik mechanism. Participation of another water molecule might improve the geometry. The proximal imidazole ring retains a single proton throughout the peroxidase reaction cycle, a point of agreement among the papers by Derat, and Shaik et al., a more recent QM/MM study (Heimdal et al., 2008), and the consensus from experimental work. It was concluded from a QM/MM study that the heme propionates play an important role in electron transfer in ascorbate peroxidase and in heme-to-heme electron transfer in bacterial diheme peroxidases (Guallar, 2008).

Other recent QM/MM studies have examined the role of bound substrate in keeping the Compound I free radical on the porphyrin of cytochrome P450 3A4 (Fishelovitch et al., 2007), and concluded that hemolytic generation of hydroxyl radicals from hydrogen peroxide occurs in heme oxygenases (Chen et al., 2008).

Molecular Dynamics Simulations

In molecular dynamics calculations the motions of atoms within a system are simulated with respect to the forces acting upon them. By heating the molecule (increasing assigned velocities of atoms) and then allowing them to cool it is possible to explore conformational space and identify low energy forms. Structures for the horseradish peroxidase-hydrogen peroxide complex were explored (Filizola and Loew, 2000).

The equilibrium behavior of COX-1 and COX-2 isoforms with arachidonic acid were explored by molecular dynamics simulations. They provided detailed information about protein channels available for oxygen and water transport (Furse et al., 2006).

An artifical cavity was created in yeast cytochrome *c* peroxidase by the W191G mutant (Fitzgerald et al., 1996). Properties of this cavity have been explored by molecular dynamics simulations (Baron and McCammon, 2007). The system contained bulk water, unlike some of the peroxidase theoretical models described earlier. One important conclusion was that water locations identified by X-ray determination of crystal structure may not be permanently occupied but represent the average position of a large number of rapidly exchanging water molecules. Another is that a single conformation for a loop is unlikely. Presence of a potassium ion in the cavity increases the ordering of water molecules.

Formation of Compound 0 of horseradish peroxidase was modeled by MD simulations (Zazza et al., 2008). It was concluded that no water was required to assist proton transfer from hydrogen peroxide to distal imidazole, and that the process is highly exothermic. Therefore, there are points of disagreement among the molecular dynamics, DFT and QM/MM treatments of Compound 0 formation.

A combination of DFT and QM/MM calculations has been used by the Barcelona group in their catalase investigations (see Chapter 15) (Rovira et al., 2006; Alfonso-Prieto et al., 2007).

Hydrogen Atom Transfers

Hydrogen atom transfers occur in every peroxidase cyclic reaction. A chemical bond has a lower zero point energy when deuterium is substituted for hydrogen. Classically, in a local mode approximation, this can lead to H/D kinetic isotope effects in the range of 6–7. If proton tunneling occurs then unusually large H/D isotope effects may be observed (Knapp and Klinman, 2002). Hydrogen atom transfers may occur as proton coupled electron transfer (PCET) or complete hydrogen atom transfers (HAT) or as completely decoupled electron transfer followed by proton transfer. A theory has been developed that connects PCET and HAT (Cukier, 2002). Which mechanism is applicable to peroxidase reactions is a problem that is just starting to be addressed. In Compounds I and II reductions by ferulic acid it was concluded that a PCET mechanism occurs, as described above (Derat and Shaik, 2006c).

REFERENCES

Alberty RA. 2007. Effects of pH in rapid-equilibrium enzyme kinetics. *J. Phys. Chem. B* **111**: 14064–14068.

Alfonso-Prieto M, Borovik A, Carpena X, Murshudov G, Melik-Adamyan W, Fita I, Rovira C, Loewen PC. 2007. The structures and electronic configuration of Compound I intermediates of *Helicobacter pylori* and *Penicillium vitale* catalases determined by X-ray crystallography and QM/MM density functional theory calculations. *J. Am. Chem. Soc.* **129**: 4193–4205.

Andreu R, Ferapontova EE, Gorton L, Calvente JJ. 2007. Direct electron transfer kinetics in horseradish peroxidase electrocatalysis. *J. Phys. Chem. B* **111**: 469–477.

Baron R, McCammon JA. 2007. Dynamics, hydration, and motional averaging of a loop-gated artificial protein cavity: the W191G mutant of cytochrome *c* peroxidase in water as revealed by molecular dynamics simulations. *Biochemistry* **46**: 10629–10642.

Beratan DN, Betts JN, Onuchic JN. 1991. Protein electron transfer rates set by the bridging secondary and tertiary structure. *Science* **252**: 1285–1288.

Beratan DN, Onuchic JN, Hopfield JJ. 1987. Electron tunneling through covalent and non-covalent pathways in proteins. *J. Chem. Phys.* **86**: 4488–4498.

Beratan DN, Onuchic JN, Winkler JR, Gray HB. 1992. Electron-tunneling pathways in proteins. *Science* **258**: 1740–1741.

Chalkias NG, Kahawong P, Giannelis EP. 2008. Activity increase of horseradish peroxidase in the presence of magnetic particles. *J. Am. Chem. Soc.* **130**: 2910–2911.

Chang Y-T, Veitch NC, Loew GH. 1998. A theoretical study of benzhydroxamic binding modes in horseradish peroxidase. *J. Am. Chem. Soc.* **120**: 5168–5178.

Chen H, Moreau Y, Derat E, Shaik S. 2008. Quantum mechanical/molecular mechanical study of the mechanisms of heme degradation by the enzyme heme oxygenase: the strategic function of the water cluster. *J. Am. Chem. Soc.* **130**: 1953–1965.

Collins JR, Du P, Loew GH. 1992. Molecular dynamics simulations of the resting and hydrogen peroxide-bound states of cytochrome *c* peroxidase. *Biochemistry* **31**: 11166–11174.

Colosi LM, Huang Q, Weber WJ, Jr. 2006. Quantitative structure-activity relationship based quantification of the impacts of enzyme-substrate binding on rates of peroxidase-mediated reactions of estrogenic phenolic chemicals. *J. Am. Chem. Soc.* **128**: 4041–4047.

Cukier RI. 2002. A theory that connects proton-coupled electron transfer and hydrogen-atom transfer reactions. *J. Phys. Chem B.* **106**: 1746–1757.

Czarnecki K, Proniewicz LM, Fujii H, Kincaid JR. 1996. Resonance Raman spectrum of a $^2A_{1u}$ ferryl porphyrin π-cation radical. *J. Am. Chem. Soc.* **118**: 4680–4685.

Dalosto SD, Prabhu NV, Vanderkooi JM, Sharp KA. 2003. A density functional theory study of conformers in the ferrous CO complex of horseradish peroxidase with distinct Fe-C-O configurations. *J. Phys. Chem. B.* **107**: 1884–1892.

de Visser SP. 2007. Preferential hydroxylation over epoxidation catalysis by a horseradish peroxidase mutant: a cytochrome P450 mimic. *J. Phys. Chem. B.* **111**: 12299–12302.

Derat E, Cohen S, Shaik S, Altun A, Thiel W. 2005. Principal active species of horseradish peroxidase, Compound i: a hybrid quantum mechanical/molecular mechanical study. *J. Am. Chem. Soc.* **127**: 13611–13621.

Derat E, Shaik S. 2006a. The Poulos-Kraut mechanism of Compound I formation in horseradish peroxidase: a QM/MM study. *J. Phys. Chem. B*, **110**: 10526–10533.

Derat E, Shaik S. 2006b. Two-state reactivity, electromerism, tautomerism, and "surprise" isomers in the formation of Compound II of the enzyme horseradish peroxidase from the principal species, Compound I. *J. Am. Chem. Soc.* **128**: 8185–8198.

Derat E, Shaik S. 2006c. An efficient proton-coupled electron-transfer process during oxidation of ferulic acid by horseradish peroxidase: coming full cycle. *J. Am. Chem. Soc.* **128**: 13940–13949.

Derat E, Shaik S, Rovira C, Vidossich P, Alfonso-Prieto M. 2007. The effect of a water molecule on the mechanism of formation of Compound 0 in horseradish peroxidase. *J. Am. Chem. Soc.* **129**: 6346–6347.

Dunford HB. 2001. How do enzymes work? Effect of electron circuits on transition state acid dissociation constants. *J. Biol. Inorg. Chem.* **6**: 819–822.

Dunford HB. 2003. Enzyme catalysis. Reversible flow of electron density. Poster presented at Enzyme Mechanisms Conference, Galveston Island, January 4–7.

Dunford HB, Hewson WD. 1977. Effect of mixed solvents on the formation of horseradish peroxidase Compound I. The importance of diffusion-controlled reactions. *Biochemistry* **16**: 2949–2957.

Dunford HB, Stillman JS. 1976. On the function and mechanism of action of peroxidases. *Coord. Chem. Rev.* **19**: 187–251.

Eigen M. 1974. Diffusion control in biochemical reactions. In: Kursunoglu B, Mintz SL, Widmayer SM, editors. *Quantum Statistical Mechanics in the Natural Sciences*. Plenum, New York.

Filizola M, Loew GH. 2000. Role of protein environment in horseradish peroxidase Compound I formation: molecular dynamics simulations of horseradish peroxidase-HOOH complex. *J. Am. Chem. Soc.* **122**: 18–25.

Fishelovitch D, Hazan C, Hirao H, Wolfson HJ, Nussinov R, Shaik S. 2007. QM/MM study of the active species of the human cytochrome P450 3A4, and the influence thereof of the multiple substrate binding. *J. Phys. Chem. B* **111**: 13822–13832.

Fitzgerald MM, Musah RA, McRee DE, Goodin D. 1996. A ligand-gated, hinged loop rearrangement opens a channel to a buried artificial protein cavity. *Nat. Struct. Biol.* **3**: 626–631.

Fujii H, Ichikawa K. 1992. Preparation and characterization of an A_{1u} oxoiron(IV) porphyrin π-cation radical complex. *Inorg. Chem.* **31**: 1110–1112.

Furse KE, Pratt DA, Porter NA, Lybrand TP. 2006. Molecular dynamics simulations of arachidonic acid complexes with COX-1 and COX-2: insights into equilibrium behavior. *Biochemistry* **45**: 3189–3205.

Gouterman M. 1978. Optical spectra and electronic structure of porphyrins and related rings. In: Dolphin D, editor. *The Porphyrins*. Volume IIIA. New York: Academic Press, pp. 1–165.

Gray HB, Winkler JR. 1996. Electron transfer in proteins. *Annu. Rev. Biochem.* **65**: 537–561.

Guallar V. 2008. Heme electron transfer in peroxidases: the propionate e-pathway. *J. Phys. Chem. B* **112**: 13460–13464.

Hanson LK, Chang CK, Davis MS, Fajer J. 1981. Electron pathways in catalase and peroxidase enzymic catalysis. Metal and macrocycle oxidations of iron porphyrins and chlorines. *J. Am. Chem. Soc.* **103**: 663–670.

Harris DL, Loew GH. 1996. Identification of putative peroxide intermediates of peroxidases by electronic structure and spectra calculations. *J. Am. Chem. Soc.* **118**: 10588–10594.

Heimdal J, Rydberg P, Ryde U. 2008. Protonation of the proximal histidine ligand in heme peroxidases. *J. Phys. Chem. B* **112**: 2501–2510.

Henriksen A, Smith AT, Gajhede M. 1999. The structures of horseradish peroxidase C-ferulic acid complex and ternary complex with cyanide suggest how peroxidases oxidize small phenolic substrates. *J. Biol. Chem.* **274**: 35005–35011.

Hopfield JJ. 1974. Electron transfer between biological molecules by thermally activated tunneling. *Proc. Natl. Acad. Sci. USA.* **71**: 3640–3644.

Hsuanyu Y, Dunford HB. 1992. Prostaglandin H synthase kinetics. The effect of substituted phenols on cyclooxygenase activity and the substituent effect on phenolic peroxidatic activity. *J. Biol. Chem.* **267**: 17649–17657.

Jensen KP, Ryde U. 2004. How O_2 binds to heme. Reasons for rapid binding and spin inversion. *J. Biol. Chem.* **279**: 14561–14569.

Jones P. 2001. Roles of water in heme peroxidase and catalase mechanisms. *J. Biol. Chem.* **276**: 13791–13796.

Jones P, Dunford HB. 2005. The mechanism of Compound I formation revisited. *J. Inorg. Biochem.* **99**: 2292–2298.

Kaposi AD, Prabhu NV, Dalosto SD, Sharp KA, Wright WW, Stavrov SS, Vanderkooi JM. 2003. Solvent dependent and independent motions of CO-horseradish peroxidase examined by infrared spectroscopy and molecular dynamics calculations. *Biophys. Chem.* **106**: 1–14.

Kaposi AD, Wright WW, Fidy J, Stavrov SS, Vanderkooi JM, Rasnik I. 2001. Carbonmonoxy horseradish peroxidase as a function of pH and substrate: influence of local electric fields on the optical and infrared spectra. *Biochemistry* **40**: 3483–3491.

Kettle AJ, Winterbourn CC. 2001. A kinetic analysis of the catalase activity of myeloperoxidase. *Biochemistry* **40**: 10204–10212.

Khajehpour M, Troxler T, Vanderkooi JM. 2003. Effect of protein dynamics upon reactions that occur in the heme pocket of horseradish peroxidase. *Biochemistry* **42**: 2672–2679.

Khopde SM, Priyadarsini KI. 2000. Application of Marcus theory of electron transfer for the reactions of HRP Compounds I and II and 2,4-disubstituted phenols. *Biophys. Chem.* **88**: 103–109.

Knapp MJ, Klinman JP. 2002. Environmentally coupled hydrogen tunneling. Linking catalysis to dynamics. *Eur. J. Biochem.* **269**: 3113–3121.

Kuramochi H, Noodleman L, Case DA. 1997. Density functional study on the electronic structures of model peroxidase Compounds I and II. *J. Am. Chem. Soc.* **119**: 11442–11451.

Laberge M, Osvath S, Fidy J. 2001. Aromatic substrate specificity of horseradish peroxidase C studied by a combined fluorescence line narrowing/energy minimization approach: the effect of localized side-chain organization. *Biochemistry* **40**: 9226–9237.

Loew G, Dupuis M. 1996. Structure of a model transient peroxide intermediate of peroxidases by ab initio methods. *J. Am. Chem. Soc.* **118**: 10584–10587.

Loew G, Dupuis M. 1997. Characterization of a resting state model of peroxidases by ab initio methods: optimized geometries, electronic structures, and relative energies of the sextet, quartet, and doublet spin states. *J. Am. Chem. Soc.* **119**: 9848–9851.

Loo S, Erman JE. 1977. The rate of reaction between cytochrome *c* peroxidase and hydrogen peroxide is not diffusion limited. *Biochim. Biophys. Acta* **481**: 279–282.

Marquez LA, Huang JT, Dunford HB. 1994. Spectral and kinetic studies on the formation of myeloperoxidase Compounds I and II: roles of hydrogen peroxide and superoxide. *Biochemistry* **33**: 1447–1454.

Marcus RA. 1993. Electron transfer reactions in chemistry: theory and experiment (Nobel lecture). *Angewandte Chemie, Int. Ed. in English* **32**: 1111–1122.

Marcus RA. 2007. Norman Sutin: a personal tribute. *J. Phys. Chem. B* **111**: 6594.

Nakatani H, Dunford HB. 1979. Meaning of diffusion-controlled association rate constants in enzymology. *J. Phys. Chem.* **83**: 2662–2665.

Nogrady T, Weaver DF. 2005. *Medicinal Chemistry. A Molecular and Biochemical Approach.* Third edition. Oxford University Press.

Onuchic JN, Beretan DN, Winkler JR, Gray HB. 1992. Pathway analysis of protein electron-transfer reactions. *Annu. Rev. Biophys. Biomol. Struct.* **21**: 349–377.

Page CC, Moser CC, Chen X, Dutton PL. 1999. Natural engineering principles of electron tunnelling in biological oxidation-reduction. *Nature* **402**: 47–52.

Picot D, Loll PJ, Garavito M. 1994. The X-ray crystal structure of the membrane protein prostaglandin H_2 synthase-1. *Nature* **367**: 243–249.

Purcell WL, Erman JE. 1976. Cytochrome *c* peroxidase catalyzed oxidations of substitution inert iron(II) complexes. *J. Am. Chem. Soc.* **98**: 7033–7037.

Rovira C, Alfonso-Prieto M, Biarnés X, Carpena X, Fita I, Loewen PC. 2006. A first principles study of the binding of formic acid in catalase complementing high resolution X-ray studies. *Chem. Phys.* **323**: 129–137.

Rovira C, Fita I. 2003. The proximal hydrogen-bonded residue controls the stability of the Compound II intermediate of peroxidases and catalases. *J. Phys. Chem. B.* **107**: 5300–5305.

Sakurada J, Sekiguchi R, Sato K, Hosoya T. 1990. Kinetic and molecular orbital studies on the rate of oxidation of monosubstituted phenols and analines by horseradish peroxidase Compound II. *Biochemistry* **29**: 4093–4098.

Siegbahn PEM, Borowski T. 2006. Modeling enzymatic reactions involving transition metals. *Acc. Chem. Res.* **39**: 729–738.

Smoluchowski Mv. 1917. Versuch einer mathematischen theorie der koagulationskinetik kolloider lösungen. *Z. physik. Chemie* **92**: 129–168.

Strickland N, Harvey JN. 2007. Spin-forbidden ligand binding to the ferrous-heme group: ab initio and DFT studies. *J. Phys. Chem. B* **111**: 841–852.

Sutin N. 2007. On the trail of the electron: bridges, tunnels, and tolls. *J. Phys. Chem. B* **111**: 6594–6603.

Terner J, Palaniappan V, Gold A, Weiss R, Fitzgerald MM, Sullivan AM, Hosten CM. 2006. Resonance Raman spectroscopy of oxoiron(IV) porphyrin π-cation and oxoiron(IV) hemes in peroxidase intermediates. *J. Inorg. Biochem.* **100**: 480–501.

Welinder KG, Bjørnholm B, Dunford HB. 1995. Functions of electrostatic potentials and conserved distal and proximal His-Asp H-bonding networks in haem peroxidases. *Biochem. Soc. Trans.* **23**: 257–262.

Wirstam M, Blomberg MRA, Siegbahn PEM. 1999. Reaction mechanism of Compound I formation in heme peroxidases: a density functional theory study. *J. Am. Chem. Soc.* **121**: 10178–10185.

Yandell JK, Yonetani T. 1983. Steady state kinetics of yeast cytochrome *c* peroxidase catalyzed oxidation of inorganic reductants by hydrogen peroxide. *Biochim. Biophys. Acta* **748**: 263–270.

Zazza C, Amadei A, Palma A, Sanna N, Tatoli S, Aschi M. 2008. Theoretical modeling of enzyme reactions: the thermodynamics of formation of Compound 0 in horseradish peroxidase, *J. Phys. Chem. B* **112**: 3184–3192.

Zelent B, Kaposi AD, Nucci NV, Sharp KA, Dalosto SD, Wright WW, Vanderkooi JM. 2004. Water channel of horseradish peroxidase studied by charge-transfer absorption band of ferric heme. *J. Phys. Chem. B.* **108**: 10317–10324.

Zerner M, Gouterman M. 1966. Porphyrins. IV. Extended Hückel calculations on transition metal complexes. *Theor. Chim. Acta* **4**: 44–63.

APPENDIX

PEROXIDASE STEADY-STATE KINETICS

MECHANISM AND STOICHIOMETRY

$$\begin{aligned}
P + H_2O_2 &\xrightarrow{k_1} \text{Compound I} + H_2O \\
\text{Compound I} + SH &\xrightarrow{k_2} \text{Compound II} + S^{\bullet} \\
\text{Compound I} + SH &\xrightarrow{k_3} P + H_2O + S^{\bullet}
\end{aligned}$$

$$H_2O_2 + 2\ SH = 2\ H_2O + 2\ S^{\bullet}$$

CONSERVATION

$$[E]_0 = [P] + [\text{Compound I}] + [\text{Compound II}]$$

RATE

$$v = -\frac{d[H_2O_2]}{dt} = k_1[H_2O_2][P]$$

STEADY STATE

$$-\frac{d[P]}{dt} = k_1[H_2O_2][P] - k_3[SH][\text{Compound II}] = 0$$

$$\frac{d[\text{Compound I}]}{dt} = k_1[H_2O_2][P] - k_2[SH][\text{Compound I}] = 0$$

$$\frac{d[\text{Compound II}]}{dt} = k_2[SH][\text{Compound I}] - k_3[SH][\text{Compound II}] = 0$$

By rearrangement of steady-state equations and substitution, the conservation relation can be put in the form

$$[P] = \frac{[E]_0}{1 + \dfrac{(k_1k_2 + k_1k_3)[H_2O_2]}{k_2k_3[SH]}}$$

Substitution into the rate equation

$$v = k[H_2O_2][P]$$

leads to

$$v = \frac{k_1[H_2O_2][E]_0}{1 + \dfrac{(k_1k_2 + k_1k_3)[H_2O_2]}{k_2k_3[SH]}}$$

Typically $k_2 > k_3$

$$k = \frac{v}{[E]_0} = \frac{k_1[H_2O_2]}{1 + \dfrac{k_1[H_2O_2]}{k_3[SH]}} = \frac{k_{cat}}{1 + \dfrac{K_M}{[SH]}}$$

Therefore, $k_{cat} = k_1[H_2O_2]$ and $K_M = k_1[H_2O_2]/k_3$.

CHAPTER 11

CLASS I: ASCORBATE PEROXIDASE

Class I peroxidases are those with no carbohydrate, no calcium ions, no cystine bridges, and no signal peptide for excretion. The monumental amount of work on yeast cytochrome *c* peroxidase is summarized in Chapters 6 and 7. In this chapter another Class I peroxidase is described.

INTRODUCTION

Identification of ascorbate peroxidase occurred in 1979, and work on highly purified enzyme began about 1989. Therefore this enzyme is one of the more recent additions to the peroxidase superfamily. Ascorbate peroxidases from three principal sources have been studied: chloroplasts, plant cytosol, and root nodules with nitrogenase activity.

The first report was an ascorbate peroxidase isolated from spinach chloroplasts (Groden and Beck, 1979; Nakano and Asada, 1981, 1987). Soluble ascorbate peroxidase activity was detected in pea leaves. Use of the enzyme in an ascorbate assay was described (Kelly and Latzko, 1979; Gerbling et al., 1984). The soluble enzyme is now known as cytosolic ascorbate peroxidase. Ascorbate peroxidase was also detected in *Euglena gracilis z* (Shigeoka et al., 1980), soybean root nodules (Puppo et al., 1980; Dalton et al., 1986, 1987), cyanobacteria (Tel-Or et al., 1986), and tea leaves (Chen and Asada, 1989).

The role of ascorbate peroxidase now appears to be clear. It is to catalyze the removal of excess hydrogen peroxide by L-ascorbate (vitamin C):

$$H_2O_2 + 2\text{Ascorbate} \longrightarrow H_2O + 2\text{Monodehydroascorbate}$$

Peroxidases and Catalases: Biochemistry, Biophysics, Biotechnology, and Physiology, Second Edition.
By H. Brian Dunford

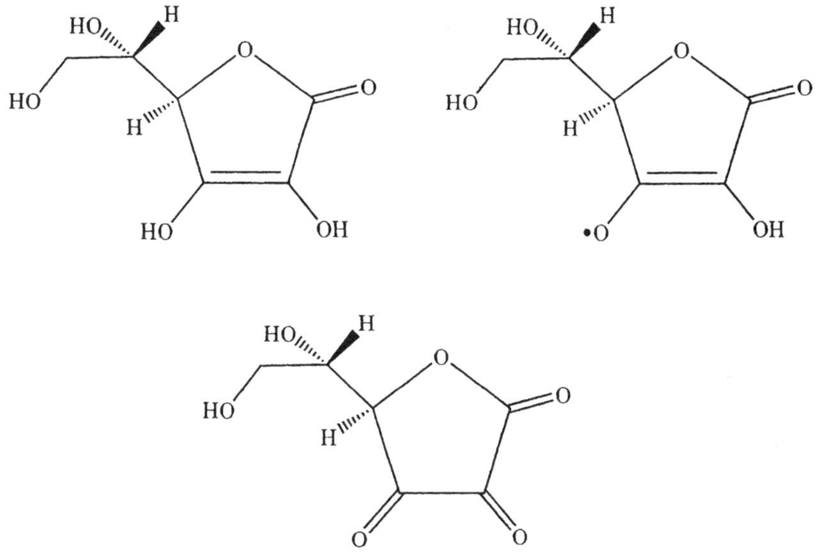

Figure 11.1. Structures of ascorbic acid, dehydroascorbic acid, and the free radical monodehydroascorbic acid.

The structures of ascorbic acid, the free radical monodehydroascorbic acid, and the final oxidation product dehydroascorbic acid, are shown in Figure 11.1.

The monodehydrocascorbate, a fairly unreactive free radical, may be further oxidized to dehydroascorbate or reduced back to ascorbate. Two free radicals can disproportionate to form one ascorbate and one dehydroascorbate. Thus L-ascorbic acid (vitamin C) is an ideal water-soluble free radical scavenger.

An alternative use of ascorbate deserves discussion, although in a system in which a role for ascorbate peroxidase has not been defined. When vitamin C is administered intraveneously rather than orally much higher concentrations are obtained in blood. Formation of ascorbate radicals can reduce metal ions leading through superoxide to hydrogen peroxide formation. This is rapidly scavenged in blood by catalase and glutathione peroxidase. Inside cells, particularly cancer cells, which rely on glycolysis rather than oxidative phosphorylation, the hydrogen peroxide may lead to cell death (Chen et al., 2007). Linus Pauling would be thrilled.

SEQUENCING AND CLONING

Tea leaf chloroplastic ascorbate peroxidase was partially sequenced and sequence similarities to yeast cytochrome *c* peroxidase were noted (Chen et al., 1992). The first major breakthrough in ascorbate peroxidase research was the purification to homogeneity of pea cytosolic ascorbate peroxidase, followed by cloning and complete amino acid sequence determination (Mittler and Zilinskas, 1991a, 1991b). This led to its classification as a Class I peroxidase (Welinder, 1992). The enzyme is a noncovalently bonded homodimer with a subunit molecular weight of 29,500.

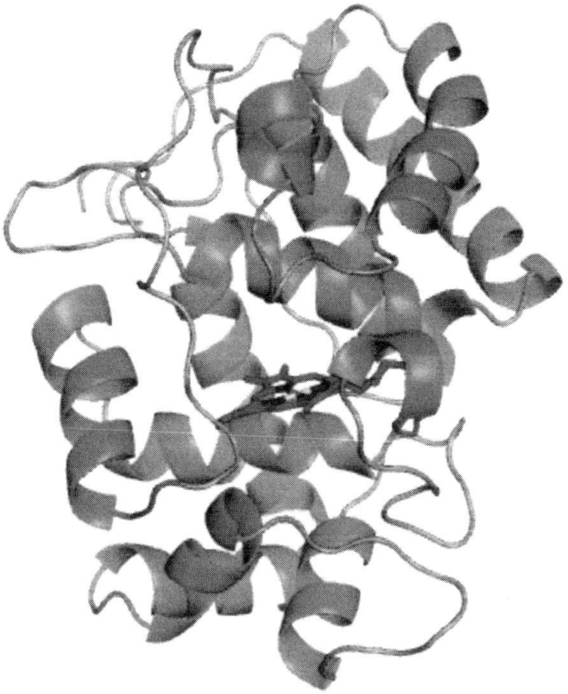

Figure 11.2. Structure of pea cytosolic ascorbate peroxidase. (Kindly supplied by Dr. Tom Poulos.)

An alternative cloning procedure was developed, followed by determination of the crystal structure (Patterson and Poulos, 1994, 1995). A representation of the ascorbate peroxidase structure is shown in Figure 11.2.

Database analyses retrieved 52 expressed sequence tags coding for ascorbate peroxidase in *Arabidopsis thaliana*. Analysis of sequence data for all known ascorbate peroxidases resulted in the following classification: five cytosolic ascorbate peroxidases, three of them membrane-bound; and two types of chloroplast ascorbate peroxidases. Key residues are conserved in the seven enzymes. A change from Phe175 to Trp175 may explain the higher ascorbate activity of the chloroplast enzymes. The proximal potassium ion binding site of the pea cytosolic enzyme is preserved in *A. thaliana*. "The data support the hypothesis that present-day peroxidases evolved by an earlier gene duplication event" (Jespersen et al., 1997, p. 305). Phylogenetic relationships were established for 60 members of peroxidase Class I (Zámocký, 2004).

PROPERTIES, REACTIONS, AND INTERMEDIATE COMPOUNDS

Ascorbate peroxidase has a tryptophan residue on the proximal side. The corresponding residue in the yeast enzyme is the site of the unpaired electron in Compound I. Surprisingly Compound I of ascorbate peroxidase has a conventional porphyrin

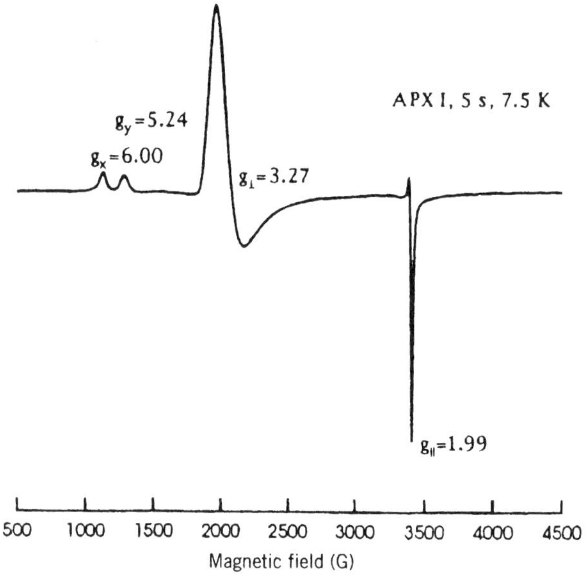

Figure 11.3. ESR spectrum of ascorbate peroxidase Compound I obtained at 7.5 K. (Reproduced with permission from Patterson WR, Poulos TL, Goodin DB. 1995. *Biochemistry* **34**: 4342–4345.)

π-cation radical, as detected by ESR spectra on the recombinant enzyme (Patterson et al., 1995), and optical spectra of the wild-type enzyme (Marquez et al., 1996). See discussion of crystal structures, below. The Compound I ESR spectrum is shown in Figure 11.3.

The rate constant for ascorbate peroxidase Compound I formation is 8.0×10^7 $M^{-1} s^{-1}$ at pH 7.8 and 20°C. The rate falls off at lower pH, with the inflection point at pH 5, somewhat higher than for the yeast enzyme. Ascorbate has markedly different reactivities toward different peroxidase Compounds I. Rate constants are 8.0×10^7 $M^{-1} s^{-1}$ for ascorbate peroxidase (Marquez et al., 1996), essentially zero for the yeast enzyme (Yonetani and Ray, 1965), and 2.3×10^5 $M^{-1} s^{-1}$ for horseradish peroxidase (Zhang, 1993).

Using the cyanide complex as a Compound I model, the binding of ascorbate to ascorbate peroxidase was studied by nuclear magnetic resonance (NMR) spectroscopy and molecular modeling. The ascorbate binding site is ~1 nm from the heme iron and near the exposed C20 heme position. Suicide inactivation using phenylhydrazine indicated reactivity at the C20 position. A possible functional role for Ala134 in ascorbate binding is implicated (Hill et al., 1997). This position is occupied by proline in the yeast enzyme, which may account for its low reactivity toward ascorbate. The modeled ascorbate binding site of ascorbate peroxidase is shown in Figure 11.4. Further evidence of the differences between cytosolic ascorbate peroxidase and yeast cytochrome *c* peroxidase was obtained from resonance Raman and electronic absorption spectroscopies. The two enzymes have markedly different

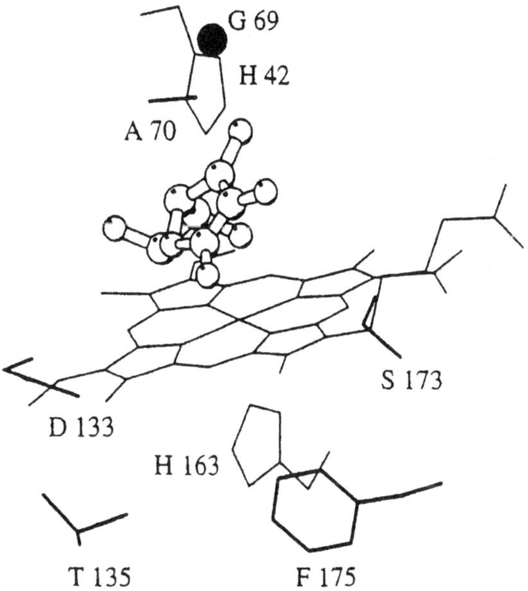

Figure 11.4. Proposed ascorbate binding site on ascorbate peroxidase, based on molecular modeling and NMR and crystallographic restraints. (Reproduced with permission from Hill AP et al., 1997. *Eur. J. Biochem.* **248**: 347–354.)

orientations of the heme vinyl substituents because of different steric constraints imposed by the protein matrix (Nissum et al., 1998).

Salicyclic acid was found to be a slow-reacting substrate for thylakoid-bound chloroplastic ascorbate peroxidase from tea leaves (Kvaratskhelia et al., 1997), and not an inhibitor as had previously been claimed (Durner and Klessig, 1995).

A large number of valuable contributions have been made by Emma Lloyd Raven and coworkers (Hill et al., 1997). The kinetics and mechanism of *p*-cresol oxidation were studied and shown to be similar to the corresponding horseradish peroxidase oxidation. Reaction between Compound II and *p*-cresol is rate-controlling. Two *p*-cresol binding sites exist (Çelik et al., 2000). Inactivation of ascorbate peroxidase by hydrogen peroxide was investigated and shown to be unique. Approximately 20% activity remains after the inactivation process. No catalase activity occurs (Hiner et al., 2000). With hydrogen peroxide as the only substrate an unstable Compound I containing a porphyrin π-cation radical decays to form a tryptophan radical (Hiner et al., 2001). The reactions of a series of thioethers with recombinant ascorbate peroxidase and the W41A mutant showed enhanced reactivity of the mutant because of a less restricted heme site. Enhanced enantioselectivity was observed for the mutant (Çelik et al., 2001).

The W41A mutation of ascorbate peroxidase causes distal His42 binding to the heme ferric ion (Badyal et al., 2006). When the iron is reduced to the ferrous state the His dissociates (Badyal et al., 2008).

Autocatalytic formation of covalent attachments to heme were demonstrated. A methionine-heme linkage led to formation of a green heme in the reaction of the

S160M mutant with hydrogen peroxide (Metcalfe et al., 2004). Covalent linkages involving Trp41 and tyrosine in the S160Y mutant were also observed (Pipirou et al., 2007a, 2007b). Oxidation of a cysteine residue, leading to enzyme inactivation, was detected in a red alga and stromal tobacco ascorbate peroxidases by use of a radical scavenger and a spin trap (Kitajima et al., 2008).

Redox potentials for wild-type and mutant ascorbate peroxidase couples were determined using an optically transparent thin layer electrode (OTTLE) using mediators to facilitate equilibria, or by a xanthine-xanthine oxidase method in which a dye of known potential is simultaneously oxidized. Reduction potentials for the wild-type couples are $E^{\circ\prime}$ (Compound I/Compound II) = 1156 mV, $E^{\circ\prime}$ (Compound II/Fe^{3+}) = 752 mV, $E^{\circ\prime}$ (Compound I/Fe^{3+}) = 954 mV. Thus Compound I has a much higher reduction potential than Compound II, unlike horseradish peroxidase where the two potentials are nearly identical (Efimov et al., 2007).

A peroxidase isolated from tea leaves has very high reactivity toward ascorbate, little reactivity with phenols, yet has more sequence analogy with Class III peroxidases (Heering et al., 2001). Germinating seeds have specialized peroxisomes known as glyoxysomes. An ascorbate peroxidase was identified in the glyoxysomal membrane of castor beans (Karyotou and Donaldson, 2005). The parasite *Trypanosoma cruzi*, which has no catalase or glutathione peroxidase, possesses ascorbate peroxidase (Wilkinson et al., 2002).

Ascorbate peroxidase accumulates in *Arabidopsis thaliana* when the plant is exposed to heat and drought stress, suggesting that the enzyme plays a role in the adaptation of the plant to stress (Koussevitzky et al., 2008).

CRYSTAL STRUCTURES

Two definitive crystal structures cast new light on the mechanism of action of ascorbate peroxidase, its differences with yeast cytochrome *c* peroxidase, and indeed on the whole field of peroxidase research. The structures of both recombinant pea and soybean ascorbate peroxidase and their ascorbate complexes show that ascorbate binds near the γ-heme edge, not the δ-heme edge (Bursey and Poulos, 2000; Sharp et al., 2004). There are five ordered water molecules directly or indirectly hydrogen bonded to Arg172 and the heme 6-propionate in the enzyme, which are displaced upon ascorbate binding. The ascorbate can readily react with the porphyrin π-cation radical because it is attached to it. It had previously been postulated that a metal ion, assumed to be potassium, near the proximal Trp prevented formation of a Trp radical in ascorbate peroxidase Compound I (Patterson and Poulos, 1995). It appears that this proposed metal ion site is occupied by water. Ascorbate is coupled to the heme group through the heme 6-propionate, providing a direct route for reaction with a porphyrin π-cation radical Compound I and eliminating the need for a Trp179 cation radical (Sharp et al., 2004). An important role for propionate side chains in electron transfer has been deduced in a theoretical study (Guallar, 2008).

The crystal structure of the recombinant soybean ascorbate peroxidase–salicylhydroxamic acid complex confirms that binding of aromatic substrates occurs near the δ-heme edge. Thus there are two binding sites for substrates in ascorbate peroxidase.

It is pointed out that proton transfer pathways are not nearly as well understood as electron transfer pathways. (There is still much to learn about the latter.) A suggestion is made that the distal arginine may be involved in proton transfer in redox reactions involving substrates bound near either of the δ- and γ-heme edges. Finally the valid point is made that yeast cytochrome *c* peroxidase is an atypical peroxidase, and that ascorbate peroxidase provides a better universal peroxidase model (Sharp et al., 2004). Yeast cytochrome *c* peroxidase has also been modified to accommodate an ascorbate binding site (Meharenna et al., 2008; Murphy et al., 2008). The ascorbate binding site was further explored by a variety of techniques, including site-directed mutants (Macdonald et al., 2006).

The crystal structures of the tuberculosis drug isoniazid complexes with ascorbate peroxidase and yeast cytochrome *c* peroxidase were obtained. These results are discussed in the following chapter on Class I catalase-peroxidase (Metcalfe et al., 2008).

REFERENCES

Badyal SK, Joyce MG, Sharp KH, Seward HE, Mewies M, Basran J, Macdonald IK, Moody PCE, Raven EL. 2006. Conformational mobility in the active site of a heme peroxidase. *J. Biol. Chem.* **281**: 24512–24520.

Badyal SK, Metcalfe CL, Basran J, Efimov I, Moody PCE, Raven EL. 2008. Iron oxidation state modulates active site structure in a heme peroxidase. *Biochemistry* **47**: 4403–4409.

Bursey EH, Poulos TL. 2000. Two substrate binding sites in ascorbate peroxidase: the role of arginine 172. *Biochemistry* **39**: 7374–7379.

Çelik A, Cullis PM, Raven EL. 2000. Catalytic oxidation of *p*-cresol by ascorbate peroxidase. *Arch. Biochem. Biophys.* **373**: 175–181.

Çelik A, Cullis PM, Sutcliffe MJ, Sangar R, Raven EL. 2001. Engineering the active site of ascorbate peroxidase. *Eur. J. Biochem.* **268**: 78–85.

Chen G-X, Asada K. 1989. Ascorbate peroxidase in tea leaves: occurrence of two isozymes and the differences in their enzymatic and molecular properties. *Plant Cell Physiol.* **30**: 987–998.

Chen G-X, Sano S, Asada K. 1992. The amino acid sequence of ascorbate peroxidase from tea has a high degree of homology to that of cytochrome *c* peroxidase from yeast. *Plant Cell Physiol.* **33**: 109–116.

Chen Q, Espey MG, Sun AY, Lee J-H, Krishna MC, Shacter E, Choyke PL, Pooput C, Kirk KL, Buettner GR, Levine M. 2007. Ascorbate in pharmacologic concentrations selectively generates ascorbate radical and hydrogen peroxide in extracellular fluid *in vivo*. *Proc. Natl. Acad. Sci. USA* **104**: 8749–8754.

Dalton DA, Hanus FJ, Russell SA, Evans HJ. 1987. Purification, properties, and distribution of ascorbate peroxidase in legume root nodules. *Plant Physiol.* **83**: 789–794.

Dalton DA, Russell SA, Hanus FJ, Pascoe GA, Evans HJ. 1986. Enzymatic reactions of ascorbate and glutathione that prevent peroxide damage in soybean root nodules. *Proc. Natl. Acad. Sci. USA* **83**: 3811–3815.

Durner J, Klessig DF. 1995. Inhibition of ascorbate peroxidase by salicyclic acid and 2,6-dichloroisonicotinic acid, two inducers of plant defence responses. *Proc. Natl. Acad. Sci. USA* **92**: 11312–11316.

Efimov I, Papadopoulou ND, McLean KJ, Badyal SK, Macdonald IK, Munro AW, Moody PCE, Raven EL. 2007. The redox properties of ascorbate peroxidase. *Biochemistry* **46**: 8017–8023.

Gerbling K-P, Kelly GJ, Fischer K-H, Latzko E. 1984. Partial purification and properties of soluble ascorbate peroxidases from pea leaves. *J. Plant Physiol.* **115**: 59–67.

Groden D, Beck E. 1979. H_2O_2 destruction by ascorbate-dependent systems from chloroplasts. *Biochim. Biophys. Acta* **546**: 426–435.

Guallar V. 2008. Heme electron transfer in peroxidases: the propionate e-pathway. *J. Phys. Chem. B* **112**: 13460–13464.

Heering HA, Jansen MAK, Thorneley RNF, Smulevich G. 2001. Cationic ascorbate peroxidase isoenzyme II from tea: structural insights into the heme pocket of a unique hybrid peroxidase. *Biochemistry* **40**: 10363–10370.

Hill AP, Modi S, Sutcliffe MJ, Turner DD, Gilfoyle DJ, Smith AT, Tam BM, Lloyd E. 1997. Chemical, spectroscopic and structural investigation of the substrate-binding site in ascorbate peroxidase. *Eur. J. Biochem.* **248**: 347–354.

Hiner ANP, Martinez JI, Arnao MB, Acosta M, Turner DD, Raven EL, Rodríguez-López JN. 2001. Detection of a tryptophan radical in the reaction of ascorbate peroxidase with hydrogen peroxide. *Eur. J. Biochem.* **268**: 3091–3098.

Hiner ANP, Rodríguez-López JN, Arnao MB, Raven EL, García-Cánovas F, Acosta M. 2000. Kinetic study of the inactivation of ascorbate peroxidase by hydrogen peroxide. *Biochem. J.* **348**: 321–328.

Jespersen HM, Kjaersgård IVH, Østergaard L, Welinder KG. 1997. From sequence analysis of three novel ascorbate peroxidases from *Arabidopsis thaliana* to structure, function and evolution of seven types of ascorbate peroxidase. *Biochem. J.* **326**: 305–310.

Karyotou K, Donaldson RP. 2005. Ascorbate peroxidase, a scavenger of hydrogen peroxide in glyoxysomal membranes. *Arch. Biochem. Biophys.* **434**: 248–257.

Kelly GJ, Latzko E. 1979. Soluble ascorbate peroxidase: detection in plants and use in vitamin C estimation. *Naturwissenschaften* **66**: 617–618.

Kitajima S, Kurioka M, Yoshimoto T, Shindo M, Kanaori K, Tajima K, Oda K. 2008. A cysteine residue near the propionate side chain of heme is the radical site in ascorbate peroxidase. *FEBS J.* **275**: 470–480.

Koussevitzky S, Suzuki N, Huntington S, Armijo L, Sha W, Cortes D, Shulaev V, Mittler R. 2008. Ascorbate peroxidase plays a key role in the response of *Arabidopsis thaliana* to stress combination. *J. Biol. Chem.* **283**: 34197–34203.

Kvaratskhelia M, George SJ, Thorneley RNF. 1997. Salicylic acid is a reducing substrate and not an effective inhibitor of ascorbate peroxidase. *J. Biol. Chem.* **272**: 20998–21001.

Macdonald IK, Badyal SK, Ghamsari L, Moody PCE, Raven EL. 2006. Interaction of ascorbate peroxidase with substrates: a mechanistic and structural analysis. *Biochemistry* **45**: 7808–7817.

Marquez LA, Quitoriano M, Zilinskas BA, Dunford HB. 1996. Kinetic and spectral properties of pea cytosolic ascorbate peroxidase. *FEBS Lett.* **389**: 153–156.

Meharenna YT, Oertel P, Bhaskar B, Poulos TL. 2008. Engineering ascorbate peroxidase activity into cytochrome *c* peroxidase. *Biochemistry* **47**: 10324–10332.

Metcalfe C, Macdonald IK, Murphy EJ, Brown KA, Raven EL. 2008. The tuberculosis prodrug isoniazid bound to activating peroxidases. *J. Biol. Chem.* **283**: 6195–6200.

Metcalfe CL, Ott M, Patel N, Singh K, Mistry SC, Goff HM, Raven EL. 2004. Autocatalytic formation of green heme: evidence for H_2O_2-dependent formation of a covalent methionine-heme linkage in ascorbate peroxidase. *J. Am. Chem. Soc.* **126**: 16242–16248.

Mittler R, Zilinskas BA. 1991a. Purification and characterization of pea cytosolic ascorbate peroxidase. *Plant Physiol.* **97**: 962–968.

Mittler R, Zilinskas BA. 1991b. Molecular cloning and nucleotide sequence analysis of a cDNA encoding pea cytosolic ascorbate peroxidase. *FEBS Lett.* **289**: 257–259.

Murphy EJ, Metcalfe CL, Basran J, Moody PCE, Raven EL. 2008. Engineering the substrate specificity and reactivity of a heme protein: creation of an ascorbate binding site in cytochrome *c* peroxidase. *Biochemistry* **47**: 13933–13941.

Nakano Y, Asada K. 1981. Hydrogen peroxide is scavenged by ascorbate-specific peroxidase in spinach chloroplasts. *Plant Cell Physiol.* **22**: 867–880.

Nakano Y, Asada K. 1987. Purification of ascorbate peroxidase in spinach chloroplasts. *Plant Cell Physiol.* **28**: 131–140.

Nissum M, Neri F, Mandelman D, Poulos TL, Smulevich G. 1998. Spectroscopic characterization of recombinant pea cytosolic ascorbate peroxidase: similarities and differences with cytochrome *c* peroxidase. *Biochemistry* **37**: 8080–8087.

Patterson WR, Poulos TL. 1994. Characterization and crystallization of recombinant pea cytosolic ascorbate peroxidase. *J. Biol. Chem.* **269**: 17020–17024.

Patterson WR, Poulos TL. 1995. Crystal structure of recombinant pea cytosolic ascorbate peroxidase. *Biochemistry* **34**: 4331–4341.

Patterson WR, Poulos TL, Goodin DB. 1995. Identification of a porphyrin π cation radical in ascorbate peroxidase Compound I. *Biochemistry* **34**: 4342–4345.

Pipirou Z, Bottrill AR, Metcalfe CM, Mistry SC, Badyal SK, Rawlings BJ, Raven EL. 2007a. Autocatalytic formation of a covalent link between tryptophan 41 and the heme in ascorbate peroxidase. *Biochemistry* **46**: 2174–2180.

Pipirou Z, Bottrill AR, Svistunenko DA, Efimov, Basran J, Mistry SC, Cooper CE, Raven EL. 2007b. The reactivity of heme in biological systems: autocatalytic formation of both tyrosine-heme and trptophan-heme covalent links in a single protein architecture. *Biochemistry* **46**: 13269–13278.

Puppo A, Rigaud J, Job D, Ricard J, Zeba B. 1980. Peroxidase content of soybean root nodules. *Biochim. Biophys Acta* **614**: 303–312.

Sharp KH, Mewies M, Moody PCE, Raven EL. 2003. Crystal structure of the ascorbate peroxidase-ascorbate complex. *Nature Struct. Biol.* **10**: 303–307.

Sharp KH, Moody PCE, Brown KA, Raven EL. 2004. Crystal structure of the ascorbate peroxidase-salicylhydroxamic acid complex. *Biochemistry* **43**: 8644–8651.

Shigeoka S, Nakano Y, Kitaoka S. 1980. Purification and some properties of L-ascorbic acid specific peroxidase in *Euglena gracilis z*. *Arch. Biochem. Biophys.* **201**: 121–127.

Tel-Or E, Huflejt ME, Packer L. 1986. Hydroperoxide metabolism in cyanobacteria. *Arch. Biochem. Biophys.* **246**: 396–402.

Welinder KG. 1992. Superfamily of plant, fungal and bacterial peroxidases. *Curr. Opinion Struct. Biol.* **2**: 388–393.

Wilkinson SR, Obado SO, Mauricio IL, Kelly JM. 2002. *Trypanosoma cruzi* expresses a plant-like ascorbate-dependent hemoperoxidase localized to the endoplasmic reticulum. *Proc. Natl. Acad. Sci. USA* **99**: 13453–13458.

Yonetani T, Ray GS. 1965. Studies on cytochrome *c* peroxidase. I. Purification and some properties. *J. Biol. Chem.* **240**: 4503–4508.

Zámocký M. 2004. Phylogenetic relationships in class I of the superfamily of bacterial, fungal, and plant peroxidases. *Eur. J. Biochem.* **271**: 3297–3309.

Zhang H. 1993. M.Sc. thesis. University of Alberta.

CHAPTER 12

CATALASE-PEROXIDASES AND *MYCOBACTERIUM TUBERCULOSIS*

INTRODUCTION

The unique bacterial catalase-peroxidases were identified in 1979 (Claiborne and Fridovich, 1979; Claiborne et al., 1979). Two enzymes were isolated from *Escherichia coli* and labeled hydroperoxidases I and II (HP-I and HP-II). HP-I is a tetramer of molecular weight 337,000 with two heme *b* (ferriprotoporphyrin IX) molecules per tetramer. It exhibited both classic catalase and peroxidase activities. HP-II is also a tetramer, molecular weight 312,000. It possessed greater catalase activity than HP-I, appeared to have a different primary structure and heme, and had essentially no peroxidase activity. The gene for HP-I was identified and labeled *KatG*, a name now used for the enzyme (Loewen et al., 1985). The nucleotide sequence followed (Triggs-Raine et al., 1988). It predicted a 726 amino acid protein, confirmed by amino acid sequencing of peptides. No homology was found with other catalases.

The term *bifunctional catalase-peroxidase hydroperoxidase I* (Loewen et al., 1985) was reduced to *catalase-peroxidase* (Nadler et al., 1986), the name that is used today interchangeably with *KatG*. The latter workers found catalase-peroxidases in *Rhodopseudomonas capsulate* and *Rhodospirillum rubrum*, similar to the catalase-peroxidase of *E. coli*. Of three catalases found in *Klebsiella pneumoniae*, one is a catalase-peroxidase (Goldberg and Hochman, 1989; Hochman and Goldberg, 1991).

The gene for a thermostable peroxidase in *Bacillus stearothermophilus* was found to have 48% homology to that for the *E. coli* catalase-peroxidase (Loprasert et al., 1989). Catalase-peroxidases were found in *Halobacterium halobium* L-33 (Fukumori et al., 1985), in the gram-negative bacterium *Vitreoscilla* (Abrams and Webster, 1990), in *Streptomyces* sp. (Youn et al., 1995), in *Mycobacterium smegmatis* (Marcinkeviciene et al., 1995), and in *Sinorhizobium meliloti* (Ardissone et al., 2004).

Peroxidases and Catalases: Biochemistry, Biophysics, Biotechnology, and Physiology, Second Edition.
By H. Brian Dunford

The gene for catalase-peroxidase from a bacterium that learned to live in very high salt concentrations, *Haloarcula marismortui*, was sequenced and the results compared to sequences for 10 other catalase-peroxidases (Cannac-Caffrey, 1998). The *Mycobacterium tuberculosis KatG* gene has been isolated and inserted via a plasmid into *E. coli*, where the catalase-peroxidase has been overexpressed (Johnsson et al., 1997; Nagy et al., 1997).

Catalase-peroxidases from seven different organisms were found to exhibit similar rates of both catalase and peroxidase activity with similar pH dependencies (Singh et al., 2004b). A molecular switch controlled by an arginine residue 20 Å from the heme has been found to facilitate the catalase activity in catalase-peroxidases (Carpena et al., 2005). Its influence on the heme is imparted by association or dissociation from a tyrosinate residue (Carpena et al., 2005). A total of five residues in *Burkholderia pseudomallei* catalase-peroxidase were established by site-directed mutagenesis to be essential for catalase but not peroxidase activity (Deemagarn et al., 2007).

STRUCTURES OF CATALASE-PEROXIDASES

There is agreement that catalase-peroxidases have a dimeric structure consisting of two very similar covalently linked monomers of molecular weight ~40,000.

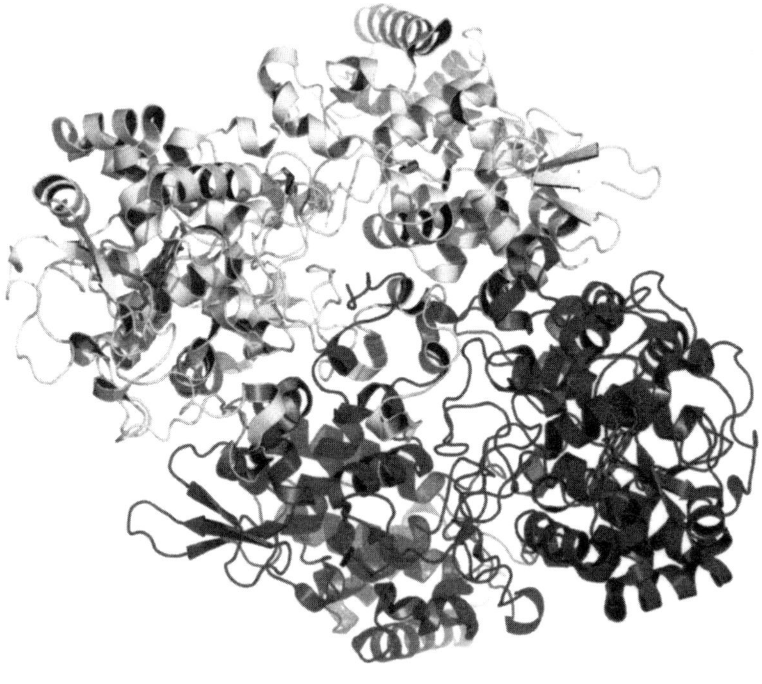

Figure 12.1. Dimeric structure of catalase-peroxidase from *Mycobacterium tuberculosis*. (Kindly supplied by Andrea Gumiero and Dr. Emma Lloyd Raven.)

A single heme *b* prosthetic group is located on the N-terminal monomer. Gene duplication is responsible for the dimer structure (Welinder, 1991). Two catalase-peroxidases of MW ~80,000 undergo homodimerization so the final MW is ~160,000. The catalase-peroxidases belong in Class I of plant, fungal, and bacterial peroxidases, a group that also contains yeast cytochrome *c* peroxidase and ascorbate peroxidase (Welinder, 1992).

The combined groups of Ignacio Fita at the Parc Científic de Barcelona and Peter Loewen at the University of Manitoba have contributed greatly to our knowledge of both catalase-peroxidases and catalases. The crystal structures of four different catalase-peroxidases have been determined: that of *Haloarcula marismortui* (Yamada et al., 2002), *Synechococcus* PCC 7942 (Wada et al., 2002), *Burkholderia pseudomallei* (Carpena et al., 2003), and of the C-terminal domain of *Escherichia coli*

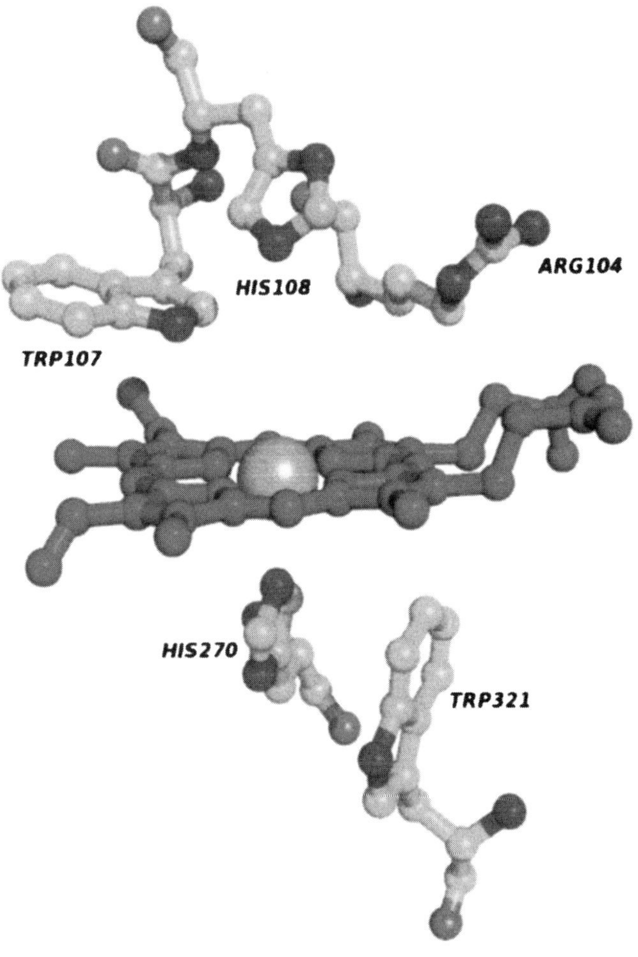

Figure 12.2. Heme environment of one monomer of *M. tuberculosis* catalase-peroxidase. (Kindly supplied by Andrea Gumiero and Dr. Emma Lloyd Raven.)

catalase-peroxidase (Carpena et al., 2004). Extra loops not found in other Class I peroxidases play a role in the main access channel to the heme. Because of its role in human health, greatest interest is shown in the structure of *Mycobacterium tuberculosis* catalase-peroxidase determined by Katherine Brown and coworkers (Bertrand et al., 2004).

A representation of the dimeric structure of *M. tuberculosis* catalase-peroxidase is shown in Figure 12.1, and of the heme active site in Figure 12.2.

Crystallization of the recombinant *M. tuberculosis* peroxidase was aided by addition of extra heme *b*, which increased the heme content to its stoichiometric level. Peptide chains, not found in monomeric catalase-peroxidases, form hooks that hold the homodimer structure together. Extra loops were also identified in the *E. coli* catalase-peroxidase (Li and Goodwin, 2004). The isoniazid binding site is located near the δ-meso heme edge, according to NMR, modeling, and computational studies on both Class I peroxidases and horseradish peroxidase (Pierattelli et al., 2004).

ISONIAZID AND OTHER REACTANTS OF CATALASE-PEROXIDASES

Because of the resurgence of tuberculosis, once thought to be obliterated, studies on the peroxidases involved have a new urgency. "Human mycobacterial pathogens . . . are phagocitized by, and can proliferate inside, macrophages, and catalase-peroxidases help to protect the bacilli from the deleterious effects of macrophage-generated hydrogen peroxide" (Marcinkeviciene et al., 1995, p. 22290). The *KatG* catalase-peroxidases of *Mycobacterium tuberculosis* are constitutively expressed, and, utilizing hydrogen peroxide, metabolize the antituberculosis drug isoniazid (a contraction of isonicotinic acid hydrazide; Hillar and Loewen, 1995). An early study of the reaction of isoniazid with horseradish peroxidase showed that the reaction can proceed without hydrogen peroxide, that oxygen is consumed, and that addition of hydrogen peroxide abolished oxygen consumption (Shoeb et al., 1985a). The same authors demonstrated that crude extracts of *Mycobacterium tuberculosis* would catalyze similar reactions, and that the isoniazid-resistant strain of *M. tuberculosis*, which has no peroxidase, did not react (Shoeb et al., 1985b). A scheme for the reaction of isoniazid with the *Mycobacterium tuberculosis* catalase-peroxidase has been published, and is reproduced in Figure 12.3 (Johnsson and Schultz, 1994). The isonicotinic acyl radical appears to be the activated reagent and its target is now known. It is the NADH in the active site of a long-chain enoyl-acyl carrier protein reductase (InhA), as shown in Figure 12.4 (Rozwarski et al., 1998; Lei et al., 2000; Rawat et al., 2003; Zhao et al., 2006). Cell wall synthesis is dependent on InhA, so its inactivation leaves the bacterium open to attack.

The hydrazide group of isoniazid reacts with peroxidases similarly to the -$NHNH_2$ group of phenylhydrazine, a known peroxidase inhibitor (Ebihara, 1939). The following reaction sequence has been proposed for phenylhydrazine (DePillis and Ortiz de Montellano, 1989):

$$\text{Ph-NHNH}_2 \longrightarrow \text{Ph-N=NH} \longrightarrow \text{Ph-N=N}^{\bullet} \longrightarrow \text{Ph}^{\bullet}$$

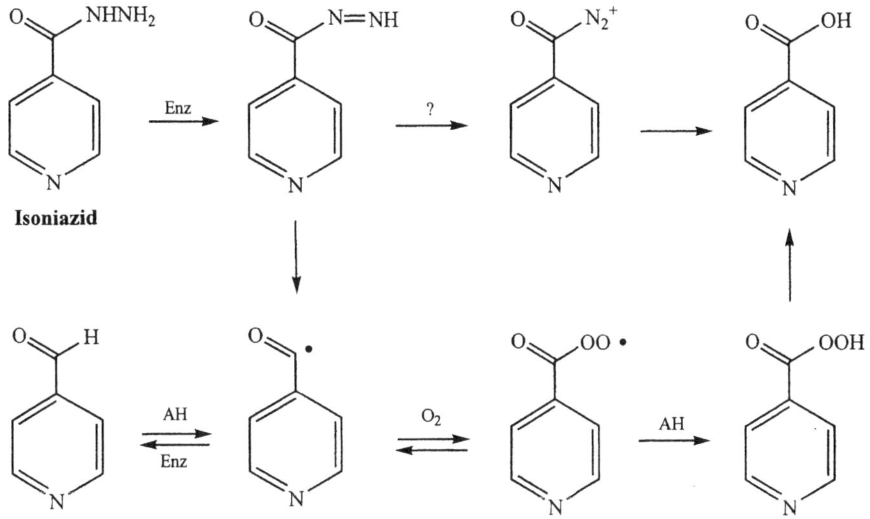

Figure 12.3. Proposed mechanism for the enzymatic oxidation of isoniazid. (Reproduced with permission from Johnsson K, Schultz PG. 1994. *J. Am. Chem. Soc.* **116**: 7425–7426.)

Thus carbon-centered radicals are produced by hydrazide-containing molecules and there is ample potential for generation of peroxidase-damaging reagents.

Reaction with NADH

Three catalase-peroxidases have been shown to react spontaneously with NADH in the presence of O_2 (Singh et al., 2004a). This is another example of the so-called peroxidase oxidase reactions, where addition of H_2O_2 is not required: it is generated within the system. Arg426 and Trp111 play a role in modulation of the NADH oxidase activity of *Burkholderia pseudomallei* (Carpena et al., 2006).

Manganese(II) enhances the activity of the catalase-peroxidase from *Mycobacteria smegmatis* (Magliozzo and Marcinkeviciene, 1997), but appears to a have limited role in the activity of the *M. tuberculosis* catalase-peroxidase (Bodiguel et al., 2001). The above studies would indicate that the reaction of *M. tuberculosis* catalase-peroxidase with isoniazid is another example of a peroxidase oxidase reaction (Chapter 5) in which either hydrogen peroxide or a suitable substitute is generated spontaneously in an aerobic environment. Mechanistic details are lacking. The activity of the catalase-peroxidase from *Bacillus stearothermophilus* is greatly increased by increasing the temperature (Kobayashi et al., 1997).

A series of papers from the group of Richard Magliozzo examined properties of recombinant and mutant forms of *M. tuberculosis* catalase-peroxidase. Resonance Raman spectra confirmed that the heme iron is mainly high-spin five-coordinate. Upon standing a small amount of six-cordinate heme with ligated water is formed (Choucane et al., 2000, 2003), which appears to be the preferred form for ligand binding (Ranguelova et al., 2008b). By use of peroxides other than H_2O_2 a porphyrin

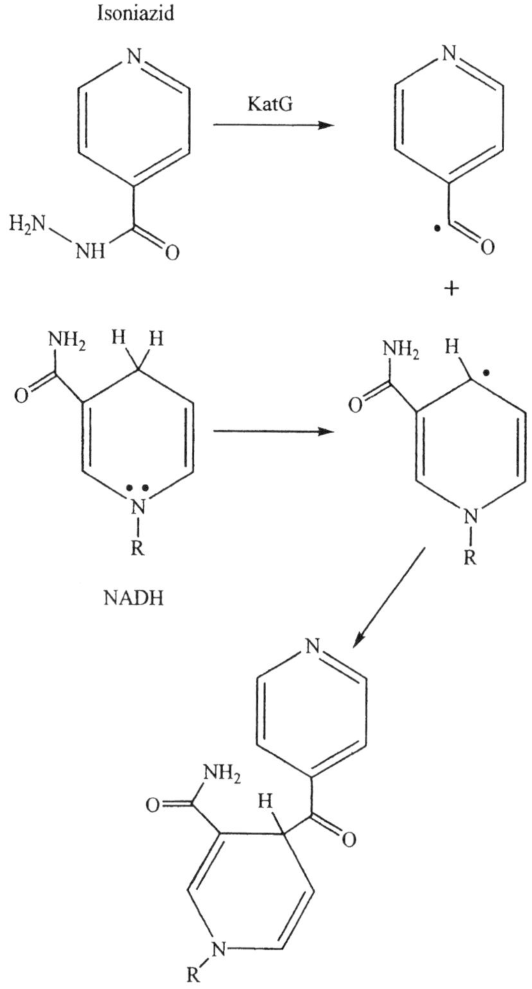

Figure 12.4. One proposed pathway for the formation of the isonicotinic acyl-NADH inhibitor of isoniazid. (Reproduced with permission from Rozwarski DA, et al., 1998. *Science* **279**: 98–102.)

π-cation radical Compound I was identified. Stopped-flow experiments showed that isoniazid reduced Compound I to native enzyme with no detectable Compound II formation, in contrast to the behavior of horseradish peroxidase (Choucane et al., 2000, 2003). Radicals produced by peroxide and isoniazid in *Mycobacterium tuberculosis* have been trapped and identified by ESR spectroscopy. Enzyme dimer formation was detected (Ranguelova et al., 2008a).

Naturally occurring mutants of *M. tuberculosis* catalase-peroxidase have been isolated from patients in Finland and the St. Petersburg region of Russia. The S315T mutant has similar peroxidase and catalase activity compared to the wild-type enzyme, but reactivity towards isoniazid is greatly reduced, indicating reduced binding of the drug (Yu et al., 2003; Kapetanaki et al., 2005; Zhao et al., 2006).

A study from the group of Paul Ortiz de Montellano examined the unique Met-Tyr-Trp cross-link near the heme active site of *M. tuberculosis* catalase-peroxidase (Ghiladi et al., 2005a, 2005b). This cross-link forms spontaneously in an oxidative reaction involving Compound I and is essential for catalase activity. The studies on both wild type and KatG(M255I) utilized peroxyacetic acid and 2-methyl-1-phenyl-2propyl hydroperoxides which allow Compound I formation, not observed with hydrogen peroxide because of the catalatic reaction. Compounds I, II, and III were characterized. Two forms of Compound II were identified, one with the extra oxidizing equivalent, compared to native enzyme, existing as a protein radical and the other as a $Fe^{IV}{=}O$ ferryl heme. The Tyr-Trp adduct was identified by mass spectrometry after tryptic digest and high performance liquid chromatography (HPLC) separation. The possible inhibition of *M. tuberculosis* growth by nitric oxide was deduced from a study of NO reactions with *M. tuberculosis* cytochrome P-450 enzymes (Ouellet et al., 2009).

Oxidized forms of *M. tuberculosis* catalase-peroxidase were examined by multifrequency ESR spectroscopy from wild-type enzyme, and the proximal W321F and distal W107F and W91F mutants (Singh et al., 2007). Two $Fe^{IV}{=}O$ $Trp^{\bullet}$ intermediates were identified. The results show that Compound I species with protein-based free radicals can be catalytically active, as in the family-related yeast cytochrome *c* peroxidase. The Compound I structure has been examined by density functional theory calculations (Vidossich et al., 2007). Redox potentials of the ferric-ferrous couple of *Synechocystis* KatG have been measured (Bellei et al., 2006).

Reactions of Compound I of recombinant catalase-peroxidase from the cyanobacterium *Synechocystis* PCC 6803 expressed in *E. coli* were studied (Regelsberger et al., 1999). Because of the fast catalatic reaction with hydrogen peroxide, peroxoacetic acid was used to produce Compound I. Compound I has about 40% hypochromocity compared to native enzyme and a Soret maximum absorbance at 406 nm. Isosbestic points with native enzyme occur at 357 and 430 nm. Rate of formation of Compound I at pH 7 is $(8.7 \pm 0.3) \times 10^3\ M^{-1}\ s^{-1}$. Substituent effects in the reactions of phenols substituted anilines fit Hammett equations: with ρ values of -3.4 ± 0.4 for phenols and -5.1 ± 0.8 for anilines.

The primary experimental approach in the following papers has been site-directed mutagenesis: of the catalase-peroxidase of *E. coli*, of *Synechocystis* PCC 6803, and of catalase-peroxidases from three different sources (Jakopitsch et al., 2007). Heme is easily lost from *E. coli* catalase-peroxidase. Variants of residues R102 and H106 reduced both catalase and peroxidase activity. W105F and W105I mutants essentially eliminated catalase activity but enhanced peroxidase reactions. The W105F mutant produced an identifiable porphyrin π-cation radical Compound I (Hillar et al., 2000). The C-terminal and N-terminal domains were expressed separately. Although the heme is not contained in the C-terminal domain, it is essential for both catalase and peroxidase activity (Baker et al., 2006).

A series of papers from the group of Christian Obinger have provided a detailed examination of the catalase-peroxidase from *Synechocystis* PCC 6803. The role of the conserved distal Arg-Trp-His triad in Compound I formation was probed with site-directed mutants. W122F and W122A completely lost their catalase activity but retained their peroxidase activity. The Trp mutants made it possible to use hydrogen peroxide to study Compound I formation where the His-Arg pair plays an important role. Compound I has a dominant conventional $Fe^{IV}{=}O$ porphyrin π-cation radical structure (Regelsberger et al., 2000; Jakopitsch et al., 2003a), although Tyr and Trp radicals were also detected by multifrequency ESR (Ivancich et al., 2003). A resonance Raman study showed that the distal Trp plays a more important role in distal H-bonding than in yeast cytochrome *c* peroxidase (Heering et al., 2002). Distal Asn153 and Asp152 also play important roles in hydrogen bonding networks (Jakopitsch et al., 2003b, 2003c). Mutations involving the unique access channel loops showed that they play an important role in the catalatic reaction (Jakopitsch et al., 2005). Comparison of catalase-peroxidases from three different sources showed common reaction intermediates in the catalatic reaction, different from those in classic catalases. In contrast the three different catalase-peroxidases all behaved differently in reactions with peroxoacetic acid. The exact nature of the intermediates remains to be determined (Jakopitsch et al., 2007).

THE OXIDATIVE DEFENSE MECHANISMS OF *MYCOBACTERIUM TUBERCULOSIS*

Although *M. tuberculosis* catalase-peroxidase has received the most attention, the bacterium has alternative defense mechanisms against hydroperoxides. *M. tuberculosis* cells, which lack functional catalase-peroxidase, have the advantage that they are immune from the effects of isoniazid, but they have lost a primary defense against oxidative stress. The cells also lack the normal bacterial sensing system against oxidative damage, OxyR. At least two other peroxidases appear to play a compensatory role. One is alkyl hydroperoxidase isoenzyme C (*ahp*C) (Chae et al., 1994b; Sherman et al., 1995, 1996; Cha et al., 2003). A thioredoxin peroxidase also plays an important role (Chae et al., 1994a; Jaeger et al., 2004; Trujillo et al., 2006). The enzyme *ahp*D is required for the reduction of *ahp*C. It has been shown to have a peroxidatic activity of its own (Koshkin et al., 2003). Thus there is an important role played by cysteine and cystine in peroxidase chemistry. These enzymes will provide a fruitful area of research for some time to come. The efficiency of cell hydrogen peroxide sensing systems, including OxyR, has been reviewed (Stone, 2004).

β-Lactam antibiotics are ineffective against *Mycobacterium tuberculosis*. However, when meropenem is complexed with the β-lactam inhibitor clavulanate, the complex exhibits potent activity against *M. tuberculosis* (Hugonnet et al., 2009).

REFERENCES

Abrams JJ, Webster DA. 1990. Purification, partial characterization, and possible role of catalase in the bacterium *Viteoscilla*. *Arch. Biochem. Biophys.* **279**: 54–59.

Ardissone S, Frendo P, Laurenti E, Jantschko W, Obinger C, Puppo A, Ferrari RP. 2004. Purification and physical-chemical characterization of the three hydroperoxidases from the symbiotic bacterium *Sinorhizobium meliloti*. *Biochemistry* **43**: 12692–12699.

Baker RD, Cook CO, Goodwin DC. 2006. Catalase-peroxidase active site restructuring by a distant and "inactive" domain. *Biochemistry* **45**: 7113–7121.

Bellei M, Jakopitsch C, Battistuzzi G, Sola M, Obinger C. 2006. Redox thermodynamics of the ferric-ferrous couple of wild-type *Synechocystis* KatG and KatG (Y249F). *Biochemistry* **45**: 4768–4774.

Bertrand T, Eady NAJ, Jones JN, Jesmin, Nagy JN, Jamart-Grégoire, Raven EL, Brown KA. 2004. Crystal structure of *Mycobacterium tuberculosis* catalase-peroxidase. *J. Biol. Chem.* **279**: 38991–38999.

Bodiguel J, Nagy JM, Brown KA, Jamart-Grégoire B. 2001. Oxidation of isoniazid by manganese and *Mycobacterium tuberculosis* catalase-peroxidase yields a new mechanism of activation. *J. Am. Chem. Soc.* **123**: 3832–3833.

Cannac-Caffrey V, Hudry-Clergeon G, Pétilloy Y, Gagnon J, Zaccai G, Franzetti B. 1998. The protein sequence of an archaeal catalase-peroxidase. *Biochimie* **80**: 1003–1011.

Carpena X, Loprasert S, Mongkolsuk S, Switala J, Loewen PC, Fita I. 2003. Catalase-peroxidase KatG of *Burkholderia pseudomallei* at 1.7 Å resolution. *J. Mol. Biol.* **327**: 475–489.

Carpena X, Melik-Adamyan W, Loewen PC, Fita I. 2004. Crystal structure of the C-terminal domain from the catalase-peroxidase KatG of *Escherichia coli*. *Acta Crystallogr.* **D60**: 1824–1832.

Carpena X, Wiseman B, Deemagarn T, Herguedas B, Ivancich A, Singh R, Loewen PC, Fita I. 2006. Roles for Arg426 and Trp111 in the modulation of NADH oxidase activity of the catalase-peroxidase KatG from *Burkholderia pseudomallei* inferred from pH-induced structural changes. *Biochemistry* **45**: 5171–5179.

Carpena X, Wiseman B, Deemagarn T, Singh R, Switala J, Ivancich A, Fita I, Loewen PC. 2005. A molecular switch and electronic circuit modulate catalase activity in catalase-peroxidases. *EMBO Rep* **6**: 1156–1162.

Cha M-K, Choi Y-S, Hong S-K, Kim W-C, No KT, Kim I-H. 2003. Nuclear thiol peroxidase as a functional alkyl-hydroperoxide reductase necessary for stationary growth of *Saccharomyces cerevisiae*. *J. Biol. Chem.* **278**: 24636–24643.

Chae HZ, Chung SJ, Rhee SG. 1994a. Thioredoxin-dependent peroxide reductase from yeast. *J. Biol. Chem.* **269**: 27670–27678.

Chae HZ, Robison K, Poole LB, Church G, Storz G, Rhee SG. 1994b. Cloning and sequencing of thiol-specific antioxidant from mammalian brain: alkyl hydroperoxide reductase and thiol-specific antioxidant define a large family of antioxidant enzymes. *Proc. Natl. Acad. Sci. USA* **91**: 7017–7021.

Choucane S, Girotto S, Kapetanaki S, Schelvis JPM, Yu S, Magliozzo RS. 2003. Analysis of heme structure heterogeneity in *Mycobacterium tuberculosis* (KatG). *J. Biol. Chem.* **278**: 8154–8162.

Choucane S, Lippai I, Magliozzo RS. 2000. Catalase-peroxidase (*Mycobacterium tuberculosis* KatG) catalysis and isoniazid activation. *Biochemistry* **39**: 9975–9983.

Claiborne A, Fridovich I. 1979. Purification of the *o*-dianisidine peroxidase from *Escherichia coli* B. Physiochemical characterization and analysis of its dual catalatic and peroxidatic activities. *J. Biol. Chem.* **254**: 4245–4252.

Claiborne A, Malinowski DP, Fridovich I. 1979. Purification and characterization of hydroperoxidase II of *Escherichia coli* B. *J. Biol. Chem.* **254**: 11664–11668.

Deemagarn T, Wiseman B, Carpena X, Ivancich A, Fita I, Loewen PC. 2007. Two alternative substrate paths for Compound I formation and reduction in catalase-peroxidase KatG from *Burkholderia pseudomallei*. *Proteins* **66**: 219–228.

DePillis GD, Ortiz de Montellano PR. 1989. Substrate oxidation by the heme edge of fungal peroxidases: reaction of *Coprinus macrorhizus* peroxidase with hydrazines and sodium azide. *Biochemistry* **28**: 7947–7952.

Ebihara T. 1939. Untersuchungen über ascorbinsäure-oxydase. III. Mitteilung: über die peroxydase in dem gereinigten ascorbinsäureoxydasepräparat. *J. Biochem. (Tokyo)* **29**: 217–223.

Fukumori Y, Fujiwara T, Okada-Takahashi Y, Mukohata Y, Yamanaka T. 1985. Purification and properties of a peroxidase from *Halobacterium halobium* L-33. *J. Biochem.* **98**: 1055–1061.

Ghiladi RA, Knudsen GM, Medzihradszky KF, Ortiz de Montellano PR. 2005a. The Met-Tyr-Trp cross-link in *Mycobacterium tuberculosis* catalase-peroxidase (KatG). *J. Biol. Chem.* **280**: 22651–22663.

Ghiladi RA, Medzihradszky KF, Ortiz de Montellano PR. 2005b. Role of the Met-Tyr-Trp cross-link in *Mycobacterium tuberculosis* catalase-peroxidase (KatG) as revealed by KatG(M255I). *Biochemistry* **44**: 15093–15105.

Goldberg I, Hochman A. 1989. Three different types of catalases found in *Klebsiella pneumoniae*. *Arch. Biochem. Biophys.* **268**: 124–128.

Heering HA, Indiani C, Regelsberger G, Jakopitsch C, Obinger C, Smulevich G. 2002. New insights into the heme cavity structure of catalase-peroxidase: a spectroscopic approach to the recombinant *Synechocystis* enzyme and selected distal cavity mutants. *Biochemistry* **41**: 9237–9247.

Hillar A, Loewen PC. 1995. Comparison of isoniazid oxidation catalyzed by bacterial catalase-peroxidases and horseradish peroxidase. *Arch. Biochem. Biophys.* **323**: 438–446.

Hillar A, Peters B, Pauls R, Loboda A, Zhang H, Mauk AG, Loewen PC. 2000. Modulation of the activities of catalase-peroxidase HPI of *Escherichia coli* by site-directed mutagenesis. *Biochemistry* **39**: 55868–5875.

Hochman A, Goldberg I. 1991. Purification and characterization of a catalase-peroxidase and a typical catalase from the bacterium *Klebsiella pneumoniae*. *Biochim. Biophys. Acta* **1077**: 299–307.

Hugonnet J-E, Tremblay LW, Boshoff HI, Barry CE, III, Blanchard JS. 2009. Meropenem-clavulanate is effective against extensively drug-resistant *Mycobacterium tuberculosis*. *Science* **323**: 1215–1218.

Ivancich A, Jakopitsch C, Auer M, Un S, Obinger C. 2003. Protein-based radicals in the catalase-peroxidase of *Synechocystis* PCC6803: a multifrequency EPR investigation of wild-type and variants on the environment of the heme active site. *J. Am. Chem. Soc.* **125**: 14093–14102.

Jaeger T, Budde H, Flohé L, Menge U, Singh M, Trujillo M, Radi R. 2004. Multiple thioredoxin-mediated routes to detoxify hydroperoxides in *Mycobacterium tuberculosis*. *Arch. Biochem. Biophys.* **423**: 182–191.

Jakopitsch C, Auer M, Ivancich A, Rüker F, Furtmüller PG, Obinger C. 2003a. Total conversion of bifunctional catalase-peroxidase (KatG) to monofunctional peroxidase by exchange of a conserved distal side tyrosine. *J. Biol. Chem.* **278**: 20185–20191.

Jakopitsch C, Auer M, Regelsberger G, Jantschko W, Furtmüller PG, Rüker F, Obinger C. 2003b. The catalytic role of the distal side asparagine-histidine couple in catalase-peroxidases. *Eur. J. Biochem.* **270**: 1006–1013.

Jakopitsch C, Auer M, Regelsberger G, Jantschko W, Furtmüller PG, Rüker F, Obinger C. 2003c. Distal side aspartate is essential in the catalase activity of catalase-peroxidases. *Biochemistry* **42**: 5292–5300.

Jakopitsch C, Droghetti E, Schmuckenschlager F, Furtmüller PG, Smulevich G, Obinger C. 2005. Role of the main access channel of catalase-peroxidases in catalysis. *J. Biol. Chem.* **280**: 42411–42422.

Jakopitsch C, Vlasits J, Wiseman B, Loewen PC, Obinger C. 2007. Redox intermediates in the catalase cycle of catalase-peroxidases from *Synechocystis* PCC 6803, *Burkholderia pseudomallei*, and *Mycobacterium tuberculosis*. *Biochemistry* **46**: 1183–1193.

Johnsson K, Froland WA, Schultz PG. 1997. Overexpression, purification, and characterization of the catalase-peroxidase KatG from *Mycobacterium tuberculosis*. *J. Biol. Chem.* **272**: 2834–2840.

Johnsson K, Schultz PG. 1994. Mechanistic studies of the oxidation of isoniazid by the catalase peroxidase from *Mycobacterium tuberculosis*. *J. Am. Chem. Soc.* **116**: 7425–7426.

Kapetanaki SM, Choucane S, Yu S, Zhao X, Magliozzo RS, Schelvis JPM. 2005. *Mycobacterium tuberculosis* KatG (S315T) catalase-peroxidase retains all active site properties for proper catalytic function. *Biochemistry* **44**: 243–252.

Kobayashi C, Suga Y, Yaamamoto K, Yomo T, Ogasahara K, Yutani K, Urabe I. 1997. Thermal conversion from low- to high-activity forms of catalase I from *Bacillus stearothermophilus*. *J. Biol. Chem.* **272**: 23011–23016.

Koshkin A, Nunn CM, Djordjevic S, Ortiz de Montellano PR. 2003. The mechanism of *Mycobacterium tuberculosis* alkylhydroperoxidase AhpD as defined by mutagenesis, crystallography, and kinetics. *J. Biol. Chem.* **278**: 29502–29508.

Lei B, Wei C-J, Tu S-C. 2000. Action mechanism of antitubercular isoniazid. Activation by *Mycobacterium tuberculosis* KatG, isolation, and characterization of InhA inhibitor. *J. Biol. Chem.* **275**: 2520–2526.

Li Y, Goodwin DC. 2004. Vital roles of an interhelical insertion in catalase-peroxidase bifunctionality. *Biochem. Biophys. Res. Commun.* **318**: 970–976.

Loewen PC, Triggs BL, George CS, Hrabarchuk BE. 1985. Genetic mapping of *katG*, a locus that affects synthesis of the bifunctional catalase-peroxidase hydroperoxidase I in *Escherichia coli*. *J. Bacteriol.* **162**: 661–667.

Loprasert S, Negoro S, Okada H. 1989. Cloning, nucleotide sequence, and expression in *Escherichia coli* of the *Bacillus stearothermophilus* peroxidase gene (*perA*). *J. Bacteriol.* **171**: 4871–4875.

Magliozzo RS, Marcinkeviciene JA. 1997. The role of Mn(II)-peroxidase activity of mycobacterial catalase-peroxidase in activation of the antibiotic isoniazid. *J. Biol. Chem.* **272**: 8867–8870.

Marcinkeviciene JA, Magliozzo RS, Blanchard JS. 1995. Purification and characterization of the *Mycobacterium smegmatis* catalase-peroxidase involved in isoniazid activation. *J. Biol. Chem.* **270**: 22290–22295.

Metcalfe C, Macdonald IK, Murphy EJ, Brown KA, Raven EL, Moody PCE. 2008. The tuberculosis prodrug isoniazid bound to activating peroxidases. *J. Biol. Chem.* **283**: 6193–6200.

Nadler V, Goldberg I, Hochman A. 1986. Comparative study of bacterial catalases. *Biochim. Biophys. Acta* **882**: 234–241.

Nagy JM, Cass AEG, Brown KA. 1997. Purification and characterization of recombinant catalase-peroxidase, which confers isoniazid sensitivity in *Mycobacterium tuberculosis*. *J. Biol. Chem.* **272**: 31265–31271.

Ouellet H, Lang J, Couture M, Ortiz de Montellano PR. 2009. Reaction of *Mycobacterium tuberculosis* cytochrome P450 enzymes with nitric oxide. *Biochemistry* **48**: 863–872.

Pierattelli R, Banci L, Eady NAJ, Bodiguel J, Jones JN, Moody PCE, Raven EL, Jamart-Grégoire B, Brown KA. 2004. Enzyme-catalyzed mechanism of isoniazid activation in Class I and Class III peroxidases. *J. Biol. Chem.* **279**: 39000–39009.

Ranguelova K, Suarez J, Magliozzo RS, Mason RP. 2008a. Spin trapping investigation of peroxide- and isoniazid-induced radicals in *Mycobacterium tuberculosis* catalase-peroxidase. *Biochemistry* **47**: 11377–11385.

Ranguelova K, Suarez J, Metlitsky L, Brejt SZ, Brejt SZ, Zhao L, Schelvis JPM, Magliozzo RS. 2008b. Impact of distal side water and residue 315 on ligand binding to ferric *Mycobacterium tuberculosis* catalase-peroxidase (KatG). *Biochemistry* **47**: 12583–12592.

Rawat R, Whitty A, Tonge PJ. 2003. The isoniazid-NAD adduct is a slow, tight-binding inhibitor of InhA, the *Mycobacterium tuberculosis* enoyl reductase: adduct affinity and drug resistance. *Proc. Natl. Acad. Sci. USA* **100**: 13881–13886.

Regelsberger G, Jakopitsch C, Engleder K, Rüker F, Peschek GA, Obinger C. 1999. Spectral and kinetic studies of the oxidation of monosubstituted phenols and analines by recombinant *Synechocystis* catalase-peroxidase Compound I. *Biochemistry* **38**: 10480–10488.

Regelsberger G, Jakopitsch C, Rüker F, Krois D, Peschek GA, Obinger C. 2000. Effect of distal cavity mutations on the formation of Compound I in catalase-peroxidases. *J. Biol. Chem.* **275**: 22854–22861.

Rozwarski DA, Grant GA, Barton DHR, Jacobs WR Jr., Sacchettini JC. 1998. Modification of the NADH of the isoniazid target (InhA) from *Mycobacterium tuberculosis*. *Science* **279**: 98–102.

Sherman DR, Mdluli K, Hickey MJ, Arain TM, Morris SL, Barry III CE, Stover CK. 1996. Compensatory *ahp*C gene expression in isoniazid-resistant *Mycobacterium tuberculosis*. *Science* **272**: 1641–1643.

Sherman DR, Sabo PJ, Hickey MJ, Arain TM, Mahairas GG, Yuan Y, Barry III CE, Stover CK. 1995. Disparate responses to oxidative stress in saprophytic and pathogenic mycobacteria. *Proc. Natl. Acad. Sci. USA* **92**: 6625–6629.

Shoeb HA, Bowman BU Jr., Ottolenghi AC, Merola AJ. 1985a. Peroxidase-mediated oxidation of isoniazid. *Antimicrob. Agents Chemother.* **27**: 399–403.

Shoeb HA, Bowman BU Jr., Ottolenghi AC, Merola AJ. 1985b. Evidence for the generation of active oxygen by isoniazid treatment of extracts of *Mycobacterium tuberculosis* H37Ra. *Antimicrob. Agents Chemother.* **27**: 399–403.

Singh R, Switala J, Loewen PC, Ivancich A. 2007. Two [Fe(IV)=O Trp$^{\bullet}$] mutants in *M. tuberculosis* catalase-peroxidase discriminated by multifrequency (9-285 GHz) EPR spectroscopy: reactivity toward isoniazid. *J. Am. Chem. Soc.* **129**: 15954–15963.

Singh R, Wiseman B, Deemagarn T, Donald LJ, Duckworth HW, Carpena X, Fita I, Loewen PC. 2004a. Catalase-peroxidases (KatG) exhibit NADH activity. *J. Biol. Chem.* **279**: 43098–43106.

Singh R, Wiseman B, Deemagarn T, Jha V, Switala J, Loewen PC. 2004b. Comparative study of catalase-peroxidases (KatGs). *Arch. Biochem. Biophys.* **471**: 207–214.

Stone JR. 2004. An assessment of proposed mechanisms for sensing hydrogen peroxide in mammalian systems. *Arch. Biochem. Biophys.* **422**: 119–124.

Triggs-Raine BL, Doble BW, Mulvey MR, Sorby PA, Loewen PC. 1988. Nucleotide sequence of *katG*, encoding catalase HPI of *Escherichia coli*. *J. Bacter.* **170**: 4415–4419.

Trujillo M, Mauri PL, Benazzi L, Comini M, De Palma A, Flohé L, Radi R, Stehr M, Singh M, Ursini F, Jaeger T. 2006. The mycobacterial thiredoxin peroxidase can act as a one-cysteine peroxiredoxin. *J. Biol. Chem.* **281**: 20555–20566.

Vidossich P, Alfonso-Prieto M, Carpena X, Loewen PC, Fita I, Rovira C. 2007. Versatility of the electronic structure of Compound I in catalase-peroxidases. *J. Am. Chem. Soc.* **129**: 13436–13446.

Wada K, Tada T, Nakamura Y, Kinoshita T, Tanoi M, Shigeoka S, Nishimura K. 2002. Crystallization and preliminary X-ray diffraction studies of catalase-peroxidase from *Synechococcus* PCC 7942. *Acta Crystallogr.* **D58**: 157–159.

Welinder KG. 1991. Bacterial catalase-peroxidases are gene duplicated members of the plant peroxidase superfamily. *Biochim. Biophys. Acta* **1080**: 215–220.

Welinder KG. 1992. Superfamily of plant, fungal and bacterial peroxidases. *Curr. Opin. Struct. Biol.* **2**: 388–393.

Yamada Y, Fujiwara T, Sato T, Igarishi N, Tanaka N. 2002. The 2.0 Å crystal structure of catalase-peroxidase from *Haloarcula marismortui*. *Nature Struct. Biol.* **9**: 691–695.

Youn H-D, Yim Y-I, Kim K, Hah YC, Kang S-O. 1995. Spectral characterization and chemical modification of catalase-peroxidase from *Streptomyces* sp. *J. Biol. Chem.* **270**: 13740–13747.

Yu S, Girotto S, Lee C, Magliozzo RS. 2003. Reduced affinity for isoniazid in the S315T mutant of *Mycobacterium tuberculosis* KatG is a key factor in antibiotic resistance. *J. Biol. Chem.* **278**: 14769–14775.

Zhao X, Yu H, Yu S, Wang F, Sacchettini JC, Magliozzo RS. 2006. Hydrogen peroxide-mediated isoniazid activation catalyzed by *Mycobacterium tuberculosis* (KatG) and its S315T mutant. *Biochemistry* **45**: 4131–4140.

CHAPTER 13

CLASS II. LIGNIN, MANGANESE, VERSATILE, AND *COPRINUS CINEREUS* PEROXIDASES

The secretory fungal peroxidases constitute Class II in the Welinder classification of heme peroxidases. Their history beginning in 1984 places them among the most newly discovered peroxidases.

LIGNIN PEROXIDASE

Lignin is second only to cellulose in terms of amount of biomass, and so plays a very important role in the carbon cycle. It is synthesized by peroxidases (see the section on *Arabidopsis thaliana* peroxidases in Chapter 14) and is degraded by peroxidases. Lignin synthesis requires a multi-enzyme system. There is stereochemical control, apparently not supplied by peroxidases. The peroxidases oxidize in their usual manner, utilizing both Compounds I and II. An authority on the total mechanism of lignin synthesis is Norman G. Lewis at Washington State University. This chapter concentrates on the lignin degrading enzymes.

Complete degradation of lignin to carbon dioxide and water is accomplished aerobically by a narrow range of microbes that have their own hydrogen peroxide-generating systems (Kirk and Farrell, 1987). These microbes are called white rot fungi because of the bleaching of wood that occurs during its degradation. Their mechanism of action has been reviewed (Kirk and Farrell, 1987; ten Have and Teunissen, 2001). The molecular biology of *P. chrysosporium* has also been reviewed (Gold and Alic, 1993).

Peroxidases and Catalases: Biochemistry, Biophysics, Biotechnology, and Physiology, Second Edition.
By H. Brian Dunford

Lignin peroxidase from the white-rot Basidiomycete *Phanerochaete chrysosporium* was discovered independently by two groups (Gold et al., 1984; Kuwahara et al., 1984; Tien and Kirk, 1984). The enzyme was originally called a hydrogen peroxide-dependent oxidase, a diarylpropane oxygenase, or ligninase. The name became lignin peroxidase when the similarities with horseradish peroxidase became apparent.

Three isoenzyme forms, labeled I to III, were isolated and shown to be glycoproteins (Renganathan et al., 1985). The dominant form is isoenzyme II. It is now known that there are at least eight isoenzymes of lignin peroxidase. The effect of glycosylation on thermal stabilities was examined. Recombinant enzymes expressed in *E. coli* had lower thermal stabilities. On the basis of chemical deglycosylation it was concluded that O-linked glycans, but not N-linked, play a role in thermal stabilitiy (Nie et al., 1999).

The enzyme contains heme *b*, and its Reinheitzahl (A_{407}/A_{280}) is 4.97. Optical spectra of oxidized forms of the enzyme were obtained (Tien et al., 1986; Andrawis et al., 1988). At pH 6 the interconversions of the oxidized forms are slow enough that a standard spectrophotometer can be used (Renganathan and Gold, 1986; Marquez et al., 1988). The iron in the native enzyme is high-spin, five-coordinate and Compound II is low-spin, six-coordinate with a $Fe^{IV}{=}O$ structure (Kuila et al., 1985; Andersson et al., 1987). Compound I is a conventional $Fe^{IV}{=}O$ porphyrin π-cation radical (Khindaria and Aust, 1996). The heme of lignin peroxidase is exceptionally sensitive to degradation by hydrogen peroxide, which hampered studies to ascertain the site of electron transfer reactions (DePillis et al., 1990; Tatarko and Bumpus, 1997).

The rate constant for Compound I formation from hydrogen peroxide and lignin peroxidase is constant from pH 2.0 to 7.5 with a value of $5.4 \times 10^5\ M^{-1}\ s^{-1}$, slow compared to most other peroxidases (Tien et al., 1986; Marquez et al., 1988; Harvey et al., 1989).

The activation energy for Compound I formation is 5.9 kcal/mole at pH 3.5. Nitrate ion has no effect at low pH, in marked contrast to horseradish peroxidase (Andrawis et al., 1988).

Reactions of lignin peroxidase and *Arthromyces ramosus* peroxidase with hydrogen peroxide, and in the absence of other substrates, have been studied in detail (Hiner et al., 2002). Catalase activity is observed as is Compound III formation, and with differences in rate constants, the same mechanism as applied to the reaction of horseradish peroxidase isoenzyme C with hydrogen peroxide is valid. Denaturation originates from a Compound I-H_2O_2 complex. No evidence was found for a second Compound III species, called Compound III* (Wariishi and Gold, 1989, 1990; Wariishi et al., 1990; Cai and Tien, 1992).

A Compound II-like species forms spontaneously from Compound I, a process that is accelerated by excess hydrogen peroxide or *p*-cresol (Marquez et al., 1988; Hiner et al., 2002). The resonance Raman spectra of Compound III are consistent with a Fe^{III}-$O_2^{\bullet -}$ structure (Mylrajan et al., 1990). Halide ion oxidations by lignin peroxidase have been explored (Shah and Aust, 1993a, 1993b).

Both lignin and manganese peroxidases have also been found in a different white-rot Basidiomycete *Trametes versicolor* (Johansson and Nyman, 1993; Johansson et al., 1993).

Veratryl Alcohol

Veratryl alcohol, a secondary metabolite of *P. chrysosporium*, is a preferred substrate for lignin peroxidase (Valli et al., 1990; Koduri and Tien, 1994; Goodwin et al., 1995). It reacts to form the veratryl alcohol cation radical (Haemmerli et al., 1987; Sheng et al., 1998), with the final principal product being veratryl aldehyde. The cation radical is the likely candidate from lignin peroxidase reactions to help to degrade lignin, either directly or as a mediator (Harvey et al., 1986). Its optical spectrum was obtained by pulse radiolysis (Candeias and Harvey, 1995).

When generated by pulse radiolysis the veratryl alcohol cation radical has a decay rate constant of $17 \pm 2\ s^{-1}$ which defines a lifetime of ~60 ms (Candeias and Harvey, 1995; Harvey and Candeias, 1995). Its product is a neutral benzyl radical which is transformed into veratryl aldehyde (Fig. 13.1).

EPR spectra of the veratryl alcohol cation radical were obtained (Khindaria et al., 1997). The cation radical appears to be stabilized by lignin peroxidase (Khindaria et al., 1996).

Veratryl alcohol reduces lignin peroxidase Compound I to II, and Compound II to native enzyme, confirming one-electron reactions (Marquez et al., 1988; Wariishi et al., 1991a). Compound III also reacts readily with veratryl alcohol (Wariishi and Gold, 1989, 1990). Mechanistic details have been discussed (Dunford, 1999). The

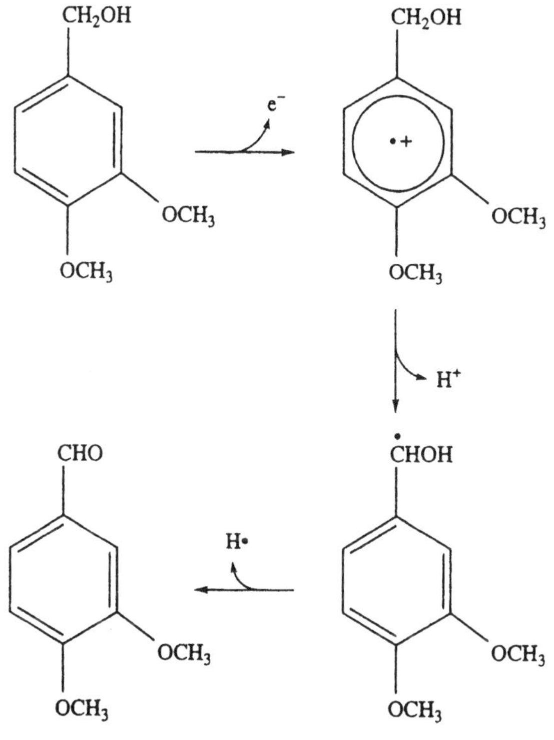

Figure 13.1. Oxidation of veratryl alcohol to veratryl aldehyde.

W171A mutant of lignin peroxidase is unable to oxidize veratryl alcohol (Gelpke et al., 2002).

Studies of reactions of lignin peroxidase with a whole host of reducing substrates have been carried out: with lignin model dimers (Hammel et al., 1985; Miki et al., 1988; ten Have et al., 2000); phenolic lignin oligomers (Banci et al., 1999b); tetrameric nonphenolic lignin models (Mester et al., 2001); 3,4-dimethoxybenzyl alcohol (Schmidt et al., 1989); 4-methoxymandelic acid (Tien and Ma, 1997); substituted phenols and 1,2-dimethoxyarenes (Ward et al., 2003); haloperoxidase reactions (Renganathan et al., 1987); ferro- and ferricytochrome *c* (Wariishi et al., 1994; Sheng and Gold, 1998); guaiacol (Koduri and Tien, 1995); and aromatic sulfides (Baciocchi et al., 2000).

Lignin peroxidase can also catalyze dimerization and polymerization as demonstrated with ferulic acid (Ward et al., 2001). Reactions in lignolytic cell cultures of *Phanaerochaete chrysosporium* were also studied (Joshi and Gold, 1994).

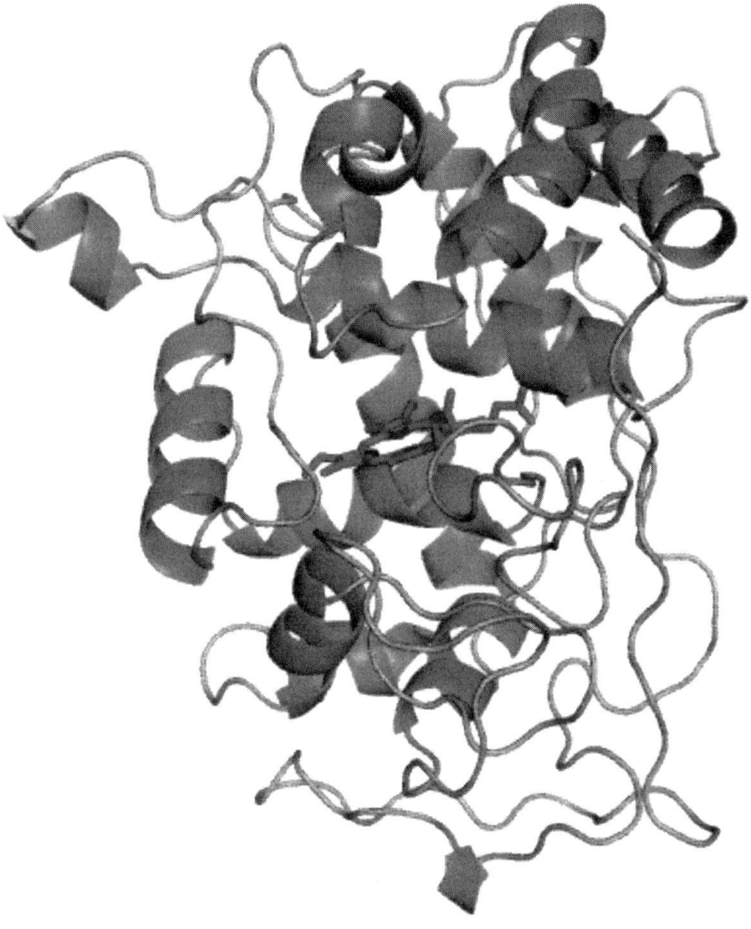

Figure 13.2. Structure of *P. chrysosporium* lignin peroxidase-2. (Kindly supplied by Dr. Tom Poulos.)

Lignin peroxidases have been isolated from other white rot fungi: *Bjerkandera* sp. strain BOS55 (ten Have et al., 1998; Mester and Field, 1998), *Pleurotus eryngii* (Camarero et al., 1999), and *Pleurotus ostreatus* (Kamitsuji et al., 2005).

Crystal Structure

Amino acid sequences were obtained for lignin peroxidase 2 (Rich et al., 1991) and lignin peroxidase 8 (Tien and Tu, 1987). The crystal structure of lignin peroxidase 2 was obtained at 2.6 Å resolution (Edwards et al., 1993), and refined to 2.0 Å (Poulos et al., 1993). This was the first Class II peroxidase structure to be determined. It has the classic distal pocket containing histidine and arginine, a proximal histidine hydrogen bonded to a buried aspartate side chain, two calcium ions, and four disulfide bonds. The tryptophans present in yeast cytochrome *c* peroxidase are replaced by phenylalanines in lignin peroxidase. Structures of *Phanerochaete chrysosporium* lignin peroxidase-2, manganese peroxidase, and versatile peroxidase are shown in Figures 13.2, 13.3, and 13.4.

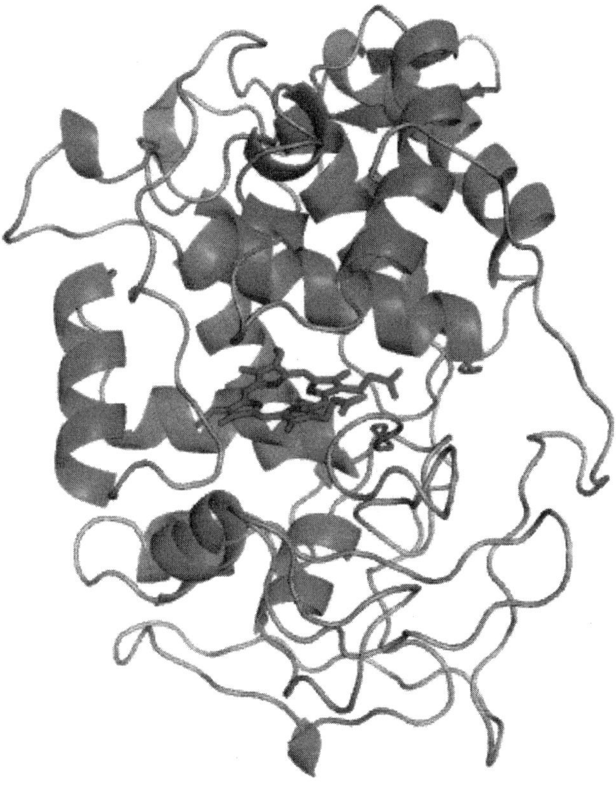

Figure 13.3. Structure of manganese peroxidase. (Kindly supplied by Dr. Tom Poulos.)

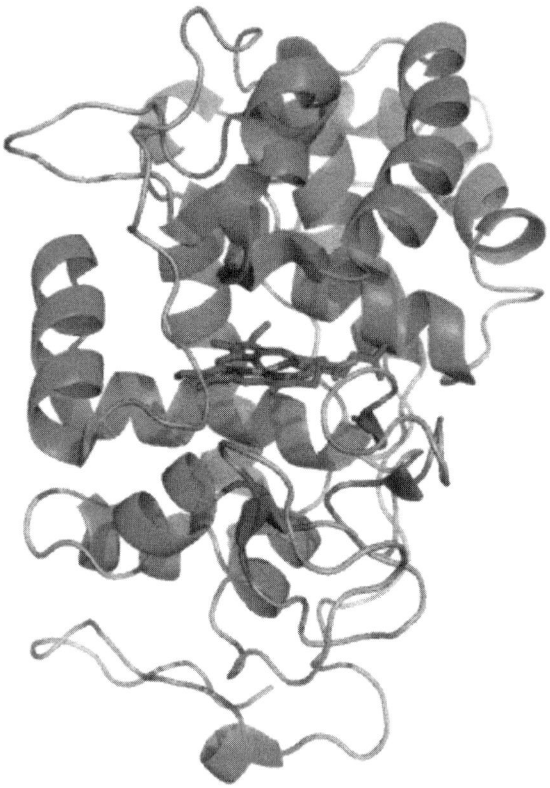

Figure 13.4. Structure of versatile peroxidase from *Pleurotus eryngii*. (Kindly supplied by Dr. Tom Poulos.)

MANGANESE PEROXIDASE

Manganese peroxidase from *P. chrysosporium* was discovered by the group of Michael Gold (Kuwahara et al., 1984). Details of its purification have been published (Glenn and Gold, 1985). Its principal function is oxidation of Mn^{II} (Glenn et al., 1986). Otherwise it is similar in its properties to lignin peroxidase (Mino et al., 1988). The enzyme exhibits typical native, Compound I, and Compound II electronic spectra. One-electron reactions occur with both Compounds I and II with Mn^{II} oxidized to Mn^{III} (Wariishi et al., 1988, 1989a). There is also a low affinity site for Mn^{II} binding (Mauk et al., 1998).

Mn peroxidase reacts more slowly with peroxides of the formula ROOH than horseradish peroxidase (Wariishi et al., 1989a). In the presence of hydrogen peroxide the enzyme oxidizes Mn^{II} to Mn^{III} in the presence of pyrophosphate (Glenn and Gold, 1985), lactate (Glenn et al., 1986), or citrate as chelators (Wariishi et al., 1989b). Oxalate and malonate are the only chelators present in significant amount in *P. chrysosporium* (Wariishi et al., 1992). Oxalate is present in 0.5 mM concentration in cultures of the organism, and at this concentration optimal enzyme activity is observed (Kuan and Tien, 1993; Kuan et al., 1993; Kishi et al., 1994). Water proton relaxation measurements show that oxalate does not form a ternary protein-Mn^{2+}-oxalate complex, but oxalate may remove the oxidized Mn^{3+} from the enzyme (Banci et al., 1998).

Depolymerization by manganese peroxidase of lignin involving degradation of ^{14}C labeled side chains was observed (Wariishi et al., 1991b). Cleavage of lignin model dimers was studied (Wariishi et al., 1989b; Tuor et al., 1992). Thiyl radicals are formed from glutathione and dithiothreitol by the lignin peroxidase-hydrogen peroxide-veratryl alcohol system and can play a role in lignin depolymerization (Wariishi et al., 1989c).

Sequencing, Expression, Active Site Probes, and Crystal Structure

The amino acid sequence of Mn peroxidase isoenzyme 1 (Pribnow et al., 1989; Godfrey et al., 1990), and isoenzyme 2 has been determined (Mayfield et al., 1994a). A series of papers examined gene expression in detail (Brown et al., 1990, 1993; Godfrey et al., 1994; Mayfield et al., 1994b). Isoenzyme 4 was expressed in *E. coli*. The same difficulties were encountered as in the expression of horseradish peroxidase in *E. coli* and were overcome in similar ways: by addition of heme, calcium ion, and oxidized glutathione. Highly active enzyme was isolated (Whitwam et al., 1995; Whitwam and Tien, 1996).

The active site of manganese peroxidase was probed using the denaturants sodium azide and alkylhydrazines. The results indicated that reaction occurs with the heme edge near the δ-*meso*-carbon, indicating this as the site of approach of reducing substrates (Harris et al., 1991). The effect of manganese replacement with cadmium was investigated (Youngs et al., 2000). Results obtained with an Asp179Asn mutant indicate that Asp179 is one of the ligands in the Mn^{II} binding site (Kusters-van Someren et al., 1995). Similarly Glu35 and Glu39 are implicated in Mn^{II} binding (Kishi et al., 1996). Further studies were made of mutants involving the Mn binding site (Whitwam et al., 1997). A Ser168Trp mutant was shown to convert Mn peroxidase into species with many similar properties to lignin peroxidase (Timofeevski et al., 2000). Mutations in the heme vicinity affect reactions of Compound II but not Compound I reduction (Whitwam et al., 1999; Ambert-Balay, 2000). Mutations on the proximal side also affect the distal side, leading to formation of a bis-histidine-iron complex and loss of calcium (Banci et al., 1999a). Redox potentials of recombinant Mn peroxidase and several mutants were measured, showing ionization of at least two groups in redox reactions (Santucci et al., 2000). The effects of engineered additional disulfide bonds were investigated (Reading and Aust, 2001).

Mutation studies of cytochrome *c* peroxidase, modified to convert it into a manganese peroxidase were undertaken. A Mn^{II} binding site was introduced (Wang and Lu, 1999), and the Trp51Phe mutation increased the rate of the rate-controlling step, oxidation of Mn^{II} by Compound II (Gengenbach et al., 1999).

The crystal structure of manganese peroxidase isoenzyme 1 was obtained at 2.06 Å resolution (Sundaramoorthy et al., 1994). The structures of all three peroxidases described in this chapter are all very similar. Manganese peroxidase differs in having five, rather than four, disulfide bonds. Asp179, Glu35, Glu39, a heme propionate, and two water molecules were identified as the Mn^{II} bonding ligands. The water molecules may be replaced by oxalate during the enzyme cycle. An electron transfer path along the sigma bonds of the propionate to the heme or iron was envisaged as the mechanism of Mn^{II} oxidation. Two calcium ions are bound to the enzyme. There is only one glycosylation site in manganese peroxidase located at Asn131. Crystal structures of manganese and ascorbate peroxidases were compared (Poulos et al., 1995).

The crystal structure of manganese peroxidase was further refined to 1.45 Å resolution, giving a more accurate picture of the Mn^{II} binding site. Structures of Cd^{II} and Sm^{III} inhibitor complexes were also obtained (Sundaramoorthy et al., 2005).

OTHER MANGANESE PEROXIDASES, VERSATILE PEROXIDASE

Manganese peroxidase has been detected in *Phlebia radiata* and *Nematoloma frowardii* (Hofrichter et al., 1998). With malonic acid as substrate, the enzyme generates its own peroxide, another example of a peroxidase oxidase reaction. *Pleurotus eryngii* and *Bjerkandera adusta* also contain a manganese peroxidase (Heinfling et al. 1998).

Both *Pleurotus* and *Bjerkandera* species contains a hybrid enzyme with both lignin peroxidase and Mn peroxidase activity (Mester and Field, 1998; Martinez, 2002; Hammel and Cullen, 2008). These enzymes have been dubbed versatile peroxidase. The crystal structure of the versatile peroxidase from *Pleurotus eryngii* has been determined, and an intensive investigation of native and mutant forms has been conducted (Ruiz-Dueñas et al., 2009). The native enzyme contains a tryptophan radical in both Compounds I and II.

COPRINUS CINEREUS (*ARTHROMYCES RAMOSUS*) PEROXIDASE

The common name for the source of this peroxidase is the black inkcap mushroom. *Coprinus cinereus* peroxidase appears to be identical to *Coprinus macrorhizus* peroxidase and *Arthromyces ramosus* peroxidase except for carbohydrate content, which varied from batch to batch (Kjalke et al., 1992; Limongi et al., 1995).

Coprinus cinereus peroxidase was purified, crystallized, and characterized. It has one heme *b* per molecule, its molecular weight is 41,600, and it forms typical Compounds I and II (Morita et al., 1988). It was further characterized by NMR (Lukat et al., 1989), kinetics and equilibria of cyanide binding (Andersen et al., 1991a), spectral and kinetic properties of Compounds I, II and III (Andersen et al., 1991b), resonance Raman spectroscopy (Smulevich et al., 1994), and in demonstration of reversible denaturation (Tams and Welinder, 1996). Whereas lignin peroxidase has at least eight isoenzymes, *C. cinereus* has only one.

Coprinus cinereus peroxidase exhibits some unusual kinetics in reactions with ferulic acid and 2,2′-azinobis(3-ethylbenzthiazoline-6-sulfonic acid) (ABTS). Apparently protonation of the proximal imidazole occurs at low pH. Reaction rates of Compounds I and II are identical at lower pH (Abelskov et al., 1997). The enzyme has a readily accessible active site, and is able to catalyze enantioselective epoxidation (Tuynman et al., 2000). Redox potentials and CO binding to ferrous wild-type *Coprinus cinereus* peroxidase and the Asp245Asn mutant were examined in detail (Ciaccio et al., 2003).

Studies have been conducted on recombinant *Coprinus cinereus* peroxidase and the site-directed mutants Gly156Phe and/or Asn157Phe. Phe156 and Phe157 have NMR spectral characteristics similar to comparable residues in horseradish peroxidase, known to be involved in aromatic donor molecule binding (Veitch et al., 1994). The Asp245Asn mutant was examined by both NMR and resonance Raman spectroscopies (Veitch et al., 1996; Smulevich et al., 1996). Protonation of the proximal histidine at low pH was confirmed, and both five- and six-coordinate heme iron were observed. The His183 hydrogen bond to residues 245 is weakened by the Asp245Asn mutant and this affects the heme coordination.

Compound I formation of lignin peroxidase has a low pH optimum of about 3, whereas the *Coprinus* enzyme optimum is from pH 3 to 9. Attempts were made to convert the *C. cinereus* peroxidase into one that mimics lignin peroxidase in Compound I formation. Mutations at Gly154 and Pro90 caused changes in electrostatic fields, causing destabilization of the oxidized compounds (Di Cerbo et al., 2001). The role of distal Arg was explored using site-directed mutants of *C. cinereus* peroxidase. The hydrogen bonding capacity of Arg51 is key to fast Compound I formation. The properties of the distal Arg differ significantly in peroxidases from *Coprinus cinereus*, yeast cytochrome *c*, and horseradish peroxidase (Schiødt et al., 2007).

Arthromyces ramosus peroxidase exhibits much higher luminescence from luminol than does horseradish peroxidase, and is therefore more suitable for luminescent assays (Kim et al., 1991). Mutants of horseradish peroxidase were prepared to improve its luminescent reaction (Tanaka et al., 1999). Hydroxylamine is a competitive inhibitor of Compound I formation in *Arthromyces ramosus* peroxidase (Wariishi et al., 2000).

Sequencing and Crystal Structures

The cDNA sequence encoding *C. cinereus* peroxidase was determined (Baunsgaard et al., 1993). The crystal structure of the recombinant enzyme was determined and refined to 2.6 Å resolution (Petersen et al., 1993, 1994). The structure of *Coprinus cinereus* peroxidase is similar to that of lignin peroxidase, but the *Coprinus* enzyme has a much larger opening to the active site. The iron was pentacoordinate with no water bound.

Arthromyces ramosus peroxidase was crystallized and characterized (Shinmen et al., 1986). It was cloned, sequenced, and expressed in yeast (Sawai-Hatanaka et al., 1995). The crystal structure was determined at 1.9 Å resolution (Kunishima et al., 1994). Water was coordinated to iron; otherwise, the structure was very similar to that of *C. cinereus* peroxidase. A later study showed pentacoordinated iron with no water ligand, and it was suggested this is the state of the enzyme under physiological conditions (Kunishima et al., 1996). Benzhydroxamic acid binds in the active site, replacing many water molecules (Itakura et al., 1997). Iodide ion binding to *Arthromyces ramosus* peroxidase was studied by ^{1}H and ^{127}I NMR spectroscopy, steady-state kinetics, and x-ray crystallographic analysis (Fukuyama et al., 1997). Binding occurs at the entrance to the access channel on the distal side of the heme. NMR spectra of the enzyme were hardly affected by iodide binding, unlike the behavior of horseradish peroxidase.

REFERENCES

Abelskov AK, Smith AT, Rasmussen CB, Dunford HB, Welinder KG. 1997. pH dependence and structural interpretation of the reactions of *Coprinus cinereus* peroxidase with hydrogen peroxide, ferulic acid, and 2,2′-azinobis(3-ethylbenzthiazoline-6-sulfonic acid). *Biochemistry* **36**: 9453–9463.

Ambert-Balay K, Dougherty M, Tien M. 2000. Reactivity of manganese peroxidase: site-directed mutagenesis of residues in proximity to the porphyrin ring. *Arch. Biochem. Biophys.* **382**: 89–94.

Andersen MB, Hsuanyu Y, Welinder KG, Schneider P, Dunford HB. 1991a. Kinetics and equilibria of cyanide binding to *Coprinus cinereus* peroxidase. *Acta Chem. Scand.* **45**: 206–211.

Andersen MB, Hsuanyu Y, Welinder KG, Schneider P, Dunford HB. 1991b. Spectral and kinetic properties of oxidized intermediates of *Coprinus cinereus* peroxidase. *Acta Chem. Scand.* **45**: 1080–1086.

Andersson LA, Renganathan V, Loehr TM, Gold MH. 1987. Lignin peroxidase: resonance Raman spectral evidence for Compound II and for a temperature-dependent coordination-state equilibrium in the ferric enzyme. *Biochemistry* **26**: 2258–2263 (1985).

Andrawis A, Johnson KA, Tien M. 1988. Studies on Compound I formation of the lignin peroxidase from *Phanerochaete chrysosporium*. *J. Biol. Chem.* **263**: 1195–1198.

Baciocchi E, Gerini MF, Harvey PJ, Lanzalunga O, Mancinelli S. 2000. Oxidation of aromatic sulfides by lignin peroxidase from *Phanerochaete chrysosporium*. *Eur. J. Biochem.* **267**: 2705–2710.

Banci L, Bertini I, Capannoli C, Del Conte R, Tien M. 1999a. Spectroscopic characterization of active mutants of manganese peroxidase: mutations on the proximal side affect calcium binding on the distal side. *Biochemistry* **38**: 9617–9625.

Banci L, Bertini I, Dal Pozzo L, Del Conte R, Tien M. 1998. Monitoring the role of oxalate in manganese peroxidase. *Biochemistry* **37**: 9009–9015.

Banci L, Ciofi-Baffoni S, Tien M. 1999b. Lignin and Mn peroxidase-catalyzed oxidation of phenolic lignin oligomers. *Biochemistry* **38**: 3205–3210.

Baunsgaard L, Dalbøge H, Houen G, Rasmussen EM, Welinder KG. 1993. Amino acid sequence of *Coprinus macrorhizus* peroxidase and cDNA sequence encoding *Coprinus cinereus* peroxidase. *Eur. J. Biochem.* **213**: 605–611.

Brown JA, Glenn JK, Gold MH. 1990. Manganese regulates expression of manganese peroxidase gene transcription in *Phanerochaete chrysosporium*. *Appl. Environ. Microbiol.* **59**: 4295–4299.

Brown JA, Li D, Alic M, Gold MH. 1993. Heat shock induction of manganese peroxidase by *Phanerochaete chrysosporium*. *J. Bacteriol.* **172**: 3125–3130.

Cai D, Tien M. 1992. Kinetic studies on the formation and decomposition of Compounds II and III. Reactions of lignin peroxidase with H_2O_2. *J. Biol. Chem.* **267**: 11149–11155.

Camarero S, Sarkar S, Ruiz-Dueñas, Martínez MJ, Martínez ÁT. 1999. Description of a versatile peroxidase involved in the natural degradation of lignin that has both manganese peroxidase and lignin peroxidase substrate interaction sites. *J. Biol. Chem.* **274**: 10324–10330.

Candeias LP, Harvey PJ. 1995. Lifetime and reactivity of the veratryl alcohol radical cation. Implications for lignin peroxidase catalysis. *J. Biol. Chem.* **270**: 16745–16748.

Ciaccio C, Rosati A, DeSanctis G, Sinibaldi F, Marini S, Santucci R, Ascenzi P, Welinder KG, Coletta M. 2003. Relationships of ligand binding, redox properties, and protonation in *Coprinus cinereus* peroxidase. *J. Biol. Chem.* **278**: 18730–18737.

DePillis GD, Wariishi H, Gold MH, Ortiz de Montellano PR. 1990. Inactivation of lignin peroxidase by phenylhydrazine and sodium azide. *Arch. Biochem. Biophys.* **280**: 217–223.

Di Cerbo P, Welinder KG, Schiødt CB. 2001. Kinetic evidence for surface residues influencing the active site of *Coprinus cinereus* peroxidase: analysis of the pH dependence of G154E, P90H and P90H-G154E substrate entrance mutants. *Biochim. Biophys. Acta* **1544**: 18–27.

Dunford HB. 1999. *Heme Peroxidases.* Wiley-VCH, New York.

Edwards SL, Raag R, Wariishi H, Gold MH, Poulos TL. 1993. Crystal structure of lignin peroxidase. *Proc. Natl. Acad. Sci. USA* **90**: 750–754.

Fukuyama K, Sato K, Itakura H, Takahashi S, Hosoya T. 1997. Binding of iodide to *Arthromyces ramosus* peroxidase investigated by X-ray crystallographic analysis, ^{1}H and ^{127}I NMR spectroscopy, and steady-state kinetics. *J. Biol. Chem.* **272**: 5752–5756.

Gelpke MDS, Lee J, Gold MH. 2002. Lignin peroxidase oxidation of veratryl alcohol: effects of mutants H82A, Q222A, W171A and F267L . *Biochemistry* **41**: 3498–3506.

Gengenbach A, Syn S, Wang X, Lu Y. 1999. Redesign of cytochrome *c* peroxidase into a manganese peroxidase: role of tryptophans in peroxidase activity. *Biochemistry* **38**: 11425–11432.

Glenn JK, Akileswaran L, Gold MH. 1986. Mn(II) oxidation is the principal function of the extracellular Mn-peroxidase from *Phanerochaete chrysosporium. Arch. Biochem. Biophys.* **251**: 688–696.

Glenn JK, Gold MH. 1985. Purification and characterization of an extracellular Mn(II)-dependent peroxidase from the lignin-degrading Basidiomycete *Phanerochaete chrysosporium. Arch. Biochem. Biophys.* **242**: 329–341.

Godfrey BJ, Akileswaran L, Gold MH. 1994. A reporter gene construct for studying the regulation of manganese peroxidase gene expression. *Appl. Environ. Microbiol.* **60**: 1353–1358.

Godfrey BJ, Mayfield MB, Brown JA, Gold MH. 1990. Characterization of a gene encoding a manganese peroxidase from *Phanerochaete chrysosporium. Gene* **93**: 119–124.

Gold MH, Alic M. 1993. Molecular biology of the lignin-degrading Basidiomycete *Phanerochaete chrysosporium. Microbiol. Rev.* **57**: 605–622.

Gold MH, Kuwahara M, Chiu AA, Glenn JK. 1984. Purification and characterization of an extracellular H_2O_2-requiring diarylpropane oxygenase from the white rot Basidiomycete, *Phanerochaete chrysosporium. Arch. Biochem. Biophys.* **234**: 353–362.

Goodwin DC, Aust SD, Grover TA. 1995. Evidence for veratryl alcohol as a redox mediator in lignin peroxidase-catalyzed oxidation. *Biochemistry* **34**: 5060–5065.

Haemmerli SD, Schoemaker HE, Schmidt HWH, Leisola MSA. 1987. Oxidation of veratryl alcohol by the lignin peroxidase of *Phanerochaete chrysosporium. FEBS Lett.* **220**: 149–154.

Hammel KE, Cullen D. 2008. Role of fungal peroxidases in biological ligninolysis. *Curr. Opinion Plant Biol.* **11**: 349–355.

Hammel KE, Tien M, Kalyanaraman B, Kirk TK. 1985. Mechanism of oxidative C_{α}-C_{β} cleavage of a lignin model dimer by *Phanerochaete chrysosporium* ligninase. *J. Biol. Chem.* **260**: 8348–8353.

Harris RZ, Wariishi H, Gold MH, Ortiz de Montellano PR. 1991. The catalytic site of manganese peroxidase: regiospecific addition of sodium azide and alkylhydrazines to the heme group. *J. Biol. Chem.* **266**: 8751–8758.

Harvey PJ, Candeias LP. 1995. Radical cation cofactors in lignin peroxidase catalysis. *Biochem. Soc. Trans.* **23**: 262–267.

Harvey PJ, Palmer JM, Schoemaker HE, Dekker HL, Wever R. 1989. Pre-steady-state kinetic study on the formation of Compound I and II of ligninase. *Biochim. Biophys. Acta* **994**: 59–63.

Harvey PJ, Schoemaker HE, Palmer JM. 1986. Veratryl alcohol as mediator and the role of radical cations in lignin biodegradation *by Phaerochaete chrysosporium. FEBS Lett.* **195**: 242–246.

Heinfling A, Ruiz-Dueñas FJ, Martínez MJ, Bergbauer M, Szewzyk U, Martínez AT. 1998. A study of the reducing substrates of manganese-oxidizing peroxidases from *Pleurotus eryngii* and *Bjerkandera adusta. FEBS Lett.* **428**: 141–146.

Hiner ANP, Hernández Ruiz J, Rodriguez Lopez JN, García Cánovas F, Brisset NC, Smith AT, Arnao MB, Acosta M. 2002. *J. Biol. Chem.* **277**: 26879–26885.

Hofrichter M, Ziegenhagen D, Vares T, Friedrich M, Jäger MG, Fritsche W, Hatakka A. 1998. Oxidative decomposition of malonic acid as a basis for the action of manganese peroxidase in the absence of hydrogen peroxide. *FEBS Lett.* **434**: 362–366.

Itakura H, Oda Y, Fukuyama K. 1997. Binding mode of benzhydroxamic acid to *Arthromyces ramosus* peroxidase shown by X-ray crystallographic analysis of the complex at 1.6 Å resolution. *FEBS Lett.* **412**: 107–110.

Johansson T, Nyman PO. 1993. Isolation of lignin peroxidase and manganese(II) peroxidase from the white rot Basidiomycete *Trametes versicolor.* I. Isolation of enzyme forms and characterization of physical and catalytic properties. *Arch. Biochem. Biophys.* **300**: 49–56.

Johansson T, Welinder KG, Nyman PO. 1993. Isolation of lignin peroxidase and manganese(II) peroxidase from the white rot Basidiomycete *Trametes versicolor.* II. Partial sequences, peptide maps, and amino acid and carbohydrate compositions. *Arch. Biochem. Biophys.* **300**: 57–62.

Joshi DK, Gold MH. 1994. Oxidation of dibenzo-p-dioxin by lignin peroxidase from the Basidiomycete *Phanerochaete chrysosporium. Biochemistry* **33**: 10969–10976.

Kamitsuji H, Watanabe T, Hond Y, Kuwahara M. 2005. Direct oxidation of polymeric substrates by multifunctional manganese peroxidase isoenzyme from *Pleurotus ostreatus* without redox mediators. *Biochem. J.* **386**: 387–393.

Kjalke M, Andersen MB, Schneider P, Christensen B, Schülein M, Welinder KG. 1992. Comparison of structure and activities of peroxidases from *Coprinus cinereus*, *Coprinus macrorhizus* and *Arthromyces ramosus. Biochim. Biophys. Acta* **1120**: 248–256.

Khindaria A, Aust SD. 1996. EPR detection and characterization of lignin peroxidase porphyrin π-cation radical. *Biochemistry* **35**: 13107–13111.

Khindaria A, Nie G, Aust SD. 1997. Detection and characterization of the lignin peroxidase Compound II-veratryl alcohol cation radical complex. *Biochemistry* **36**: 14181–14185.

Khindaria A, Yamazaki I, Aust SD. 1996. Stabilization of the veratryl alcohol cation radical by lignin peroxidase. *Biochemistry* **35**: 6418–6424.

Kim BB, Pisarev VV, Egorov AM. 1991. A comparative study of peroxidases from horse radish and *Arthromyces ramosus* as labels in luminol-mediated chemiluminescent assays. *Anal. Biochem.* **199**: 1–6.

Kirk TK, Farrell RL. 1987. Enzymatic "combustion": the microbial degradation of lignin. *Annu. Rev. Microbiol.* **41**: 465–505.

Kishi K, Kusters-van Someren M, Mayfield MB, Sun J, Loehr TM, Gold MH. 1996. Characterization of manganese(II) binding site mutants of manganese peroxidase. *Biochemistry* **35**: 8986–8994.

Kishi K, Wariishi H, Marquez L, Dunford HB, Gold MH. 1994. Mechanism of manganese Compound II reduction: effect of organic acid chelators and pH. *Biochemistry* **33**: 8694–8701.

Koduri RS, Tien M. 1994. Kinetic analysis of lignin peroxidase: explanation for the mediation phenomenon by veratryl alcohol. *Biochemistry* **33**: 4225–4230.

Koduri RS, Tien M. 1995. Oxidation of guaiacol by lignin peroxidase. Role of veratryl alcohol. *J. Biol. Chem.* **270**: 22254–22258.

Kuan I-C, Johnson KA, Tien M. 1993. Kinetic analysis of manganese peroxidase: the reaction with manganese complexes. *J. Biol. Chem.* **268**: 20064–20070.

Kuan I-C, Tien M. 1993. Stimulation of Mn peroxidase activity: a possible role for oxalate in lignin biodegradation. *Proc. Natl. Acad. Sci. USA* **90**: 1242–1246.

Kuila D, Tien M, Fee JA, Ondrias MR. 1985. Resonance Raman spectra of extracellular ligninase: evidence for a heme active site similar to those of peroxidases. *Biochemistry* **24**: 3394–3397.

Kunishima N, Amada F, Fukuyama K, Kawamoto M, Matsunaga T, Matsubara H. 1996. Pentacoordination of the heme iron of *Arthromyces ramosus* peroxidase shown by a 1.8 Å resolution crystallographic study at pH 4.5. *FEBS Lett.* **378**: 291–294.

Kunishima N, Fukuyama K, Matsubara H, Hatanaka H, Shibano Y, Amachi T. 1994. Crystal structure of the fungal peroxidase from *Arthromyces ramosus* at 1.9 Å resolution. *J. Mol. Biol.* **235**: 331–344.

Kusters-van Someren M, Kishi K, Lundell T, Gold MH. 1995. The manganese binding site of manganese peroxidase: characterization of an Asp179Asn site-directed mutant protein. *Biochemistry* **34**: 10620–10627.

Kuwahara M, Glenn JK, Morgan MA, Gold MH. 1984. Separation and characterization of two extracellular H_2O_2-dependent oxidases from ligninolytic cultures of *Phanerochaete chrysosporium*. *FEBS Lett.* **169**: 247–250.

Limongi P, Kjalke M, Vind J, Tams JW, Johansson T, Welinder KG. 1995. Disulfide bonds and glycosylation in fungal peroxidases. *Eur. J. Biochem.* **227**: 270–276.

Lukat GS, Rodgers KR, Jabro MN, Goff HM. 1989. Magnetic resonance spectral characterization of the heme active site of peroxidase. *Biochemistry* **28**: 3338–3345.

Marquez L, Wariishi H, Dunford HB, Gold MH. 1988. Spectroscopic and kinetic properties of the oxidized intermediates of lignin peroxidase from *Phanerochaete chrysosporium*. *J. Biol. Chem.* **263**: 10549–10552.

Martinez AT. 2002. Molecular biology and structure-function of lignin-degrading heme peroxidases. *Enzyme Microb. Technol.* **30**: 425–444.

Mauk MR, Kishi K, Gold MH, Mauk AG. 1998. pH-linked binding of Mn(II) to manganese peroxidase. *Biochemistry* **37**: 6767–6771.

Mayfield MB, Godfrey BJ, Gold MH. 1994a. Characterization of the *mnp2* gene encoding manganese peroxidase isozyme 2 from the basidiomycete *Phanerochaete chrysosporium*. *Gene* **142**: 231–235.

Mayfield MB, Kishi K, Alic M, Gold MH. 1994b. Homologous expression of recombinant manganese peroxidase in *Phanerochaete chrysosporium*. *Appl. Environ. Microbiol.* **60**: 4303–4309.

Mester T, Ambert-Balay K, Ciofi-Baffoni S, Banci L, Jones AD, Tien M. 2001. Oxidation of a tetrameric nonphenolic lignin model compound by lignin peroxidase. *J. Biol. Chem.* **276**: 22985–22990.

Mester T, Field JA. 1998. Characterization of a novel manganese peroxidase-lignin peroxidase hybrid isozyme produced by *Bjerkandera* species strain BOS55 in the absence of manganese. *J. Biol. Chem.* **273**: 15412–15417.

Miki K, Kondo R, Renganathan V, Mayfield MB, Gold MH. 1988. Mechanism of aromatic ring cleavage of a β-phenylyl ether dimer catalyzed by lignin peroxidase of *Phanerochaete chrysosporium*. *Biochemistry* **27**: 4787–4794.

Mino Y, Wariishi H, Blackburn NJ, Loehr TM, Gold MH. 1988. Spectral characterization of manganese peroxidase, an extra cellular heme enzyme from the lignin-degrading Basidiomycete, *Phanerochaete chrysosporium*. *J. Biol. Chem.* **263**: 7029–7036.

Morita Y, Yamashita H, Mikami B, Iwamoto H, Aibara S, Terada M, Minami J. 1988. Purification, crystallization, and characterization of peroxidase from *Coprinus cinereus*. *J. Biochem.* **103**: 693–699.

Mylrajan M, Valli K, Wariishi H, Gold MH, Loehr TM. 1990. Resonance Raman spectroscopic characterization of Compound III of lignin peroxidase. *Biochemistry* **29**: 9617–9623.

Nie G, Reading NS, Aust SD. 1999. Relative stability of recombinant versus native peroxidases from *Phanerochaete chrysosporium*. *Arch. Biochem. Biophys.* **365**: 328–334.

Petersen JFW, Tams JW, Vind J, Svensson A, Dalbøge H, Welinder KG, Larsen S. 1993. Crystallization and X-ray diffraction analysis of recombinant *Coprinus cinereus* peroxidase. *J. Mol. Biol.* **232**: 989–991.

Petersen JFW, Tams JW, Vind J, Svensson A, Dalbøge H, Welinder KG, Larsen S. 1994. Three-dimensional structure of a recombinant peroxidase from *Coprinus cinereus* at 2.6 Å resolution. *J. Mol. Biol.* **232**: 989–991.

Poulos TL, Edwards SL, Wariishi H, Gold MH. 1993. Crystallographic refinement of lignin peroxidase at 2 Å. *J. Biol. Chem.* **268**: 4429–4440.

Poulos TL, Patterson WR, Sundaramoorthy M. 1995. The crystal structure of ascorbate and manganese peroxidases: the role of non-haem metal in the catalytic mechanism. *Biochem. Soc. Trans.* **23**: 228–232.

Pribnow D, Mayfield MB, Nipper VJ, Brown JA, Gold MH. 1989. Characterization of a cDNA encoding a manganese peroxidase, from the lignin-degrading Basidiomycete *Phanerochaete chrysosporium*. *J. Biol. Chem.* **264**: 5036–5040.

Reading NS, Aust SD. 2001. Role of disulfide bonds in the stability of recombinant manganese peroxidase. *Biochemistry* **40**: 8161–8168.

Renganathan V, Gold MH. 1986. Spectral characterization of the oxidized states of lignin peroxidase, an extracellular heme enzyme from the white rot basidiomycete *Phanerochaete chrysosporium*. *Biochemistry* **25**: 1626–1631.

Renganathan V, Miki K, Gold MH. 1985. Multiple molecular forms of diarylpropane oxygenase an H_2O_2-requiring, lignin-degrading enzyme *Phanerochaete chrysosporium*. *Arch. Biochem. Biophys.* **241**: 304–314.

Renganathan V, Miki K, Gold MH. 1987. Haloperoxidase reactions catalyzed by lignin peroxidase, an extracellular enzyme from the Basidiomycete *Phanerochaete chrysosporium*. *Biochemistry* **26**: 5127–5132.

Ritch TG Jr, Nipper VJ, Akileswaran L, Smith AJ, Pribnow DG, Gold MH. 1991. Lignin peroxidase from the Basidiomycete *Phanerochaete chrysosporium* is synthesized as a preproenzyme. *Gene* **107**: 119–126.

Ruiz-Dueñas FJ, Pogni R, Morales M, Giansanti S, Mate MJ, Romero A, Martinez MJ, Basosi R, Martinez AT. 2009. Protein radicals in fungal versatile peroxidase: catalytic tryptophan radical in both Compound I and Compound II and studies on W164Y, W164H, and W164S variants. *J. Biol. Chem.* **284**: 7986–7994.

Santucci R, Bongiovanni C, Marini S, Del Conte R, Tien M, Banci L, Coletta M. 2000. Redox equilibria of manganese peroxidase from *Phanerochaete chrysosporium:* functional role of residues on the proximal side of the haem pocket. *Biochem. J.* **349**: 85–90.

Sawai-Hatanaka H, Ashikari T, Tanaka Y, Asada Y, Nakayama T, Minakata H, Kunishima N, Fukuyama K, Yamada H, Shibano Y, Amachi T. 1995. Cloning, sequencing, and heterologous expression of a gene coding for *Arthromyces ramosus* peroxidase. *Biosci. Biotech. Biochem.* **59**: 1221–1228.

Schiødt CB, Veitch NC, Welinder KG. 2007. Roles of distal arginine in activity and stability of *Coprinus cinereus* peroxidase elucidated by kinetic and NMR analysis of Arg51Gln, -Asn, -Leu, and -Lys mutants. *J. Inorg. Biochem.* **101**: 336–347.

Schmidt HWH, Haemerli SD, Schoemaker HE, Leisola MSA. 1989. Oxidative degradation of 3,4-dimethoxybenzyl alcohol and its methyl ester by the lignin peroxidase of *Phanerochaete chrysosporium*. *Biochemistry* **28**: 1776–1783.

Shah MM, Aust SD. 1993a. Iodide as mediator for the reductive reactions of peroxidases. *J. Biol. Chem.* **268**: 8503–8506.

Shah MM, Aust SD. 1993b. Oxidation of halides by peroxidases and their subsequent reductions. *Arch. Biochem. Biophys.* **300**: 253–257.

Sheng D, Gold MH. 1998. Irreversible oxidation of ferricytochrome *c* by lignin peroxidase. *Biochemistry* **37**: 2029–2036.

Sheng D, Joshi DK, Gold MH. 1998. Nitration of veratryl alcohol by lignin peroxidase and tetranitromethane. *Arch. Biochem. Biophys.* **352**: 121–129.

Shinmen Y, Asami S, Amachi T, Shimizu S, Yamada H. 1986. Crystallization and characterization of an extracellular fungal peroxidase. *Agric. Biol. Chem.* **50**: 247–249.

Smulevich G, Feis A, Focardi C, Tams J, Welinder KG. 1994. Resonance Raman study of the active site of *Coprinus cinereus* peroxidase. *Biochemistry* **33**: 15425–15432.

Smulevich G, Neri F, Marzocchi MP, Welinder KG. 1996. Versatility of heme coordination in a fungal peroxidase: absorption and resonance Raman studies of *Coprinus cinereus* peroxidase and the Asp245 → Asn mutant at various pH values. *Biochemistry* **33**: 15425–15432.

Sundaramoorthy M, Kishi K, Gold MH, Poulos TL. 1994. The crystal structure of manganese peroxidase from *Phanerochaete chrysosporium* at 2.06-Å resolution. *J. Biol. Chem.* **269**: 32759–32767.

Sundaramoorthy M, Youngs HL, Gold MH, Poulos TL. 2005. High-resolution crystal structure of manganese peroxidase: substrate and inhibitor complexes. *Biochemistry* **44**: 6463–6470.

Tams JW, Welinder KG. 1996. Unfolding and refolding of *Coprinus cinereus* peroxidase at high pH, in urea, and at high temperature: effect of organic and ionic additives on these processes. *Biochemistry* **35**: 7573–7579.

Tanaka M, Ishimori K, Morishima I. 1999. Luminol activity of horseradish peroxidase mutants mimicking a proposed binding site for luminol in *Arthromyces ramosus* peroxidase. *Biochemistry* **38**: 10463–10473.

Tatarko M, Bumpus JA. 1997. Further studies on the inactivation by sodium azide of lignin peroxidase from *Phanerochaete chrysosporium*. *Arch. Biochem. Biophys.* **339**: 200–209.

ten Have R, Franssen MCR, Field JA. 2000. Lignin peroxidase initiates O_2-dependent self-propagating chemical reactions which accelerate the consumption of 1-(3′,4′-dimethoxyphenyl)propene. *Biochem. J.* **347**: 585–591.

ten Have R, Hartmans S, Teunissen PJM, Field JA. 1998. Purification and characterization of two lignin peroxidase isozymes produced by *Bjerkandera* sp. strain BOS55. *FEBS Lett.* **422**: 391–394.

ten Have R, Teunissen PJM. 2001. Oxidative mechanisms involved in lignin degradation by white-rot fungi. *Chem. Rev.* **101**: 3397–3413.

Tien M, Kirk TK. 1984. Lignin-degrading enzyme from *Phanerochaete* chrysosporium: purification, characterization, and catalytic properties of a unique H_2O_2-requiring oxygenase. *Proc. Natl. Acad. Sci. USA* **81**: 2280–2284.

Tien M, Kirk TK, Bull C, Fee JA. 1986. Steady-state and transient-state kinetic studies on the oxidation of 3,4-dimethoxybenzyl alcohol catalyzed by the ligninase of *Phanerochaete chrysosporium* burds. *J. Biol. Chem.* **261**: 1687–1693.

Tien M, Ma D. 1997. Oxidation of 4-methoxymandelic acid by lignin peroxidase: mediation by veratryl alcohol. *J. Biol. Chem.* **272**: 8912–8917.

Tien M, Tu C-P D. 1987. Cloning and sequencing of a cDNA for a ligninase from *Phanerochaete chrysosporium*. *Nature* **326**: 520–523.

Timofeevski SL, Nie G, Reading NC, Austs SD. 2000. Substrate specificity of lignin peroxidase and a S168W variant of manganese peroxidase. *Arch. Biochem. Biophys.* **373**: 147–153.

Tuor U, Wariishi H, Schoemaker HE, Gold MH. 1992. Oxidation of phenolic arylglycerol β-aryl ether lignin model compounds by manganese peroxidase from *Phanerochaete chrysosporium:* oxidative cleavage of an α-carbonyl model compound. *Biochemistry* **31**: 4986–4995.

Tuynman A, Spelberg JL, Kooter IM, Schoemaker HE, Wever R. 2000. Enantioselective epoxidation and carbon-carbon bond cleavage catalyzed by *Coprinus cinereus* peroxidase and myeloperoxidase. *J. Biol. Chem.* **275**: 3025–3030.

Valli K, Wariishi H, Gold MH. 1990. Oxidation of monomethoxylated aromatic compounds by lignin peroxidase: role of veratryl alcohol in lignin biodegradation. *Biochemistry* **29**: 8535–8539.

Veitch NC, Gao Y, Welinder KG. 1996. The Asp245 → Asn mutant of *Coprinus cinereus* peroxidase: characterization by ^{1}H-NMR spectroscopy and comparison with the wild-type enzyme. *Biochemistry* **35**: 14370–14380.

Veitch NC, Tams JW, Vind J, Dalbøge H, Welinder KG. 1994. NMR studies of recombinant *Coprinus* peroxidase and three site-directed mutants: implications for peroxidase substrate binding. *Eur. J. Biochem.* **222**: 909–918.

Wang X, Lu Y. 1999. Proton NMR investigation of the heme active site structure of an engineered cytochrome *c* peroxidase that mimics manganese peroxidase. *Biochemistry* **38**: 9146–9157.

Ward G, Hadar Y, Bilkis I, Dosoretz CG. 2003. Mechanistic features of lignin peroxidase-catalyzed oxidation of substituted phenols and 1,2-dimethoxyarenes. *J. Biol. Chem.* **278**: 39726–39734.

Ward G, Hadar Y, Bilkis I, Konstantinovsky L, Dosoretz CG. 2001. Initial steps of ferulic acid polymerization by lignin peroxidase. *J. Biol. Chem.* **276**: 18734–18741.

Wariishi H, Akileswaran L, Gold MH. 1988. Manganese peroxidase from the Basidiomycete *Phanerochaete chrysosporium:* spectral characterization of the oxidized states and the catalytic cycle. *Biochemistry* **27**: 5365–5370.

Wariishi H, Dunford HB, MacDonald ID, Gold MH. 1989a. Manganese peroxidase from the lignin-degrading Basidiomycete *Phanerochaete chrysosporium*: transient state kinetics and reaction mechanism. *J. Biol. Chem.* **264**: 3335–3340.

Wariishi H, Gold MH. 1989. Lignin peroxidase Compound III: formation, inactivation, and conversion to the native enzyme *FEBS Lett.* **243**: 165–168.

Wariishi H, Gold MH. 1990. Lignin peroxidase Compound III: mechanism of formation and decomposition. *J. Biol. Chem.* **265**: 2070–2077.

Wariishi H, Huang J, Dunford HB, Gold MH. 1991a. Reactions of lignin peroxidase Compounds I and II with veratryl alcohol: transient-state kinetic characterization. *J. Biol. Chem.* **266**: 20694–20699.

Wariishi H, Marquez L, Dunford HB, Gold MH. 1990. Lignin peroxidase Compounds II and III: spectral and kinetic characterization of reactions with peroxides. *J. Biol. Chem.* **265**: 2070–2077.

Wariishi H, Nonaka D, Johjima T, Nakamura N, Naruta Y, Kubo S, Fukuyama K. 2000. Direct binding of hydroxylamine to the heme iron of *Arthromyces ramosus* peroxidase. *J. Biol. Chem.* **275**: 32919–32924.

Wariishi H, Sheng D, Gold MH. 1994. Oxidation of ferrocytochrome *c* by lignin peroxidase. *Biochemistry* **33**: 5545–5552.

Wariishi H, Valli K, Gold MH. 1989b. Oxidative cleavage of a phenolic diarylpropane model dimer by manganese peroxidase from *Phanerochaete chrysosporium. Biochemistry* **28**: 6017–6023.

Wariishi H, Valli K, Gold MH. 1991b. In vitro depolymerization of lignin by manganese peroxidase of *Phanerochaete chrysosporium. Biochem. Biophys. Res. Commun.* **176**: 269–275.

Wariishi H, Valli K, Gold MH. 1992. Manganese(II) oxidation by manganese peroxidase from the Basidiomycete *Phanerochaete chrysosporium*: kinetic mechanism and role of chelators. *J. Biol. Chem.* **267**: 23688–23695.

Wariishi H, Valli K, Renganathan V, Gold MH. 1989c. Thiol-mediated oxidation of nonphenolic lignin model compounds by manganese peroxidase of *Phanerochaete chrysosporium. J. Biol. Chem.* **264**: 14185–14191.

Whitwam RE, Brown KR, Musick M, Natan MJ, Tien M. 1997. Mutagenesis of the Mn^{2+}-binding site of manganese peroxidase affects oxidation of Mn^{2+} by both Compound I and Compound II. *Biochemistry* **36**: 9766–9773.

Whitwam RE, Gazarian IG, Tien M. 1995. Expression of fungal Mn peroxidase in *E. coli* and refolding to yield active enzyme. *Biochem. Biophys. Res. Commun.* **218**: 1013–1017.

Whitwam RE, Koduri RS, Natan M, Tien M. 1999. Role of axial ligands in the reactivity of Mn peroxidase from *Phanerochaete chrysosporium*. *Biochemistry* **38**: 9608–9616.

Whitwam RE, Tien M. 1996. Heterologous expression and reconstitution of fungal Mn peroxidase. *Arch. Biochem. Biophys.* **333**: 439–446.

Youngs HL, Sundaramoorthy M, Gold MH. 2000. Effects of cadmium on manganese peroxidase. *Eur. J. Biochem.* **267**: 1761–1769.

CHAPTER 14

OTHER CLASS III PEROXIDASES

The Class III peroxidases are the classical secretory plant enzymes. The best known example is horseradish peroxidase, so well studied that it is now used for comparative purposes to elucidate the behavior of other peroxidases. The properties and behavior of horseradish peroxidase, mainly isoenzyme C, are the topics of Chapters 3 to 5. Ironically HRP was not the first Class III peroxidase to have its crystal structure determined. That honor belongs to peanut peroxidase.

Peroxidases are ubiquitous in plant species. They exert a control mechanism over indole-3-acetic acid, auxin, the plant growth hormone; and they play an important role in cell wall and lignin synthesis, and in wound healing. A large literature on the biochemistry and physiology of plant peroxidases is compiled in two books by Thomas Gaspar, Hubert Greppin, and Claude Penel (Gaspar et al., 1982; Penel et al., 1992), a series of conference proceedings involving the same threesome (Greppin et al., 1986; Lobarzewski et al., 1991; Welinder et al., 1993; Obinger et al., 1996), and one led by Manuel Acosta (Acosta et al., 2003). Two papers illustrate results obtained on the correlation of peroxidase activity and plant development (Gaspar et al., 1985, 1992).

In this chapter peroxidases from *Arabidopsis thaliana*, barley, peanut, soybean, tobacco, and turnip are discussed.

ARABIDOPSIS THALIANA PEROXIDASE

In a joint U.S.-European Union project, the *Arabidopsis* Genome Initiative 2000, the first plant genome, that of *Arabidopsis thaliana*, was sequenced. *A. thaliana* belongs to the Crucifereae family that also contains horseradish peroxidase. Initially ~30 peroxidase clones were identified and many of them were sequenced (Simon et al., 1996). There are now known to be 73 Class III peroxidase genes in *A. thaliana* (Tognolli

Peroxidases and Catalases: Biochemistry, Biophysics, Biotechnology, and Physiology, Second Edition.
By H. Brian Dunford

et al., 2002; Welinder et al., 2002), and a phylogenetic tree has been constructed (Duroux and Welinder, 2003). A similar study of rice peroxidase genes has been published (Hiraga et al., 2000).

A predominant anionic peroxidase of pI 3.5 was purified from *A. thaliana* and partially sequenced. The cDNA clone was completely sequenced and shown to encode a 30-residue signal peptide and a 305-residue mature protein. Its sequence is 95% identical to the anionic horseradish peroxidase A2; the latter has never been successfully cloned. The new enzyme was designated as *A. thaliana* A2 (Østergaard et al., 1996). Two novel new peroxidases were discovered in *A. thaliana*, and their many entries in the database of expressed sequence tags indicate that they have essential roles. They lack the asparagine residue linked to the distal histidine and possess an N-glycosylation site at the entrance to the heme channel (Kjaersgård et al., 1997).

Lignification and Crystal Structure

The *A. thaliana* isoenzyme A2 was isolated from a rapidly lignifying cell culture and its cDNA cloned. Substrate specificity was analyzed by x-ray crystallography and docking of lignin precursors. Structure at 1.45 Å resolution was obtained at 100 K. The lignin precursors *p*-coumaryl and coniferyl alcohols were preferred substrates, while oxidation of sinapyl alcohol would be sterically hindered by Pro139 (Østergaard et al., 2000). Later the structure was obtained at room temperature. By a variety of techniques the recombinant *A. thaliana* A2 was shown to be identical to glycosylated horseradish peroxidaseA2. It showed some differences in reactivity with HRP A2, attributable to the lack of glycosylation (Nielsen et al., 2001).

BARLEY PEROXIDASE

Barley peroxidase has been shown to have properties that, so far, are unique for Class III peroxidases. The dominant isoenzyme, barley peroxidase 1, BP1, has a messenger RNA that is expressed only in the barley seed endosperm, and only 15 to 20 days after flowering (Rasmussen et al., 1991). A C-terminal peptide targets it for vacuoles. It has less than 50% sequence identity with other plant peroxidases. Using antibodies towards BP1, a cDNA clone was isolated. The pI of BP1 is 8.5 and molecular weight 37,000. BP1 contains 309 amino acid residues and has only one site of glycosylation. It is the first characterized plant peroxidase that is not blocked by pyroglutamate. Its unusual properties suggest a unique biological role (Johansson et al., 1992). Large-scale production of BP1 has been accomplished with an improved isolation protocol. Physical and spectroscopic characteristics were documented (Rasmussen et al., 1997).

The reactivity of barley peroxidase as originally isolated was very low (Rasmussen et al., 1993, 1995). The crucial factors in controlling the reactivity are now known to be pH and calcium ion concentration (Rasmussen et al., 1998). The enzyme as isolated is inactive at neutral pH. A decrease to pH 3.7 converts it into a form that reacts slowly with hydrogen peroxide. Calcium ion binding, with a dissociation constant of 4 mM, fully activates the enzyme. It then exhibits a pH-independent reaction with hydrogen peroxide over the low pH range of 3 to 5. The fully active form binds the usual ligands for peroxidase iron(III): fluoride, azide, cyanide, and (weakly) chloride. In the presence of hydrogen peroxide and the reducing substrates ferulic acid, caffeic acid, and coniferyl alcohol, the fully active enzyme exhibits reaction rates typical of other plant peroxidases, but only at low pH.

The crystal structure of the inactive form of barley peroxidase 1 has been published (Henriksen et al., 1998). The causes of lower reactivity at neutral and higher pH appear to be the following: disruption of a hydrogen bond between the distal His and Arg, wrong orientation of distal His, substitution of sodium for calcium at the distal binding site, and a dramatic structural rearrangement to the loop region between helices B and C. Calcium ion binding at lower pH restores the reactivity. See the discussion of Class I ascorbate peroxidase for another example of problems with a metal binding site.

Low temperature studies of electronic absorption, resonance Raman, and ESR spectra of barley peroxidase 1 (BP1) led to the conclusion that the enzyme exists in a quantum mechanically mixed-spin state. Some saddling of the heme structure may be required to attain this unique mixed-spin state. The vinyl stretching modes are split, indicating that the vinyl groups are conjugated differently to the porphyrin. The mixed-spin state and splitting of vinyl stretching modes appear to be unique to Class III plant peroxidases (Howes et al., 1999).

PEANUT PEROXIDASE

Work on peanut peroxidase and the career of Robert van Huystee are inextricably interwoven. His initial research was on the effect of ionizing radiation on cell growth for which he chose cultured peanut cells (*Arachis hypogaea* L.) from cotyledon tissue (Verma and van Huystee, 1970, 1971a, 1971b; van Huystee and Turcon, 1973). An anionic peroxidase was isolated from radioactively labeled culture (van Huystee, 1976), and later a cationic excreted peroxidase (Stephan and van Huystee, 1981). Both the anionic and cationic peroxidases were purified (Chibbar and van Huystee, 1983). The cationic peanut peroxidase was expressed in transgenic tobacco (Lige et al., 1998).

Differences in carbohydrate content were detected (van Huystee and Maldonado, 1982; O'Donnell et al., 1992; van Huystee et al., 1992; Wan and van Huystee, 1993). Unlike horseradish peroxidase, the carbohydrate chains of cationic peanut peroxidase appear necessay for structural integrity and enzyme activity (van Huystee and Wan, 1994). The same carbohydrate chains are present in the cationic peroxidase expressed in tobacco plants (Lige et al., 2000). The role of glycans has received scant attention, a situation that is being corrected (van Huystee and McManus, 1998). The glycan chain was analyzed by NMR spectroscopy (Shaw et al., 2000), and by electrospray ionization mass spectrometry (Zhang et al., 2004). Stability in relation to calcium and glycan content has been reviewed (van Huystee et al., 2004).

By use of ^{14}C-labeling it was demonstrated that the peroxidase heme is derived from glutamic acid and synthesized in mitochondria (Chibbar and van Huystee, 1983). The heme is ferriprotoporphyrin IX and there is one heme per molecule of peanut peroxidase (Chibbar et al., 1984; Chibbar and van Huystee, 1984). Two calcium ions are part of the enzyme structure and are necessary for optimal enzyme activity (Hu et al., 1987; Marañón et al., 1993; Barber et al., 1995). Tunicamycin inhibits peanut peroxidase release and glycosylation at N-linked sites (Ravi et al., 1986). Accurate spectra of cationic peanut peroxidase Compounds I, II, and III were obtained. The enzyme is similar in optical, Magnetic Circular Dichroism (MCD), and kinetic properties to horseradish peroxidase (Lambeir et al., 1985; Marañón and van Huystee, 1994; Marañón et al., 1994).

Production of monoclonal antibodies suggested a structural nonidentity between the anionic and cationic forms of peroxidase (van Huystee, 1987; Hu et al., 1987, 1990; Xu et al., 1990, 1991).

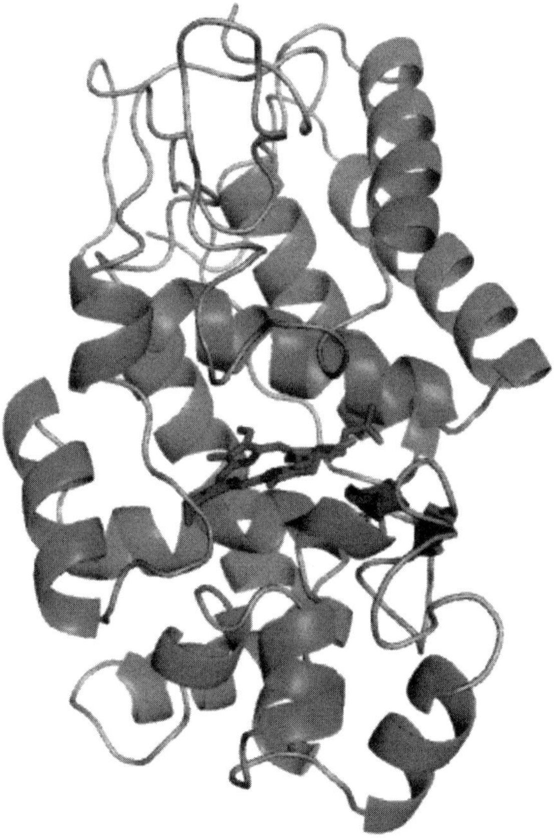

Figure 14.1. Structure of peanut peroxidase. (Kindly supplied by Dr. Tom Poulos.)

Crystal Structure

Two cDNAs for cationic peanut peroxidase were isolated, cloned, and characterized (Buffard et al., 1990). The dominant one corresponds to a 294 amino acid sequence. Crystals of native enzyme satisfactory for x-ray structural determination were obtained, overcoming the long held view of the prevention of crystallization by carbohydrate side chains (Ban et al., 1992). This was followed by determination of the crystal structure at 2.7 Å resolution (Schuller et al., 1996) (Fig. 14.1). The latter article contains an excellent comparison of the structures of Class I, II, and III peroxidases.

The Class III peroxidases, of which cationic peanut peroxidase is typical, have additional helical structures, D′, F′, and F″, that direct access to the δ-meso heme edge. There are no β-sheet structures. Water is present in the heme pocket, too distant to be a fifth iron ligand. Subsequently the peanut peroxidase structure used as a reference led to that of horseradish peroxidase.

SOYBEAN PEROXIDASE

The major source of anionic soybean (*Glycine max*) peroxidase is in the seed coat. It is a heterogeneous glycoprotein with a pI of 4.1 and a molecular weight of 37,000 (Gillikin and Graham, 1991). The enzyme has unusual stability, even for a peroxidase (McEldoon and Dordick, 1996). It brominates and iodinates veratryl alcohol, but is incapable of chlorination (Munir and Dordick, 2000). Its reaction with hydrogen peroxide and veratryl alcohol was further examined (Nissum et al., 2001). The recombinant enzyme was expressed in *E. coli* and its crystal structure determined. It has one of the most accessible δ-meso heme edges of any peroxidase. There is a contact between the heme C8 vinyl group and the sulfur atom of Met37, which might affect the stability of a porphyrin π-cation radical Compound I structure (Henriksen et al., 2001). Thermal stability and unfolding of soybean peroxidase has been examined (Kamal and Behere, 2003).

TOBACCO PEROXIDASES

There are at least 12 isoenzymes of tobacco peroxidase. Cationic peroxidases (pI 8 to 11), located in the central vacuole, have indole-3-acetic acid activity that does not require hydrogen peroxide (Mäder, 1980; Mäder et al., 1980; Grambow and Langenbeck-Schwich, 1983). Moderately anionic peroxidases (pI 4.5 to 6.5) are highly expressed in wounded stem tissue and may have a role in wound healing or suberization.

The highly anionic tobacco peroxidases, the best understood, have been purified from tobacco leaves (Mäder, 1980). They are associated with the cell wall, catalyze the polymerization of cinnamyl alcohols in vitro, and function in lignification (Mäder et al., 1977).

The roles of tobacco isoperoxidases in growth, development, and physiological state have been investigated (Trinh et al., 1981). Peroxidase activity was found to be an indissociable factor from auxin and polyamine metabolisms, and in the induction of rooting and flowering (Gaspar et al., 1992).

Extensive work by Mark Lagrimini and coworkers on tobacco peroxidase is described. The anionic tobacco peroxidase A has been cloned and sequenced (Lagrimini et al., 1987; Diaz-De-Leon et al., 1993). Different tissues were examined for their peroxidase isoenzyme distribution, and changes induced by wounding and tobacco mosaic virus infection were studied (Lagrimini and Rothstein, 1987). Tobacco plants were transformed with a chimeric anionic tobacco peroxidase gene that caused all tissues in the plants to produce high levels of the peroxidase. The subsequent effect of wounding was examined (Lagrimini et al., 1990; Lagrimini, 1991). A nearly twofold greater synthesis of lignin was observed. Plant browning was correlated with peroxidase activity. The studies on transgenic plants were extended to transformed tomato plants into which the tobacco anionic peroxidase gene was introduced (Lagrimini et al., 1993).

The anionic tobacco peroxidase was isolated and purified from leaves of transgenic plants. An order-of-magnitude better yield was obtained. The enzyme is similar to horseradish peroxidase isoenzyme A. Its reaction with indole-3-acetic acid was studied (Gazaryan and Lagrimini, 1996a, 1996b).

TURNIP PEROXIDASES

Studies on turnip peroxidases were initiated in the laboratory of Jacques Ricard (Mazza et al., 1968; Ricard, 1969). Seven isoenzymes were isolated, designated P_1 to P_7. Isoenzyme P_1 appears closely related to horseradish peroxidase isoenzyme C, whereas isoenzyme P_7, found in wintering turnips, has markedly different properties (Ricard et al., 1972). Amino acid sequences were determined for peptide fragments of P_1, P_2, and P_3 (Welinder and Mazza, 1975, 1977), and the complete sequence of P_7 (Mazza and Welinder, 1980).

Ligand binding was studied (Job and Ricard, 1975). The alkaline transition of P_1 is similar to that of horseradish peroxidase. It is now known that this results in a hydroxyl group occupying the fifth-coordination position of iron. The transition occurs at pH 10.1. A striking difference is observed with turnip peroxidase P_7 (Job et al., 1977). This less reactive isoenzyme appears to have water coordinated to iron, and the following reactions occur.

$$Fe^{III}\text{-}H_2O + H_2O \rightleftharpoons Fe^{III}\text{-}OH^- + H_3O^+$$

$$Fe^{III}\text{-}H_2O + OH^- \rightleftharpoons Fe^{III}\text{-}OH^- + H_2O$$

Thus, acid catalysis by $H_2O\text{-}H_3O^+$ occurs at lower pH, with the net rate decreasing with increasing pH, until base catalysis by $OH^-\text{-}H_2O$ takes over, with the net rate increasing with increasing pH. The switch from acid to alkaline catalysis is governed by the relevant rate constants and the value of K_w, the ionization constant of water.

Thus the value of K_w can be calculated from the kinetic results. It was found to be $(9.3 \pm 0.4) \times 10^{-15}\ M^2$, an example of how thermodynamic data can be deduced from kinetics experiments (Job et al., 1977). The converse is not true. The pH-dependent kinetics of Compound I formation from reaction with hydrogen peroxide was determined for P_1 and P_7 (Job et al., 1978).

REFERENCES

Acosta M, Rodriguez-López JN, Pedreño MA. 2003. *Plant Peroxidases: Biochemistry and Physiology*. Universidad de Murcia, Murcia, Spain.

Ban N, van Huystee RB, Day J, Greenwood A, Larson S, Esnault R, McPherson A. 1992. Preliminary crystallographic study of peanut peroxidase. *Acta Crystallogr. B* **48**: 109–111.

Barber KR, Marañón MJR, Shaw GS, van Huystee RB. 1995. Structural influence of calcium on the heme cavity of cationic peanut peroxidase as determined by ^{1}H-NMR spectroscopy. *Eur. J. Biochem.* **232**: 825–833.

Buffard D, Breda C, van Huystee RB, Asemota O, Pierre M, Ha DBD, Esnault R. 1990. Molecular cloning of complementary DNAs encoding two cationic peroxidases from cultivated peanut cells. *Proc. Natl. Acad. Sci. USA* **87**: 8874–8878.

Chibbar RN, Cella R, van Huystee RB. 1984. The heme moiety in peanut peroxidase. *Can J. Biochem. Cell Biol.* **62**: 1046–1050.

Chibbar RN, van Huystee RB. 1983. Characterization of peroxidase in plant cells. *Plant Physiol.* **75**: 956–958.

Chibbar RN, van Huystee RB. 1984. Site of haem synthesis in cultured peanut cells. *Phytochemistry* **25**: 585–587.

Diaz-De-Leon F, Klotz KL, Lagrimini LM. 1993. Nucleotide sequence of the tobacco (*Nicotiana tabacum*) anionic peroxidase gene. *Plant Physiol.* **101**: 1117–1119.

Duroux L, Welinder KG. 2003. The peroxidase gene family in plants: a phylogenetic overview. *J. Mol. Evol.* **57**: 397–407.

Gaspar T, Kevers C, Hausman JF, Berthon JY, Ripetti V. 1992. Practical uses of peroxidase activity as a predictive marker of rooting performance of micropropagated shoots. *Agronomie* **12**: 757–765.

Gaspar T, Penel C, Castillo FJ, Greppin H. 1985. A two-step control of basic and acidic peroxidases and its significance for growth and development. *Physiol. Plant.* **64**: 418–423.

Gaspar T, Penel CL, Thorpe T, Greppin H. 1982. *Peroxidases. 1970–1980*. University of Geneva, Switzerland.

Gazaryan IG, Lagrimini LM. 1996a. Purification and unusual kinetic properties of a tobacco anionic peroxidase. *Phytochemistry* **41**: 1029–1034.

Gazaryan IG, Lagrimini LM. 1996b. Tobacco anionic peroxidase overexpressed in transgenic plants: aerobic oxidation of indole-3-acetic acid. *Phytochemistry* **42**: 1271–1278.

Gillikin JW, Graham JS. 1991. Purification and developmental analysis of the major anionic peroxidase from the seed coat of *Glycine max*. *Plant Physiol.* **96**: 214–220.

Grambow HJ, Langenbeck-Schwich B. 1983. The relationship between oxidase activity, peroxidase activity, hydrogen peroxide, and phenolic ccompounds in the degredation of indole-3-acetic acid in vitro. *Planta* **157**: 131–137.

Greppin H, Penel C, Gaspar Th. 1986. *Molecular and Physiological Aspects of Plant Peroxidases*. University of Geneva, Switzerland.

Henriksen A, Mirza O, Indiani C, Teilum K, Smulevich G, Welinder KG, Gajhede M. 2001. Structure of soybean seed coat peroxidase: a plant peroxidase with unusual stability and haem-apoprotein interactions. *Protein Sci.* **10**: 108–115.

Henriksen A, Welinder KG, Gajhede M. 1998. Structure of barley grain peroxidase refined at 1.9-Å resolution: a plant peroxidase reversibly inactivated at neutral pH. *J. Biol. Chem.* **273**: 2241–2248.

Hiraga S, Sasaki T, Ito H, Ohashi Y, Matsui H. 2001. A large family of class III peroxidases. *Plant Cell Physiol.* **42**: 462–468.

Hiraga S, Yamamoto K, Ito H, Sasaki T, Ohashi Y. 2000. Diverse expression profiles of 21 rice peroxidase genes. *FEBS Lett.* **471**: 245–250.

Howes BD, Schiødt CB, Welinder KG, Marzocchi MP, Ma J-G, Zhang J, Shelnutt JA, Smulevich G. 1999. The quantum mixed-spin heme state of barley peroxidase: a paradigm for Class III peroxidases. *Biophys. J.* **77**: 478–492.

Hu C, Carbonera D, van Huystee RB. 1987. Production and preliminary characterization of monoclonal antibodies against cationic peanut peroxidase. *Plant Physiol.* **85**: 299–303.

Hu C, Krol M, van Huystee RB. 1990. Comparison of anionic with cationic peroxidase from cultured peanut cells. *Plant Cell, Tissue, Organ Culture* **22**: 65–70.

Hu C, Lee D, Chibbar RN, van Huystee RB. 1987. Ca^{2+} and peroxidase derived from cultured peanut cells. *Physiol. Plant.* **70**: 99–102.

Job D, Ricard J. 1975. Kinetic and equilibrium studies of cyanide and fluoride binding to turnip peroxidases. *Arch. Biochem. Biophys.* **170**: 427–437.

Job D, Ricard J, Dunford HB. 1977. The alkaline transition of turnip peroxidases. *Arch. Biochem. Biophys.* **179**: 95–99.

Job D, Ricard J, Dunford HB. 1978. Kinetics of formation of the primary compound (compound I) from hydrogen peroxide and turnip peroxidases. *Can. J. Biochem.* **56**: 702–707.

Johansson A, Rasmussen SK, Hartill JE, Welinder KG. 1992. cDNA, amino acid and carbohydrate sequence of barley seed-specific peroxidase BP1. *Plant Mol. Biol.* **18**: 1151–1161.

Kamal JKA, Behere DV. 2003. Thermal and conformational stability of seed coat soybean peroxidase. *Biochemistry* **41**: 9034–9042.

Kizhakkedathu AKJ, Nazeerunnisa M, Behere DV. Thermal unfolding of soybean peroxidase. *J. Biol. Chem.* **277**: 40717–40721.

Kjaersgård IVH, Jespersen HM, Rasmussen SK, Welinder KG. 1997. Sequence and RT-PCR expression analysis of two peroxidases from *Arabidopsis thaliana* belonging to a novel evolutionary branch of plant peroxidases. *Plant Mol. Biol.* **33**: 699–708.

Lagrimini LM. 1991. Wound-induced deposition of polyphenols in transgenic plants overexpressing peroxidase. *Plant Physiol.* **96**: 577–583.

Lagrimini LM, Bradford S, Rothstein S. 1990. Peroxidase-induced wilting in transgenic tobacco peroxidase. *The Plant Cell.* **2**: 7–18.

Lagrimini LM, Burkhart W, Moyer M, Rothstein S. 1987. Molecular cloning of complementary DNA encoding the lignin-forming peroxidase from tobacco: molecular analysis and tissue-specific expression. *Proc. Natl. Acad. Sci. USA* **84**: 7542–7546.

Lagrimini LM, Rothstein S. 1987. Tissue specificity of tobacco peroxidase isozymes and their induction by wounding and tobacco mosaic virus infection. *Plant Physiol.* **84**: 438–442.

Lagrimini LM, Vaughn J, Erb WA, Miller SA. 1993. Peroxidase overexpression in tomato: wound-induced polyphenol deposition and disease resistance. *Hort. Sci.* **28**: 218–221.

Lambeir A-M, Dunford HB, van Huystee RB, Lobarzewski J. 1985. Spectral and kinetic properties of a cationic peroxidase secreted by cultured peanut cells. *Can. J. Biochem. Cell Biol.* **63**: 1086–1092.

Lige B, Ma S, van Huystee RB. 2000. Glycosylation of the cationic peanut peroxidase: gene expressed in transgenic tobacco. *Plant Sci.* **136**: 55–63.

Lige B, Ma S, Zhao D, van Huystee RB. 1998. Cationic peanut peroxidase: expression and characterization in transgenic tobacco and purification of the histidine-tagged protein. *Plant Sci.* **136**: 159–168.

Lobarzewski J, Greppin H, Penel C, Gaspar Th. 1991. *Biochemical, Molecular, and Physiological Aspects of Plant Peroxidases*. University of Geneva, Switzerland.

Mäder M. 1980. Origin of the heterogeneity of peroxidase isoenzyme group G_I from *Nicotiana tabacum*. I. Conformation. Z. *Pflanzenphysiol.* **96**: 283–296.

Mäder M, Nessel A, Bopp M. 1977. Über die physiologische bedeutung der peroxidase-isoenzym-gruppen des tabaks anhand einiger biochemischer eigenschaften. II. pH-optima, Michaelis-konstanten, maximale oxidationsraten. Z. *Pflanzenphysiol.* **82**: 247–260.

Mäder M, Ungemach J, Schloss P. 1980. The role of peroxidase isozyme groups of *Nicotiana tabacum* in hydrogen peroxide formation. *Planta* **147**: 467–470.

Marañón MJR, Mercier D, van Huystee RB, Stillman MJ. 1994. Analysis of the optical absorption and magnetic-circular-dichroism spectra of peanut peroxidase: electronic structure of a peroxidase with biochemical properties similar to those of horseradish peroxidase. *Biochem. J.* **301**: 335–341.

Marañón MJR, Stillman MJ, van Huystee RB. 1993. Co-dependency of calcium and porphyrin for an integrated molecular structure of peanut peroxidase: a circular dichroism analysis. *Biochem. Biophys. Res. Commun.* **194**: 326–332.

Marañón MJR, van Huystee RB. 1994. Plant peroxidases: interactions between their prosthetic groups. *Phytochemistry* **37**: 1217–1225.

Mazza G, Charles C, Bouchet M, Ricard J, Raynaud J. 1968. Isolement, purification et propriétés physico-chimiques des peroxydases de navet. *Biochim. Biophys. Acta* **167**: 89–98.

Mazza G, Welinder KG. 1980. Covalent structure of turnip peroxidase 7. *Eur. J. Biochem.* **108**: 473–479.

McEldoon JP, Dordick JS. 1996. Unusual thermal stability of soybean peroxidase. *Biotechnol. Prog.* **12**: 555–558.

Munir IZ, Dordick JS. 2000. Soybean peroxidase as an effective bromination catalyst. *Enz. Microb. Tech.* **26**: 337–341.

Nielsen KL, Indiani C, Henriksen A, Feis A, Becucci M, Gajhede M, Smulevich G, Welinder KG. 2001. Differential activity and structure of highly similar peroxidases: spectroscopic, crystallographic, and enzymatic analyses of lignifying *Arabidopsis thaliana* peroxidase A2 and horseradish peroxidase A2. *Biochemistry* **40**: 11013–11021.

Nissum M, Schiødt CB, Welinder KG. 2001. Reactions of soybean peroxidase and hydrogen peroxide pH 2.4-12.0, and veratryl alcohol pH 2.4. *Biochim. Biophys. Acta* **1545**: 339–348.

Obinger C, Burner U, Ebermann R, Penel C, Greppin H, editors. 1996. *Plant Peroxidases: Biochemistry and Physiology*. University of Geneva, Switzerland.

O'Donnell JP, Wan L, van Huystee RB. 1992. Characterization of two forms of cationic peroxidase from cultured peanut cells. *Biochem. Cell Biol.* **70**: 166–169.

Østergaard L, Abelskov AK, Mattson O, Welinder KG. 1996. Structure and organ specificity of an anionic peroxidase from *Arabidopsis thaliana* cell suspension culture. *FEBS Lett.* **398**: 243–247.

Østergaard L, Teilum K, Mirza O, Mattson O, Petersen M, Welinder KG, Mundy J, Gajhede M, Henriksen A. 2000. *Arabidopsis* ATP A2 peroxidase: expression and high-resolution structure of a plant peroxidase with implications for lignification. *Plant Mol. Biol.* **44**: 231–243.

Penel C, Gaspar Th, Greppin H. 1992. *Plant Peroxidases. 1980–1990.* University of Geneva, Switzerland.

Rasmussen CB, Bakovic M, Welinder KG, Dunford HB. 1993. Unique reaction of barley peroxidase with hydrogen peroxide. *FEBS Lett.* **321**: 1151–1161.

Rasmussen CB, Henriksen A, Abelskov AK, Jensen RB, Rasmussen SK, Hejgaard J, Welinder KG. 1997. Purification, characterization and stability of barley grain peroxidase BP1, a new type of plant peroxidase. *Physiol. Plant.* **100**: 102–110.

Rasmussen CB, Hiner ANP, Smith AT, Welinder KG. 1998. Effect of calcium, other ions, and pH on the reactions of barley peroxidase with hydrogen peroxide and fluoride: control of activity through conformational change. *J. Biol. Chem.* **273**: 2232–2240.

Rasmussen CB, Welinder KG, Dunford HB. 1995. Rate enhancement of Compound I formation of barley peroxidase by ferulic acid, caffeic acid, and coniferyl alcohol. *Biochemistry* **34**: 4022–4029.

Rasmussen SK, Welinder KG, Hejgaard J. 1991. cDNA cloning, characterization and expression of an endosperm-specific barley peroxidase. *Plant Mol. Biol.* **16**: 317–327.

Ravi K, Hu C, Reddi PS, van Huystee RB. 1986. Effect of tunicamycin on peroxidase release by cultured peanut cell suspensions. *J. Exp. Bot.* **37**: 1708–1715.

Ricard J. 1969. Les peroxydases des végétaux supérieurs. *Bull. Soc. Fr. Physiol. Veg.* **15**: 321–362.

Ricard J, Mazza G, Williams RJP. 1972. Oxidation-reduction potentials and ionization states of two turnip peroxidases. *Eur. J. Biochem.* **28**: 566–578.

Schuller DJ, Ban N, van Huystee RB, McPherson A, Poulos TL. 1996. The crystal structure of peanut peroxidase. *Structure* **4**: 311–321.

Shaw GS, Sun Y, Barber KR, van Huystee RB. 2000. Sequence specific analysis of the heterogeneous glycan chain from peanut peroxidase by ^{1}H-NMR spectroscopy. *Phytochemistry* **53**: 135–144.

Simon P, Capelli N, Flach J, Overney S, Tognolli M, Penel C, Greppin H. 1996. The peroxidase gene family of *Arabidopsis thaliana*. In: Obinger C, Burner U, Ebermann R, Penel C, Greppin H, editors. *Plant Peroxidases: Biochemistry and Physiology*, University of Geneva, Switzerland, pp. 179–183.

Stephan D, van Huystee RB. 1981. Some aspects of peroxidase synthesis by cultured peanut cell. *Zeit. Pflanzenphysiol.* **101**: 313–321.

Tognolli M, Penel C, Greppin H, Simon P. 2002. Analysis and expression of the Class III peroxidase large gene family in *Arabidopsis thaliana*. *Gene* **288**: 129–138.

Trinh TH, Gaspar T, Tran Than Van K, Marcotte JL. 1981. Genotype, ploidy and physiological state in relation to isoperoxidases in *Nicotiana*. *Physiol. Plant.* **53**: 153–157.

van Huystee RB. 1976. A study of peroxidase synthesis by means of double labeling and affinity chromatography. *Can. J. Bot.* **54**: 876–880.

van Huystee RB. 1987. Some molecular aspects of plant peroxidase biosynthetic studies. *Annu. Rev. Plant Physiol.* **38**: 205–219.

van Huystee RB, Maldonado BA. 1982. Some physico-chemical properties of a major cationic peroxidase from cultured peanut cells. *Physiol. Plant.* **56**: 88–92.

van Huystee RB, McManus MT. 1998. Glycans of higher plant peroxidases: recent observations and future speculations. *Glycoconjugate J.* **15**: 101–106.

van Huystee RB, Roig MG, Shnyrov VL, Sakharov IY. 2004. *Phytochem. Rev.* **3**: 19–28.

van Huystee RB, Sesto PA, O'Donnell JP. 1992. Number and size of oligosaccharides linked to peanut peroxidase. *Plant Physiol. Biochem.* **30**: 147–152.

van Huystee RB, Turcon G. 1973. Rapid release of peroxidase by peanut cells in suspension culture. *Can. J. Bot.* **51**: 1169–1175.

van Huystee RB, Wan L. 1994. Carbohydrate moiety of peanut peroxidase necessary for enzyme activity. Rôle des chaînes glycosidiques de peroxydases d'arachide: conformation moléculaire et activité enzymatique. *C.R. Acad. Sci. Paris* **317**: 789–794.

Verma DPS, van Huystee RB. 1970. Cellular differentiation and peroxidase isozymes in cell culture of peanut cotyledons. *Can. J. Bot.* **48**: 429–431.

Verma DPS, van Huystee RB. 1971a. Derivation, characteristics and large scale culture of a cell line from *Arachis hypogaea*, L. cotyledons. *Exp. Cell Res.* **69**: 402–408.

Verma DPS, van Huystee RB. 1971b. Induction of giant cells in suspension cultures of *Arachis hypogaea*, L. by massive irradiation. *Radiat. Res.* **48**: 518–530.

Wan L, van Huystee RB. 1993. A study on glycosylation of cationic peanut peroxidase. *Biochem. Biophys. Res. Commun.* **194**: 1398–1405.

Welinder KG, Justesen AF, Kjaergård IV, Rassmussen SK, Jespersen HM, Duroux L. 2002. Structural diversity and transcription of Class III peroxidases from *Arabidopsis thaliana*. *Eur. J. Biochem.* **269**: 6063–6081.

Welinder KG, Mazza G. 1975. Similarities and differences of five peroxidases from turnip and horseradish: peptide mapping studies on glycoproteins. *Eur. J. Biochem.* **57**: 415–424.

Welinder KG, Mazza G. 1977. Amino-acid sequences of heme-linked, histidine-containing peptides of five peroxidases from horseradish and turnip. *Eur. J. Biochem.* **73**: 353–358.

Welinder KG, Rasmussen SK, Penel C, Greppin H. 1993. *Plant Peroxidases: Biochemistry and Physiology*. University of Geneva, Switzerland.

Xu Y, Hu C, van Huystee RB. 1990. Quantitative determination of a peanut anionic peroxidase by specific monoclonal antibodies. *J. Exp. Botany* **41**: 1479–1488.

Xu Y, van Huystee RB. 1991. Identification of an antigenic determinant on anionic peanut peroxidase by monoclonal antibodies. *J. Exp. Botany* **42**: 935–945.

Zhang C, Doherty-Kirby A, van Huystee RB, Lajoie G. 2004. Investigation of cationic peanut peroxidase glycans by electrospray ionization mass spectrometry. *Phytochemistry* **65**: 1575–1588.

CHAPTER 15

CATALASES

PETER JONES

INTRODUCTION

The activity of catalase was observed some 190 years ago (long before the concepts of catalysis and enzymes emerged). The enzyme was named 100 years ago and first crystallized 70 years ago. It may seem surprising that there are so many ongoing studies of the chemistry, biochemistry, and biology of these enzymes but unpredictable behavior and remarkable complexity have been consistent findings. Over the years the catalases have often been applauded as superlative catalysts, labeled as "monofunctional" enzymes (certainly an over-simplification and perhaps a distortion of their roles), and once considered to be "fossil" enzymes (Nicholls and Schonbaum, 1963). More recently the enzymes have been described as "perplexing" (Percy, 1984) and "enigmatic" (Kalko et al., 2001); terms that neatly encompass the current levels of understanding in both chemical and biological contexts.

Reviews with different emphases have appeared frequently; perhaps the most influential (certainly the most detailed accounts of earlier work) have been those of Nicholls and Schonbaum (1963) and Schonbaum and Chance (1976). Jones and Wilson (1978) list earlier catalase reviews and also give a comparative account of the reactivities of catalases and water-soluble ferrihemes [iron(III)-porphyrins]; this chapter extends this approach to more recent work. Nicholls et al. (2001) have presented a detailed review of the enzymology and structures of catalases and comparisons with the catalase-like properties of chloroperoxidase and the catalase-peroxidases. The diversity of structures and properties of catalases from different sources has been discussed by Chelikani et al. (2004).

Peroxidases and Catalases: Biochemistry, Biophysics, Biotechnology, and Physiology, Second Edition.
By H. Brian Dunford

PERSPECTIVE

Following his discovery of hydrogen peroxide in 1818, Louis-Jacques Thenard made a comprehensive survey of its reactivity, particularly noting decompositions occurring without alteration of the decomposing agent. He found that animal matter—blood and tissues of the organs—was effective in decomposing many successive portions of hydrogen peroxide (see Schumb et al., 1955). Thenard could not explain his results but they provided an important part of the data used by Berzelius in formulating the concept of catalysis in 1836. A century later the catalases had been characterized as a group of heme peroxidases that display a catalytic activity with such remarkable (at the time unique) substrate selectivity that the Enzyme Commission assigned a distinct classification (EC 1.11.1.6 hydrogen peroxide: hydrogen peroxide oxido-reductase). Thus, the enzymes display "catalatic" activity, that is, they catalyze the dismutation reaction:

$$2H_2O_2 = 2H_2O + O_2 \tag{15.1}$$

The evolved dioxygen is usually assumed to be in its lowest energy triplet state; more on that later.

Perhaps Thenard's greatest achievement was his preparation of anhydrous hydrogen peroxide—no chemist for many years after was able to match the purity and stability of his material. In contemporary terms he demonstrated that hydrogen peroxide, although highly unstable thermodynamically with respect to dismutative decomposition, is kinetically stable. Reports of uncatalyzed decomposition at ambient temperatures in alkaline solution proved to be unfounded when tested with rigorously purified reagents (Koubek et al., 1963). This lack of reactivity is less surprising when the nature of Equation (15.1) is considered in more detail; the process formally involves the breaking of three covalent bonds and the formation of two new bonds. A reaction of such complexity is not likely to occur simply, for example, in a single bimolecular collision. Dismutation is, however, susceptible to catalysis by a wide variety of species. The processes are often very complex and mechanistic interpretations usually speculative.

The most intensively studied "simple" system has been that involving ferric ions [Fe^{3+}; Fe(III) aqua ions]. A widely (but by no means unequivocally) held mechanism was initially proposed by Haber and Weiss and later modified by Barb et al. (1951), which envisages a chain reaction, initiated by a reductive action of perhydroxyl anion on Fe(III):

$$Fe^{3+} + HO_2^{\bullet} \longrightarrow Fe^{2+} + HO_2 \tag{15.2}$$

and carried by Fe^{2+} and the oxygen free radicals $HO_2^{\bullet}$ ($O_2^{\bullet -}$) and $^{\bullet}OH$. Although Weiss proposed a similar scheme for the "catalatic" reaction of catalases it received little support, more especially when radiation, chemical, and biological studies showed that hydroxyl radicals ($^{\bullet}OH$), formed from the aqueous matrix are a major component of the indirect action of ionizing radiation on biological systems. Chance and Oshino (1971) estimated the rate of hydrogen peroxide formation in purified rat liver microsomal fraction and, assuming microsomal protein to be 1% of the total liver weight, the rate of formation can be expressed as 9 μmol H_2O_2/min/kg liver. In 1989 the maximum dose rate obtainable from the Newcastle $^{60}Co\gamma$ source was ~30 Gy/min, so that [taking G(OH) = 0.3 μmol J^{-1}] the rate of •OH radical formation in water exposed to this dose rate would be 9 μmol •OH/min/kg water. This comparison (Jones, 1989) suggests that the avoidance of hydroxyl free radical formation in hydrogen peroxide metabolism is, indeed, highly desirable. For the evolution of a heme-based catalase it would be advantageous: (1) to select a structure that does not permit reduction of native ferricatalase to ferrocatalase and (2) to select a mechanism that avoids the formation of reactive oxygen species (ROS). Both criteria are substantially met by the heme catalases.

In their attempts to uncover the mechanism of catalase action the early researchers had few techniques available; activity assays required very low enzyme concentrations (<10 pM) and were prone to major problems arising from enzyme inactivation in both the absence and presence of substrate. Chance decided to attempt to detect catalytic intermediates using UV-vis spectrophotometry, an approach that had found success in (more straightforward) peroxidase studies. Some major workers in the field were discouraging; Keilin and Hartree (1945, p. 148) wrote, "the reaction with hydrogen peroxide consists in a violent, almost explosive, decomposition of the latter, during which it is impossible to observe any changes in colour and absorption spectrum of the enzyme." Although studies were initiated in 1938, "due to a large number of technical difficulties," the first results were reported nine years later (Chance, 1947). At the same time, important technical improvements in classical kinetic assays were described (Bonnichsen et al., 1947) using higher enzyme concentrations (~1 to 10 nM) and rapid titration techniques.

With these and other careful studies it became possible to begin to develop a mechanism of catalase action. Before 1981 there was, however, an important gap in the knowledge base. Although it had been established that the active enzymes are tetramers of identical subunits (each containing a ferriheme center), which act independently during catalysis, the enzyme molecular structures were unknown. The triumphs of protein x-ray crystallography have brought a molecular context to the discussion of mechanism in the post-1981 period—although the results have often led to a more sophisticated posing of questions, without directly providing answers.

PROGRESS

1947–1981, The Peroxidatic Mechanism

The spectroscopically distinct intermediate first observed by Chance (1947), was shown to be kinetically capable of fulfilling the role of a catalytic intermediate and has become known as "Catalase Compound I." In the steady state only 25% to 30% of the heme centers are converted to Compound I when hydrogen peroxide is the

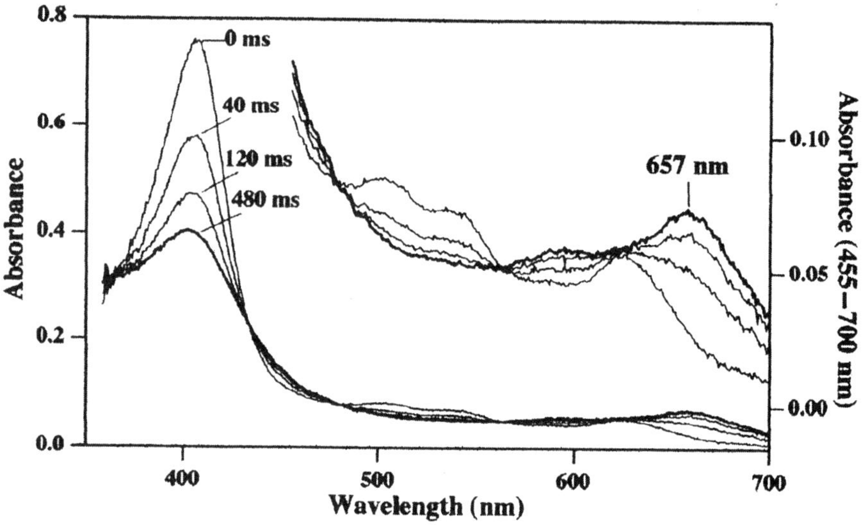

Figure 15.1. Stopped-flow rapid-scan study of the formation of *Exiguobacterium oxidotolerans* (EKTA) catalase Compound I by reaction with methyl hydroperoxide. (Reproduced with permission from Hara I et al., 2007. *Biochemistry* **46**: 11–22.)

oxidant. Complete formation of Compound I using methyl hydroperoxide is illustrated from a recent example in Figure 15.1 (Hara et al., 2007).

The kinetic and spectrophotometric data obtained with hydrogen peroxide were rationalized by the basic peroxidatic mechanism (the rate constant numbering is traditional):

$$\text{Catalase} + H_2O_2 \xrightarrow{k_1} \text{Compound I} \quad (15.3)$$

$$\text{Compound I} + H_2O_2 \xrightarrow{k_4} \text{Catalase} + \text{Products} \quad (15.4)$$

Although it was first thought that Compound I was an enzyme-substrate adduct (albeit a very peculiar one), it became clear that the intermediate is an oxidized form of the enzyme (later shown to be a hypervalent iron ($Fe^{IV}{=}O$, ferryl) species, surrounded by a porphyrin π-cation radical ($Por^{\bullet+}$) (Dolphin et al., 1971; Benecky et al., 1993), which contains both oxidizing equivalents of its parent hydrogen peroxide molecule. The peroxidatic mechanism can be rewritten to express the heme redox chemistry involved:

$$\text{Catalase } Fe^{III}\text{ Por} + H_2O_2 \xrightarrow{k_1} \text{Compound I } Fe^{IV}{=}O\text{ Por}^{\bullet+} + H_2O \quad (15.5)$$

$$\text{Compound I } Fe^{IV}{=}O\text{ Por}^{\bullet+} + H_2O_2 \xrightarrow{k_4} \text{Catalase } Fe^{III}\text{ Por} + H_2O + O_2 \quad (15.6)$$

The overall catalytic rate constant, k_0, and the individual rate constants are very large (typically $>10^7\ M^{-1}\ s^{-1}$), are independent of pH over a wide range ($\sim$4 to 10), and have very low activation energies (less than diffusion control; Job et al., 1993). Jones and Dunford (1977) argued that these data imply that a mechanistically significant, reversible enzyme-substrate complex acts as a Compound I precursor in the reactions of both catalase and horseradish peroxidase. In the latter reaction a reversible adduct has been observed (Baek and van Wart, 1989) at low temperatures. Ogura (1955) used a double mixer, quenched flow technique to study the overall kinetics at high hydrogen peroxide concentrations but, because of the high exothermicity of the reaction and the essentially adiabatic nature of the technique, the experiments were nonisothermal and extrapolation to zero time was necessary. In a later study (Jones and Suggett, 1968) the temperature rise was restricted to 1°C. Both studies reported Michaelis–Menten kinetics with $K_M = 1.1$ M. Jones and Suggett (1968) suggested an extended peroxidatic mechanism:

$$\text{Catalase} + H_2O_2 \rightleftharpoons \underset{\text{Compound 0}}{\text{Catalase} - H_2O_2} \tag{15.7}$$

$$\underset{\text{Compound 0}}{\text{Catalase} - H_2O_2} \longrightarrow \text{Compound I} + H_2O \tag{15.8}$$

$$\text{Compound I} + H_2O_2 \longrightarrow \text{Catalase} + H_2O + O_2 \tag{15.9}$$

Ogura (1955) suggested the formation of a reversible adduct by Compound I.

$$\text{Compound I} + H_2O_2 \rightleftharpoons \underset{\text{Ogura Complex}}{\text{Compound I} - H_2O_2} \tag{15.10}$$

$$\underset{\text{Ogura Complex}}{\text{Compound I} - H_2O_2} \longrightarrow \text{Catalase} + H_2O + O_2 \tag{15.11}$$

It seems likely that the most satisfactory extended peroxidatic mechanism is a combination of both concepts (i.e. replacing Eq. 15.9 with 15.10 and 15.11).

Compound I can also be formed by reaction of a range of small alkyl and acyl hydroperoxides with the enzyme (see Schonbaum and Chance, 1976). Complete Compound I formation can be achieved in these systems but only with an excess of hydroperoxide, implying a measure of complexity. Although with the alkyl hydroperoxides the reactions are pH independent (but see below), with peracetic acid (Jones and Middlemiss, 1972) the rate of reaction decreases as the acid ionizes with increasing pH. The size of the end group of these pseudo-substrates is critically important for reactivity; hydrogen peroxide typically reacts at least an order of magnitude faster than methyl hydroperoxide and two to four orders of magnitude faster than ethyl and hydroxymethyl hydroperoxides and peracetic acid (depending on the particular enzyme), suggesting an important steric constraint on access to the active site. Compound I generated by these methods, and also by using hydrogen peroxide generating oxidase systems, has been used extensively in studies of reactions of donor molecules other than hydrogen peroxide. Of major interest are those donors that react by $2e^-$ equivalent redox processes, so that the half reduced Compound II species is not observed. These donors include formic acid (but not formate), nitrous acid (but not nitrite), and the lower alcohols. Nitrous acid is a particularly active donor (its reactivity approaching that of hydrogen peroxide) but its maximum activity is only evident at low pH. Ethanol has a much lower reactivity (more than three orders of magnitude lower than hydrogen peroxide) but its reaction is pH independent over a wide range. Competition between ethanol and hydrogen peroxide for Compound I has been demonstrated at very high ethanol concentrations (Kremer, 1970).

The requirement for neutral substrates appears to extend to the reactions of catalase with ligands. For example, in the catalase-cyanide system, HCN has been regarded as the reactant. Nevertheless, since FeCN seems to be the only possible product (Schonbaum and Chance, 1976) it is important to know where and how the proton is retained. There is, as yet, no definite answer to this question. A study of chloride binding to beef liver catalase by Litchfield (1977) was interpreted as an interaction of molecular HCl with the enzyme but is, more reasonably, accounted for by binding of chloride to a protonated enzyme residue, the base form of which controls enzyme activity and the acid form has $pK_a < 4$. When chloride is bound the apparent pK_a controlling enzyme inactivation shifts to ~ 6. This behavior shows very close parallels with the formation of horseradish peroxidase Compound I, which has been extensively studied and interpreted by Dunford (see Chapter 3) who has argued compellingly that the rate-controlling base residue is the essential distal histidine-imidazole.

Catalase Compound II, a ferryl catalase species formed by the $1e^-$ reduction of the porphyrin π-cation radical of Compound I, brings some intriguing complications into catalase chemistry. In marked contrast to horseradish peroxidase, where the Compound II derivative is an important catalytic intermediate, catalase Compound II is catalatically inactive. When native catalase reacts with a hydroperoxide pseudo-substrate, the Compound I formed converts to Compound II at a rate that is the sum of two kinetic terms: (1) a reaction involving a mysterious endogenous donor, which is presumed to be a component of the enzyme protein but remains unidentified; and (2) a process which, when hydroperoxide is present in excess, proceeds at a rate proportional to the excess peroxide (Fig. 15.2; Marklund, 1972). The formation of Compound II during the catalatic reaction with hydrogen peroxide complicates the reaction kinetics.

Jones and Suggett (1968) separated the catalytic kinetics and the enzyme inactivation kinetics over a wide range of hydrogen peroxide concentration and showed that both processes follow Michaelis–Menten kinetics with the same K_M value. This behavior can be rationalized if catalase Compound I can react with hydrogen peroxide in two distinct ways: (1) a $2e^-$-equivalent redox reaction as in Equation 15.6 and (2) a $1e^-$ process leading to Compound II and perhydroxyl radical (or its conjugate base) as coproduct:

$$\text{Compound I Fe}^{IV}\!=\!\text{O Por}^{\bullet +} + H_2O_2 \longrightarrow \text{Compound II Fe}^{IV}\!=\!\text{O Por} + H^+ + HO_2^{\bullet} \tag{15.12}$$

The data for *M. luteus* (formerly *lysodeikticus*) catalase at pH 7, 20°C, imply that reaction (15.12) occurs only once in every 10^8 catalatic cycles. Thus catalase meets the

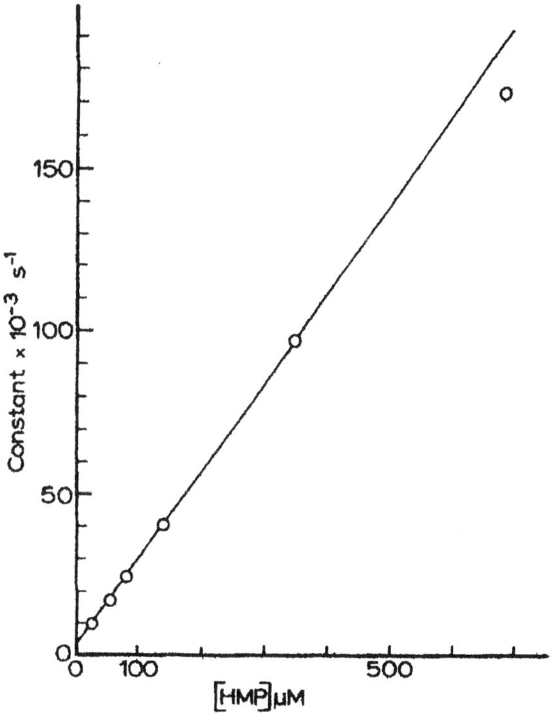

Figure 15.2. An example of complex catalase kinetics. Native beef liver catalase is converted to Compound II by reaction with hydroxymethyl hydroperoxide. (Reproduced with permission from Marklund S. 1972. *Biochim. Biophys. Acta* **289**: 269–275.)

criteria for avoiding oxygen free radical formation with remarkable efficiency. The situation may be even better in biological systems where the hydrogen peroxide concentration is low so that the "endogenous donor" pathway dominates, particularly if there is an additional pathway for repair of the endogenous donor.

Post 1981

The immediate consequence of the determination of the high resolution crystal structure of beef liver catalase (Fig. 15.3; Murthy et al., 1981; Reid et al., 1981), was to make it evident that earlier speculation about the co-catalytic role(s) of the enzyme protein in the catalatic reaction was largely unsound. The evidence from studies of the irreversible specific inhibitor 3-aminotriazole, which strongly implicated an important histidine-imidazole in the active site, was upheld but the (postulated histidine-imidazole) proximal ligand to ferriheme iron was shown to be a tyrosine-phenolate anion. The postulated water or hydroxide distal ligand to iron is not present (although some later studies of other catalases suggest water may be weakly bound). The assignment of a molecular form to the peroxidatic mechanism now began (Fita and Rossmann, 1985a) and more detailed studies of the molecular properties of the enzymes were stimulated.

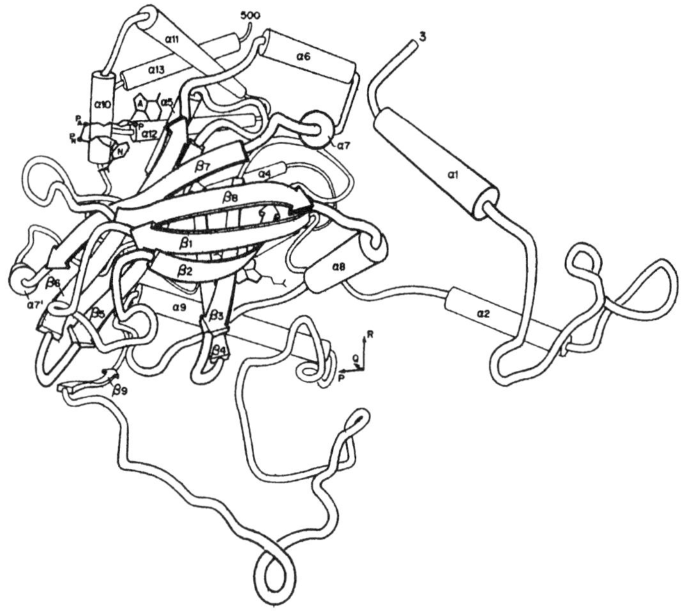

Figure 15.3. Crystal structure of one subunit of beef liver catalase showing the NADPH binding site. (Reproduced with permission from Fita I, Rossmann MG. 1985b. *Proc. Natl. Acad. Sci. USA* **82**: 1604–1608.)

It has now become clear that the ferriheme catalases are a much more complex family of enzymes than had been supposed; see Nicholls et al. (2001) and Chelikani et al. (2004). The active tetrameric enzyme had been considered to be fairly uniform in size (MW $\sim$250 kDa) but two groups are now recognized: small subunit (55 to 69 kDa) enzymes containing heme *b* and large subunit (75 to 84 kDa) enzymes in which the heme group has been flipped through 180° and modified to heme *d*. Analysis of more than 200 catalase sequences has revealed three genetically distinct groups (termed clades) which all arise from a single progenitor catalase (Klotz et al., 1997; Klotz and Loewen, 2003). It has been shown that proteolytic cleavage of a large subunit catalase can convert it fairly successfully into a catalase with small subunits (Chelikani et al., 2005).

The formation of tetrameric catalase involves a very complex folding and interweaving process in which subunit–subunit association must begin before subunit folding is complete. Figure 15.4 shows a graphic representation of the intimate embraces of pairs of subunits in which the N-terminal arms are interwoven through loops in the adjacent subunit. The resultant structures have variable thermal stabilities, which reflect the extent and nature of the network of interactions. Alkaline unfolding and salt-induced folding have been examined (Prajapati et al., 1998). A view of the beef liver catalase dimer structure is shown in Figure 15.5 and of the HP-II catalase tetramer from *Escherichia coli* in Figure 15.6 (Bravo et al., 1995).

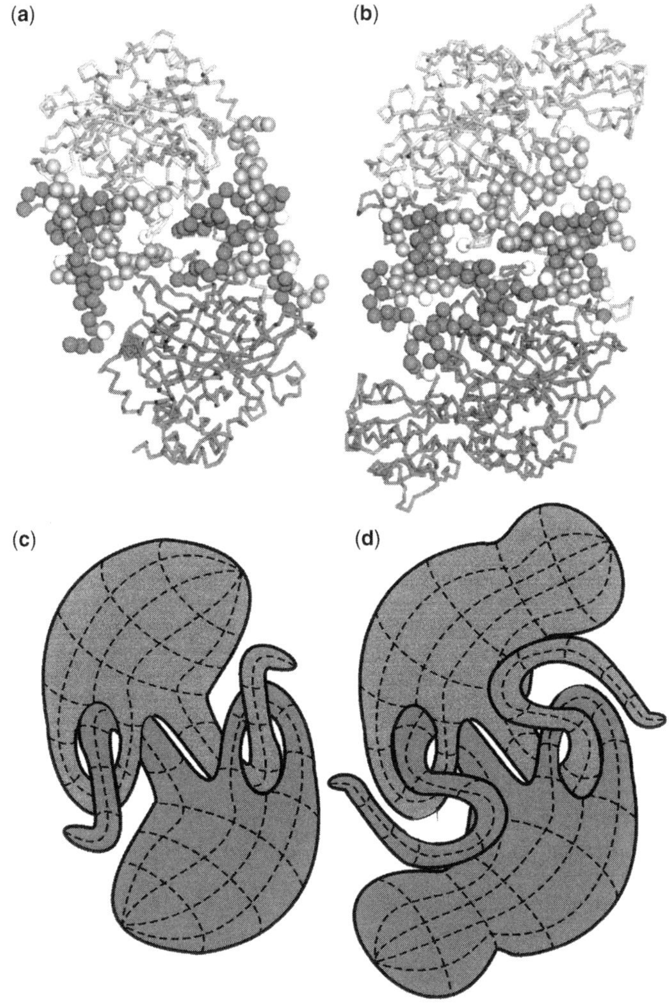

Figure 15.4. Crystal structures and schematic representations of the interweaving of two subunits for a small subunit enzyme (beef liver catalase) and a large susbunit enzyme (*E. coli* HP II). (Kindly supplied by Dr. Peter Loewen.)

In the post-1981 years the crystal structures of a dozen catalases in all three clades have been described (for a list of references see Hara et al., 2007). In the monomer units of the active tetramers the ferriheme groups are deeply buried (the depth varying from 25 to 55 Å from the protein surface). The most evident common feature, and probable route for substrate access to the heme, is provided by a channel (the main or vertical channel) from the surface perpendicular to the distal face of the heme plane. About 15 Å from the heme iron the channel narrows, forming a bottleneck the interior of which is lined mainly with hydrophobic residues. Studies of this

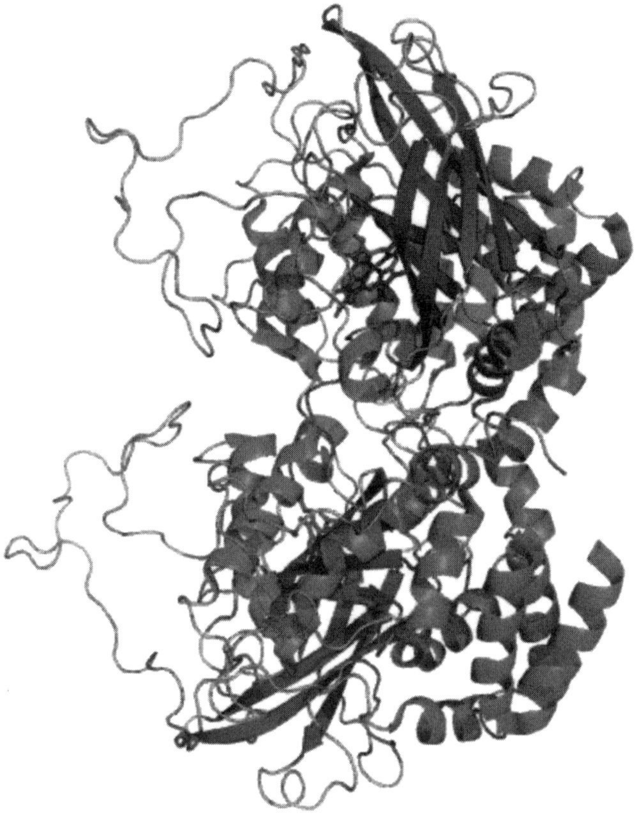

Figure 15.5. A structural view of the beef liver catalase dimer. (Kindly supplied by Dr. Tom Poulos.)

region by site-directed mutagenesis have been productive. Enlarging the channel by replacement of larger with smaller hydrophobic residues causes a decrease in catalatic activity but an increase in peroxidatic activity. Changing a conserved aspartate near the top of the bottleneck for any uncharged residue caused a loss of enzyme activity (Chelikani et al., 2003). In the "active site" surrounding the heme iron, the distal pendant histidine-imidazole is truly "essential" for catalatic activity; asparagine and serine residues are reaction "facilitators" but not essential. Mutation of the proximal tyrosine ligand to iron abolished accumulation of any protein.

These results have led to some intriguing suggestions about protein co-catalysis. On the basis of the dimensions and nature of the bottleneck it has been proposed that it acts by a "molecular ruler" effect to select and orient hydrogen peroxide for entry into the active site (Putnam et al., 2000). Extending this proposal, Chelikani et al. (2003) noted the influence of the negatively charged aspartate residue and suggested that its effect could arise from an electrical potential field between the negative charge and the positively charged heme iron. Hara et al. (2007) reported a

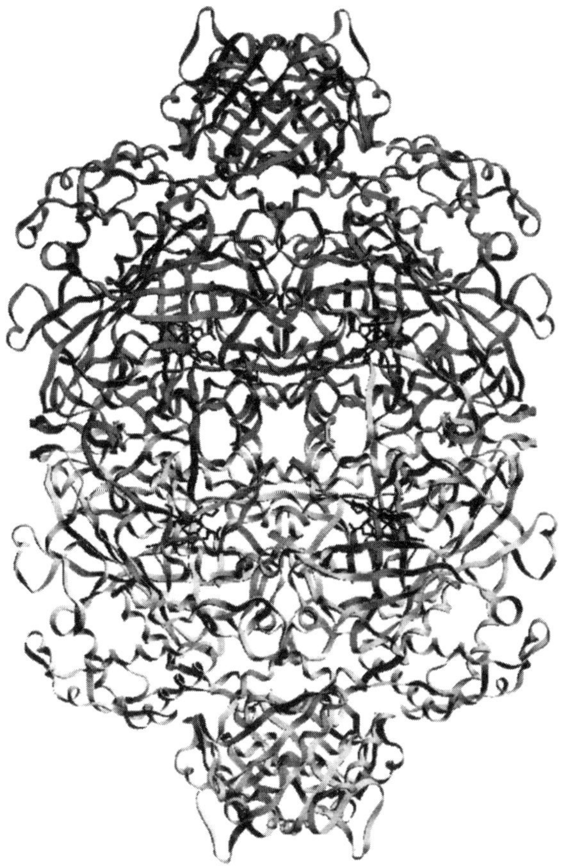

Figure 15.6. Structure of the catalase tetramer HP-II from *E. coli.* (Kindly supplied by Dr. Peter Loewen.)

comparative study of the rate constants for the formation of Compound I with hydrogen peroxide and a range of hydroperoxides reacting with beef liver, *Micrococcus luteus*, and EKTA catalases. (EKTA catalase is a clade 1 catalase from a new microorganism, *Exiguobacterium oxidotolerans* T-2-2^{T}, found in a drain pool of a fish processing plant in which hydrogen peroxide is used as a bleaching agent.) The diameters of the bottleneck entrances vary significantly: 2.7A (EKTA); 2.3A (BLC) and 2.1A (MLC). The rate constants with methyl hydroperoxide, peracetic acid, and *t*-butyl hydroperoxide all increase markedly and systematically with increasing size of the bottleneck. These results match the pattern noted by Schonbaum and Chance (1976, p. 395), who concluded that this: "strongly suggests that enzyme-peroxide association is the rate-determining step and, as such, is rather irrelevant to the elucidation of the redox transformations." In this case the exclusion of peroxyanions from the active site (Palcic and Dunford, 1980; Jones and Middlemiss, 1972), must occur at the entrance to the bottleneck, either by the hydrophobicity of the lower channel, or by the aspartate anion acting as an "electrostatic gate" (Davies et al., 1976) or by a combination of the two effects. It might be argued that this is not functionally relevant for catalatic action since $pK_a(H_2O_2) \sim 11.6$, but such a feature could be important in preventing the escape of a perhydroxyl anion unless neutralized by its proton partner, hence preserving the pH independence of the reaction. Since the rate constants for reaction with hydrogen peroxide are much higher, and not significantly influenced by bottleneck dimensions, it seems that access of the natural substrate is not seriously restrained and reaction is determined by the nature and topography of the essential groups in the active site and their co-catalytic effects on the mechanism, kinetics, and energetics of the redox process.

The construction of a plausible catalatic reaction mechanism on the basis of the structure of the native enzyme only is a formidable problem and the molecular graphics study of Fita and Rossmann (1985a) was a considerable feat (Fig. 15.7).

A notable feature was that the model showed how the constraints on access to the beef liver catalase active site impressed the observed stereoselectivity on the oxidation of ethanol by Compound I. Nevertheless, the model is essentially geometric rather than dynamic and this is largely because there is a remarkable lack of

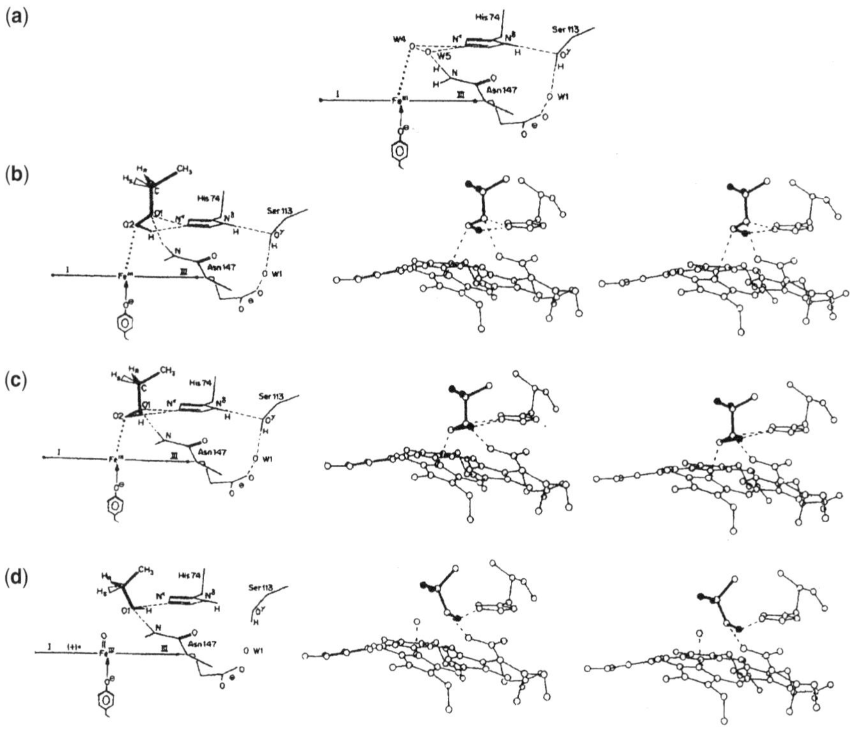

Figure 15.7. Formation of catalase Compound I using ethyl hydroperoxide: a molecular graphics study. (Reproduced with permission from Fita I, Rossmann MG. 1985a. *J. Mol. Biol.* **185**: 21–37.)

information concerning the acid-base properties of the essential distal histidine-imidazole. Schonbaum and Chance (1976, p. 395) noted that: "The co-catalytic role of the apoprotein in facilitating Compound I generation is increasingly regarded as that of a general acid–general base." For the formation of horseradish peroxidase Compound I, detailed studies (see Chapter 3) have made a compelling case that the imidazolium form of the essential distal histidine is highly acidic ($pK_a = 2.5$) and that only the base form is active. A serious complication is the potential for distortion of the pH effects by acid induced anion binding, which has only been quantitatively resolved for horseradish peroxidase. For yeast cytochrome *c* peroxidase, the controlling active site ionization is less well defined ($pK_a < 4$). For catalases the most useful study is that of Litchfield (1977) since it extends measurements to lower pH and separates pH and anion binding effects. The lack of detailed studies is unsurprising since the possibilities of acid-induced protein unfolding and heme release suggest a "can of worms" situation. However, Palcic and Dunford (1980) have shown a way forward in which a fast reaction (human erythrocyte catalase + MeOOH) can be studied separately from very fast processes (acid-base reactions of amino acid residues) and much slower processes (protein unfolding) using pH jump-stopped flow-with rapid scanning technology. Detailed studies for the reactions of well-characterized catalases at low pH would be extremely valuable.

Poulos and Kraut (1980) proposed (originally for the yeast cytochrome *c* peroxidase reaction) that hydrogen peroxide entering the active site donated a proton to the base form of distal His-imidazole as the first step in Compound I formation. This would be highly unfavorable, since $pK_a(H_2O_2) \sim 11.6$, but becomes more favorable if Compound 0 is first formed, since the pK_a of an Fe^{III}porphyrin(H_2O_2) complex can be estimated as 3.2 to 4.0 (Jones and Dunford, 2005). The situation becomes even more favorable as formation of Compound I initiates since, in the product, the corresponding $pK_a = 5.0$. As noted above, the proton, when transferred, cannot be released to the external milieu. A scheme to address this has been suggested in which Compound 0 comprises a hydrogen bonding network (involving ferriheme, substrate, imidazole and, possibly, a water molecule) which remains intact as the necessary proton transfer occurs to form a second intermediate, Compound 0*. It was argued (Jones and Dunford, 2005) that a scheme of this type avoids the reaction energetic problem that the high enthalpy of ionization of hydrogen peroxide provides (see Job et al., 1993).

The complex structures of the catalases show a second (minor or lateral) channel, which runs from the enzyme surface, approaching the ferriheme essentially in the porphyrin plane. The question of the possible role of this channel in catalysis has been debated but not resolved (Nicholls et al., 2001). Two further observations have suggested that this channel may be involved in more complex catalase chemistry, at least in mammalian and other clade 3 catalases, perhaps providing enzyme protection mechanisms. By an elegant piece of classical biochemical detective work, Kirkman and Gaetani (1984) showed that each monomer unit of tetrameric beef liver catalase and human erythrocyte catalase tightly binds a molecule of NADPH. Cystallographic analysis showed that the NADPH binds at the entrance to the minor channel, some 13.7 Å from the ferriheme (Fita and Rossmann, 1985b). Subsequent kinetic and stoichiometric studies found that NADPH protects the enzyme mainly by preventing the formation of Compound II (Kirkman et al., 1999), although a broader action was also suspected. Ivancich et al. (1996, 1997) found that decay of mammalian catalase Compound I resulted in the formation of a "Compound II-like species" containing a stable tyrosine radical (Tyr•). The site of this radical is not the proximal ligand, and Putnam et al. (2000) support Tyr 370 (a rather remote proximal residue) in human erythrocyte catalase. They propose that Tyr· and NADPH serve as alternative reduction pathways for catalase trapped in the Compound I state at low peroxide concentrations. The crystal structure of *Helicobacter pylori* catalase has been determined with and without formic acid bound at 1.6 Å resolution (Loewen et al., 2004). It is the only known clade 3 catalase that does not bind NADPH. The formic acid binding has been examined theoretically using density functional molecular dynamics simulations (Rovira et al., 2006). The heme environments in Compound I (Green, 2001) and Compound II (Rovira and Fita, 2003) have been examined theoretically.

It is likely that catalases have developed a variety of other approaches to the problem of enzyme inactivation during catalysis. The unusual covalent bonds between amino acid residues in some clade 2 catalases may contribute to enzyme protection. The clade 1 EKTA catalase is a small subunit enzyme, has no NADPH binding site, and shows no posttranslational modification of proximal tyrosine—and yet, EKTA catalase Compound I is very stable and decays to native enzyme with very little formation of Compound II (Hara et al., 2007). Catalases possess some oxidase activity (Vetrano et al., 2005).

Ferrihemes as Catalase Models

The nomenclature of hemes (iron-porphyrin complexes) is chaotic in the literature. For most purposes the 1964 recommendations of the International Union of Biochemistry (IUB) remains the most useful approach to systematic trivial names. The generic term heme (or haem) denotes an iron-porphyrin complex and, when the oxidation state of iron is to be specified, an appropriate prefix should be used (e.g. ferriheme). Additional prefixes, following traditional usage, differentiate various hemes. However, older trivial names continue to be used and inorganic chemists commonly employ different sets of trivial names based on traditions of coordination chemistry. Thus, protoferriheme chloride (or other counter-anion) ≡ hemin chloride (chlorohemin or often simply hemin, sometimes hematin) ≡ protoporphyrinIX-iron(III)chloride (or variants thereof!).

It has long been known that protoferriheme possesses modest "catalatic" activity. However, ferriheme solutions show evidence of heme aggregation (dimerisation and, at high pH, polydispersity) and this complexity provided a deterrent against systematic studies of catalytic properties. The observation that deuteroferriheme (in which the vinyl groups of protoferriheme have been removed) is up to 100 time more active than protoferriheme at last provoked more detailed studies. The difference in activity is substantially accounted for by the differences in the dimerization constants for the two ferrihemes, and the monomeric ferrihemes are the predominant reactive species (Fig. 15.8). Similar experiments with an extended range of ferrihemes have confirmed this conclusion and studies up to 1978 have been reviewed (Jones and Wilson, 1978).

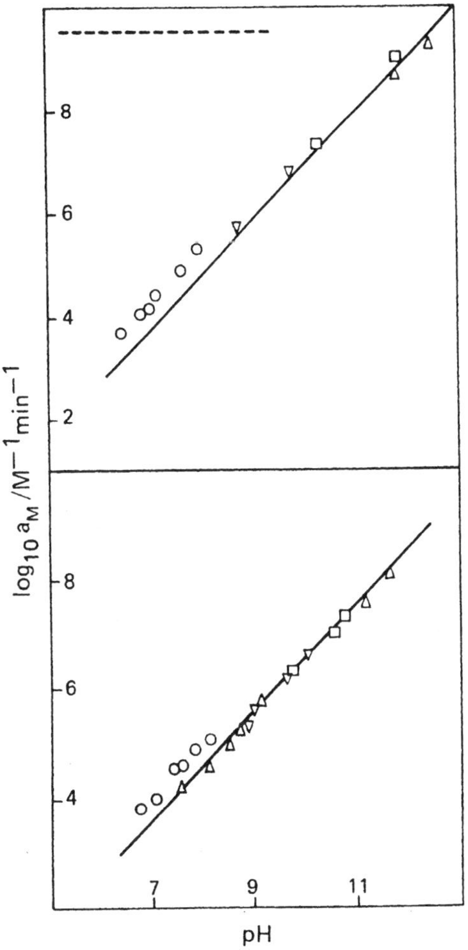

Figure 15.8. Dependence on pH of the catalactic activity of ferriheme monomers. Upper curve, protoferriheme monomer; lower curve, deuteroferriheme monomer. The same solid line of unit positive slope is shown in both plots. The dashed line represents the pH-independent activity of *M. luteus* catalase. (Reproduced with permission from Jones P, Wilson I. 1978. In: Sigel H, editor. *Metal Ions in Biological Systems*, Vol. 7. Marcel Dekker, New York, pp. 185–240.)

Dimerization constants have been determined by measurements of deviations from Beer's law and the nature and dynamics of the processes by temperature-jump relaxation measurements (Jones et al., 1974; Brown and Hatzikonstantinou, 1978a, 1978b). The very mobile equilibria observed under these conditions strongly suggest that the dimers are of the face-to-face type characterized in the NMR studies of uroferriheme by Satterlee and Shelnutt (1984, 1985), rather than the μ-oxo bridged species formed at very high pH. Figure 15.7 suggests, and extended studies confirm, that, although the influence of porpyrin peripheral structure on dimerization is substantial, the influence on monomer activity is small. This conclusion has also been supported by a peroxidatic oxidation study (Zipplies et al., 1986), using a highly monomeric, water soluble derivative of iron(III) tetraphenylporphyrin.

Figure 15.8 also compares the catalatic activity of ferriheme monomers with that of catalase; the ferriheme activity at pH $\sim$12 is comparable to the maximal activity of catalase. Of course, at pH 12 the enzyme activity is greatly reduced because the tetramer dissociates, so that, under these conditions, the ferriheme monomer is much more active than the enzyme. As Figure 15.8 also makes clear the activity of catalase is pH independent (in the active range), whereas the ferriheme monomer activity decreases with $[H^+]$, so that the enzyme is very much superior as a catalyst at physiological pH. Thus, the statement of Nicholls et al. (2001, p. 52) that "Heme alone can reportedly elicit a catalatic reaction but at a much reduced, almost negligible, rate compared to the catalatic proteins containing heme," needs to be taken with a pinch, not of salt, but of hydrogen ion.

Pre-steady-state studies searching for oxidized (Compound I-like) reaction intermediates using hydrogen peroxide yielded results only with deuteroferriheme (Portsmouth and Beal, 1971). The intermediate can be formed by reaction of the ferriheme monomer with a range of pseudo-substrates, including hydroperoxides (Jones et al., 1977; Kelly et al., 1977), with the nonperoxidic oxidants chlorite and hypochlorite (Kelly et al., 1981) and, surprisingly, by the 1e$^-$ oxidant, Br_2^- produced by pulse radiolysis or γ-radiolysis (Wilson et al., 1985). With highly monomeric coproferriheme a Compound I intermediate is formed with peracids in a reaction of 1:1 stoichiometry (Bretscher and Jones, 1988), but with deuteroferriheme the Compound I intermediate dimerizes with unoxidized deuteroferriheme to an extent which, at pH 7, simulates an irreversible comproportionation reaction. Low temperature studies have reported sequential formation of intermediates in the deuteroferriheme-*m*-chloroperoxybenzoic acid reaction (Imper and Van Wart, 1996). That the intermediate dimerization reaction is reversible is revealed by studies of the oxidation of phenol, a donor that reacts only with the monomeric intermediate, and shows competitive inhibition by unoxidized deuteroferriheme (Jones et al., 1982). Some other donors are less discriminating, for example, anilines (Frew and Jones, 1983), where the oxidation rates are remarkably similar to those with horseradish peroxidase Compound I. Thus the ferrihemes act as undifferentiated "catalatic" and "peroxidatic" catalysts, producing highly reactive oxidized heme intermediates but showing little donor selectivity. However, the kinetics of intermediate formation severely restrict overall catalytic efficiency because the proton released from bound hydrogen peroxide acceptor substrate equilibrates with the milieu. Effectively the substrate is perhydroxyl anion, which is present at very low concentration at physiological pH, and the activation energy is high (higher than the ionization enthalpy of hydrogen peroxide (Job et al., 1993). A different perspective on mechanisms is offered by setting to music heme model and catalase reactions (Jones, 1985).

CATALASES IN BIOLOGY

The roles of catalases in biology remain insecurely defined, especially in higher organisms. For example, although human erythrocytes contain large amounts of catalase, the conclusion of Cohen and Hochstein (1963) that catalase is unimportant in protecting erythrocytes from endogenous hydrogen peroxide, and that the important pathway is via glutathione peroxidase, was widely accepted. However, a more recent comparison of normal and acatalasemic erythrocytes concluded that the catalase and glutathione peroxidase pathways are equally involved in the removal of hydrogen peroxide (Gaetani et al., 1989). Since both mechanisms depend (in different ways) on the generation of NADPH, a failure of this process, as in genetic deficiency of glucose-6-phosphate dehydrogenase, makes the erythrocytes very susceptible to oxidative hemolysis.

The pros and cons of the possible role of catalase in ethanol metabolism have long been debated. Physicochemically it is not a straightforward matter to demonstrate the precise conditions for competition between the donor's hydrogen peroxide and ethanol for Compound I (Kremer, 1970). It is possible that relatively slow release of product acetaldehyde may attenuate input of substrates under competition conditions. Nicholls et al. (2001) suggest that the amount of catalase expressed in cells under most conditions will keep the maximum peroxide concentration during a pulse of production below 0.1 to 0.2 μM. The maximum permitted blood alcohol concentration for driving a car in the United Kingdom is $\sim$17 mM. At these levels ethanol should compete effectively for Compound I; much higher alcohol levels are encountered in addicted individuals. Acetaldehyde formed in the brain in this way (Hamby-Mason et al., 1997) is implicated in the neurological effects of alcohol in humans (Hunt, 1996). This biochemistry represents an aberrant side reaction of catalase induced by a particular behavior of individuals.

The formation of hydrogen peroxide in respiring organisms, either directly or via superoxide, is both a burden (implicated in a wide range of ROS-mediated conditions; Halliwell and Gutteridge, 1984), and a necessity (see earlier chapters on specialized peroxidases). Studies of the protective "caretaker" roles of catalases are fragmented but increasing. Far from the evolutionary dead end concept of a "fossil" enzyme, it seems likely that catalases play vital but usually unspectacular roles in coping with the excess by-products of oxygen metabolism. It is plausible that the duality of catalases and glutathione peroxidase represents an evolutionary "belt and braces" approach to a persistent problem.

Undoubtedly the most spectacular and well-documented example of the catalatic reaction in biology (although it appears not to be widely recognized) is the defensory discharge of the Bombardier beetle (Fig. 15.9). The bioengineering, chemistry, thermodynamics, kinetics and enzymology of this process have been described in a series of masterly papers (Aneshanley et al., 1969; Dean et al., 1990; Eisner and Aneshanley, 1999; Schildknecht and Holoubek, 1961; Schildknecht et al., 1970).

When attacked the insects fire a defensive spray from glands in the abdominal tip with an audible report. The insect can aim the spray at an assailant from any direction and can fire more than 20 times in rapid succession, each discharge lasting <30 ms. The spray consists of a 25% hydrogen peroxide solution, containing 10% hydroquinone and methyl hydroquinone, which is squeezed into a vestibule where admixture with catalases and peroxidases induces decomposition of hydrogen peroxide and coupled oxidation of the hydroquinones with explosive violence. The reaction occurs essentially adiabatically; the heat of reaction converts the water to steam and

Figure 15.9. Explosive discharge of the Bombardier beetle: a spectacular biological application of the catalatic reaction. (Photo kindly supplied by Drs. T. Eisner and D. Aneshanley, Cornell University, Ithaca, New York.)

this, together with the oxygen generated, propels the spray (which also contains residual hot peroxide and repellent quinones) over the predator at 100°C. In order to avoid anal self-harm the insect employs a pulse-jet mechanism! The major enzymes recovered from the explosion chambers were a small subunit homotetramer heme catalase and a glycoprotein monomeric heme peroxidase, which possesses substantial diphenol peroxidase activity.

PROSPECTS

The elegant crystallographic methods developed by Hajdu and colleagues for the study of peroxidase intermediates (Berglund et al., 2002) have also been applied to catalase intermediates (Gouet et al., 1996). The only significant structural changes observed are small movements of heme iron, from slightly below the porphyrin plane (towards the proximal ligand) to slightly above in Compound I and back into the plane in Compound II. Are these movements mechanistically significant?

There are remarkably close parallels between all aspects of the formation of the Compound I intermediates of the catalases and peroxidases. Indeed, the similarities are so close as to suggest that the complex architecture of the catalases has little to do with the *formation* of Compound I but is almost entirely concerned with selection, kinetics, and mechanism of *reduction* of Compound I by donor hydrogen peroxide. The steric constraints of the access channel ensure that catalases are not good general peroxidases; but how is efficient $2e^-$ oxidation of donor hydrogen peroxide achieved? A classical peroxidase intermediate, such as horseradish peroxidase Compound I, is capable only of $1e^-$ oxidations, except when the essential His-imidazole becomes protonated (see Chapter 3) and $2e^-$ oxidation of, for example, iodide becomes possible. Ortiz de Montellano (1992) has argued that the $1e^-$ oxidations involve electron transfer into the heme edge of the porphyrin radical and that the iron atom is not readily accessible to substrates. But hydrogen peroxide, as an acceptor substrate, has facile access to heme iron in peroxidase and, with horseradish peroxidase, peroxybenzoic acids react even faster. So what is the origin of the constraint that prevents horseradish peroxidase Compound I displaying significant catalatic properties? It has been speculated (Jones, 2001) that active site water may be of importance in controlling the accessibility of $1e^-$ and $2e^-$ oxidation pathways in Compound I species, blocking access to the $2e^-$ oxidation site in the horseradish intermediate. For catalase, a corollary is that the Compound I active site should be "dry" to allow $2e^-$ oxidation of hydrogen peroxide. A mechanism for drying out the active site by disposal of water reaction product is necessary if efficient throughput of substrate is to be maintained. Kalko et al. (2001) have considered this problem in a theoretical study and suggest that the lateral channel and another small channel, which runs from the heme to the monomer–monomer interface, may act as routes for product (water and oxygen) release rather than for substrate entry. However, Nicholls et al. (2001), consider that, at least in NADPH binding enzymes, the lateral channel cannot have either an inlet or an exhaust function. An encouraging aspect of the theoretical approach of Kalko et al. (2001) is a prediction, based on an analysis of competition between water and hydrogen peroxide for active site access, that the K_M for the catalatic reaction should be in the molar range, in reasonable agreement with the experimental results of Ogura (1955) and Jones and Suggett (1968). It begins to seem probable that the roles of water in the actions of the enzymes are both subtle and significant.

For improved understanding of catalase mechanisms direct studies of the properties of the essential active site His-imidazole and the important bottleneck aspartate are necessary, in the native enzyme and in the intermediates. Extension of native enzyme studies to low pH, of the type and in the detail pioneered by Palcic and Dunford (1980), and which are mindful of the problems of acid-induced anion binding noted by Litchfield (1977), would offer a useful beginning. Similar, although undoubtedly more complicated, studies of Compound I reactions would be very worthwhile.

Catalases react with hydrogen peroxide in at least six different ways and none of these reactions is adequately understood. The evolved oxygen has been claimed to be in an excited singlet state, a claim that appears unlikely (Alfonso-Prieto et al., 2008, and references therein). Adding to the mystery, these hemoproteins continue to show quite unpredictable and remarkable behavior; see, for example, Heck et al. (2003) and Alfonso-Prieto et al. (2007). The catalases remain perplexing, enigmatic, and utterly fascinating enzymes.

REFERENCES

Alfonso-Prieto M, Borovik A, Carpena X, Murshudov G, Melik-Adamyan W, Fita I, Rovira C, Loewen PC. 2007. The structures and electronic configuration of Compound I intermediates of *Helicobacter pylori* and *Penicillium vitale* catalases determined by X-ray crystallography and QM/MM density functional theory calculations. *J. Am. Chem. Soc.* **129**: 4193–4205.

Alfonso-Prieto M, Vidossich P, Rodriguez-Fortea A, Carpena X, Fita I, Loewen PC, Rovira C. 2008. Electronic state of the molecular oxygen released by catalase. *J. Phys. Chem. A* **112**: 12842–12848.

Aneshanley DJ, Eisner T, Widom JM, Widom B. 1969. Biochemistry at 100°C: explosive secretory discharge of Bombardier beetles (*Brachinus*). *Science* **165**: 61–63.

Baek HK, van Wart HE. 1989. Elementary steps in the formation of horseradish peroxidase Compound I: direct observation of Compound 0, a new intermediate with a hyperporphyrin spectrum. *Biochemistry* **28**: 5714–5719.

Barb WG, Baxendale JH, George P, Hargrave KR. 1951. Reactions of ferrous and ferric ions with hydrogen peroxide. II. The ferric ion reaction. *Trans. Faraday Soc.* **47**: 591–616.

Benecky MJ, Frew JE, Scowen N, Jones P, Hoffman BM. 1993. EPR and ENDOR detection of Compound I from *Micrococcus lysodeikticus* catalase. *Biochemistry* **32**: 11929–11933.

Berglund GI, Carlsson GA, Smith AT, Szöke H, Henrickson A, Hadju J. 2002. The catalytic pathway of horseradish peroxidase at high resolution. *Nature* **417**: 463–468.

Bonnichsen RK, Chance B, Theorell H. 1947. Catalase activity. *Acta Chem. Scand.* **1**: 685–709.

Bravo J, Verdaguer N, Tormo J, Betzel C, Switala J, Loewen PC, Fita I. 1995. Crystal structure of catalase HPII from *Escherichia coli*. *Structure* **3**: 491–502.

Bretscher KR, Jones P. 1988. Formation and peroxidatic properties of coprohaem Compound I. *J. Chem. Soc. Dalton Trans.* 2267–2272.

Brown SB, Hatzikonstantinou H. 1978a. The dimerization of ferrihaems. II. Equilibrium and kinetic studies of mesoferrihaem. *Biochim. Biophys. Acta* **539**: 352–363.

Brown SB, Hatzikonstaninou H. 1978b. The dimerization of ferrihaems. III. Equilibrium and kinetic studies of the dimerisation of coproferrihaem. *Biochim. Biophys. Acta* **544**: 407–417.

Chance B. 1947. An intermediate compound in the catalase-hydrogen peroxide reaction. *Acta Chem. Scand.* **1**: 236–267.

Chance B, Oshino N. 1971. Kinetics and mechanism of catalase in peroxisomes of the mitochondrial fraction. *Biochem. J.* **122**: 225–239.

Chelikani P, Carpena X, Fita I, Loewen PC. 2003. An electrical potential in the access channel of catalase enhances catalysis. *J. Biol Chem.* **278**: 31290–31296.

Chelikani P, Carpena X, Perez-Luque R, Donald LJ, Duckworth HW, Switala J, Fita I, Loewen PC. 2005. Characterization of a large subunit catalase truncated by proteolytic cleavage. *Biochemistry* **44**: 5597–5605.

Chelikani P, Fita I, Loewen PC. 2004. Diversity of structures and properties among catalases. *Cell Mol. Life Sci.* **61**: 192–208.

Cohen G, Hochstein P. 1963. Glutathione peroxidase: the primary agent for the elimination of hydrogen peroxide in erythrocytes. *Biochemistry* **2**: 1420–1428.

Davies DM, Jones P, Mantle D. 1976. The kinetics of formation of horseradish peroxidase Compound I by reaction with peroxobenzoic acids: pH and peroxoacid substituent effects. *Biochem. J.* **157**: 247–253.

Dean J, Aneshanley DJ, Edgerton HE, Eisner T. 1990. Defensive spray of the Bombardier beetle: a biological pulse jet. *Science* **248**: 1219–1221.

Dolphin D, Forman A, Borg DC, Fajer J, Felton RH. 1971. Compounds I of catalase and horse radish peroxidase: π-cation radicals. *Proc. Natl. Acad. Sci. USA* **68**: 614–618.

Eisner T, Aneshanley DJ. 1999. Spray aiming of the Bombardier beetle: photographic evidence. *Proc. Natl. Acad. Sci. USA* **96**: 9705–9709.

Fita I, Rossmann MG. 1985a. The active centre of catalase. *J. Mol. Biol.* **185**: 21–37.

Fita I, Rossmann MG. 1985b. The NADPH binding site of beef liver catalase. *Proc. Natl. Acad. Sci. USA* **82**: 1604–1608.

Frew JE, Jones P. 1983. Peroxidase-like activities of iron-III porphyrins: kinetics of the reduction of a peroxidatically active derivative of deuteroferriheme by anilines. *J. Inorg. Biochem.* **18**: 33–39.

Gaetani GF, Galiano S, Canepa L, Ferraris AM, Kirkman HN. 1989. Catalase and glutathione peroxidase are equally active in detoxification of hydrogen peroxide in human erythrocytes. *Blood* **73**: 334–339.

Gouet P, Jouve HM, Williams PA, Andersson I, Andreoletti P, Nussaumi L, Hadju J. 1996. Ferryl intermediates of catalase captured by time resolved Weissenberg, crystallography and UV-VIS spectroscopy. *Nature Struct. Biol.* **3**: 951–956.

Green MT. 2001. The structure and spin coupling of catalase Compound I: a study of noncovalent effects. *J. Am. Chem. Soc.* **123**: 9218–9219.

Halliwell B, Gutteridge JMC. 1984. Oxygen toxicity, oxygen radicals, transition metals and disease. *J. Biochem.* **219**: 1–14.

Hamby-Mason R, Schenker S, Chen JJ, Perez A, Henderson GI. 1997. Catalase mediates acetaldehyde formation from ethanol in fetal and neo-natal rat brain. *Alcohol Clin. Exp. Res.* **21**: 1063–1072.

Hara I, Ichise N, Kojima K, Kondo H, Ohgiya S, Matsuyama H, Yumoto I. 2007. Relationship between the size of the bottleneck 15 Å from iron in the main channel and the reactivity of catalase corresponding to the molecular size of substrates. *Biochemistry.* **46**: 11–22.

Heck DE, Vetrano AM, Mariano TM, Laskin JD. 2003. UV light stimulates production of reactive oxygen species: unexpected role for catalase. *J. Biol. Chem.* **278**: 22432–22436.

Hunt WA. 1996. Role of acetaldehyde in the actions of ethanol on the brain: a review. *Alcohol* **13**: 147–151.

Imper V, Van Wart HE. 1996. Catalytic intermediates in the reaction of deuteroferriheme with *m*-chloroperoxybenzoic acid. *Croat. Chem. Acta* **69**: 997–1006.

Ivancich A, Jouve HM, Gaillard J. 1996. EPR evidence for a tyrosyl radical intermediate in bovine liver catalase. *J. Am. Chem. Soc.* **118**: 12852–12853.

Ivancich A, Jouve HM, Sartor B, Gaillard J. 1997. EPR investigation of Compound I in *Proteus mirabilis* and bovine liver catalases: formation of porphyrin and tyrosyl radical intermediates. *Biochemistry* **36**: 9356–9364.

Job D, Jones P, Dunford HB. 1993. Evolution of enzyme activity: is diffusion control important? Activation parameters in the reactions of ferric heme species with hydrogen peroxide. *J. Phys. Chem.* **97**: 9259–9262.

Jones P. 1985. Musical mechanisms. *J. Chem. Ed.* **62**: 1093–1095.

Jones P. 1989. Radicals – free and otherwise in iron-complex peroxide reactions. In: Beaumont PC, Deeble DJ, Parsons BJ, Rice-Evans C, editors. *Free Radicals, Metal Ions and Biopolymers*. Richelieu Press, London, pp 45–53.

Jones P. 2001. Roles of water in heme peroxidase and catalase mechanisms. *J. Biol. Chem.* **21**: 13791–13796.

Jones P, Dunford HB. 1977. On the mechanism of Compound I formation from peroxidases and catalases. *J. Theor. Biol.* **69**: 457–470.

Jones P, Dunford HB. 2005. The mechanism of Compound I formation revisited. *J. Inorg. Biochem.* **99**: 2292–2298.

Jones P, Mantle D, Davies DM, Kelly HK. 1977. Hydroperoxidase activities of ferrihemes: heme analogs of peroxidase enzyme intermediates. *Biochemistry* **16**: 3974–3978.

Jones P, Mantle D, Wilson I. 1982. Peroxidase-like activities of iron(III)-porphyrins: kinetics of the reduction of a peroxidatically active derivative of deuteroferriheme by phenols. *J. Inorg. Biochem.* **18**: 33–39.

Jones P, Middlemiss DN. 1972. Formation of Compound I by reaction of catalase with peroxoacetic acid. *Biochem. J.* **130**: 411–415.

Jones P, Prudhoe K, Brown SB. 1974. Equilibrium and kinetic studies of deuteroferrihaem {chloro[dihydrogen-3,7,12,17-tetramethyl-2,18-porphinedipropionato(2^-)]iron(III)} dimerization. *J. Chem. Soc. Dalton Trans.* 911–913.

Jones P, Suggett A. 1968. The catalase-hydrogen peroxide system: kinetics of catalatic action at high substrate concentrations. *Biochem. J.* **110**: 5714–5719.

Jones P, Wilson I. 1978. Catalases and catalase-like iron complexes. In: Sigel H, editor. *Metal Ions in Biological Systems*, Vol. 7. Marcel Dekker, New York, pp. 185–240.

Kalko SG, Gelpi JL, Fita I, Orozco M. 2001. Theoretical study of the mechanisms of substrate recognition by catalase. *J. Am. Chem. Soc.* **123**: 9665–9672.

Keilin D, Hartree EF. 1945. Properties of azide-catalase. *Biochem. J.* **39**: 148–157.

Kelly HC, Davies DM, King MJ, Jones P. 1977. Pre-steady state kinetics of intermediate formation in the deuteroferriheme-hydrogen peroxide system. *Biochemistry* **16**: 3543–3549.

Kelly HC, Parigi KJ, Wilson I, Jones P, Roettger LJ. 1981. Chlorite ion oxidation of the iron(III) complex of deuteroporphyrin IX. *Inorg. Chem.* **20**: 1086–1090.

Kirkman HN, Gaetani GF. 1984. Catalase: a tetrameric enzyme with four tightly bound molecules of NADPH. *Proc. Natl. Acad. Sci. USA* **81**: 4343–4348.

Kirkman HN, Rolfo M, Ferraris AM, Gaetani GF. 1999. Mechanism of protection of catalase by NADPH. *J. Biol. Chem.* **274**: 13908–13914.

Klotz MG, Klassen GR, Loewen PC. 1997. Phylogenetic relationships among prokaryotic and eukaryotic catalases. *Mol. Biol. Evol.* **14**: 951–958.

Klotz MG, Loewen PC. 2003. The molecular evolution of catalatic hydroperoxidases: evidence for multiple lateral transfers of genes between prokaryota and from bacteria into eukaryota. *Mol. Biol. Evol.* **20**: 1089–1112.

Koubek E, Haggett ML, Battaglia CJ, Ibne-Rasa KM, Pyun HY, Edwards JO. 1963. Kinetics and mechanisms of the spontaneous decomposition of some peroxoacids, hydrogen peroxide and *t*-butyl hydroperoxide. *J. Am. Chem. Soc.* **85**: 2263–2268.

Kremer ML. 1970. Peroxidatic activity of catalase. *Biochim. Biophys. Acta* **198**: 199–209.

Litchfield WJ. 1977. Inactivation of catalase by chloride. *FEBS Lett.* **83**: 281–284.

Loewen PC, Carpena X, Rovira C, Ivancich A, Perez-Luque R, Haas R, Odenbreit S, Nicholls P, Fita I. 2004. Structure of *Helicobacter pylori* catalase, with and without formic acid bound, at 1.6 Å resolution. *Biochemistry* **43**: 3089–3103.

Marklund S. 1972. Interactions between hydroxymethyl hydroperoxide and catalase. *Biochim. Biophys. Acta* **289**: 269–275.

Murthy MR, Reid TJ III, Sicignano A, Tanaka N, Rossmann MG. 1981. Structure of beef liver catalase. *J .Mol .Biol.* **152**: 465–499.

Nicholls P, Fita I, Loewen PC. 2001. Enzymology and structure of catalases. *Adv. Inorg. Chem.* **51**: 51–106.

Nicholls P, Schonbaum GR. 1963. Catalases. In: Boyer PD, Lardy H, Myrback K, editors. *The Enzymes*, Vol. 8. Academic Press, New York, pp. 147–225.

Ogura Y. 1955. Catalase activity at high concentrations of hydrogen peroxide. *Arch. Biochem. Biophys.* **57**: 288–300.

Ortiz de Montellano PR. 1992. Catalytic sites of hemoprotein peroxidases. *Annu. Rev. Pharmacol. Toxicol.* **12**: 89–107.

Palcic MM, Dunford HB. 1980. The reaction of human erythrocyte catalase with hydroperoxides to form Compound I. *J. Biol. Chem.* **255**: 6128–6132.

Percy ME. 1984. Catalase: an old enzyme with a new role? *Can. J. Biochem. Cell Biol.* **62**: 1006–1014.

Portsmouth D, Beal EA. 1971. The peroxidase activity of deuterohemin. *Eur. J. Biochem.* **19**: 479–487.

Poulos TL, Kraut J. 1980. The stereochemistry of peroxidase catalysis. *J. Biol. Chem.* **255**: 8199–8205.

Prajapati S, Bhakuni V, Babu KR, Jain SK. 1998. Alkaline folding and salt-induced folding of bovine liver catalase at high pH. *Eur. J. Biochem.* **255**: 178–184.

Putnam CD, Arvai AS, Bourne Y, Tainer JA. 2000. Active and inhibited human catalase studies: ligand and NADPH binding and catalytic mechanism. *J. Mol. Biol.* **296**: 295–309.

Reid TJ III, Murthy MR, Sicignano A, Tanaka N, Musick WD, Rossmann MG. 1981. Structure and heme environment of beef liver catalase at 2.5 Å resolution. *Proc. Natl. Acad. Sci. USA* **78**: 4767–4771.

Rovira C, Alfonso-Prieto M, Biarnés X, Carpena X, Fita I, Loewen PC. 2006. A first principles study of the binding of formic acid in catalase complemeneting high resolution X-ray structures. *Chem. Phys.* **323**: 129–137.

Rovira C, Fita I. 2003. The proximal hydrogen-bonded residue controls the stability of the Compound II intermediate of peroxidases and catalases. *J. Phys. Chem. B* **107**: 5300–5305.

Satterlee JD, Shelnutt JA. 1984. Studies of urohemin: thermodynamics of self-association and electronic properties of two species detected by NMR spectroscopy. *J. Phys. Chem.* **88**: 5487–5492.

Satterlee JD, Shelnutt JA. 1985. Characterisation of pH dependent axial ligation changes of monomer and dimer forms of iron(III) uroporphyrin I in aqueous solution. *Inorg. Chim. Acta* **106**: 165–170.

Schildknecht H, Holoubek K. 1961. Die Bombardierkäfer und ihre explsionschimie. *Angew. Chem.* **73**: 1–7.

Schildknecht H, Maschwitz E, Maschwitz U. 1970. Die explosionschemie der Bombardierkäfer: struktur und eigenschaften der brennkammerenzyme. *J. Insect. Physiol.* **16**: 749–789.

Schonbaum GR, Chance B. 1976. Catalases. In: Boyer PD, editor. *The Enzymes*, Vol. 13. Academic Press, New York, pp. 363–408.

Schumb WC, Satterfield CN, Wentworth RL. 1955. *Hydrogen Peroxide*. American Chemical Society Monograph Series, No. 128. Reinhold, New York.

Vetrano AM, Heck DE, Mariano TM, Mishin V, Laskin DL, Laskin JD. 2005. Characterization of the oxidase activity in mammalian catalase. *J. Biol. Chem.* **280**: 35372–35381.

Wilson I, Bretscher KE, Jones P. 1985. One-electron oxidation of iron(III) porphyrins by pulse radiolysis. In: Bensasson RV, Jori G, Land EJ, Truscott TG, editors. *Primary Photo-Processes in Biology and Medicine*. NATO-ASI Series A: Life Sciences, Vol. 85. Plenum Press, New York, pp. 53–55.

Zipplies MF, Lee WA, Bruice TC. 1986. Influence of hydrogen ion activity and general acid-base catalysis on the rate of decomposition of hydrogen peroxide by a novel nonaggregating water-soluble iron(III) tetraphenylporphyrin derivative. *J. Am. Chem. Soc.* **108**: 4433–4445.

CHAPTER 16

MYELOPEROXIDASE: ENZYMOLOGY

JEFFREY HENDERSON and JAY HEINECKE

INTRODUCTION

This chapter addresses the enzymology of two mammalian peroxidases: myeloperoxidase and, in less detail, eosinophil peroxidase. Oxidation of guaiacol by pus was demonstrated in the nineteenth century and by leukocytes early in the twentieth century. The molecules responsible for the reaction were not clearly identified until much later. The enzyme, called verdoperoxidase because of its green color, was isolated from the sputum of tuberculosis patients (Agner, 1941). Later the same enzyme was isolated from leukocytes of a patient with myeloid leukemia (Agner, 1943), and from infected dog uteri pus (Agner, 1958; Harrison et al., 1977). The name verdoperoxidase was changed later to myeloperoxidase, because Hugo Theorell had discovered that lactoperoxidase was also green.

Leukocytes, or white blood cells, can be subdivided into those that contain a large number of granules, *granulocytes*, and those that contain a small number, originally thought to be nongranular, the *agranulocytes*. The granulocytes were subdivided into three categories based on their response to histochemical staining: *neutrophils*, *eosinophils*, and *basophils*. A property of granulocytes, particularly neutrophils, is that they possess an irregularly shaped nucleus, one with many lobes. For this reason neutrophils are also called *polymorphonuclear (PMN) leukocytes*. The precursor of a neutrophil is called a myelocyte, hence the choice of the name myeloperoxidase.

Phagocytes are leukocytes capable of *phagocytosis*, the ingestion of foreign cells. The major types of phagocytes are the neutrophils, eosinophils, and monocytes. The monocytes are immature agranulocytes, which become macrophages upon maturation. These relationships are summarized in Figure 16.1. The myeloperoxidases from neutrophils and monocytes are identical, but that from eosinophils is distinctly different and is called eosinophil peroxidase. The enzyme found in basophils is also different (Klebanoff, 1991). There is a comprehensive review of neutrophils (Klebanoff and

Peroxidases and Catalases: Biochemistry, Biophysics, Biotechnology, and Physiology, Second Edition.
By H. Brian Dunford

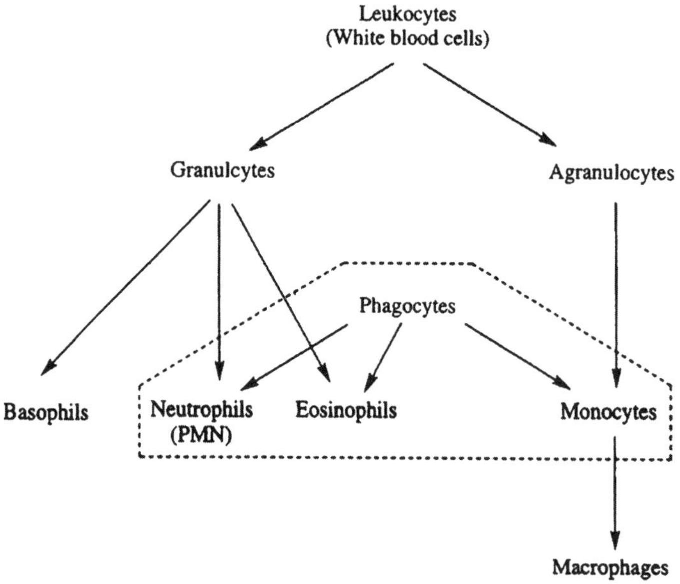

Figure 16.1. Classification of various types of leukocytes.

Clark, 1978), and a review of all phagocytic cells (Klebanoff, 1999). For an introduction to medical aspects of mammalian peroxidases see (Metodiewa and Dunford, 1993).

A green spleen heme protein was proven to be identical to neutrophil myeloperoxidase (Davis and Averill, 1981; Ikeda-Saito, 1985; Ikeda-Saito et al., 1989). Beef spleen is an excellent source of myeloperoxidase. A combination of three isolation procedures for the spleen enzyme was used (Marquez et al., 1990). The spleen enzyme is apparently obtained as a waste product from dead leukocytes. Improvements have been made on the isolation of myeloperoxidase from leukocytes (Odajima and Yamazaki, 1970; Bakkenist et al., 1978; Matheson et al., 1981; Andrews and Krinsky, 1981, 1986; Morita et al., 1986; Marquez et al., 1990). Myeloperoxidases from dog, pig, beef, and human sources are very similar.

Three isoenzymatic forms of human myeloperoxidase were isolated (Pember et al., 1983; Miyasaki et al., 1991). Some differences in reactivity were detected but they have identical EPR (electron paramagnetic resonance) spectra (Wright et al., 1990). The differences may be caused by variable glycosylation, and at any rate are lost in pooled samples.

The biosynthesis and processing pathway of myeloperoxidase has been outlined: from the endoplasmic reticulum, where interactions with molecular chaperones and heme incorporation occurs, to the Golgi, and finally to azurophil granules (Hansson et al., 2006).

PROPERTIES OF MYELOPEROXIDASE

The molecular weight of myeloperoxidase is 149,000 with two irons per molecule (Ehrenberg and Agner, 1958), and two calciums (Booth et al., 1989). The enzyme contains 3% to 4% carbohydrate, which consists of mannose, glucose, and *N*-acetylglucosamine (Bakkenist et al., 1978). Because of the covalent attachment of the heme, the usual primary standard of the pyridine hemochromogen spectra was replaced by iron determination. The secondary standard is the molar absorptivity of the Soret peak of the native enzyme at 430 nm, with values ranging from 89 to 95 $mM^{-1} cm^{-1}$ per heme group (Agner, 1958; Odajima and Yamazaki, 1970). I suggest use of the latest literature value, 91 $mM^{-1} cm^{-1}$ (Jantschko et al., 2004). The active site structure of myeloperoxidase compared to other mammalian peroxidases has been reviewed (Furtmüller et al., 2006).

Myeloperoxidase was cleaved in half by reduction of one connecting disulfide bridge. Each hemi-myeloperoxidase consists of a heavy and light chain, molecular weights 57,500 and 14,000 (Andrews and Krinsky, 1981, 1986; Andrews et al., 1984; Olsen and Little, 1984; Wright et al., 1987). Hemi-myeloperoxidase appears virtually identical in enzymatic properties to the intact enzyme. The linking of two monomers may be to provide a high enough molecular weight to minimize excretion from the cell.

The antibacterial properties of myeloperoxidase-halide-hydrogen peroxide systems have been described (Klebanoff, 1968), as has a potential role in defense of viral infections (Belding et al., 1970). The myeloperoxidase-hydrogen peroxide system will oxidize chloride to hypochlorous acid (Harrison and Schultz, 1976).

A series of studies on myeloperoxidase were conducted in the laboratory of Ron Wever. Halide complexes (Bakkenist et al., 1980), formation of ferrous nitrosyl compounds (Bolscher and Wever, 1984b), and stimulation of chlorinating activity by vitamin C (Bolscher et al., 1984) were investigated. The kinetics of cyanide binding was studied on different sources of myeloperoxidase with similar results (Bolscher and Wever, 1984a; Ikeda-Saito, 1987; Marquez and Dunford, 1989). A pK_a value of ~4 is indicated for an active site group, with HCN binding to the basic form of the enzyme. At neutral pH the association rate constant is in the range 1 to $4 \times 10^6 M^{-1} s^{-1}$, faster by at least an order of magnitude than with plant peroxidases. The dissociable group on the enzyme has a ΔH value of ~8 kcal/mol and was assigned to the imidazole group of distal histidine (Ikeda-Saito, 1987). The effects of cyanide binding were examined using ESR and MCD (Eglington et al., 1982) and resonance Raman spectroscopy (López-Garriga et al., 1990).

Halide ions bind rapidly to the acidic form of the enzyme with the pK_a value of ~4 (Bakkenist et al., 1980; Ikeda-Saito and Prince, 1985; Ikeda-Saito et al., 1985). Whereas cyanide binds in the form of HCN to basic enzyme, it is Cl^- that binds to acidic enzyme. Nitrite (Cooper and Odell, 1992) and thiocyanate (Bolscher and Wever, 1984a) also bind. These small molecules/anions bind in the heme pocket, perhaps directly to iron(III). Bromide binds in close proximity to distal His95 (Fiedler et al., 2000).

Aromatic substrate binding was examined by ESR and model building, utilizing the 3 Å resolution structure of canine myeloperoxidase (Hori et al., 1994). Phenol, *p*-cresol, resorcinol, and 4-aminosalicylate generate a high-spin ESR signal slightly different from native enzyme. In contrast, salicylhydroxamic acid and benzhydroxamic acid generate a low-spin signal that is strongly temperature dependent. All substrates bind in a hydrophobic pocket on the distal side of the heme. The crystal structure of the salicylhydroxamic acid complex, shown in Figure 16.2, shows that

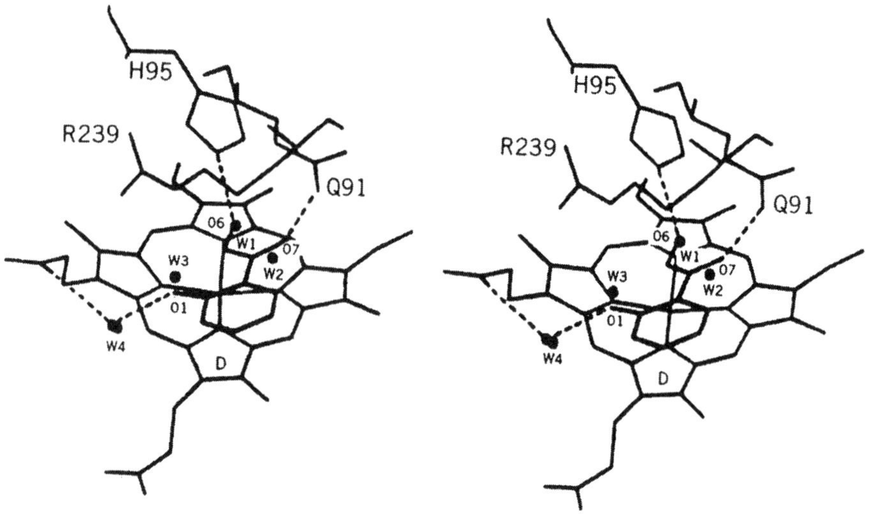

Figure 16.2. Model for salicylhydroxamic acid bound to human myeloperoxidase. The positions of the three oxygen atoms of the acid are close to those formerly occupied by the three displaced water molecules (W1, W2, and W3). (Reproduced with permission from Davey CA, Fenna RE. 1996. *Biochemistry* **35**: 10967–10973.)

the hydroxamic acid side chain is not bound to the iron, but displaces three heme cavity water molecules (Davey and Fenna, 1996).

Inhibitors of myeloperoxidase include acetaminophen (van Zyl et al., 1989b), diclofenac (Zuurbier et al., 1990a), anti-inflammatory drugs (Kettle and Winterbourn, 1991), guanidine HCl (Wever et al., 1991), isoniazid (van Zyl et al., 1989a), propylthiouracil (Lee et al., 1990), and salicylhydroxamic acid (Davies and Edwards, 1989; Davey and Fenna, 1996).

THE COMPOUNDS OF MYELOPEROXIDASE

Myeloperoxidase can form the same Compounds I, II, and III, as other peroxidases. With a very pure myeloperoxidase sample a 20-fold excess of hydrogen peroxide is required to obtain a high yield of Compound I (Harrison et al., 1980; Marquez et al., 1994). The Soret maximum of Compound I is 44 $mM^{-1}\,cm^{-1}$ at 430 nm. With a 40-fold excess complete formation of Compound I is not attained as it is rapidly converted to Compound II. Spectra are shown in Figure 16.3.

Formation of Compound I is reversible, with a rate constant for formation using hydrogen peroxide of $1.8 \times 10^7\ M^{-1}\ s^{-1}$, and a reverse dissociation rate of 58 s^{-1} (Bolscher and Wever, 1984a; Marquez et al., 1994). The reversibility of Compound I formation plus the reaction of hydrogen peroxide with Compound I place the restrictions on the excess of H_2O_2 to obtain a good yield of Compound I. The rate constant for the reduction of Compound I by H_2O_2 is $8 \times 10^4\ M^{-1}\ s^{-1}$. Both reactions of H_2O_2

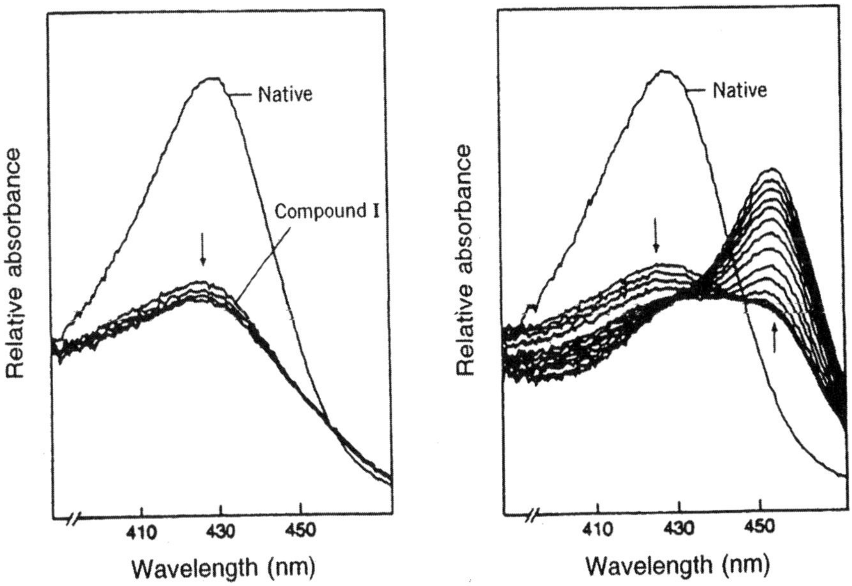

Figure 16.3. Left panel: conversion of myeloperoxidase into its Compound I using a 20-fold excess of hydrogen peroxide. Spectra were taken 4 to 7 ms after flow was stopped. Right panel: spontaneous conversion of Compound I into Compound II, using a 40-fold excess of hydrogen peroxide. Spectra were recorded starting 3 ms after flow had stopped, then in 1-ms intervals. (Reproduced with permission from Marquez LA, Huang JT, Dunford HB. 1994. *Biochemistry* **33**: 1447–1454.)

are pH dependent, with reaction occurring with a basic form of the enzyme. For Compound I the pK_a value is 4.0, and for Compound II 4.6. The conversion of Compound I to II is biphasic, because the hydroperoxyl radical can also reduce Compound I. With a large excess of hydrogen peroxide the reduction of Compound I becomes monophasic, because hydrogen peroxide reacts faster and is present in higher concentration than hydroperoxyl radical. These results are summarized in the following equations. The reaction of hydroperoxyl or superoxide radical is suppressed by the addition of superoxide dismutase.

$$\mathrm{MPO} + \mathrm{H_2O_2} \rightleftharpoons \mathrm{MPO\text{-}I} + \mathrm{H_2O}$$
$$\mathrm{MPO\text{-}I} + \mathrm{H_2O_2} \longrightarrow \mathrm{MPO\text{-}II} + \mathrm{HO_2}^{\bullet}$$
$$\mathrm{MPO\text{-}I} + \mathrm{HO_2}^{\bullet} \longrightarrow \mathrm{MPO\text{-}II} + \mathrm{O_2}$$

Organic hydroperoxides react with myeloperoxidase to form Compound I, and like H_2O_2 they will also reduce Compound I to Compound II. The result is formation of free radicals, which has been observed in human polymorphonuclear neutrophils. The cytotoxicity of organic hydroperoxides leads to membrane damage (Chamulitrat et al., 1991). Myeloperoxidase has been implicated in oxidative damage to both low-density and high-density lipoproteins (Malle et al., 2006).

Four methods of preparation of Compound III have been described. In the first, Compound II is reacted with excess hydrogen peroxide followed by catalase addition (Hoogland et al., 1987, 1988). In the second, xanthine-xanthine oxidase is used as a source of superoxide, which reacts with native enzyme to form Compound III (Winterbourn et al., 1985; Metodiewa and Dunford 1989; Marquez and Dunford, 1990). In the third, aerobic oxidation of dihydroxyfumarate is used to generate superoxide (Marquez, 1990), a method first used with horseradish peroxidase (HRP) (Yamazaki and Piette, 1963). Finally, in the fourth, oxygen was reacted with ferrous myeloperoxidase (Abu-Soud et al., 2004; Jantschko et al., 2004). Well-defined peaks of Compound III at 450 and 625 nm were obtained with beef spleen myeloperoxidase, with an isosbestic point between native enzyme and Compound III at 440 nm (Marquez and Dunford, 1990). Compound III, prepared using dihydroxyfumarate, is stable for 15 min at pH 4.7. At higher pH it decays more rapidly to native enzyme, a process accelerated by the addition of superoxide dismutase. Ascorbic acid reduces Compound III to native enzyme with a rate constant of $(4.0 \pm 0.1) \times 10^2\ M^{-1}\ s^{-1}$.

Is Singlet Oxygen Formed by Myeloperoxidase?

The $^1\Delta_g$ state of dioxygen is the lowest energy excited state, lying 1.0 ev (23 kcal/mol) above the ground $^3\Sigma_g^-$ state. Because of spin conservation rules, as well as the excitation energy, singlet oxygen is highly reactive, whereas triplet ground state oxygen is not. This reactivity was exploited brilliantly in organic syntheses by the late Christopher Foote. It was demonstrated that singlet oxygen could be produced in an electric discharge and also by reaction of hydrogen peroxide with chlorine in alkaline solution (hypochlorite) (Khan and Kasha, 1963; Arnold et al., 1964; Corey and Taylor, 1964; Foote and Wexler, 1964). Therefore, the necessary components for singlet oxygen production, hydrogen peroxide and hypochlorite, are present in a myeloperoxidase system. The question is whether they are present in high enough concentration. In the first attempt to detect singlet oxygen positive results were reported (Rosen and Klebanoff, 1977), but the interpretation of results was criticized (Harrison et al., 1978). Strong evidence was presented that singlet oxygen is produced in the lactoperoxidase-hydrogen peroxide-bromide system (Kanofsky, 1983), and later in the myeloperoxidase system (Kanofsky et al., 1984). Bromide is much more effective than chloride, and great enhancement in red luminescence is observed in D_2O. The optimum pH is 5. Singlet oxygen production was demonstrated in stimulated human eosinophils (Kanofsky et al., 1988). The short answer to the question of singlet oxygen production by myeloperoxidase is yes. With bromide not available in the in vivo myeloperoxidase systems, the amount of singlet oxygen produced does not appear to be of major importance.

Some singlet oxygen may also be generated in the reaction

$$HOCl + O_2^{\bullet -} \longrightarrow {}^{\bullet}OH + O_2 + Cl^-$$

The latter reaction appears to generate hydroxyl radicals (Long and Bielski, 1980). The rate constant for the reaction is $(7.5 \pm 0.4) \times 10^6\ M^{-1}\ s^{-1}$. Hydroxyl radicals were detected by spin trapping in stimulated human neutrophils and monocytes (Ramos et al., 1992). Thus, hypochlorous acid combined with other oxidants found in myeloperoxidase systems is capable of producing singlet oxygen and hydroxyl radicals. There is strong evidence, however, that the latter oxidants will be produced in negligible quantities because the concentration of HOCl is so low. HOCl reacts rapidly with native enzyme to form Compound I (Floris and Wever, 1992).

$$\mathrm{HOCl + MPO \longrightarrow MPO\text{-}I + H^+ + Cl^-}$$

At pH 7.2 the reaction is too fast to be measured, but at pH 8.3 the rate constant is $2 \times 10^7\ \mathrm{M^{-1}\,s^{-1}}$ and at pH 9.2 it is $2.3 \times 10^6\ \mathrm{M^{-1}\,s^{-1}}$. Thus, in the myeloperoxidase-hydrogen peroxide-chloride system Compound I is a dominant enzyme species compared to native enzyme. The implications are further explored below.

Ferrous Myeloperoxidase

Ferrous myeloperoxidase has a red-shifted spectrum with the Soret peak at 472 nm and the α-band at 636 nm (Floris et al., 1994). The redox potentials of the $\mathrm{Fe^{III}/Fe^{II}}$ couple, both free and cyanide-bound, have been measured (Battistuzzi et al., 2006). Two independent studies of the reaction of dioxygen with ferrous myeloperoxidase were published about the same time, with rather different results (Abu-Soud et al., 2004; Jantschko et al., 2004). The study by Jantschko et al. was conducted at 25°C. They found that carefully prepared anaerobic ferrous myeloperoxidase still decayed spontaneously to ferric enzyme with rate constants of $0.04\ \mathrm{s^{-1}}$ at pH 8 and $1.4\ \mathrm{s^{-1}}$ at pH 5. These rates were too slow to affect dioxygen binding to ferrous enzyme, with rate constants of $(1.1 \pm 0.1) \times 10^4\ \mathrm{M^{-1}\,s^{-1}}$ at pH 7 and double this rate at pH 5. The binding fits single exponential curves. It is reversible with a dissociation equilibrium constant of $(170 \pm 10)\ \mu\mathrm{M}$ at pH 7. The rate of oxygen dissociation from Compound III to form ferrous enzyme is faster than conversion of Compound III to ferric enzyme. Spectral peaks occur at the following wavelengths: native MPO peaks at 430 and 570 nm; ferrous MPO peaks at 473 and 638 nm; and Compound III peaks at 449 and 625 nm.

The study of oxygen binding to ferrous MPO by Abu-Soud et al. (2004) was conducted at the lower temperature of 10°C. At times up to 0.2 s the ferrous MPO peak at 473 nm is replaced by one at 434 nm, and at later times, up to 6 s, the 434 nm peak is replaced by that of Compound III at 450 nm. Thus the species at 434 nm appears to be a detectable ferrous MPO—O_2 intermediate formed prior to Compound III formation. Compared to other peroxidases the binding of dioxygen to ferrous MPO is relatively slow. This has been attributed to collapse of the heme pocket upon ferrous MPO formation, thus slowing the binding of other diatomic substrates CO and NO (Abu-Soud and Hazen, 2001).

Compound III thus can be formed either from ferric MPO and hydroperoxyl radical, or from ferrous enzyme and dioxygen. It is reduced by ascorbate to ferric enzyme with a rate constant of $(4.0 \pm 0.1) \times 10^2\ \mathrm{M^{-1}\,s^{-1}}$ at 25°C, a process in the phagocyte that would enable the enzyme to reenter the ferric-Compound I cycle for the two-electron oxidation of chloride (Marquez and Dunford, 1990).

Assays for Myeloperoxidase

A sensitive assay for myeloperoxidase was developed using a luciferin analog excited by singlet oxygen in a bromide-containing system (Uehara et al., 1991). The relative merits of other assays have been reviewed (Kettle and Winterbourn, 1994).

REACTIONS OF MYELOPEROXIDASE

Some of the studies of reactions of a wide variety of myeloperoxidase substrates are the following: 4-chloroaniline (Bakkenist et al., 1981), thiocyanate (Wever et al., 1982), chlorpromazine (van Zyl et al., 1990), and polyhydroxyphenols (substituted catechols) (Metodiewa and Dunford, 1990a). Butylated hydroxyanisole (BHA), widely used as an antioxidant in processed foods and cosmetics, has harmful as well as beneficial effects. It and its dimer are substrates for myeloperoxidase and lactoperoxidase (Metodiewa et al., 1991). Myeloperoxidase, structurally related to thyroid peroxidase, was shown to catalyze iodination of thyroglobulin and coupling of diiodotyrosyl residues (Taurog and Dorris, 1992). Myeloperoxidase along with *Coprinus cinereus* peroxidase can catalyze the epoxidation of styrene (Tuynman et al., 2000). Transient-state kinetic studies on myeloperoxidase include reactions of Compounds II and III with ascorbic acid (Marquez and Dunford 1990; Marquez et al., 1990), acetaminophen (Marquez and Dunford, 1993), tyrosine and dityrosine (Marquez and Dunford, 1995), and 3,5,3′,5′-tetramethyl benzidine (Marquez and Dunford, 1997).

Nitrite is oxidized in one-electron steps by Compounds I and II. The Compound II reaction is slow but the cycle is enhanced by the presence of a good Compound II reductant (Burner et al., 2000).

Rates were measured for the reactions of five indole derivatives: tryptamine, *N*-acetyltryptamine, tryptophan, melatonin, and serotonin (5-hydroxy-tryptamine), with Compounds I and II of myeloperoxidase, lactoperoxidase, and horseradish peroxidase. The mammalian peroxidase Compounds I react much faster than HRP Compound I (Jantschko et al., 2002). Nitroxide radicals were shown to be good scavengers of glutathionyl radicals formed in myeloperoxidase reactions (Borisenko et al., 2004).

Tyrosine residues are nitrated by myeloperoxidase, hydrogen peroxide, and nitrite, a reaction inhibited when cysteine residues are available (Zhang et al., 2005). Thiocyanate is called a pseudo-halide because of its properties. Its oxidation by myeloperoxidase is enhanced by the presence of nitric oxide (Galijasevic et al., 2006). Protein tyrosine nitration is catalyzed by myeloperoxidase in the presence of hydrogen peroxide and nitrite (Sampson et al., 1998).

The inhibitory effect of the antioxidant tempol (4-hydroxy-2,2,6,6-tetramethyl-1-piperiddinyloxy) on ribonuclease nitration was examined (Vaz and Augusto, 2008).

Myeloperoxidase and Peroxynitrite

Both superoxide and nitric oxide are present in stimulated neutrophils. They may react to form peroxynitrite or its acid form, peroxynitrous acid. The latter species reacts with myeloperoxidase, and also with lactoperoxidase and horseradish peroxidase. No Compound I could be detected, only Compound II (Floris et al., 1993).

Oscillatory Reactions

NADH is oxidized by myeloperoxidase (Odajima, 1971). Oscillatory reactions involving myeloperoxidase Compound III, NADH, and dioxygen have been observed (Olsen et al., 2003a; Gabdoulline et al., 2003; Brasen et al., 2004). The results may be important for an understanding of oxidant production in neutrophils. The reaction has been modeled using NADPH oxidase, present in the leukocyte, and melatonin as a cofactor. The system "can be considered a model system for the nonrespiratory oxygen metabolism in activated neutrophils and other similar cells participating in the defence against invading pathogens" (Olsen et al., 2003b, p. 287).

Role of Superoxide

In the respiratory burst in neutrophils NADPH oxidase converts dioxygen to superoxide. Disproportionation of superoxide yields hydrogen peroxide, but Compound III is formed in high yield from addition of superoxide to native enzyme (Winterbourn et al., 1985; Kettle and Winterbourn, 1988, 1989). Small reducing substrates reduce both Compounds II and III, dead ends for chlorination reactions, back to native enzyme. It has never been proven, but a very sensible reaction would be that of Compound III with superoxide to generate hydrogen peroxide and dioxygen, implying that myeloperoxidase could be its own superoxide dismutase. The latter enzyme apparently is not present in leukocytes. An examination of the catalase reaction of myeloperoxidase was flawed by the arbitrary assumption of a major role for endogenous donor (Kettle and Winterbourn, 2001).

What Is the Role of Hypochlorous Acid?

Bleach is the commonly used solution for scrubbing down hospital operating rooms. The ability of HOCl to kill microorganisms is well established. HOCl was used as a model for the myeloperoxidase chlorinating system to examine injury to target cells (Schraufstätter et al., 1990). The two-electron oxidation of chloride by myeloperoxidase to produce HOCl is almost universally accepted (Spalteholz et al., 2006). Thus, it made sense to assume that HOCl is the killing reagent generated by myeloperoxidase in the phagosome. However, evidence is accumulating that this is not the case. First, HOCl is an indiscriminant reagent and it is capable of inactivating myeloperoxidase. The rapid reaction of HOCl to regenerate Compound I shows that the myeloperoxidase system is avoiding excess HOCl like the plague (Floris and Wever, 1992). Second, HOCl will react with amino acid residues to form chloramines, which are usually unstable (Zgliczyński et al., 1968, 1971). The research in the group of Edwin Thomas has shown that chloramines are the likely bactericidal reagent ultimately formed by myeloperoxidase. The chloramines are interchangeable (Thomas and Learn, 1990). Third, in the chlorination of taurine by myeloperoxidase an enzyme intermediate species formed from Compound I and chloride was detected (Marquez and Dunford, 1994). The data excluded HOCl as the taurine chlorinating reagent. Taurine is a unique β-amino acid with a sulfonic acid rather than carboxlate group. Myeloperoxidase could avoid harming itself by chlorinating one of its own amino acid residues, which in turn chlorinates taurine, which becomes a mobile chlorinating reagent. The role of chloramines in protein fragmentation has been examined by the group of Michael Davies (Hawkins and Davies, 1999; Pattison and Davies, 2005).

Chlorination of monochlorodimedone by myeloperoxidase has been studied (Zuurbier et al., 1990). Both bromination and chlorination reactions of myeloperoxidase have been examined (Senthilmohan and Kettle, 2006).

CLONING OF MYELOPEROXIDASE: SITE-DIRECTED MUTAGENESIS

The gene for human myeloperoxidase has been cloned, characterized (Chang et al., 1986; Morishita et al., 1987; Yamada et al., 1987; Weil et al., 1987), and expressed in a baculovirus system (Taylor et al., 1992). An 84 kDa recombinant enzymatically active promyeloperoxidase containing heme has been produced in Chinese hamster ovary cells, and its spectral and enzymatic properties have been documented (Jacquet et al., 1991; Moguilevsky et al., 1991). Mutants of the 84 kDa heavy chain confirmed that His95, Asp94, Glu242, and Met243 are residues responsible for maintaining the correct orientation of the heme (Jacquet et al., 1994). See discussion of the heme structure, below. The Glu242Gln mutant was intensively studied. Its resonance Raman spectrum was drastically altered, and it was no longer able to oxidize chloride (Floris et al., 1995).

THE CRYSTAL STRUCTURE AND THE PROSTHETIC GROUP OF MYELOPEROXIDASE

The green color of native myeloperoxidase provided an early lead that the heme chromophoric group is not ferriproptoporphyrin IX (heme *b*). The large peak in the visible spectral region, centered at 571 nm, is responsible for the green color. Since the heme is covalently bound, its structure could not be determined by the usual chemical techniques. A variety of spectroscopic techniques provided important but limited structural information (Ehrenberg, 1962; Sibbett and Hurst, 1984; Andersson et al., 1984; Babcock et al., 1985; Dugad et al., 1990). A photochemical reduction of myeloperoxidase produced an active enzyme with prosthetic group similar to that in lactoperoxidase (Hori and Ikeda-Saito, 1990). The covalent binding of the heme in myeloperoxidase is accomplished in an autocatalytic reaction. Work on horseradish peroxidase mutants showed that a single appropriately placed carboxylate is all that is required to cause autocatalytic covalent heme binding (Colas and Ortiz de Montellano, 2004).

The crystal structure of canine myeloperoxidase at 3 Å resolution (Zeng and Fenna, 1992), was refined to 2.28 Å (Fenna et al., 1995). The higher resolution results provided sufficient detail to provide a structure for the heme group. Asp94 and Glu242 were identified as participating in ester bond formation with heme, and Met243 in a sulfonium ion linkage. Further structural information was obtained with Fourier transform infrared spectra, combined with site-directed mutagenesis, confirming that Glu242 and Asp94 are involved in ester bond formation (Kooter et al., 1997b, 1999b). The corrected structure of the myeloperoxidase heme, shown in Figure 16.4, was confirmed in a later crystal structure obtained at 1.8 Å resolution (Fiedler et al., 2000). Similar ester bonds are found in lactoperoxidase and eosinophil peroxidase. The Met243 sulfonium ion heme linkage was also investigated by a variety of techniques (Kooter et al, 1997a, 1999a, 1999c), and believed to involve the α-carbon atom of a vinyl group. A vinyl sufonium ion was proposed in earlier work (Taylor et al., 1992a, 1995).

The crystal structures show that His336, the proximal ligand, is hydrogen bonded to Asn421. Distal His95 and Arg239 are positioned to participate in catalysis. Although water is present in the heme cavity, it is not coordinated to the iron. In addition to the disulfide bridge between the two hemi-myeloperoxidases, each hemi-myeloperoxidase possesses an additional five disulfide bridges in its large polypeptide chain and one in its small. The structure is dominated by 19 α-helices.

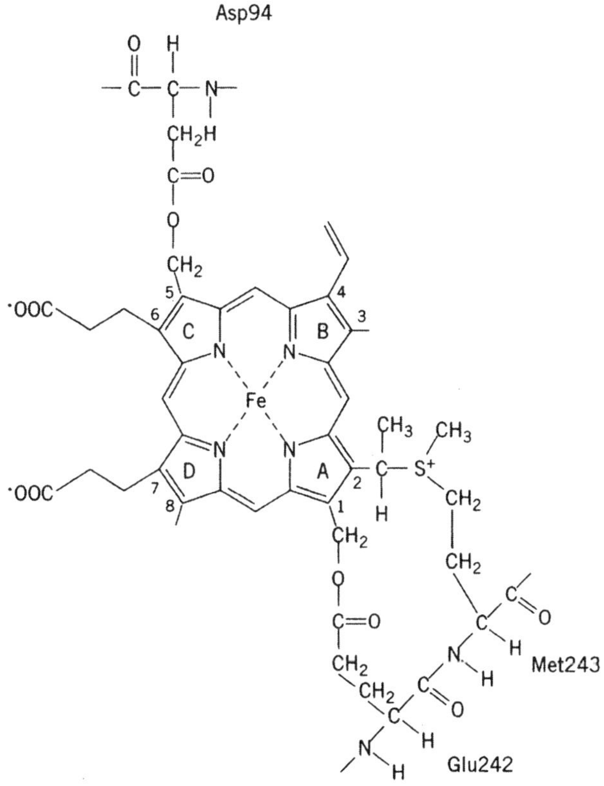

Figure 16.4. Heme structure of myeloperoxidase. (Reproduced with permission from Kooter IM, et al., 1997b. *J. Am. Chem. Soc.* **119**: 11542–11543.)

The inhibitor salicylhydroxamic acid (SHA) binds in the active site of myeloperoxidase. The aromatic ring of the inhibitor binds between heme pyrrole ring D and the side chain of Arg239. The hydroxamic acid moiety is hydrogen bonded to distal His95. Three SHA oxygens closely duplicate the hydrogen bonding pattern of three displaced water molecules (Davey and Fenna, 1996).

EOSINOPHIL PEROXIDASE

Little did the goddess Eos realize that she would have a dye, a leukocyte, and a peroxidase named in her honor. Her name was used for the dye eosin, which is a particularly good stain for the type of leukocyte known as eosinophil. The eosinophils, much less abundant than the neutrophils, can be detected and quantified by flow cytometry (Visser, 1990). Persons who suffer from hereditary myeloperoxidase deficiency have normal eosinophil peroxidase content. There is a distinct eosinophil peroxidase deficiency (Romano et al., 1994). Eosinophils are involved in the detoxification of foreign proteins and other substances that appear to have tumoricidal activity. They have bactericidal activity (Jong et al., 1980; Klebanoff et al., 1989). For reviews see Henderson (1991) and Wang and Slungaard (2006).

Eosinophil peroxidase has been isolated from humans (Wever et al., 1981; Bolscher et al., 1984), guinea pig (Jong et al., 1980), and horse (Klebanoff et al., 1989).

The enzyme with hydrogen peroxide can oxidize iodide, bromide, chloride, and thiocyanate (Jong et al., 1980; Wever et al., 1981; Weiss et al., 1986; Slungaard and Mahoney, 1991a, 1991b; Arlandson et al., 2001; van Dalen and Kettle, 2001; Spalteholz et al., 2006). Thus eosinophil peroxidase is similar to myeloperoxidase in the reactions it can catalyze.

Bromide oxidation was shown to be consistent with HOBr formation that reacts with amine compounds to form bromamines (Thomas et al., 1995). Bromotyrosine compounds were detected (Wu et al., 1999b). 5-Bromouracil is also generated (Henderson et al., 2001). Free nucleosides and double-stranded DNA are brominated (Shen et al., 2001).

Nitrite is also oxidized by eosinophil peroxidase, leading to nitration of proteins (Wu et al., 1999a; van Dalen et al., 2006). By use of eosinophil peroxidase and myeloperoxidase knockout mice, nitration by these enzymes in vivo was demonstrated (Brennan et al., 2002). Nitric oxide was shown to be a one-electron competitor for two-electron bromide as an eosinophil peroxidase substrate (Galijasevic et al., 2007).

Human eosinophil peroxidase was estimated to have a molecular weight of $71{,}000 \pm 1{,}000$. It is composed of a light (14,000) and heavy (58,000) subunit, both of which contain carbohydrate. The Soret maximum absorbance is $110 \pm 4\,mM^{-1}\,cm^{-1}$ at 412 nm, the same as for lactoperoxidase (Bolscher et al., 1984). From the cloned gene the amino acid sequence was obtained, and a molecular weight of 81,036 was calculated (Sakamaki et al., 1989; Ten et al., 1989).

Both myeloperoxidase and eosinophil peroxidase are inhibited by dapsone (Bozeman et al., 1992). An assay at high pH using homovanillic acid was shown to react with eosinophil peroxidase with a barely detectable myeloperoxidase reaction (Menegazzi et al., 1991). Hydroxyl radicals generated by eosinophil peroxidase were detected by spin trapping (McCormick et al., 1994).

Asp93 and Glu241, one from the light chain and the other from the heavy, were shown to form ester linkages with eosinophil peroxidase heme (Oxvig et al., 1999).

Human eosinophil peroxidase has 14 half-cystine residues. The disulfide bridges were identified (Thomsen et al., 2000). The Compound I/native enzyme redox potential is 1.10 ± 0.01 volts for eosinophil peroxidase, very similar to that for myeloperoxidase (Arnhold et al., 2001). The spectra and kinetics of eosinophil peroxidase Compound I formation and reactions with halides and thiocyanate were measured (Furtmüller et al., 2000).

Like myeloperoxidase, eosinophil peroxidase is oxidized rapidly to Compound I by hypochlorous acid. Eosinophil peroxidase Compound I reacts slowly with chloride, rapidly with bromide, and even more rapidly with iodide.

REFERENCES

Abu-Soud HM, Hazen SL. 2001. Interrogation of heme pocket environment of mammalian peroxidases with diatomic ligands. *Biochemistry* **40**: 10747–10755.

Abu-Soud HM, Raushel FM, Hazen SL. 2004. A novel multistep mechanism for oxygen binding to ferrous hemoproteins: rapid kinetic analysis of ferrous-dioxy myeloperoxidase (Compound III) formation. *Biochemistry* **43**: 11589–11595.

Agner K. 1941. Verdoperoxidase: a ferment isolated from leukocytes. *Acta Physiol. Scand.* **2** (Suppl. 8): 1–62.

Agner K. 1943. Verdoperoxidase. *Adv. Enzymol.* **3**: 137–148.

Agner K. 1958. Crystalline myeloperoxidase. *Acta Chem. Scand.* **12**: 89–94.

Andersson LA, Loehr TM, Lim AR, Mauk AG. 1984. Sulfmyoglobin. Resonance Raman evidence for an iron-chlorin prosthetic group. *J. Biol. Chem.* **259**: 15340–15349.

Andrews PC, Krinsky NI. 1981. The reductive cleavage of myeloperoxidase in half, producing enzymically active hemi-myeloperoxidase. *J. Biol. Chem.* **256**: 4211–4218.

Andrews PC, Krinsky NI. 1986. Human myeloperoxidase and hemi-myeloperoxidase. *Methods Enzymol.* **132**: 369–378.

Andrews PC, Parnes C, Krinsky NI. 1984. Comparison of myeloperoxidase and hemi-myeloperoxidase with respect to catalysis, regulation and bactericidal activity. *Arch. Biochem. Biophys.* **228**: 439–442.

Arlandson M, Decker T, Roongta VA, Bonilla L, Mayo KH, MacPherson JC, Hazen SL, Slungaard A. 2001. Eosinophil peroxidase oxidation of thiocyanate: characterization of major reaction products and a potential sulfhydryl-targeted cytotoxicity system. *J. Biol. Chem.* **276**: 215–224.

Arnhold J, Furtmüller PG, Regelsberger G, Obinger C. 2001. Redox properties of the couple Compound I/native enzyme of myeloperoxidase and eosinophil peroxidase. *Eur. J. Biochem.* **268**: 5142–5148.

Arnold SJ, Ogryzlo EA, Witke H. 1964. Some new emission bands of molecular oxygen. *J. Chem. Phys.* **40**: 1769–1770.

Babcock GT, Ingle RT, Oertling WA, Davis JC, Averill BA, Hulse CL, Stufkins DJ, Bolscher BGJM, Wever R. 1985. Raman characterization of human leukocyte myeloperoxidase and bovine spleen green haemoprotein: insight into chromophore structure and evidence that the chromophores of myeloperoxidase are equivalent. *Biochim. Biophys. Acta* **828**: 58–66.

Bakkenist ARJ, De Boer JEG, Plat H, Wever R. 1980. The halide complexes of myeloperoxidase and the mechanism of halogenation reactions. *Biochim. Biophys. Acta* **613**: 337–348.

Bakkenist ARJ, Plat H, Wever R. 1981. Oxidation of 4-chloroaniline catalyzed by human myeloperoxidase. *Bioorganic Chem.* **10**: 324–328.

Bakkenist ARJ, Wever R, Vulsma T, Plat H, Van Gelder BF. 1978. Isolation procedure and some properties of myeloperoxidase from human leukocytes. *Biochim. Biophys. Acta* **524**: 45–54.

Battistuzzi G, Bellei M, Zederbauer M, Furtmüller PG, Sola M, Obinger C. 2006. Redox thermodynamics of the Fe(III)/Fe(II) couple of human myeloperoxidase in its high-spin and low-spin forms. *Biochemistry* **45**: 12750–12755.

Belding ME, Klebanoff SJ, Ray CG. 1970. Peroxidase-mediated virucidal systems. *Science* **167**: 195–196.

Bolscher BGJM, Plat H, Wever R. 1984. Some properties of human eosinophil peroxidase, a comparison with other peroxidases. *Biochim. Biophys. Acta* **784**: 177–186.

Bolscher BGJM, Wever R. 1984a. A kinetic study of the reaction between human myeloperoxidase, hydroperoxides and cyanide. Inhibition by chloride and thiocyanate. *Biochim. Biophys. Acta* **788**: 1–10.

Bolscher BGJM, Wever R. 1984b. The nitrosyl compounds of ferrous animal haloperoxidases. *Biochim. Biophys. Acta* **791**: 75–81.

Bolscher BGJM, Zoutberg GR, Cuperus RA, Wever R. 1984. Vitamin C stimulates the chlorinating activity of human myeloperoxidase. *Biochim. Biophys. Acta* **791**: 75–81.

Booth KS, Kimura S, Lee HC, Ikeda-Saito M, Caughey WS. 1989. Bovine myeloperoxidase and lactoperoxidase each contain a high affinity site for calcium. *Biochem. Biophys. Res. Commun.* **160**: 897–902.

Borisenko GG, Martin I, Zhao Q, Amoscato AA, Kagan VE. 2004. Nitroxides scavenge myeloperoxidase-catalyzed thiyl radicals in model systems and in cells. *J. Am. Chem. Soc.* **126**: 9221–9232.

Bozeman PM, Learn DB, Thomas EL. 1992. Inhibition of the human leukocyte enzymes myeloperoxidase and eosinophil peroxidase by dapsone. *Biochem. Pharmacol.* **44**: 553–563.

Brasen JC, Lunding A, Olsen LF. 2004. Human myeloperoxidase catalyzed an oscillating peroxidase-oxidase reaction. *Arch. Biochem. Biophys.* **431**: 55–62.

Brennan M-L, Wu W, Fu X, Shen Z, Song W, Frost H, Vadseth C, Narine L, Lenkiewicz E, Borchers MT, Lusis AJ, Lee JJ, Lee NA, Abu-Soud HM, Ischiropoulos H, Hazen SL. 2002. A tale of two controversies: defining both the role of peroxidases in nitrotyrosine formation in vivo using eosinophil peroxidase and myeloperoxidase-deficient mice, and the nature of peroxidase-generated reactive nitrogen species. *J. Biol. Chem.* **277**: 17415–17427.

Burner U, Furtmüller PG, Kettle AJ, Koppenol WH, Obinger C. 2000. Mechanism of reaction of myeloperoxidase with nitrite. *J. Biol. Chem.* **275**: 20597–20601.

Chamulitrat W, Cohen MS, Mason RP. 1991. Free radical formation from organic hydroperoxides in isolated human polymorphonuclear neutrophils. *Free Radical Biol. Med.* **11**: 439–445.

Chang KS, Trujillo JM, Cook RG, Stass SA. 1986. Human myeloperoxidase gene: molecular cloning and expression in leukemic cells. *Blood* **68**: 1411–1414.

Colas C, Ortiz de Montellano PR. 2004. Horseradish peroxidase mutants that autocatalytically modify their prosthetic heme group. Insights into mammalian peroxidase heme-protein covalent bonds. *J. Biol. Chem.* **279**: 24131–24140.

Cooper CE, Odell E. 1992. Interaction of human myeloperoxidase with nitrite. *FEBS Lett.* **314**: 58–60.

Corey EJ, Taylor WC. 1964. A study of the peroxidation of organic compounds by externally generated singlet oxygen molecules. *J. Am. Chem. Soc.* **86**: 3881–3882.

Davey CA, Fenna RE. 1996. 2.3 Å resolution X-ray crystal structure of the bisubstrate analogue inhibitor salicylhydroxamic acid bound to human myeloperoxidase: a model for a prereaction complex with hydrogen peroxide. *Biochemistry* **35**: 10967–10973.

Davis JC, Averill BA. 1981. Isolation from bovine spleen of a green heme protein with properties of myeloperoxidase. *J. Biol. Chem.* **256**: 5992–5996.

Davies B, Edwards SW. 1989. Inhibition of myeloperoxidase by salicylhydroxamic acid. *Biochem. J.* **258**: 801–806.

Dugad LB, La Mar GN, Lee HC, Ikeda-Saito M, Booth KS, Caughey WS. 1990. A nuclear Overhauser effect study of the active site of myeloperoxidase: structural similarity of the prosthetic group to that on lactoperoxidase. *J. Biol. Chem.* **265**: 7173–7179.

Eglington DG, Barber D, Thomson AJ, Greenwood C, Segal AW. 1982. Studies of cyanide binding to myeloperoxidase by electron paramagnetic resonance and magnetic circular dichroism spectroscopies. *Biochim. Biophys. Acta* **703**: 187–195.

Ehrenberg A. 1962. Electron spin resonance absorption by some hemoproteins. *Arkiv Kemi* **19**: 119–127.

Ehrenberg A, Agner K. 1958. The molecular weight of myeloperoxidase. *Acta Chem. Scand.* **12**: 95–100.

Fenna R, Zeng J, Davey C. 1995. Structure of the green heme in myeloperoxidase. *Arch. Biochem. Biophys.* **316**: 653–656.

Fiedler TJ, Davey CA, Fenna RE. 2000. X-ray crystal structure and characterization of halide-binding sites of human myeloperoxidase at 1.8 Å resolution. *J. Biol. Chem.* **275**: 11964–11971.

Floris R, Kim Y, Babcock GT, Wever R. 1994. Optical spectrum of myeloperoxidase: origin of the red shift. *Eur. J. Biochem.* **222**: 677–685.

Floris R, Moguilevsky N, Puppels G, Jacquet A, Renirie R, Bollen A, Wever R. 1995. Heme-protein interaction in myeloperoxidase: modification of spectroscopic properties and catalytic activity by single residue mutation. *J. Am. Chem. Soc.* **117**: 3907–3912.

Floris R, Piersma SR, Yang G, Jones P, Wever R. 1993. Interaction of myeloperoxidase with peroxynitrite: a comparison with lactoperoxidase, horseradish peroxidase and catalase. *Eur. J. Biochem.* **215**: 767–775.

Floris R, Wever R. 1992. Reaction of myeloperoxidase with its product HOCl. *Eur. J. Biochem.* **207**: 697–702.

Foote CS, Wexler S. 1964. Olefin oxidations with excited singlet molecular oxygen. *J. Am. Chem. Soc.* **86**: 3879–3880.

Furtmüller PG, Burner U, Regelsberger G, Obinger C. 2000. Spectral and kinetic studies on the formation of eosinophil peroxidase Compound I and its reaction with halides and thiocyanate. *Biochemistry* **39**: 15578–15584.

Furtmüller PG, Zederbauer M, Jantschko W, Helm J, Bogner M, Jakopitsch C, Obinger C. 2006. Active site structures and catalytic mechanisms of human peroxidases. *Arch. Biochem. Biophys.* **445**: 199–213.

Gabdoulline RR, Kummer U, Olsen LF, Wade RC. 2003. Concerted simulations reveal how peroxidase Compound III formation results in cellular oscillations. *Biophys. J.* **85**: 1421–1428.

Galijasevic S, Proteasa G, Abdulhamid I, Abu-Soud HM. 2007. The potential role of nitric oxide in substrate switching in eosinophil peroxidase. *Biochemistry* **46**: 406–415.

Galijasevic S, Saed GM, Hazen SL, Abu-Soud HM. 2006. Myeloperoxidase metabolizes thiocyanate in a reaction driven by nitric oxide. *Biochemistry* **45**: 1255–1262.

Hansson M, Olsson I, Nauseef WM. 2006. Biosynthesis, processing, and sorting of human myeloperoxidase. *Arch. Biochem. Biophys.* **445**: 214–224.

Harrison JE, Araiso T, Palcic MM, Dunford HB. 1980. Compound I of myeloperoxidase. *Biochem. Biophys. Res. Commun.* **94**: 34–40.

Harrison JE, Pabalan S, Schultz J. 1977. The subunit structure of crystalline canine myeloperoxidase. *Biochim. Biophys. Acta* **493**: 247–259.

Harrison JE, Schultz J. 1976. Studies on the chlorinating activity of myeloperoxidase. *J. Biol. Chem.* **251**: 1371–1374.

Harrison JE, Watson BD, Schultz J. 1978. Myeloperoxidase and singlet oxygen: a reappraisal. *FEBS Lett.* **92**: 327–332.

Hawkins CL, Davies MJ. 1999. Hypochlorite-induced oxidation of proteins in plasma: formation of chloraamines and nitrogen-centred radicals and their role in protein fragmentation. *Biochem. J.* **340**: 539–548.

Henderson JP. 1991. Eosinophil peroxidase: occurrence and biological function. In: Everse J, Everse KE, Grisham MB, editors. *Peroxidases in Chemistry and Biology*, Vol. 1. CRC Press, Boca Raton, FL, pp. 105–121.

Henderson JP, Byun J, Mueller DM, Heinecke JW. 2001. The eosinophil peroxidase-hydrogen peroxide-bromide system of human eosinophils generates 5-bromouracil, a mutagenic thymine analogue. *Biochemistry* **40**: 2052–2059.

Hoogland H, Dekker HL, van Riel C, van Kuilenburg A, Muijsers AO, Wever R. 1988. A steady-state study on the formation of Compounds II and III of myeloperoxidase. *Biochim. Biophys. Acta* **955**: 337–345.

Hoogland H, van Kuilenburg A, van Riel C, Muijsers AO, Wever R. 1987. Spectral properties of myeloperoxidase Compounds II and III. *Biochim. Biophys. Acta* **916**: 76–82.

Hori H, Fenna RE, Kimura S, Ikeda-Saito M. 1994. Aromatic substrates bind at the distal heme pocket of myeloperoxidase. *J. Biol. Chem.* **269**: 8388–8392.

Hori H, Ikeda-Saito M. 1990. Photochemically modified myeloperoxidase, with optical spectral properties analogous to those of lactoperoxidase, retains its original catalytic activity. *Biochemistry* **29**: 7106–7112.

Ikeda-Saito M. 1985. Spectroscopic, ligand binding, and enzymatic properties of the spleen green hemeprotein: a comparison with myeloperoxidase. *J. Biol. Chem.* **260**: 11688–11696.

Ikeda-Saito M. 1987. A study of ligand binding to spleen myeloperoxidase. *Biochemistry* **26**: 4344–4349.

Ikeda-Saito M, Argade PV, Rousseau DL. 1985. Resonance Raman evidence of chloride binding to the heme iron in myeloperoxidase. *FEBS Lett.* **184**: 52–55.

Ikeda-Saito M, Lee HC, Adachi K, Eck HS, Prince RC, Booth KS, Caughey WS, Kimura S. 1989. Demonstration that spleen green hemeprotein is identical to granulocyte myeloperoxidase. *J. Biol. Chem.* **264**: 4559–4563.

Ikeda-Saito M, Prince RC. 1985. The effect of chloride on the redox and EPR properties of myeloperoxidase. *J. Biol. Chem.* **260**: 8301–8305.

Jacquet A, Deby C, Mathy M, Moguilevsky N, Deby-Dupont G, Thirion A, Goormaghtigh E, Garcia-Quintana L, Bollen A, Pincemail J. 1991. Spectral and enzymatic properties of human recombinant myeloperoxidase: comparison with the mature enzyme. *Arch. Biochem. Biophys.* **291**: 132–138.

Jacquet A, Garcia-Quintana L, Deleersnyder V, Fenna R, Bollen A, Moguilevsky N. 1994. Site-directed mutagenesis of human myeloperoxidase: further identification of residues involved in catalytic activity and heme interaction. *Biochem. Biophys. Res. Commun.* **202**: 73–81.

Jantschko W, Furtmüller PG, Allegra M, Livrea MA, Jakopitsch C, Regelsberger G, Obinger C. 2002. Redox intermediates of plant and mammalian peroxidases: a comparative transient-state study of their reactivity toward indole derivatives. *Arch. Biochem. Biophys.* **398**: 12–22.

Jantschko W, Furtmüller PG, Zederbauer M, Jakopitsch C, Obinger C. 2004. Kinetics of oxygen binding to ferrous myeloperoxidase. *Arch. Biochem. Biophys.* **426**: 91–97.

Jong EC, Henderson WR, Klebanoff SJ. 1980. Bactericidal activity of eosinophil peroxidase. *J. Immunol.* **124**: 1378–1382.

Kanofsky JR. 1983. Singlet oxygen production by lactoperoxidase: evidence from 1270 nm chemiluminescence. *J. Biol. Chem.* **258**: 5991–5993.

Kanofsky JR, Hoogland H, Wever R, Weiss SJ. 1988. Singlet oxygen production by human eosinophils. *J. Biol. Chem.* **263**: 9692–9696.

Kanofsky JR, Wright J, Miles-Richardson GE, Tauber AI. 1984. Biochemical requirements for singlet oxygen production by purified human myeloperoxidase. *J. Clin. Invest.* **74**: 1489–1495.

Kettle AJ, Winterbourn CC. 1988. Superoxide modulates the activity of myeloperoxidase and optimizes the production of hypochlorous acid. *Biochem. J.* **252**: 529–536.

Kettle AJ, Winterbourn CC. 1989. Influence of superoxide on myeloperoxidase kinetics measured with a hydrogen peroxide electrode. *Biochem. J.* **263**: 823–828.

Kettle AJ, Winterbourn CC. 1991. Mechanism of inhibition of myeloperooxidase by anti-inflammatory drugs. *Biochem . Pharmacol.* **41**: 1485–1492.

Kettle AJ, Winterbourn CC. 1994. Assays for the chlorination activity of myeloperoxidase. *Methods Enzymol.* **233**: 502–512.

Kettle AJ, Winterbourn CC. 2001. A kinetic analysis of the catalase activity of myeloperoxidase. *Biochemistry* **40**: 10204–10212.

Khan AU, Kasha M. 1963. Red chemiluminescence of molecular oxygen in aqueous solution. *J. Chem. Phys.* **39**: 2105–2106.

Klebanoff SJ. 1968. Myeloperoxidase-halide-hydrogen peroxide antibacterial system. *J. Bacteriol.* **95**: 2131–2138.

Klebanoff SJ. 1991. Myeloperoxidase: occurrence and biological function. In Everse J, Everse KE, Grisham MB, editors, *Peroxidases in Chemistry and Biology*, Vol. 1, CRC Press, Boca Raton, FL, pp. 1–35.

Klebanoff SJ. 1999. Oxygen metabolites from phagocytes. In: Gallin JI, Snyderman R, editors. *Inflammation: Basic Principles and Clinical Correlates*, 3rd edition. Lippincott Williams & Wilkins, Philadelphia, pp. 391–444.

Klebanoff SJ, Agosti JM, Jörg A, Waltersdorph AM. 1989. Comparative toxicity of the horse eosinophil peroxidase-H_2O_2-halide system and granule basic proteins. *J. Immunol.* **143**: 239–244.

Klebanoff SJ, Clark RA. 1978. *The Neutrophil: Function and Clinical Disorders*, North-Holland, Amsterdam.

Kooter IM, Koehler BP, Moguilevsky N, Bollen A, Wever R, Johnson MK. 1999a. The Met243 sulfonium ion linkage is responsible for the anomalous magnetic circular dichroism and optical spectral properties of myeloperoxidase. *J. Biol. Inorg. Chem.* **4**: 684–691.

Kooter IM, Moguilevsky N, Bollen A, Sijtsema NM, Otto C, Dekker HL, Wever R. 1999b. Characterization of the Asp94 and Glu242 mutants in myeloperoxidase, the residues linking the heme group via ester bonds. *Eur. J. Biochem.* **264**: 211–217.

Kooter IM, Moguilevsky N, Bollen A, Sitjsema NM, Otto C, Wever R. 1997a. Site-directed mutagenesis of Met243, a residue of myeloperoxidase involved in binding of the prosthetic group. *J. Biol. Inorg. Chem.* **2**: 191–197.

Kooter IM, Moguilevsky N, Bollen A, van der Veen LA, Otto C, Dekker HL, Wever R. 1999c. The sulfonium ion linkage in myeloperoxidase: direct spectroscopic detection by isotopic labeling and effect of mutation. *J. Biol. Chem.* **274**: 26794–26802.

Kooter IM, Pierik AJ, Merkz M, Averill BA, Moguilevsky N, Bollen A, Wever R. 1997b. Difference Fourier transform infrared evidence for ester bonds linking the heme group in myeloperoxidase, lactoperoxidase, and eosinophil peroxidase. *J. Am. Chem. Soc.* **119**: 11542–11543.

Lee E, Miki Y, Katsura H, Kariya K. 1990. Mechanism of inactivation of myeloperoxidase by propylthiouracil. *Biochem. Pharmacol.* **39**: 1467–1471.

Long CA, Bielski BHJ. 1980. Rate of reaction of superoxide radical with chloride-containing species. *J. Phys. Chem.* **84**: 555–557.

López-Garriga JJ, Oertling AO, Kean RT, Hoogland H, Wever R, Babcock GT. 1990. Metal-ligand vibrations of cyanoferric myeloperoxidase and cyanoferric horseradish peroxidase: evidence for a constrained heme pocket in myeloperoxidase. *Biochemistry* **29**: 9387–9395.

Malle E, Marsche G, Panzenboeck U, Sattler W. 2006. Myeloperoxidase-mediated oxidation of high-density lipoproteins: fingerprints of newly recognized potential proaatherogenic lipoproteins. *Arch. Biochem. Biophys.* **445**: 245–255.

Marquez LA. 1990. Spleen myeloperoxidase: role of ascorbic acid. Ph.D. thesis, University of Alberta.

Marquez L, Dunford HB. 1989. Cyanide binding to canine myeloperoxidase. *Biochem. Cell Biol.* **67**: 187–191.

Marquez L, Dunford HB. 1990. Reaction of Compound III of myeloperoxidase with ascorbic acid. *J. Biol. Chem.* **265**: 6074–6078.

Marquez LA, Dunford HB. 1993. Interaction of acetaminophen with myeloperoxidase intermediates: optimum stimulation of enzyme activity. *Arch. Biochem. Biophys.* **305**: 414–420.

Marquez L, Dunford HB. 1994. Chlorination of taurine by myeloperoxidase: kinetic evidence for an enzyme-bound intermediate. *J. Biol. Chem.* **269**: 7950–7956.

Marquez LA, Dunford HB. 1995. Kinetics of oxidation of tyrosine and dityrosine by myeloperoxidase Compounds I and II. *J. Biol. Chem.* **270**: 30434–30440.

Marquez LA, Dunford HB. 1997. Mechanism of the oxidation of 3,5,3′,5′-tetramethylbenzidine by myeloperoxidase determined by transient- and steady-state kinetics. *Biochemistry* **36**: 9349–9355.

Marquez LA, Dunford HB, Van Wart H. 1990. Kinetic studies on the reaction of Compound II of myeloperoxidase with ascorbic acid: role of ascorbic acid in myeloperoxidase function. *J. Biol. Chem.* **265**: 5666–5670.

Marquez LA, Huang JT, Dunford HB. 1994. Spectral and kinetic studies on the formation of myeloperoxidase Compounds I and II: roles of hydrogen peroxide and superoxide. *Biochemistry* **33**: 1447–1454.

Matheson NR, Wong PS, Travis J. 1981. Isolation and properties of human neutrophil peroxidase. *Biochemistry* **20**: 325–330.

McCormick ML, Roeder TL, Railsback MA, Britigan BE. 1994. Eosinophil peroxidase-dependent hydroxyl radical generation by human eosinophils. *J. Biol. Chem.* **269**: 27914–27919.

Menegazzi R, Zabucchi G, Zuccato P, Cramer R, Piccinini C, Patriarca P. 1991. Oxidation of homovanillic acid as a selective assay for eosinophil peroxidase in eosinophil peroxidase-myeloperoxidase mixture and its use in detecting human eosinophil peroxidase deficiency. *J. Immunol. Methods* **137**: 55–63.

Metodiewa D, Dunford HB. 1989. The reactions of horseradish peroxidase, lactoperoxidase and myeloperoxidase with enzymatically generated superoxide. *Arch. Biochem. Biophys.* **272**: 245–253.

Metodiewa D, Dunford HB. 1990a. The role of myeloperoxidase in the oxidation of biologically active polyhydroxyphenols (substituted catechols). *Eur. J. Biochem.* **193**: 445–448.

Metodiewa D, Dunford HB. 1990b. Medical aspects and techniques for peroxidases and catalases. In: Scott G, editor. *Atmospheric Oxidation and Antioxidants*, Vol. 3. Elsevier, Amsterdam, pp. 287–332.

Metodiewa D, Dunford HB. 1993. Medical aspects and techniques for peroxidases and catalases. In: Scott G, editor. *Atmospheric Oxidation and Antioxidants*, Vol. III, Chapter 11, pp. 287–332.

Metodiewa D, Marquez LA, Dunford HB. 1991. The activity of mammalian peroxidases (lactoperoxidase and myeloperoxidase and their Compound III toward 2-t-butyl-4-methoxyphenol (butylated hydroxyanisole) and its dimer (2,2′-dihydroxy-3,3′-di-t-butyl-5,5′-dimethoxydiphenyl). *Biochem. Int.* **23**: 281–290.

Miyasaki KT, Song J-P, Murthy ARK. 1991. Secretion of myeloperoxidase isoforms by human neutrophils. *Anal. Biochem.* **193**: 38–44.

Moguilevsky N, Garcia-Quintana L, Jacquet A, Tournay C, Fabry L, Piérard L, Bollen A. 1991. Structural and biological properties of human recombinant myeloperoxidase produced by Chinese hamster ovary cell lines. *Eur. J. Biochem.* **197**: 605–614.

Morishita K, Kubota N, Asano S, Kaziro Y, Nagata S. 1987. Molecular cloning and characterization of cDNA for human myeloperoxidase. *J. Biol. Chem.* **262**: 3844–3851.

Morita Y, Iwamoto H, Aibara S, Kobayashi T, Hasegawa E. 1986. Crystallization and properties of myeloperoxidase from normal human leukocytes. *J. Biochem.* **99**: 761–770.

Odajima T. 1971. Myeloperoxidase of the leukocyte of normal blood. II. The oxidation-reduction reaction mechanism of the myeloperoxidase system. *Biochim. Biophys. Acta* **235**: 52–60.

Odajima T, Yamazaki I. 1970. Myeloperoxidase of the leukocyte of normal blood. I. Reaction of myeloperoxidase with hydrogen peroxide. *Biochim. Biophys. Acta* **206**: 71–77.

Olsen LF, Kummer U, Kindzelsksii AL, Petty HR. 2003a. A model for the oscillatory metabolism of activated neutrophils. *Biophys. J.* **84**: 69–81.

Olsen LF, Lunding A, Kummer U. 2003b. Mechanism of melatonin-induced oscillations in the peroxidase-oxidase reaction. *Arch. Biochem. Biophys.* **410**: 287–295.

Olsen RL, Little C. 1984. Studies on the subunits of human myeloperoxidase. *Biochem. J.* **222**: 701–709.

Oxvig C, Thomsen AR, Overgaard MT, Sørensen ES, Højrup P, Bjerrum MJ, Gleich GJ, Sottrup-Jensen L. 1999. Biochemical evidence for heme linkage through esters with Asp-93 and Glu-241 in human eosinophil peroxidase. *J. Biol. Chem.* **274**: 16953–16958.

Pattison DI, Davies MJ. 2005. Kinetic analysis of the role of histidine chloramines in hypochlorous acid mediated protein oxidation. *Biochemistry* **44**: 7378–7387.

Pember SO, Shapira R, Kinkade JM Jr. 1983. Multiple forms of myeloperoxidase from human neutrophilic granulocytes: evidence for differences in compartmentalization, enzymatic activity, and subunit structure. *Arch. Biochem. Biophys.* **221**: 391–403.

Ramos CL, Pou S, Britigan BE, Cohen MS, Rosen GM. 1992. Spin trapping evidence for myeloperoxidase-dependent hydroxyl radical formation by human neutrophils and monocytes. *J. Biol. Chem.* **267**: 8307–8312.

Romano M, Patriarca P, Melo C, Baralle FE, Dri P. 1994. Hereditary eosinophil peroxidase deficiency: immunochemical and spectroscopic studies and evidence for a compound heterozygosity of the defect. *Proc. Natl. Acad. Sci. USA* **91**: 12496–12500.

Rosen H, Klebanoff SJ. 1977. Formation of singlet oxygen by the myeloperoxidase-mediated antimicrobial system. *J. Biol. Chem.* **252**: 4803–4810.

Sakamaki K, Tomonaga M, Tsukui K, Nagata S. 1989. Molecular cloning and characterization of a chromosomal gene for human eosinophil peroxidase. *J. Biol. Chem.* **264**: 16828–16836.

Sampson JB, Ye YZ, Rosen H, Beckman JS. 1998. Myeloperoxidase and horseradish peroxidase catalyze tyrosine nitration in proteins from nitrite and hydrogen peroxide. *Arch. Biochem. Biophys.* **356**: 207–213.

Schraufstätter IU, Browne K, Harris A, Hyslop PA, Jackson JH, Quehenberger O, Cochrane CG. 1990. Mechanisms of hypochlorite injury of target cells. *J. Clin. Invest.* **85**: 554–562.

Senthilmohan R, Kettle AJ. 2006. Bromination and chlorination reactions of myeloperoxidase at physiological concentrations of bromide and chloride. *Arch. Biochem. Biophys.* **445**: 235–244.

Shen Z, Mitra SN, Wu W, Chen Y, Yang Y, Qin J, Hazen SL. 2001. Eosinophil peroxidase catalyzes bromination of free nucleosides and double-stranded DNA. *Biochemistry* **40**: 2041–2051.

Sibbett SS, Hurst JK. 1984. Structural analysis of myeloperoxidase by resonance Raman spectroscopy. *Biochemistry* **23**: 3007–3013.

Slungaard A, Mahoney JR Jr. 1991a. Bromide-dependent toxicity of eosinophil peroxidase for endothelium and isolated working rat hearts: a model for eosinophilic endocarditis. *J. Exp. Med.* **173**: 117–126.

Slungaard A, Mahoney JR Jr. 1991b. Thiocyanate is the major substrate for eosinophil peroxidase in physiologic fluids: implications for cytotoxicity. *J. Biol. Chem.* **266** 4903–4910.

Spaltcholz H, Panasenko OM, Arnhold J. 2006. Formation of reactive halide species by myeloperoxidase and eosinophil peroxidase. *Arch. Biochem. Biophys.* **445**: 225–234.

Taurog A, Dorris ML. 1992. Myeloperoxidase-catalyzed iodination and coupling. *Arch. Biochem. Biophys.* **296**: 239–246.

Taylor KL, Pohl J, Kinkade JM Jr. 1992a. Unique autolytic cleavage of human myeloperoxidase: implications for the involvement of active site Met409. *J. Biol. Chem.* **267**: 25282–25288.

Taylor KL, Strobel F, Yue KT, Ram P, Pohl J, Woods AS, Kincade JM Jr. 1995. Isolation and identification of a protoheme IX derivative released during autolytic cleavage of human myeloperoxidase. *Arch. Biochem. Biophys.* **316**: 635–642.

Taylor KL, Uhlinger DJ, Kincade Jr JM. 1992b. Expression of recombinant myeloperoxidase using a baculovirus expression system. *Biochem. Biophys. Res. Commun.* **187**: 1572–1578.

Ten RM, Pease LR, McKean DJ, Bell MP, Gleich GJ. 1989. Molecular cloning of the human eosinophil peroxidase: evidence for the existence of a peroxidase multigene family. *J. Exp. Med.* **169**: 1757–1769.

Thomas EL, Bozeman PM, Jefferson MM, King CC. 1995. Oxidation of bromide by the human leukocyte enzymes myeloperoxidase and eosinophil peroxidase: formation of bromamines. *J. Biol. Chem.* **270**: 2906–2913.

Thomas EL, Learn DB. 1990. Myeloperoxidase-catalyzed oxidation of chloride and other halides: the role of chloramines. In: Everse J, Everse KE, Grisham MB, editors. *Peroxidases in Chemistry and Biology*, Vol. 1. CRC Press, Boca Raton, FL, pp. 83–103.

Thomsen AR, Sottrup-Jensen L, Gleich GJ, Oxvig C. 2000. The status of half-cystine residues and locations of N-glycosylated asparagine residues in human eosinophil peroxidase. *Arch. Biochem. Biophys.* **379**: 147–152.

Tuynman A, Spelberg JL, Kooter IM, Schoemaker HE, Wever R. 2000. Enantioselective epoxidation and carbon-carbon bond cleavage catalyzed by *Coprinus cinereus* peroxidase and myeloperoxidase. *J. Biol. Chem.* **275**: 3025–3030.

Uehara K, Hori K, Nakano M, Koga S. 1991. Highly sensitive chemiluminescence method for determining myeloperoxidase in human polymorphonuclear leukocytes. *Anal. Biochem.* **199**: 191–196.

van Dalen CJ, Kettle AJ. 2001. Substrates and products of eosinophil peroxidase. *Biochem. J.* **358**: 233–239.

van Dalen CJ, Winterbourn CC, Kettle AJ. 2006. Mechanism of nitrite oxidation by eosinophil peroxidase: implications for oxidant production and nitration by eosinophils. *Biochem. J.* **394**: 707–713.

van Zyl JM, Basson K, Kriegler A, van der Walt BJ. 1990. Activation of chlorpromazine by the myeloperoxidase system of the human neutrophil. *Biochem. Pharmacol.* **40**: 947–954.

van Zyl JM, Basson K, Uebel RA, van der Walt BJ. 1989a. Isoniazid-mediated irreversible inhibition of the myeloperoxidase antimicrobial system of the human neutrophil and the effect of thyronines. *Biochem. Pharmacol.* **38**: 2363–2373.

van Zyl JM, Basson K, van der Walt BJ. 1989b. The inhibitory effect of acetaminophen on the myeloperoxidase-induced antimicrobial system of the polymorphonuclear leukocyte. *Biochem. Pharmacol.* **38**: 161–165.

Vaz SM, Augusto O. 2008. Inhibition of myeloperoxidase-mediated protein nitration by tempol: kinetics, mechanism, and implications. *Proc. Natl. Acad. Sci. USA* **105**: 8191–8196.

Visser JWM. 1990. Analysis and sorting of blood and bone marrow cells. In: Melamed MR, Lindmo T, Mendelsohn ML, editors. *Flow Cytometry and Sorting*, 2nd edition. Wiley-Liss, New York, pp 669–683.

Wang J, Slungaard A. 2006. Role of eosinophil peroxidase in host defence and disease pathology. *Arch. Biochem. Biophys.* **445**: 256–260.

Weil SC, Rosner GL, Reid MS, Chisholm RL, Farber NM, Spitznagel JK, Swanson MS. 1987. cDNA cloning of human myeloperoxidase: decrease in myeloperoxidase mRNA upon induction of HL-60 cells. *Proc. Natl. Acad. Sci. USA* **84**: 2057–2061.

Weiss SJ, Test ST, Eckmann CM, Roos D, Regiani S. 1986. Brominating oxidants generated by human eosinophils. *Science* **234**: 200–203.

Wever R, Kast WM, Kasinoedin JH, Boelens R. 1982. The peroxidation of thiocyanate catalyzed by myeloperoxidase and lactoperoxidase. *Biochim. Biophys. Acta* **709**: 212–219.

Wever R, Oertling WA, Hoogland H, Bolscher BGJM, Kim Y, Babcock GT. 1991. Denaturation and renaturation of myeloperoxidase. *J. Biol. Chem.* **266**: 24308–24313.

Wever R, Plat H, Hamers MN. 1981. Human eosinophil peroxidase: a novel isolation procedure, spectral properties and chlorinating activity. *FEBS Lett.* **123**: 327–331.

Winterbourn CC, Garcia RC, Segal AW. 1985. Production of the superoxide adduct of myeloperoxidase (compound III) by stimulated human neutrophils and its reactivity with hydrogen peroxide and chloride. *Biochem. J.* **228**: 583–592.

Wright J, Bastian N, Davis TA, Zuo C, Yoshimoto S, Orme-Johnson WH, Tauber AI. 1990. Structural characterization of the isoenzymatic forms of human myeloperoxidase: evaluation of the iron-containing prosthetic group. *Blood* **75**: 238–241.

Wright J, Yoshimoto S, Offner GD, Blanchard RA, Troxler R, Tauber AI. 1987. Structural characterization of the isoenzymatic forms of human myeloperoxidase. *Biochim. Biophys. Acta* **915**: 68–76.

Wu W, Chen Y, d'Avignon A, Hazen SL. 1999b. 3-Bromotyrosine and 3,5-dibromotyrosine are major products of protein oxidation by eosinophil peroxidase: potential markers for eosinophil-dependent tissue injury in vivo. *Biochemistry* **38**: 3538–3548.

Wu W, Chen Y, Hazen SL. 1999a. Eosinophil peroxidase nitrates protein tyrosyl residues: implications for oxidative damage by nitrating intermediates in eosinophilic inflammatory disorders. *J. Biol.Chem.* **274**: 25933–25944.

Yamada M, Hur SJ, Hashinaka K, Tsuneoka K, Saeki T, Nishio C, Sakiyama F, Tsunasawa S. 1987. *Arch. Biochem. Biophys.* **255**: 147–155.

Yamazaki I, Piette LH. 1963. The mechanism of aerobic oxidase reaction catalyzed by peroxidase. *Biochim. Biophys. Acta* **77**: 47–64.

Zeng J, Fenna RE. 1992. X-ray crystal structure of canine myeloperoxidase at 3 Å resolution. *J. Mol. Biol.* **226**: 185–207.

Zgliczyński JM, Stelmaszyńska T, Domański J, Ostrowski W. 1971. Chloramines as intermediates of oxidation reaction of amino acids by myeloperoxidase. *Biochim. Biophys. Acta* **235**: 419–424.

Zgliczyński JM, Stelmaszyńska T, Ostrowski W, Naskalski J, Sznajd J. 1968. Myeloperoxidase of human leukaemic leucocytes: oxidations of amino acids in the presence of hydrogen peroxide. *Eur. J. Biochem.* **4**: 540–547.

Zhang H, Xu Y, Joseph J, Kalyanaraman B. 2005. Intramolecular electron transfer between tyrosyl radical and cysteine residue inhibits tyrosine nitration and induces thiyl radical formation in model peptides treated with myeloperoxidase, H_2O_2, and NO_2^-. *J. Biol. Chem.* **280**: 40684–40698.

Zuurbier KWM, Bakkenist ARJ, Fokkens RH, Nibbering NMM, Wever R, Muijsers AO. 1990a. Interaction of myeloperoxidase with diclofenac. *Biochem. Pharmacol.* **40**: 1801–1808.

Zuurbier KWM, Bakkenist ARJ, Wever R, Muijsers AO. 1990b. The chlorinating activity of human myeloperoxidase: high initial activity at neutral pH value and activation by electron donors. *Biochim. Biophys. Acta* **1037**: 140–146.

CHAPTER 17

BIOMEDICAL ASPECTS OF MYELOPEROXIDASE: HALOGENATION REACTIONS IN CARDIOVASCULAR DISEASE, INFECTION, AND CANCER

JEFFREY P. HENDERSON and JAY. W. HEINECKE

INTRODUCTION

Because of its abundance in neutrophils and ability to produce microbicidal oxidants, the heme protein myeloperoxidase (MPO) has classically been considered to be a host innate immune effector (Kettle and Winterbourn, 1997; Nauseef, 1988; Klebanoff, 1980). Accordingly, many studies of MPO have focused on its role in host defense mechanisms. The oxidizing molecular species generated by MPO during the inflammatory response, however, can be indiscriminate and will react with host as well as pathogen molecules. In this chapter, we discuss the redox chemistry executed by MPO-derived oxidants and the potential consequences of this for different pathophysiological processes.

OXIDANTS PRODUCED BY MPO IN HUMANS

Phagocytes contain a membrane-associated NADPH oxidase (Kettle and Winterbourn, 1997; Nauseef, 1988; Klebanoff, 1980) that generates superoxide ($O_2^{\bullet-}$), which spontaneously or enzymatically dismutates to form hydrogen peroxide (H_2O_2).

$$NADPH + O_2 \longrightarrow NADP^+ + O_2^{\bullet-} \longrightarrow H_2O_2$$

Peroxidases and Catalases: Biochemistry, Biophysics, Biotechnology, and Physiology, Second Edition.
By H. Brian Dunford

Hydrogen peroxide is relatively unreactive, but activated phagocytes also secrete MPO, which transforms H_2O_2 into potent microbicidal oxidants. The enzyme reacts with H_2O_2 to form compound I, a π-cation radical complex that is reduced back to its native oxidation state by electron donors (Kettle and Winterbourn, 1997; Hurst and Barrette, 1989). In vitro, the MPO-H_2O_2 system can use many electron donors, including halides, reduced pyridine nucleotides, and phenols. Studies of mice deficient in MPO indicate that the enzyme generates reactive intermediates that play key roles in host defenses against invading microbes (Brennan et al., 2001; Gaut et al., 2001).

MPO can catalyze the formation of a variety of reactive species in vitro (Kettle and Winterbourn, 1997; Nauseef, 1988; Klebanoff, 1980), where controlled conditions and supraphysiologic substrate concentrations make it possible to directly detect these molecules. The chemical complexity of the in vivo inflammatory microenvironment, however, makes direct detection of these evanescent molecules a challenge. Study of the in vivo reactions of MPO is therefore aided significantly by developing assays for stable oxidation products that can serve as biomarkers of specific reactive species (Fig. 17.1). Performing and interpreting these studies requires sensitive, specific, and quantitative chemical analysis, as well as detailed knowledge of redox chemistry.

Does MPO Produce Halogenating Species in Vivo?

One of MPO's hallmark reactions is the two-electron oxidation of chloride ion by compound I to hypochlorous acid (Harrison and Schultz, 1976; Foote et al., 1981).

$$Cl^- + H_2O_2 + H^+ \longrightarrow HOCl + H_2O$$

At physiologic pH, hypochlorous acid rapidly reacts with amines to form chloramines, which in turn react with other organic substrates (Zgliczynski et al., 1971; Thomas, 1979). When chloramines of amino acids are formed, the resulting amino acid undergoes oxidative decomposition to form aldehydes that may, in turn react with primary amines of proteins or lipids to form aldehyde conjugates (Heller et al., 2000; Hazen et al., 1996, 1997a, 1998; Anderson et al., 1997, 1999). Protein and lipid adducts of *para*-hydroxyphenylacetaldehyde, a major product of hypochlorous acid and free tyrosine, have been detected in human tissue, using mass spectrometry (Heller et al., 2000; Hazen et al., 2000), suggesting that chloramine formation occurs in humans and that this process may generate an array of additional reactive species.

Perhaps the most specific markers of reactive chlorinating species are products containing a carbon-chlorine bond. For example, aromatic substitution of tyrosine or pyrimidine nucleobases yields acid-stable products that can be isolated from tissue and detected by mass spectrometry (Hazen and Heinecke, 1997; Hazen et al., 1997b; Henderson et al., 2003; Takeshita et al., 2006). Indeed, this approach has detected 3-chlorotyrosine in human tissue hydrolysates representing a variety of pathophysiologic processes (Gaut et al., 2001; Hazen and Heinecke, 1997; Hazen et al., 1997b; Lamb et al., 1999; Van Der Vliet et al., 2000; Wu et al., 2000;

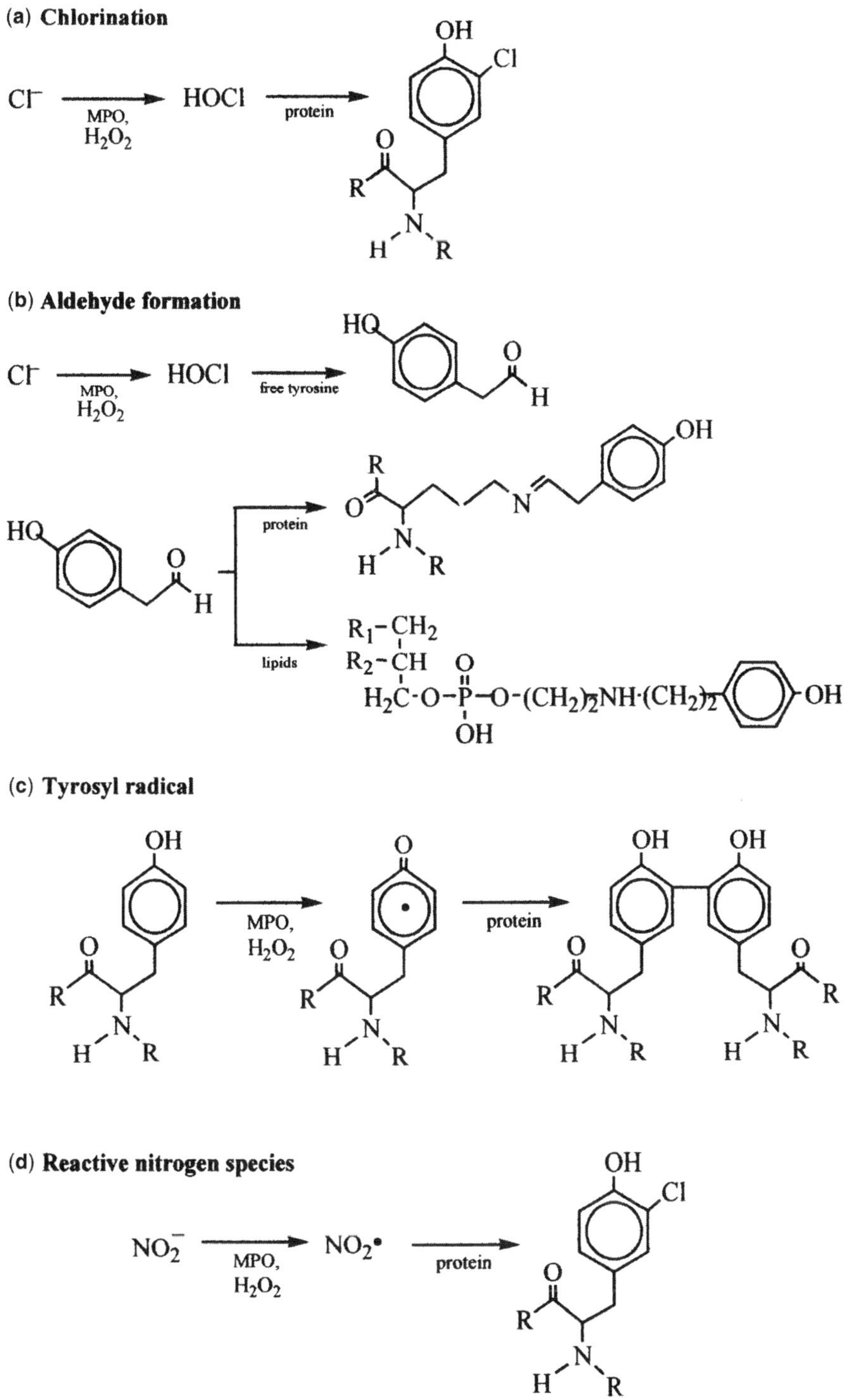

Figure 17.1. Oxidation of proteins and amino acids by MPO. Oxidation products include 3-chlorotyrosine (1), *p*-hydroxyphenylacetaldehyde (2), and its Schiff base adduct with lysine (3), tyrosyl radical (4), *o-o*′-dityrosine (5), and 3-nitrotyrosine (6).

Himmelfarb et al., 2001; Buss et al., 2003; Bergt et al., 2004; Higashi et al., 2004; Choi et al., 2005). These data strongly support an in vivo chlorinating role for MPO. They also raise the possibility that relevant quantities of chlorine gas are generated at sites of inflammation through the reaction:

$$HOCl + Cl^- + H^+ \longrightarrow Cl_2 + H_2O$$

A role for molecular chlorine was suggested by studies of the reaction between hypochlorous acid and free tyrosine in which chloride ion was required to generate 3-chlorotyrosine (Hazen and Heinecke, 1997). Direct detection of chlorine gas was also achieved with analysis of headspace gas from a reaction mixture of MPO, hydrogen peroxide, and sodium chloride (Hazen and Heinecke, 1997). The existence of molecular chlorine in vivo would be of significance because of the high redox potential of the gas and its ability to cross cell membranes. Extrapolation of in vitro data suggests that chlorine production should be favored in acidic environments.

While chloride is the most abundant halide in human plasma ($\sim$100 mM), bromide may also play a significant role in biohalogenation reactions in humans. Despite its relatively low physiologic concentrations (20 to 100 μM; Holzbecher and Ryan, 1980), bromide has a lower oxidation potential and participates readily in halogenation reactions. There are two potential routes for bromide oxidation by MPO in vivo (Henderson et al., 2001c). In one, bromide interacts directly with MPO (Thomas et al., 1995):

$$Br^- + H_2O_2 + H^+ \longrightarrow HOBr + H_2O$$

The other involves oxidation of bromide by MPO-derived hypochlorous acid to form the interhalogen gas BrCl, which may undergo spontaneous hydrolysis to hypobromous acid and chloride ion (Kumar and Margerum, 1987; Wang et al., 1994):

$$Br^- + HOCl + H^+ \longrightarrow BrCl + H_2O \longrightarrow HOBr + Cl^-$$

This second route for generating brominating agents has been observed with both purified MPO and isolated neutrophils (Henderson et al., 2001c). Like 3-chlorotyrosine, which serves as a marker of chlorinating oxidants, 3-bromotyrosine is a specific biomarker of brominating oxidants (Wu et al., 1999). While other mammalian peroxidases, such as eosinophil peroxidase (Thomas et al., 1995) and lactoperoxidase (Buchberger, 1988), may also directly oxidize bromide, evidence for an MPO origin of brominating agents in vivo consists of studies demonstrating the coexistence of both chlorinated and brominated uracil and tyrosine in infected human tissue (Henderson et al., 2003; Wu et al., 2000). Further study in a mouse model of peritoneal infection revealed a significant reduction in 3-bromotyrosine levels in MPO-null mice versus wild-type mice (Gaut et al., 2001). Whether MPO produces 3-bromotyrosine in vivo through direct oxidation of bromide or via a bromine chloride intermediate cannot be distinguished through 3-bromotyrosine analysis alone.

Does MPO Produce Nitrating Oxidants in Vivo?

While nitrating oxidants in vivo ultimately derive from nitric oxide ($NO^{\cdot}$) produced by one of several nitric oxide synthases, the precise pathways leading to nitration reactions have been the subject of multiple studies. MPO has been proposed as one source of nitrating intermediates. In addition to the two-electron oxidation of chloride ion, MPO can also catalyze one-electron oxidation of nitrite (NO_2^-), a nitric oxide decomposition product commonly found in the inflammatory milieu. The oxidation product of this reaction is nitrogen dioxide ($NO_2^{\bullet}$), a free radical gas more commonly recognized as an air pollutant (Eiserich et al., 1996, 1998).

$$NO_2^{\bullet} - + \text{MPO Compound I} \longrightarrow NO_2^{\bullet} + \text{MPO Compound II}$$

Nitrogen dioxide may subsequently oxidize other substrates, such as lipids (Byun et al., 1999) and aromatic compounds (Eiserich et al., 1998; Yeh et al., 2003; Singh et al., 1998; Yamakura and Ikeda, 2006), or it may rapidly react with tyrosyl radical to produce stable 3-nitrotyrosine, a biomarker of nitrating oxidants (Eiserich et al., 1998). MPO, as well as activated human neutrophils, was found to form 3-nitrotyrosine as well as 8-nitroxanthine in a manner consistent with a role for MPO activity (Eiserich et al., 1998; Yeh et al., 2003).

Because 3-nitrotyrosine is a common endpoint of multiple physiologically plausible nitrating pathways, mouse models have been used to determine if MPO is capable of executing nitration reactions in vivo. In a mouse model of *Klebsiella* infection, increased levels of 3-nitrotyrosine and 3-chlorotyrosine were observed in the peritoneum, suggesting the joint presence of chlorinating and nitrating intermediates (Gaut et al., 2002). Furthermore, the increase in 3-nitrotyrosine observed in this model of inflammation was abrogated in mice lacking MPO (Gaut et al., 2002). Together, these data strongly support a role for MPO in generating nitrating intermediates at sites of infection.

MPO and Tyrosyl Radical Formation

Certain peroxidases oxidize tyrosine to *o,o'*-dityrosine in vitro (Gross and Sizer, 1959). The reaction also occurs in vivo, as when the ovoperoxidase of nascent sea urchin embryos generates *o,o'*-dityrosine cross-links in the protective fertilization envelope by a reaction requiring H_2O_2 (Foerder and Shapiro, 1977). It has not been established whether these reactions involve tyrosyl radical.

1. To determine whether MPO might catalyze the formation of tyrosyl radical, we used an electron paramagnetic resonance (EPR) spin-trapping system containing the nitroso spin trap MNP (McCormick et al., 1998). MNP solutions typically demonstrate a small, 3-line background spectrum ($a^N = 16.87$ G) due to the light-induced formation of di-*tert*-butyl nitroxide. Exposing L-tyrosine to the MPO-H_2O_2 system in the presence of MNP yielded a larger, slightly narrower ($a^N = 15.55$) 3-line spectrum, which is consistent with formation of an MNP-tyrosyl radical spin adduct. Removing either L-tyrosine or H_2O_2 from

the system yielded only the background di-*tert*-butyl nitroxide spectrum. Including the MPO inhibitor azide also prevented tyrosyl radical formation, demonstrating the requirement for active peroxidase. Importantly, adding 150 mM NaCl failed to appreciably decrease the MNP-tyrosyl radical signal, suggesting that MPO can still form tyrosyl radical in the presence of physiological levels of chloride ion (one of its known substrates). These EPR spin-trapping studies demonstrated that MPO can oxidize tyrosine to tyrosyl radical by a process dependent on H_2O_2 and active peroxidase.

2. Substituting ^{13}C ring-labeled L-tyrosine for unlabeled L-tyrosine confirmed that tyrosyl radical was the species responsible for the EPR spectrum generated by MPO (^{13}C has a spin of $\frac{1}{2}$, whereas ^{12}C has a spin of 0). When L-[$^{13}C_6$]tyrosine was substituted for unlabeled L-tyrosine, a unique 8-line spectrum appeared. This 8-line spectrum was also seen when lactoperoxidase or horseradish peroxidase replaced MPO in the presence of L-[$^{13}C_6$]tyrosine. Similar results using horseradish peroxidase reacted with H_2O_2 were reported previously (Barr et al., 1996). As with MPO, these 8-line spectra were not seen if L-[$^{13}C_6$]tyrosine or H_2O_2 was omitted from the reaction mixture.
3. The experiments with L-[$^{13}C_6$]tyrosine provide strong evidence that tyrosyl radical is produced when the MPO and other peroxidase-H_2O_2 systems oxidize L-tyrosine.

MPO AND CORONARY ARTERY DISEASE

Low density lipoprotein (LDL) and high density lipoprotein (HDL) are the major carriers of cholesterol in human blood. Epidemiological, clinical, and genetic studies indicate that elevated levels of LDL greatly increase the risk for atherosclerosis, the leading cause of cardiovascular disease (CAD) in humans (Brown and Goldstein, 1992). Paradoxically, LDL fails to exert potentially atherogenic effects in vitro, suggesting that it must be modified to promote vascular disease. Indeed, many lines of evidence suggest that oxidative damage to LDL is one important mechanism for rendering lipoproteins atherogenic (reviewed in Berliner and Heinecke, 1996; Heinecke, 2001). Although this oxidation hypothesis raises the possibility that antioxidants might prevent or retard human atherosclerosis, recent clinical trials of the lipid-soluble antioxidant vitamin E have yielded mixed results (reviewed in Heinecke, 2001), leaving major questions about the role of LDL oxidation in human vascular disease. However, it is important to note that there is no compelling evidence that vitamin E is an antioxidant in vivo, or that vitamin E scavenges reactive intermediates produced by MPO (Heinecke, 2001).

MPO activity and protein have been detected in human atherosclerotic tissue (Daugherty et al., 1994; Sugiyama et al., 2001). In intermediate lesions, the enzyme colocalizes with macrophages; in advanced lesions, it associates closely with cholesterol clefts in extracellular lipid deposits. A similar pattern of immunostaining of protein-bound lipid oxidation products has been seen in rabbit atherosclerotic lesions

(Witztum and Steinberg, 1991). Moreover, LDL exposed to hypochlorous acid, MPO's major product, aggregates and promotes the transformation of macrophages into foam cells (Hazell and Stocker, 1993). Thus, MPO might be important for lipoprotein oxidation in vivo.

We obtained further evidence that MPO oxidizes artery wall constituents by quantifying levels of 3-chlorotyrosine. Chlorinated biomolecules should be specific markers of oxidative damage by activated phagocytes because, at plasma concentrations of halide, MPO is the only human enzyme known to generate hypochlorous acid (Harrison and Schultz, 1976). In vitro, 3-chlorotyrosine forms when LDL is oxidized by the MPO-peroxide-chloride system (Hazen and Heinecke, 1997). In contrast, it is undetectable when LDL is oxidized by hydroxyl radical, copper, iron, horseradish peroxidase, hemin, glucose, or peroxynitrite.

When we used isotope dilution GC/MS to analyze LDL isolated from surgical samples of human atherosclerotic lesions, we detected dramatically elevated levels of 3-chlorotyrosine (Hazen and Heinecke, 1997). By using MPO-deficient mice, we also obtained evidence that normal neutrophils use the MPO system to chlorinatc tyrosine residues in vivo (Hazen and Heinecke, 1997). These results provide strong evidence that oxidative damage to LDL—and perhaps to other macromolecules involved in plaque formation—can be ascribed at least partly, to MPO in the human artery wall.

In contrast to LDL, high density lipoprotein (HDL) is cardioprotective (Gordon and Rifkind, 1989). For example, it removes cholesterol from artery wall macrophages. To determine whether MPO might oxidize HDL in vivo, we isolated HDL from plasma of subjects with established CAD and from age- and sex-matched healthy subjects (Bergt et al., 2004). Isotope dilution gas chromatography-mass spectrometry detected a 13-fold higher level of protein-bound 3-chlorotyrosine in the HDL from the CAD patients than in that from the controls. These observations strongly support the hypothesis that HOCl derived from MPO contributes to HDL oxidation in vivo.

One important mechanism for HDL's cardioprotective effects involves the ABCA1 pathway. By interacting with lipid-free apoA-I, HDL's major protein, ABCA1 promotes the initial step in cholesterol efflux from macrophages. We therefore determined whether oxidation impaired apoA-I's ability to activate the ABCA1 pathway. Oxidation by H_2O_2 had no effect. In striking contrast, chlorination of apoA-I by MPO or HOCl impaired the protein's ability to promote cholesterol efflux (Bergt et al., 2004). Interestingly, nitration of apoA-I was much less inhibitory (Shao et al., 2005). These observations, together with those of other investigators (Zheng et al., 2004), indicate that oxidation of apoA-I by HOCl generated by MPO selectively impairs the protein's ability to remove cholesterol from cells by a pathway requiring ABCA1.

These observations suggest that MPO could promote atherosclerosis by two distinct pathways. One involves oxidation of LDL, which would increase the delivery of cholesterol to macrophages. Another involves damage to HDL, which might prevent removal of cholesterol from macrophage foam cells by impairing the ABCA1 pathway. It is important to note that vitamin E, the major so-called "antioxidant" investigated in human clinical trials, fails to inhibit LDL or HDL oxidation by MPO-generated HOCl. Thus, the failure of antioxidants to prevent CAD in human clinical trials might reflect the vitamin's inability to inhibit pathophysiologically relevant pathways for lipoprotein damage.

The observation that elevated plasma levels of MPO associate strongly with increased risk of coronary artery disease (Zhang et al., 2001) was not confirmed by other investigators (Kubala et al., 2008; Baldus et al., 2006). One potential problem with such studies is that the vast majority of MPO in blood is carried by neutrophils. Thus, differences in neutrophil activation (which causes the cells to degranulate MPO into plasma) during blood collection and storage might produce large, artifactual differences in plasma MPO levels.

MPO AND CARCINOGENESIS

Rudolf Virchow proposed over a century ago that persistent tissue inflammation is a risk factor for carcinogenesis. Indeed, chronic inflammation appears to be associated with up to one half of human cancers (Triolo, 1965; Beckman and Ames, 1997; Weitzman and Gordon, 1990). These inflammatory conditions derive from both infectious (such as *Helicobacter pylori* and hepatitis C virus) and noninfectious (such as cigarette smoking and asbestos exposure) etiologies (Valinluck and Sowers, 2007; Rosin et al., 1994; Ames et al., 1995). The broad range of inflammatory conditions that precede many cancers have led researchers to consider general features of inflammation that may contribute to human carcinogenesis. The production of reactive oxygen species at sites of inflammation has been one such feature, and a number of studies have implicated MPO as playing a potentially important carcinogenic role.

Recently, studies of a polymorphism in the MPO promoter region that relate to transcription have revealed correlations with a variety of human cancers. One such polymorphism, G463A, decreases the transcription and activity of MPO (Piedrafita et al., 1996; Van Schooten et al., 2004). Multiple genetic studies have linked G463A to a reduction in lung cancer risk among smokers (Van Schooten et al., 2004; London et al., 1997; Le Marchand et al., 2000; Dally et al., 2002; Kantarci et al., 2002; Schabath et al., 2002a, 2002b; Cajas-Salazar et al., 2003; Zhu et al., 2006), a conclusion supported by a large meta-analysis (Taioli et al., 2007). Smaller studies have pointed to similar effects (higher risk with G at 463, lower risk with A) in gastric, pancreatic, and bladder cancer (Zhu et al., 2006; Steenport et al., 2007; Hsu et al., 2008; Wheatley-Price et al., 2008; Hung et al., 2004).

MPO's ability to modify nucleic acids represents one potential carcinogenic mechanism. For example, it can transform polyaromatic hydrocarbons (PAH) to species capable of forming adducts with nucleic acids (Mallett et al., 1991; Petruska et al., 1992). PAHs are ubiquitous in the environment as combustion products and have been hypothesized to contribute to the dramatic increase in cancer risk seen with cigarette smoking. Association of the high expression polymorphism of MPO with lung cancer is in keeping with this hypothesized activity. Indeed, smokers who are heterozygous or homozygous for the low-expression polymorphism have lower levels of both MPO and bulky DNA adducts in lung lavages (Van Schooten et al., 2004).

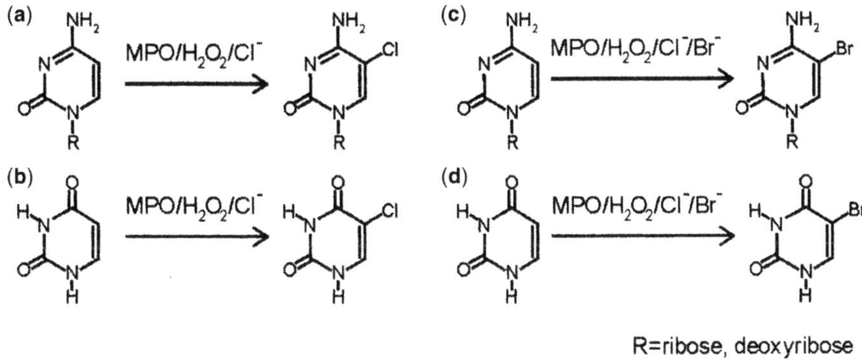

Figure 17.2. Halogenation of nucleic acids by MPO. Cytidine and deoxycytidine (a, c), as well as uracil (b, d) are susceptible to chlorination or bromination by myeloperoxidase-derived oxidants.

Another possible procarcinogenic activity of MPO relates to the enzyme's ability to generate halogenating intermediates in vivo. In vitro studies of nucleosides exposed to hypohalous acids have revealed that the carbon five position of the pyrimidines cytosine and uracil are susceptible to halogen substitution reactions by oxidants from MPO and eosinophil peroxidase (Henderson et al., 1999, 2001a, 2003; Fig. 17.2). Specifically, the MPO-hydrogen peroxide-chloride system and activated humanneutrophils can produce 5-chlorodeoxycytidine and 5-chlorouracil (Henderson et al., 1999, 2003). In the presence of physiologically plausible bromide concentrations, the MPO system can also generate 5-bromodeoxycytidine and 5-bromouracil (Henderson et al., 2001b). Analysis of human tissue using gas chromatography-mass spectrometry has demonstrated the presence of both 5-chlorouracil and 5-bromouracil during bacterial infection (Henderson et al., 2003). These halogenated bases were not detectable in plasma or urine from uninfected patients, and the use of isotope-labeled uracil as a control revealed no artifactual product formation in these studies. Thus, pyrimidine chlorination and bromination can occur at sites of inflammation.

A particularly notable property of chlorine or bromine substituents at the five-carbon position of pyrimidine is their ability to sterically mimic physiologic methylation (Henderson et al., 1999, 2001b). Thus, 5-chloro- and 5-bromouracil are able to mimic thymine (5-methyluracil) and may be incorporated from extranuclear sources into mammalian DNA. These halogenated bases are mutagenic, either because they interfere with deoxynucleotide precursor pools or because the electron-withdrawing effect of halogens stabilizes a tautomeric form of uracil that pairs more effectively with guanine than with adenine (O'Neill et al., 1983; Morris, 1991, 1993; Penman et al., 1976; Freese, 1959; Yu et al., 1993; Davidson et al., 1988).

Because genomic DNA of living cells is remarkably well protected, the ability of MPO to generate modified DNA precursors outside the nucleus—or even outside the cell—suggests a plausible scenario in which procarcinogenic DNA modification occurs as the result of redox chemistry in a more permissive extranuclear environment (Henderson et al., 1999). Peroxidase-modified deoxycytidine was taken up by mammalian cells and incorporated into DNA (Henderson et al., 2001b). Incorporation of halogenated pyrimidines might be favored if halogenated deoxyuridines also diminished the size of competing thymidine pools, for example, by inhibiting thymidylate synthase, as has been observed in vitro (Brandon et al., 2000). Thymidine pools may also be diminished with common micronutrient deficiencies in folic acid or vitamin B12, which are believed to predispose to malignancy (Fenech, 2001).

The presence of halogenated cytosine in genomic DNA may similarly mimic the normal methylation of cytosine that plays a role in gene silencing. This could contribute to the observed perturbations in cytosine methylation patterns that are widely observed in human tumors (Valinluck et al., 2005). Such perturbations may arise because the presence of halogenated cytosine in a parent DNA strand during replication may stimulate methyltransferases to act on the daughter strand, perpetuating a false methylation signal (reviewed in Valinluck and Sowers, 2007). Thus, if halogenated cytosine were incorporated into a tumor suppressor or DNA repair gene, it could have a procarcinogenic effect on replicating cells.

PROSPECTS

Oxidants generated by the MPO system of phagocytes are essential for host defense mechanisms. However, reactive intermediates generated by MPO may damage tissue during acute and chronic inflammation, and this damage might promote diseases such as atherosclerosis and cancer. The ability to quantify distinct amino acid and nucleic acid oxidation products in biological fluids, proteins, and tissues is invaluable for exploring the roles of various oxidation pathways in the pathogenesis of atherosclerosis and other diseases. This information would allow interventions that specifically inhibit physiologically relevant pathways to be tested in clinical trials which, to date, have used only nonspecific antioxidants, such as vitamin E. Future trials should be conducted on subjects with high levels of oxidative stress; such individuals could be identified by quantifying levels of oxidized amino acids, nucleobases, or other markers in plasma and urine. Such a rational approach to therapy has worked well in other fields, and is also likely to aid progress against human disorders promoted by oxidative stress.

REFERENCES

Ames BN, Gold LS, Willett WC. 1995. The causes and prevention of cancer. *Proc. Natl. Acad. Sci. USA* **92**: 5258–5265.

Anderson MM, Hazen SL, Hsu FF, Heinecke JW. 1997. Human neutrophils employ the myeloperoxidase-hydrogen peroxide-chloride system to convert hydroxy-amino acids into glycolaldehyde, 2-hydroxypropanal, and acrolein: a mechanism for the generation of highly reactive alpha-hydroxy and alpha,beta-unsaturated aldehydes by phagocytes at sites of inflammation. *J. Clin. Invest.* **99**: 424–432.

Anderson MM, Requena JR, Crowley JR, Thorpe SR, Heinecke JW. 1999. The myeloperoxidase sytem of human phagocytes generates N-(carboxymethyl)lysine on proteins: a mechanism for producing advanced glycation end products at sites of inflammation. *J. Clin. Invest.* **104**: 103–113.

Baldus S, Rudolph V, Roiss M, Ito WD, Rudolph TK, et al. 2006. Heparins increase endothelial nitric oxide bioavailability by liberating vessel-immobilized myeloperoxidase. *Circulation* **113**: 1871–1878.

Barr DP, Gunther MR, Deterding LJ, Tomer KB, Mason RP. 1996. ESR spin-trapping of a protein-derived tyrosyl radical from the reaction of cytochrome *c* with hydrogen peroxide. *J. Biol. Chem.* **271**: 15498–15503.

Beckman KB, Ames BN. 1997. Oxidative decay of DNA. *J. Biol. Chem.* **272**: 19633–19636.

Bergt C, Pennathur S, Fu X, Byun J, O'Brien K, et al. 2004. The myeloperoxidase product hypochlorous acid oxidizes HDL in the human artery wall and impairs ABCA1-dependent cholesterol transport. *Proc. Natl. Acad Sci. USA* **101**: 13032–13037.

Berliner JA, Heinecke JW. 1996. The role of oxidized lipoproteins in atherogenesis. *Free Radical Biol. Med.* **20**: 707–727.

Brandon ML, Mi L, Chaung W, Teebor G, Boorstein RJ. 2000. 5-Chloro-2′-deoxyuridine cytotoxicity results from base excision repair of uracil subsequent to thymidylate synthase inhibition. *Mutation Res.* **459**: 161–169.

Brennan ML, Anderson MM, Shih DM, Qu XD, Wang X, et al. 2001. Increased atherosclerosis in myeloperoxidase-deficient mice. *J. Clin. Invest.* **107**: 419–430.

Brown MS, Goldstein JL. 1992. Koch's postulates for cholesterol. *Cell* **71**: 187–188.

Buchberger W. 1988. Investigations into lactoperoxidase-catalysed bromination of tyrosine and thyroglobulin. *J Chromatogr.* **432**: 57–63.

Buss IH, Senthilmohan R, Darlow BA, Mogridge N, Kettle AJ, et al. 2003. 3-Chlorotyrosine as a marker of protein damage by myeloperoxidase in tracheal aspirates from preterm infants: association with adverse respiratory outcome. *Pediatr. Res.* **53**: 455–462.

Byun J, Mueller DM, Fabjan JS, Heinecke JW. 1999. Nitrogen dioxide radical generated by the myeloperoxidase-hydrogen peroxide-nitrite system promotes lipid peroxidation of low density lipoprotein. *FEBS Lett.* **455**: 243–246.

Cajas-Salazar N, Sierra-Torres CH, Salama SA, Zwischenberger JB, Au WW. 2003. Combined effect of MPO, GSTM1 and GSTT1 polymorphisms on chromosome aberrations and lung cancer risk. *Int. J. Hyg. Environ. Health* **206**: 473–483.

Choi DK, Pennathur S, Perier C, Tieu K, Teismann P, et al. 2005. Ablation of the inflammatory enzyme myeloperoxidase mitigates features of Parkinson's disease in mice. *J. Neurosci.* **25**: 6594–6600.

Dally H, Gassner K, Jager B, Schmezer P, Spiegelhalder B, et al. 2002. Myeloperoxidase (MPO) genotype and lung cancer histologic types: the MPO-463 A allele is associated with reduced risk for small cell lung cancer in smokers. *Int. J. Cancer* **102**: 530–535.

Daugherty A, Dunn JL, Rateri DL, Heinecke JW. 1994. Myeloperoxidase, a catalyst for lipoprotein oxidation, is expressed in human atherosclerotic lesions. *J. Clin. Invest.* **94**: 437–444.

Davidson RL, Broeker P, Ashman CR. 1988. DNA base sequence changes and sequence specificity of bromodeoxyuridine-induced mutations in mammalian cells. *Proc. Natl. Acad. Sci. USA* **85**: 4406–4410.

Eiserich JP, Cross CE, Jones AD, Halliwell B, van der Vliet A. 1996. Formation of nitrating and chlorinating species by reaction of nitrite with hypochlorous acid: a novel mechanism for nitric oxide-mediated protein modification. *J. Biol. Chem.* **271**: 19199–19208.

Eiserich JP, Hristova M, Cross CE, Jones AD, Freeman BA, et al. 1998. Formation of nitric oxide-derived inflammatory oxidants by myeloperoxidase in neutrophils. *Nature* **391**: 393–397.

Fenech M. 2001. The role of folic acid and vitamin B12 in genomic stability of human cells. *Mutat. Res.* **475**: 57–67.

Foerder CA, Shapiro BM. 1977. Release of ovoperoxidase from sea urchin eggs hardens the fertilization membrane with tyrosine crosslinks. *Proc. Natl. Acad. Sci. USA* **74**: 4214–4218.

Foote CS, Goyne TE, Lehrer RI. 1981. Assessment of chlorination by human neutrophils. *Nature* **301**: 715–716.

Freese E. 1959. The specific mutagenic effect of base analogues on Phage T4. *J. Mol. Biol.* **1**: 87–105.

Gaut JP, Byun J, Tran HD, Lauber WM, Carroll JA, et al. 2002. Myeloperoxidase produces nitrating oxidants in vivo. *J. Clin. Invest.* **109**: 1311–1319.

Gaut JP, Yeh GC, Tran HD, Byun J, Henderson JP, et al. 2001. Neutrophils employ the myeloperoxidase system to generate antimicrobial brominating and chlorinating oxidants during sepsis. *Proc. Natl. Acad. Sci. USA* **98**: 11961–11966.

Gordon DJ, Rifkind BM. 1989. High-density lipoprotein: the clinical implications of recent studies. *New Engl. J. Med.* **321**: 1311–1316.

Gross AJ, Sizer IW. 1959. The oxidation of tyramine, tyrosine, and related compounds by peroxidase. *J. Biol. Chem.* **234**: 1611–1614.

Harrison JE, Schultz J. 1976. Studies on the chlorinating activity of myeloperoxidase. *J. Biol. Chem.* **251**: 1371–1374.

Hazell LJ, Stocker R. 1993. Oxidation of low-density lipoprotein with hypochlorite causes transformation of the lipoprotein into a high-uptake form for macrophages. *Biochem. J.* **290**: 165–172.

Hazen SL, d'Avignon A, Anderson MM, Hsu FF, Heinecke JW. 1998. Human neutrophils employ the myeloperoxidase-hydrogen peroxide-chloride system to oxidize alpha-amino acids to a family of reactive aldehydes: mechanistic studies identifying labile intermediates along the reaction pathway. *J. Biol. Chem.* **273**: 4997–5005.

Hazen SL, Gaut JP, Crowley JR, Hsu FF, Heinecke JW. 2000. Elevated levels of protein-bound *p*-hydroxyphenylacetaldehyde, an amino-acid-derived aldehyde generated by myeloperoxidase, are present in human fatty streaks, intermediate lesions and advanced atherosclerotic lesions. *Biochem. J.* **352**: 693–699.

Hazen SL, Crowley JR, Mueller DM, Heinecke JW. 1997a. Mass spectrometric quantification of 3-chlorotyrosine in human tissues with attomole sensitivity: a sensitive and specific marker for myeloperoxidase-catalyzed chlorination at sites of inflammation. *Free Radical Biol. Med.* **23**: 909–916.

Hazen SL, Gaut JP, Hsu FF, Crowley JR, d'Avignon A, et al. 1997b. *p*-Hydroxyphenylacetaldehyde, the major product of L-tyrosine oxidation by the myeloperoxidase-H_2O_2-chloride system of phagocytes, covalently modifies epsilon-amino groups of protein lysine residues. *J. Biol. Chem.* **272**: 16990–16998.

Hazen SL, Heinecke JW. 1997. 3-Chlorotyrosine, a specific marker of myeloperoxidase-catalyzed oxidation, is markedly elevated in low density lipoprotein isolated from human atherosclerotic intima. *J. Clin. Invest.* **99**: 2075–2081.

Hazen SL, Hsu FF, Heinecke JW. 1996. *p*-Hydroxyphenylacetaldehyde is the major product of L-tyrosine oxidation by activated human phagocytes: a chloride-dependent mechanism for the conversion of free amino acids into reactive aldehydes by myeloperoxidase. *J. Biol. Chem.* **271**: 1861–1867.

Heinecke JW. 2001. Is the emperor wearing clothes? Clinical trials of vitamin E and the LDL oxidation hypothesis. *Arterioscler. Thromb. Vasc. Biol.* **21**: 1261–1264.

Heller JI, Crowley JR, Hazen SL, Salvay DM, Wagner P, et al. 2000. *p*-Hydroxyphenylacetaldehyde, an aldehyde generated by myeloperoxidase, modifies phospholipid amino groups of low density lipoprotein in human atherosclerotic intima. *J. Biol. Chem.* **275**: 9957–9962.

Henderson JP, Byun J, Heinecke JW. 1999. Molecular chlorine generated by the myeloperoxidase-hydrogen peroxide-chloride system of phagocytes produces 5-chlorocytosine in bacterial RNA. *J. Biol. Chem.* **274**: 33440–33448.

Henderson JP, Byun J, Mueller DM, Heinecke JW. 2001a. The eosinophil peroxidase-hydrogen peroxide-bromide system of human eosinophils generates 5-bromouracil, a mutagenic thymine analogue. *Biochemistry* **40**: 2052–2059.

Henderson JP, Byun J, Williams MV, McCormick ML, Parks WC, et al. 2001b. Bromination of deoxycytidine by eosinophil peroxidase: a mechanism for mutagenesis by oxidative damage of nucleotide precursors. *Proc. Natl. Acad. Sci. USA* **98**: 1631–1636.

Henderson JP, Byun J, Williams MV, Mueller DM, McCormick ML, et al. 2001c. Production of brominating intermediates by myeloperoxidase: a transhalogenation pathway for generating mutagenic nucleobases during inflammation. *J. Biol. Chem.* **276**: 7867–7875.

Henderson JP, Byun J, Takeshita J, Heinecke JW. 2003. Phagocytes produce 5-chlorouracil and 5-bromouracil, two mutagenic products of myeloperoxidase, in human inflammatory tissue. *J. Biol. Chem.* **278**: 23522–23528.

Higashi N, Mita H, Taniguchi M, Turikisawa N, Higashi A, et al. 2004. Urinary eicosanoid and tyrosine derivative concentrations in patients with vasculitides. *J Allergy Clin. Immunol.* **114**: 1353–1358.

Himmelfarb J, McMenamin ME, Loseto G, Heinecke JW. 2001. Myeloperoxidase-catalyzed 3-chlorotyrosine formation in dialysis patients. *Free Radical Biol. Med.* **31**: 1163–1169.

Holzbecher J, Ryan DE. 1980. The rapid determination of total bromine and iodine in biological fluids by neutron activation. *Clin. Biochem.* **13**: 277–278.

Hsu PI, Jwo JJ, Yang CL, Hsu PN, Yang HB, et al. 2008. Association of the myeloperoxidase polymorphism with the risk of gastric cancer. *Anticancer Res.* **28**: 1317–1323.

Hung RJ, Boffetta P, Brennan P, Malaveille C, Gelatti U, et al. 2004. Genetic polymorphisms of MPO, COMT, MnSOD, NQO1, interactions with environmental exposures and bladder cancer risk. *Carcinogenesis* **25**: 973–978.

Hurst JK, Barrette WC, Jr. 1989. Leukocytic oxygen activation and microbicidal oxidative toxins. *CRC Crit. Rev. Biochem. Mol. Biol.* 271–328.

Kantarci OH, Lesnick TG, Yang P, Meyer RL, Hebrink DD, et al. 2002. Myeloperoxidase-463 (G → A) polymorphism associated with lower risk of lung cancer. *Mayo Clin. Proc.* **77**: 17–22.

Kettle AJ, Winterbourn CC. 1997. Myeloperoxidase: a key regulator of neutrophil oxidant production. *Redox Report* **3**: 3–15.

Klebanoff SJ. 1980. Oxygen metabolism and the toxic properties of phagocytes. *Ann. Intern. Med.* **93**: 480–489.

Kubala L, Lu G, Baldus S, Berglund L, Eiserich JP. 2008. Plasma levels of myeloperoxidase are not elevated in patients with stable coronary artery disease. *Clin. Chim. Acta* **394**: 59–62.

Kumar K, Margerum DW. 1987. Kinetics and mechanism of general-acid-assisted oxidation of bromide by hypochlorite and hypochlorous acid. *Inorg. Chem.* **26**: 2706–2711.

Lamb NJ, Gutteridge JM, Baker C, Evans TW, Quinlan GJ. 1999. Oxidative damage to proteins of bronchoalveolar lavage fluid in patients with acute respiratory distress syndrome: evidence for neutrophil-mediated hydroxylation, nitration, and chlorination. *Crit. Care Med.* **27**: 1738–1744.

Le Marchand L, Seifried A, Lum A, Wilkens LR. 2000. Association of the myeloperoxidase-463G → A polymorphism with lung cancer risk. *Cancer Epidemiol Biomarkers Prevention* **9**: 181–184.

London SJ, Lehman TA, Taylor JA. 1997. Myeloperoxidase genetic polymorphism and lung cancer risk. *Cancer Res.* **57**: 5001–5003.

Mallet WG, Mosebrook DR, Trush MA. 1991. Activation of (+-)-trans-7,8-dihydroxy-7,8-dihydrobenzo[a]pyrene to diolepoxides by human polymorphonuclear leukocytes or myeloperoxidase. *Carcinogenesis* **12**: 521–524.

McCormick ML, Gaut JP, Lin T, Britigan BE, Buettner GR, et al. 1998. Electron paramagnetic resonance detection of free tyrosyl radical generated by myeloperoxidase, lactoperoxidase, and horseradish peroxidase. *J. Biol. Chem.* **273**: 32030–32037.

Morris SM. 1991. The genetic toxicology of 5-bromodeoxyuridine in mammalian cells. *Mutation Res.* **258**: 161–188.

Morris SM. 1993. The genetic toxicology of 5-fluoropyrimidines and 5-chlorouracil. *Mutation Res.* **297**: 39–51.

Nauseef WM. 1988. Myeloperoxidase deficiency: phagocytic defects. I. Abnormalities outside of the respiratory burst. *Hematol. Oncol. Clin. North Am.* **2**: 135–138.

O'Neill JP, Heartlein MW, Preston RJ. 1983. Sister-chromatid exchanges and gene mutations are induced by the replication of 5-bromo- and 5-chloro-deoxyuridine substituted DNA. *Mutation Res.* **109**: 259–270.

Penman BW, Wong M, Thilly WG. 1976. Mutagenicity of 5-halodeoxyuridines to diploid human lymphoblasts. *Life Sciences* **19**: 563–568.

Petruska JM, Mosebrook DR, Jakab GJ, Trush MA. 1992. Myeloperoxidase-enhanced formation of (+-)-trans-7,8-dihydroxy-7,8-dihydrobenzo[a]pyrene-DNA adducts in lung tissue in vitro: a role of pulmonary inflammation in the bioactivation of a procarcinogen. *Carcinogenesis* **13**: 1075–1081.

Piedrafita FJ, Molander RB, Vansant G, Orlova EA, Pfahl M, et al. 1996. An Alu element in the myeloperoxidase promoter contains a composite SPI-thyroid hormone-retinoic acid response element. *J. Biol. Chem.* **271**: 14412–14420.

Rosin MP, Anwar WA, Ward AJ. 1994. Inflammation, chromosomal instability, and cancer: the schistosomiasis model. *Cancer Res.* **54** (Suppl.): 1929s–1933s.

Schabath MB, Spitz MR, Delclos GL, Gunn GB, Whitehead LW, et al. 2002a. Association between asbestos exposure, cigarette smoking, myeloperoxidase (MPO) genotypes, and lung cancer risk. *Am. J. Ind. Med.* **42**: 29–37.

Schabath MB, Spitz MR, Hong WK, Delclos GL, Reynolds WF, et al. 2002b. A myeloperoxidase polymorphism associated with reduced risk of lung cancer. *Lung Cancer* **37**: 35–40.

Shao B, Bergt C, Fu X, Green P, Voss JC, et al. 2005. Tyrosine 192 in apolipoprotein A-I is the major site of nitration and chlorination by myeloperoxidase, but only chlorination markedly impairs ABCA1-dependent cholesterol transport. *J. Biol. Chem.* **280**: 5983–5993.

Singh RJ, Goss SP, Joseph J, Kalyanaraman B. 1998. Nitration of gamma-tocopherol and oxidation of alpha-tocopherol by copper-zinc superoxide dismutase/H2O2/NO2: role of nitrogen dioxide free radical. *Proc. Natl. Acad. Sci. USA* **95**: 12912–12917.

Steenport M, Eom H, Uezu M, Schneller J, Gupta R, et al. 2007. Association of polymorphisms in myeloperoxidase and catalase genes with precancerous changes in the gastric mucosa of patients at inner-city hospitals in New York. *Oncol. Rep.* 235–240.

Sugiyama S, Okada Y, Sukhova GK, Virmani R, Heinecke JW, et al. 2001. Macrophage myeloperoxidase regulation by granulocyte macrophage colony-stimulating factor in human atherosclerosis and implications in acute coronary syndromes. *Am. J. Pathol.* **158**: 879–891.

Taioli E, Benhamou S, Bouchardy C, Cascorbi I, Cajas-Salazar N, et al. 2007. Myeloperoxidase G-463A polymorphism and lung cancer: a HuGE genetic susceptibility to environmental carcinogens pooled analysis. *Genet. Med.* **9**: 67–73.

Takeshita J, Byun J, Nhan TQ, Pritchard DK, Pennathur S, et al. 2006. Myeloperoxidase generates 5-chlorouracil in human atherosclerotic tissue: a potential pathway for somatic mutagenesis by macrophages. *J. Biol. Chem.* **281**: 3096–3104.

Thomas EL. 1979. Myeloperoxidase, hydrogen peroxide, chloride antimicrobial system: nitrogen-chlorine derivatives of bacterial components in bactericidal action against *Escherichia coli. Infect. Immun.* **23**: 522–531.

Thomas EL, Bozeman PM, Jefferson MM, King CC. 1995. Oxidation of bromide by the human leukocyte enzymes myeloperoxidase and eosinophil peroxidase: formation of bromamines. *J. Biol. Chem.* **270**: 2906–2913.

Triolo VA. 1965. Nineteenth century foundations of cancer research: advances in tumor pathology, nomenclature, and theories of oncogenesis. *Cancer Res.* **25**: 75–105.

Valinluck V, Liu P, Kang JI, Jr., Burdzy A, Sowers LC. 2005. 5-halogenated pyrimidine lesions within a CpG sequence context mimic 5-methylcytosine by enhancing the binding of the methyl-CpG-binding domain of methyl-CpG-binding protein 2 (MeCP2). *Nucleic Acids Res.* **33**: 3057–3064.

Valinluck V, Sowers LC. 2007. Inflammation-mediated cytosine damage: a mechanistic link between inflammation and the epigenetic alterations in human cancers. *Cancer Res.* **67**: 5583–5586.

van der Vliet A, Eiserich JP, Halliwell B, Cross CE. 1997. Formation of reactive nitrogen species during peroxidase-catalyzed oxidation of nitrite: a potential additional mechanism of nitric oxide-dependent toxicity. *J. Biol. Chem.* **272**: 7617–7625.

van der Vliet A, Nguyen MN, Shigenaga MK, Eiserich JP, Marelich GP, et al. 2000. Myeloperoxidase and protein oxidation in cystic fibrosis. *Am. J. Physiol. Lung Cell Mol. Physiol.* **279**: L537–L546.

Van Schooten FJ, Boots AW, Knaapen AM, Godschalk RW, Maas LM, et al. 2004. Myeloperoxidase (MPO) -463G- > A reduces MPO activity and DNA adduct levels in bronchoalveolar lavages of smokers. *Cancer Epidemiol. Biomarkers Prev.* **13**: 828–833.

Wang TX, Kelley MD, Cooper JN, Beckwith RC, Margerum DW. 1994. Equilibrium, kinetic, and UV-spectral characteristics of aqueous bromine and chlorine species. *Inorg. Chem.* **33**: 5872–5878.

Weitzman SA, Gordon LI. 1990. Inflammation and cancer: role of phagocyte-generated oxidants in carcinogenesis. *Blood* **76**: 655–663.

Wheatley-Price P, Asomaning K, Reid A, Zhai R, Su L, et al. 2008. Myeloperoxidase and superoxide dismutase polymorphisms are associated with an increased risk of developing pancreatic adenocarcinoma. *Cancer* **112**: 1037–1042.

Witztum JL, Steinberg D. 1991. Role of oxidized low density lipoprotein in atherogenesis. *J. Clin. Invest.* **88**: 1785–1792.

Wu W, Samoszuk MK, Comhair SA, Thomassen MJ, Farver CF, et al. 2000. Eosinophils generate brominating oxidants in allergen-induced asthma. *J. Clin. Invest.* **105**: 1455–1463.

Wu W, Chen Y, d'Avignon A, Hazen SL. 1999. 3-Bromotyrosine and 3,5-dibromotyrosine are major products of protein oxidation by eosinophil peroxidase: potential markers for eosinophil-dependent tissue injury in vivo. *Biochemistry* **38**: 3538–3548.

Yamakura F, Ikeda K. 2006. Modification of tryptophan and tryptophan residues in proteins by reactive nitrogen species. *Nitric Oxide* **14**: 152–161.

Yeh GC, Henderson JP, Byun J, Andre d'Avignon D, Heinecke JW. 2003. 8-Nitroxanthine, a product of myeloperoxidase, peroxynitrite, and activated human neutrophils, enhances generation of superoxide by xanthine oxidase. *Arch. Biochem. Biophys.* **418**: 1–12.

Yu H, Eritja R, Bloom LB, Goodman MF. 1993. Ionization of bromouracil and fluorouracil stimulates base mispairing frequencies with guanine. *J. Biol. Chem.* **268**: 15935–15943.

Zgliczynski JM, Stelmaszynska T, Domanski J, Ostrowski W. 1971. Chloramines as intermediates of oxidation reaction of amino acids by myeloperoxidase. *Biochim. Biophys. Acta* **235**: 419–424.

Zhang R, Brennan ML, Fu X, Aviles RJ, Pearce GL, et al. 2001. Association between myeloperoxidase levels and risk of coronary artery disease. *J. Am. Med. Assoc.* **286**: 2136–2142.

Zheng L, Nukuna B, Brennan ML, Sun M, Goormastic M, et al. 2004. Apolipoprotein A-I is a selective target for myeloperoxidase-catalyzed oxidation and functional impairment in subjects with cardiovascular disease. *J. Clin. Invest.* **114**: 529–541.

Zhu H, Yang L, Zhou B, Yu R, Tang N, et al. 2006. Myeloperoxidase G-463A polymorphism and the risk of gastric cancer: a case-control study. *Carcinogenesis* **27**: 2491–2496.

CHAPTER 18

PROSTAGLANDIN H SYNTHASE

INTRODUCTION

The structures and physiology of the prostaglandins, leukotrienes, and thromboxanes are closely interrelated. Discussion here is restricted to the prostaglandin enzymes. Sune Bergström, Bengt Samuelsson, and John Vane won the Nobel Prize in Medicine in 1982 for their discovery of the prostaglandins and related substances, which have a wide range of functions including smooth muscle stimulating and blood pressure depressing activity. Investigations of the biological synthesis of the prostaglandins were first published in 1964 (Van Dorp et al., 1964; Bergström et al., 1964; Samuelsson 1972), and prostaglandin synthase was isolated (Hemler et al., 1976). The isoenzyme prostaglandin H synthase-1 catalyzes the conversion of arachidonic acid into prostaglandin H_2.

In much of this earlier work the reaction of a trienoic acid (20:3), not the tetraenoic arachidonic acid (20 : 4), was studied (Samuelsson, 1965; Hamberg and Samuelsson, 1967; Miyamoto et al., 1976; Ogino et al., 1978; Ohki et al., 1979). *cis,cis*-Eicosa-11,14-dienoic acid (20 : 2) also reacts with PGH synthases. One molecule of oxygen is added per molecule of the 20:2 acid (Hemler et al., 1978). The eicosadienoic acid is the only substrate to form conjugated diene products readily detected spectrophotometrically. Its basic mechanism up to the addition of oxygen is identical to that of arachidonic acid (Koshkin and Dunford, 1998). Recent work has examined the mechanism of action of prostacyclin synthase, the enzyme that converts prostaglandin H_2 to prostacyclin (Yeh et al., 2008, 2009).

This chapter will focus on the properties and mechanism of action of the most intensively investigated enzymes, the prostaglandin H synthases: the constitutive prostaglandin H synthase-1 and the later discovered inducible prostaglandin H synthase-2. These enzymes are also known as COX-1 and COX-2. The cyclooxgenase (COX) nomenclature is misleading since the enzymes are peroxidases, not oxygenases, but is now firmly embedded in the literature. (See note on nomenclature in Chapter 24.)

Peroxidases and Catalases: Biochemistry, Biophysics, Biotechnology, and Physiology, Second Edition.
By H. Brian Dunford

Immunochemical evidence was obtained using monoclonal antibodies to precipitate cyclooxygenase and peroxidase activities from solubilized ram seminal vesicle microsomes (Pagels et al., 1983). The most reactive substrate is arachidonic acid (5,8,11,14-eicosatetraenoic acid; Needleman et al., 1986). This molecule provides the start of a chain of syntheses known as the arachidonic acid cascade, which produces the prostaglandins, thromboxane, and prostacyclin (Zeldin, 2001). The first product is prostaglandin G_2 (15-hydroperoxy-9,11-peroxidoprosta-5,13-dienoic acid). Prostaglandin G_2, PGG_2, reacts with the enzyme as a peroxidase substrate to produce prostaglandin H_2 (15-hydroxy-9,11-peroxidoprosta-5,13-dienoic acid). The structures of these compounds are shown in Figure 18.1. Detection and isolation of PGG_2 and PGH_2 has been described (Graff, 1982). PGH synthase reacts readily with large fatty acid hydroperoxides such as PGG_2, 15-hydroperoxy-5,8,11,14-eicosatetraenoic acid (15-HPETE), and 5-phenyl-4-pentenyl hydroperoxide (PPHP). It reacts more slowly with hydrogen peroxide.

Minor products different from PGG_2 and PGH_2 are also formed (Hecker et al., 1987). Substrate metabolites were shown to attach themselves to PGH synthase but at a slower rate than that for inactivation (Kulmacz, 1987). Lysyl adducts of PGH synthase were characterized (Boutard et al., 1999, 2001).

The enzyme that catalyzes the conversion of arachidonic acid to prostaglandins was isolated (Hemler et al., 1976), and shown to be a glycoprotein of total molecular weight of 129,000, consisting of two identical subunits (van der Ouderaa et al.,

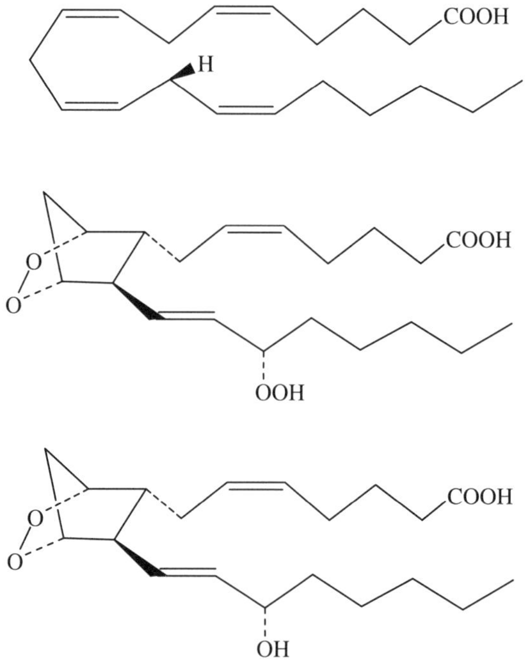

Figure 18.1. Structures of arachidonic acid, prostaglandin G_2, and prostaglandin H_2.

1977). The possibility of non-heme iron in the enzyme was excluded (van der Ouderaa et al., 1979). The prosthetic group is ferriprotoporphyrin IX (heme *b*) with one heme per subunit. The heme is easily lost during purification but can be titrated back. Both bovine and ovine PGH synthase have been isolated, with the latter appearing to be the preferred source.

Radiation inactivation was used to determine that the monomeric unit of about 72 kDa contains both the cyclooxygenase and peroxidase activities (Ruf et al., 1992). The suicide inhibitor, methimazole for lacto- and thyroid peroxidase, inhibits prostaglandin H synthase by a different mechanism, reducing free-radical-derived metabolites (Petry and Eling, 1987).

CRYSTAL STRUCTURES

The biggest single breakthrough in the prostaglandin H synthase story was the determination of the ram seminal vesicle isoenzyme-1 crystal structure at 3.5 Å resolution by Michael Garavito and coworkers (Picot et al., 1994). This required first the determination of the primary sequence (DeWitt and Smith, 1988; Merlie et al., 1988). The crystals were grown in the presence of the nonsteroidal anti-inflammatory drug (NSAID) flurbiprofen.

Each identical monomer of the dimeric structure, consisting largely of helical structures, has two distinct folding units not found in other peroxidases. The first folding unit, residues 34 to 72, similar to the epidermal growth factor, contains two small double-stranded β-sheets containing three disulfide linkages. This unit is covalently bound to the main body of the enzyme by a Cys13-Cys135 bridge. The second distinct unit, residues 73 to 116, consists of a group of four helices, labeled A to D, which lie along one side of the monomer. Helices A to D are largely hydrophobic which facilitates their insertion into the membrane bilayer. The enzyme is described as monotopic since it does not bridge the membrane but reaches from the cytoplasm halfway across. The interior of these helices provide the arachidonic acid binding site that extends from the exterior of the protein into the center near the heme. The enzyme has been described as similar to a Portugese man-of-war, skimming the surface of the membrane looking for its prey, which is arachidonic acid.

A third domain, the largest, includes helices numbered 1 to 19 and residues 117 to 587. It is a large globular region containing the heme binding site. The hydrophobic channel, which is the arachidonic acid binding site, is on the proximal side of the heme. This channel intrudes into the third domain, causing the proximal side of the heme to differ significantly from that of other peroxidases. The proximal side location of the arachidonic acid binding site came as a huge surprise to many researchers, including myself, causing them to rethink how the enzyme operates. His388 is the proximal iron ligand. The distal side of the third domain has structural homologies with myeloperoxidase, and the numbering of the helices in the third domain is based on that for the latter enzyme. A large part of the heme on the distal side is exposed to solvent. His207 on the distal side is $\sim$5 Å from the heme iron and is not coordinated with it.

The crystal structure of cobalt-substituted PGH synthase-1 with bound arachidonic acid was determined so that details of productive binding were obtained (Malkowski et al., 2000). L-shaped binding occurs so the 13-*pro-S* hydrogen is positioned for H abstraction by Tyr385. A conformational rearrangement appeared necessary for attack by the second oxygen molecule.

Dihomo-γ-linolenic acid also binds to PGH synthases-1 and -2 in L-shaped configurations. Differences in the positioning of linolenic and arachidonic acids were explored using mutant enzymes (Thuresson et al., 2001a).

Crystallographic studies on a series of mutants were carried out to examine the arachidonic acid reaction mechanism (Thuresson et al., 2001b, p. 10347). They found

> "five functional categories as follows: (a) residues directly involved in hydrogen abstraction from C-13 of arachidonate (Tyr-385); (b) residues essential for positioning C-13 of arachidonate for hydrogen abstraction (Gly-533 and Tyr-348); (c) residues critical for high affinity arachidonate binding; (Arg-120); (d) residues critical for positioning arachidonate in a conformation so that when hydrogen abstraction does occur the molecule is optimally arranged to yield PGG_2 versus monohydroperoxy acid products (Val-349, Trp-387, and Leu-534); and (e) all other active site residues, which individually make less but measurable contributions to optimal catalytic efficiency."

PROSTAGLANDIN H SYNTHASE-2

Evidence for an inducible form of PGH synthase was first published by Philip Needleman and coworkers (Masferer et al., 1990, 1992). Its gene was isolated and expressed (Xie et al., 1991, 1992; Kujubu et al., 1991; Fletcher et al., 1992; Sirois and Richards, 1992; Hla and Neilson, 1992; Barnett et al., 1994; Yamamoto et al., 1995; Morris and Richards, 1996), and the name PGH synthase-2 (COX-2) was established. Crystal structures rapidly followed (Luong et al., 1996; Kurumball et al., 1996). The structure is very similar to that of the constitutive PGH synthase-1. A major difference is that the arachidonic acid binding site is considerably larger, estimated to be 394 $Å^3$ compared to 316 $Å^3$ for PGH synthase-1 (Luong et al., 1996). The major physiological difference is that production of PGH synthase-2 is induced by cytokines, mitogens, and endotoxins in cells that are already inflamed, which can lead to gastric lesions and renal toxicity (Kurumball et al., 1996). The larger arachidonic acid binding site offers a target for unique NSAID binding that does not occur in PGH synthase-1. PGH synthase-2 is also produced in macrophages (O'Sullivan et al., 1992). The following order of reactivity was found for both PGH synthase-1 and -2: arachidonate > dihomo-γ-linolenate > linolenate > α-linolenate (Laneuville et al., 1995).

In a later crystallographic study of PGH synthase-2, the distal His was replaced with Ala so the peroxidase reaction, required to initiate the cyclooxygenase reaction, could not occur. This permitted a study of the structures of the enzyme with either arachidonic acid or prostaglandin bound in the proximal channel (Kiefer et al., 2000). Arachidonic acid was bound in an orientation that would be nonproductive for reaction.

In a second study the structure of the wild-type COX-2 apoenzyme was crystallized with arachidonic acid. Apparently traces of heme were present because reaction occurred and the detected substrate orientation was not consistent with arachidonic acid, which is achiral, but with the presence of PGG_2 or PGH_2, which have five chiral centers. On the basis of these results "we propose a working model for the mechanistically competent conformation of AA that leads to prostaglandin synthesis" (Kiefer et al., 2000, p. 100). This model is shown in Figure 18.2.

Inhibition of PGH synthase-1 was shown to be markedly dependent on arachidonic acid concentration, which is not the case for PGH synthase-2 (Swinney et al., 1997). A reduced iron porphyrin was detected in an aspirin-treated PGH synthase-2 (Tang et al., 1997). However, the Fe^{III}/Fe^{II} reduction potential is well below that required for cyclooxygenase activation (Goodwin et al., 2000).

New COX enzymes have been discovered in the laboratory of Daniel Simmons, labeled COX-3 and partial pCOX-1. They are made from the COX-1 gene but retain intron 1 in their mRNAs (Chandrasekharan et al., 2002). The medical implications of these new enzymes have been discussed (Warner and Mitchell, 2002).

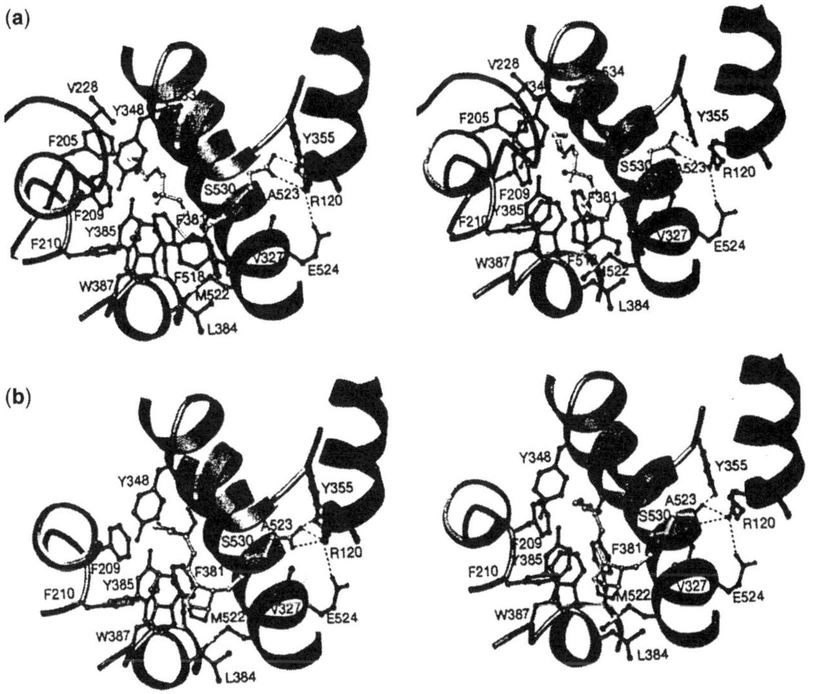

Figure 18.2. Stereo diagram of the models of arachidonic acid (a) and prostaglandin H_2 (b) bound in the cycloxoygenase active site. (Reproduced with permission from Kiefer JR, et al., 2000. *Nature* **405**: 97–101.)

PRELIMINARY MECHANISTIC STUDIES

The beginning of investigations of prostaglandin synthases may be likened to the fable of the blind men describing an elephant. A fairly detailed picture of the beast has emerged, but not universal agreement on its likely mechanism of action. Important mechanistic details are still missing.

In a paper with valuable experimental results, the effects of cyanide, carbon monoxide, and hydroperoxide additions to PGH synthase-1 were determined (Hemler and Lands, 1980). Cyanide was found to bind loosely and to inhibit cyclooxygenase activity at high concentrations. Competition from glutathione peroxidase showed that hydroperoxide was required not only to initiate but to maintain the cyclooxygenase reaction. Carbon monoxide failed to inhibit. It was also shown that phenol at low concentration accelerated and at high concentration inhibited the cyclooxygenase reaction. These experiments provided evidence that (1) the resting enzyme is in the ferric state, (2) there is no ferrous peroxide intermediate, and (3) the peroxidase reaction must be repeated for the cyclooxygenase reaction to continue. In other words, there is tight coupling between the peroxidase and cyclooxygenase reactions. However, the results were interpreted in terms of a reduction of ferric to ferrous for the cyclooxygenase reaction. A later study showed that cyanide binding is much tighter, with a dissociation constant of 65 μM. HCN binds to a basic form of the enzyme with an active site pK_a value 4.15 (MacDonald and Dunford, 1989a).

Because of the dependence of the prostaglandin H synthase reactions on hydroperoxide, a search began for peroxidase-like oxidized intermediate species. A species that was called Compound I (Nastainczyk et al., 1984) was later identified as Compound II (Lambeir et al., 1985). The conversion of PGH synthase from native enzyme to Compound I to Compound II was demonstrated. The Compound II spectrum of PGH synthase is remarkably similar to that of horseradish peroxidase, and the sequence of events, discovery of Compound II prior to Compound I, parallels the history of horseradish peroxidase research. Electronic spectral intermediates were assigned to Compound I and a tyrosyl radical-containing species (Intermediate II; Dietz et al., 1988). It was shown that either phenol or diethyldithiocarbamate (DDC) would prevent enzyme bleaching by acting as reducing substrates, and that the optical spectra observed during the peroxidase cycle only and the cyclooxygenase cycle were similar (MacDonald and Dunford, 1989b).

Reductions of Compound II were studied. An order of reactivity of *N,N,N′,N′*-tetramethyl-phenylenediamine > phenol > indomethacin > phenylbutazone was found. Aspirin had no effect on Compound II reactions. Bromfenac appears to act as both a protective agent and inhibitor (MacDonald et al., 1989).

Resonance Raman Spectra

It was concluded from two studies that the ferric heme iron is hexacoordinated with water in the sixth coordination position (Gaspard et al., 1996; Lou et al., 2000). It is most likely an indication that the known native five-coordinate state is rather readily converted to six coordination, a general feature of peroxidases.

Resonance Raman spectra of PGH synthase-1 were consistent with a neutral His proximal heme iron ligand (Seibold et al., 2000).

DETECTION OF FREE RADICALS: ROLE OF ESR SPECTROSCOPY

Work in the laboratory of Ronald Mason, Graham Palmer, Hans Ruf, and more recently Wilfred van der Donk, established the crucial roles of arachidonic acid and tyrosyl radicals in the cyclooxygenase reaction. A carbon-centered radical on arachidonic acid was detected by spin trapping (Mason et al., 1980; Schreiber et al., 1986). ESR signals at $g = 6.6$ and 5.4 develop at liquid helium temperatures when apoprostaglandin H synthase-1 is titrated with heme, and indicate a high-spin ferric state (Karthein et al., 1987).

The ESR spectrum of a transient radical induced upon hydroperoxide was detected (Kulmacz et al., 1987). This was proposed to be a tyrosyl radical on the enzyme, the reagent that abstracted the 13-*pro-S* hydrogen atom from arachidonic acid (Karthein et al., 1988). Although some questions were raised (Smith et al., 1992), the role of the tyrosyl radical has proven to be correct. Two tyrosyl radical signals can be detected by ESR, a doublet and narrow singlet. A broad singlet seen at intermediate times is a composite of the two other signals (DeGray et al., 1992). The doublet signal was correlated with cyclooxygenase activity (Tsai et al., 1992). Tyrosyl radical signals were monitored in the lipoxygenase activity of human PGH synthase-2 (Xiao et al., 1997).

It was proposed that the arachidonyl radical is a pentadienyl radical spanning positions C11 to C15 of arachidonic acid (Tsai et al., 1995, 1998). This was confirmed by the ESR spectra obtained by reaction of deuterated arachidonic acid (Peng et al., 2001).

ESR and electron nuclear double resonance (ENDOR) spectra were obtained of tyrosyl radicals in PGH synthase-1. It was shown that a different tyrosyl radical is generated in the indomethacin-treated enzyme, indicating redirection of radical formation to a catalytically inactive side chain (Shi et al., 2000).

Radicals formed in COX-2 reactions have been characterized. Acetylation of Ser530 of COX-1 abolishes all cycloxygenase activity, but the corresponding acetylation of COX-2 does not. ESR spectra indicate that both pentadienyl and a strained allyl radical are formed (Tsai et al., 2002).

When nitric oxide is bound to the ferrous enzyme two different ESR spectra are observed. First, an initial six-coordinate ferrous complex and later a five-coordinate complex in which the proximal enzyme ligand is removed (Karthein et al., 1987). Nitric oxide will trap the tyrosyl radical of PGH synthase-2 leading to tyrosine iminoxyl radical and nitrotyrosine formation (Gunther et al., 1997). In an earlier study NO was found to stimulate peroxidase activity but to have no effect on cyclooxygenase activity in murine macrophages (Curtis et al., 1996). 3-Morpholinosydnonimine (SIN-1), a peroxynitrite generator, was found to stimulate purified PGH synthase-1 and also smooth muscle cells in the presence of arachidonic acid (Upmacis et al., 1999). NO was found to stimulate platelet formation and its consumption was stimulated by the peroxidase substrate phenol (O'Donnell et al., 2000).

The aminopyrine cation radical is formed by the peroxidase activity of prostaglandin H synthase-1 (Eling et al., 1985). Glutathione is oxidized to a thiyl radical by both prostaglandin H synthase and horseradish peroxidase. The radical forms a conjugate with styrene. Radical intermediates were detected by spin trapping (Stock et al., 1986).

THE ROLE OF ASPIRIN AND RELATED SUBSTANCES: CONTRIBUTIONS OF VANE AND SMITH

It was proposed by John Vane that the role of aspirin in its anti-inflammatory action is to inhibit prostaglandin synthesis (Vane, 1971, 1994). That same year an informative paper was published by William L. Smith and William E. M. Lands. Using acetone powder preparations of sheep vesicular glands they demonstrated time-dependent aspirin and indomethacin inhibition, with indomethacin 100 times as effective. *o*-Phenanthroline blocked inhibition. Reaction was activated by phenol (Smith and Lands, 1971). Further outstanding contributions have been made by William Smith, whose brief biography was recently published (Kresge et al., 2008). The PGH synthase-1 amino acid sequence was determined from DNA analysis (DeWitt and Smith, 1988). The decisive paper established the location of aspirin reaction in the cyclooxygenase binding site, reaction with Ser530 of PGH synthase-1 (DeWitt et al., 1990). Effects of Ser530 replacement were examined (Shimokawa and Smith, 1992). The effects of aspirin-like molecules on PGH synthase-1 and -2 were examined (Meade et al., 1993). Mini- and comprehensive reviews have been published (Smith et al., 1996; Smith et al., 2000).

A new class of compounds have been developed that show preference for inhibition of COX-2 (Garg et al., 2003). A single amino acid mutation was shown to reverse the selectivities of COX-1 and COX-2 (Gierse et al., 1996). Kinetic parameters were measured for the slow inhibition of COX-1 and COX-2 by indomethacin and flurbiprofen (Callan et al., 1996).

Crystal structures of slow tight-binding (alclofenac and flurbiprofen) and rapidly reversible (ibuprofen and methyl flurbiprofen) NSAID inhibitors to PGH synthase-1 were obtained (Selinsky et al., 2001). The results show clearly that the sites and positioning of binding are similar. The differences in behavior may be due to the speed and efficiency with which hydrogen bonding networks around Arg120 and Tyr355 can be perturbed.

A better understanding of the roles of NSAIDs has been obtained with hopes of complete clarification in the not-too-distant future. It was shown that aspirin, indomethacin, and ibuprofen were more potent inhibitors of COX-1 than COX-2. "Some inhibitors had higher activity in intact cells than against purified enzymes, suggesting that pure enzyme preparations may not be predictive of therapeutic action" (Mitchell et al., 1994, p. 11693). Aspirin inhibits COX-2 gene transcription. "The beneficial actions of nonsteroid antiinflammatory drugs ... can be associated with inhibition of cyclooxygenase (COX)-2 whereas their harmful side effects are associated with inhibition of COX-1" (Xu et al., 1999, p. 5296). Selectivities of over 40 NSAIDs for COX-1 correlated with the risk of gastrointestinal complications

(Warner et al., 1999). The kinetics of cyclooxygenase inhibition has been studied (Gierse et al., 1999; Walker et al., 2001).

WORK OF MARNETT AND COWORKERS

A large body of work on the prostaglandin synthases has been conducted in the laboratory of Lawrence Marnett and in collaborations with other groups. 5-Phenyl-4-pentenyl-hydroperoxide (PPHP) reacts with horseradish peroxidase as well as PGH synthase and is reduced to 5-phenyl-4-pentenyl-alcohol (PPA). Both PPHP and PPA are readily detected by liquid chromatography. The procedure is suitable for detecting peroxide-reducing enzymes and evaluating the relative reactivities of oxidizable compounds as reducing substrates for peroxidases (Weller et al., 1985). Significant differences were found between cyclooxygenase inhibitory sites and peroxidase oxidation sites (Markey et al., 1987). The wide variety of reducing substrates includes 13-*cis*-retinoic acid (Samokyszyn and Marnett, 1987), alkylaryl sulfides (Plé and Marnett, 1989), and peroxynitrite (Landino et al., 1996). Inhibitors were found to be selective between prostaglandin H synthase-1 and -2. Fluorescence energy transfer from inhibitor to heme was observed and distances from inhibitor to heme of 19 ± 1 Å were calculated (Lanzo et al., 1998).

Mini- and major reviews have been published (Marnett et al., 1999; Rouzer and Marnett, 2003). A series of studies were undertaken on mutants of COX-2 expressed in insect cells. An L-shaped binding conformation for arachidonic acid to COX-2 was proposed (Rowlinson et al., 1999). Tyrosine-385 is critical for acetylation of cyclooxygenase-2 by aspirin (Hochgesang et al., 2000). Replacement of the proximal His by Tyr greatly diminished peroxidase activity in COX-2 but did not affect cyclooxygenase activity, supporting a branching chain mechanism for COX-2 (Goodwin et al., 2000). Two amino acids, Ser530 and Val349 (COX-1 numbering), control the stereochemistry of the 15*S*-hydroxyl group formation (Schneider et al., 2002). Indomethacin is a time-dependent inhibitor of cyclooxygenases. Binding of the 2′-methyl of indomethacin is an important part of its inhibition mechanism (Prusakiewicz et al., 2004).

COX-2 has been shown to metabolize esters and amides of arachidonic acid, most important of which are endocannabinoids. A hypothesis of a selective COX-2 signaling pathway has been proposed (Kozak et al., 2001; Rouzer and Marnett, 2008).

WORK OF KULMACZ, TSAI, AND COWORKERS

Extensive contributions have been made by Richard Kulmacz, Ah-Lim Tsai, and coworkers and in cooperation with Graham Palmer and Wilfred van der Donk (van der Donk et al., 2002). Reaction and inactivation steady-state kinetics were studied (Kulmacz, 1986). The results were interpreted in terms of one Compound I formation leading to 1000 hydroperoxide molecules synthesized from arachidonic acid—in other words, that a branching chain reaction is occurring. The controversy of branching chain versus tightly coupled mechanisms has continued to the present day. See detailed discussion of mechanism below. Heme coordination was examined using ESR, MCD, resonance Raman, and optical spectroscopies (Tsai et al., 1993). The branching chain and tightly coupled mechanisms were compared using several criteria. It was concluded that the branching chain mechanism more correctly describes the action of PGH synthase (Wei et al., 1995). At low temperature or high pH it was found that cosubstrates such as phenol had little accelerating effect on the cyclooxygenase reaction of COX-1, again supporting the branching chain mechanism (Bambai and Kulmacz 2000). In the absence of added reducing cosubstrates, inactivation of cycloxyxgenase activity was examined. It was concluded that inactivation of Intermediate II can occur either at the distal heme site or the proximal tyrosyl radical site (Wu et al., 2003).

Self-inactivation of the peroxidase activity of PGH synthase-1 was examined. It was concluded that inactivation begins with Compound I and Intermediate II. Intermediate II is converted in two steps to a terminal species (Wu et al., 1999).

Relative sensitivities of COX-1 and COX-2 reactions to hydroperoxide levels were examined. COX-2 requires less hydroperoxide for full activity, an indication that its mechanism is more likely to be that of a branching chain (Kulmacz and Wang, 1995). The relative roles of saturable kinetics and cooperative responses in PGHS-1 and -2 cycloxygenase activity were examined. A "linkage between activation efficiency and cyclooxgenase cooperativity indicates an interdependence between fatty acid and hydroperoxide levels in controlling the synthesis of potent prostanoid mediators" (Chen et al., 1999, p. 20301). These experiments were conducted in the presence of phenol.

It was found that the rate of conversion of Compound I (Intermediate I) to Intermediate II (Fe^{IV} plus a tyrosyl radical) was faster in COX-2 (Lu et al., 1999).

MANGANESE PROSTAGLANDIN SYNTHASES

The original report on Mn^{III}-porphyrin-substituted COX-1 indicated that it possessed full cyclooxygenase activity to produce PGG_2 but no peroxidase activity (Ogino et al., 1978). This would validate the branching chain mechanism and indicate that it is completely independent of the peroxidase reaction. Later studies showed that the peroxidase reaction is required to initiate the cyclooygenase reaction in the manganese-substituted enzyme (Odenwaller et al., 1992; Strieder et al., 1992), showing that this behavior is the same as in the native enzyme (Hemler and Lands, 1980). Both the peroxidase reaction and initiation of the cyclooxygenase reaction are slower in the Mn-substituted enzyme but the cyclooxygenase reaction lasts longer. This was taken as evidence for the branching chain mechanism. If the cyclooxygenase reaction remains rate controlling in the Mn-substituted enzyme, then the slower peroxidase cycle would not influence the duration of cyclooxygenase activity. Phenol, known to prolong cyclooxygenase activity, was present in the reaction mixtures. Is this evidence for a branching chain mechanism or a tightly coupled mechanism?

A Mn^{V} species was detected (Tsai et al., 1997). It appears that the manganese enzyme does not form a porphyrin π-cation radical. However, formation of an active tyrosyl radical was demonstrated (Kulmacz et al., 1994; Tsai et al., 1998). Self-inactivation was examined in PGH synthase-1 and the corresponding mangano enzyme (Wu et al., 2001).

Iron replacement by another metal ion can be regarded as creation of a new different enzyme, just as isoenzymes synthesized from different genes are chemically different. Properties of one may have much in common with the related enzyme but differences in behavior do occur. Thus much of the results on the manganese enzyme are in common with that of the iron PGH synthase, but not all. For a more detailed discussion of Mn-PGH synthase see Rouzer and Marnett (2003).

MECHANISTIC DETAILS

Two very different mechanisms of action of PGH synthase have been proposed: the branching chain mechanism and the tightly coupled mechanism. Different PGH synthases may behave differently. Now that the number of COX enzymes has been expanded to COX-1, pCOX-1a, COX-2, and COX-3 much work remains to be done (Chandrasekharan et al., 2002). The following discussion is restricted to COX-1 where a great deal of effort has already been expended.

The Branching Chain Mechanism

This mechanism was originally proposed by Hans Ruf and coworkers before the crystal structure of PGH synthase-1 was published (Dietz et al., 1988). These researchers deserve a great deal of credit for showing the role of tyrosyl radicals in the cyclooxygenase reaction. There was the usual lag as the dramatically new idea was subjected to intense scrutiny, but it is now accepted as fact. Their branching chain mechanism now has broad support (Wei et al., 1995; Smith et al., 1996, 2000; Rouzer and Marnett, 2003), but is not yet universally accepted. It is shown in Figure 18.3.

Electron donors are also hydrogen atom donors and are shown as such (e^- and H^+). In the original paper Compound I was called Intermediate I, but I adhere to the conventional peroxidase notation. The species with the neutral tyrosyl radical (hydrogen atom lost) is named Intermediate II. The conversion of Compound I to Intermediate II was shown as a single step. However, on the basis of behavior of other peroxidases that have a tendency to act as endogenous donors, an electron jump to the porphyrin cation radical forming a tyrosyl cation radical would appear to be the initial elementary step, followed by phenolic proton loss to form the electronically neutral tyrosyl radical of Intermediate II.

Dietz et al. (1988) also indicate as a tentative path the conversion of Intermediate II to Compound II, a step that would partially couple the branching chain mechanism to the peroxidase cycle. This step requires not only the addition of one electron but two protons, since Compound II has its distal imidazole protonated. Therefore, the conversion would appear to require more than one elementary step. In the original paper, reductions are shown simply as electron transfers (Dietz et al., 1988). I have added

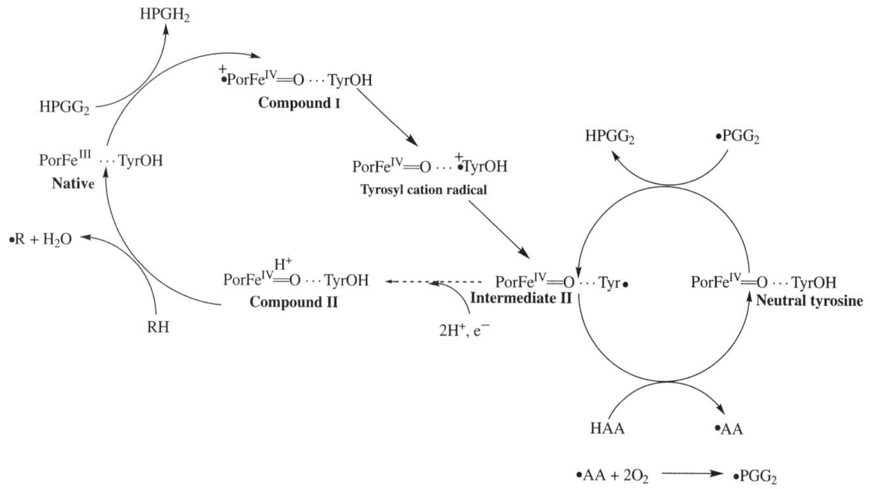

Figure 18.3. The branching chain mechanism of action of prostaglandin H synthase based on the model of Dietz et al. (1988). Necessary proton, as well as electron, balances have been added. The conversion of Intermediate II to Compound II is unlikely to occur in a single step, yet a rapid rate is assigned to it in modeling studies.

accompanying transferred hydrogen atoms in Figure 18.3, where these are known. The fate of the proton lost from the tyrosyl cation radical is unknown but a likely acceptor is solvent water.

Once the neutral tyrosyl radical is formed the resultant Intermediate II starts a branching chain reaction. An arachidonic acid molecule loses a proton to the tyrosyl radical, reacts with two molecules of O_2 to form the PGG_2 radical which in turn abstracts the tyrosyl proton, and the cycle is repeated. Without a chain breaking step it would go on as long as reactants are present. The branching chain mechanism thus shows that Compound I disappears through Intermediate II formation, and subsequent steps of the normal peroxidase cycle are not required. The rationale for this mechanism may be summarized as follows. "The initiation of cyclooxygenase activity in a single synthase molecule can result in the synthesis of over a thousand molecules of hydroperoxide. This burst of activator would make the subsequent initiation of cyclooxygenase catalysis in nearby synthase molecules much more likely" (Kulmacz, 1986, p. 284).

A free radical mechanism consists of initiation, propagation, and termination. It is established that initiation of cyclooxygenase activity occurs by peroxide forming Compound I which is converted by an intramolecular reaction into Intermediate II. The proposed propagation steps in the branching chain have been described above. What are the chain termination steps? Plausible termination steps are the following.

$$RH + {}^{\bullet}AA \longrightarrow R^{\bullet} + HAA$$

$$RH + {}^{\bullet}PGG_2 \longrightarrow R^{\bullet} + HPGG_2$$

The small cosubstrates are free radical scavengers and removal of the fatty acid free radicals would terminate the chain. Could this occur in arachidonic acid binding site? Small molecules such as O_2, aspirin, salicylate, and ascorbate readily diffuse into the binding channel. There is sufficient adjacent room for O_2 to react with the newly formed arachidonate free radical but a rearrangement is necessary for reaction of the second O_2 (Malkowski et al., 2000). Thus the PGG_2 radical may be more available for scavenging.

The Tightly Coupled Mechanism

I am indebted to Lawrence Marnett for getting my group started on work on PGH synthase-1. His student Christine Markey flew from Detroit to Edmonton with a sample of PGH synthase-1, and we obtained rapid-scan optical spectra showing that Compound I and Compound II spectra were very similar to those of horseradish peroxidase (Lambeir et al., 1985).

The papers that led to the conclusion in my group that the cyclooxygenase activity of PGH synthase-1 is tightly coupled to the peroxidase cycle are described below. At the time of this work the role of the proximal tyrosyl radical was still controversial. Following the dictum of using the minimum number of parameters to fit the kinetic data, we assumed all activity was on the distal side of the enzyme. The role of the proximal tyrosyl radical is now universally accepted and so the tightly coupled mechanism was adapted to it (Dunford, 1999, 2005).

The peroxidase cycle of PGH synthase-1 was confirmed using PPHP as oxidant and phenol and hydroquinone as reducing substrates. These are metabolites of benzene so the work is relevant to benzene toxicity. Indomethacin was shown to inhibit the peroxidase cycle as well as the cyclooxygenase cycle. The stabilizing agent diethyldithiocarbamate (DDC) was shown to act as a reducing substrate and corrections were applied for its contribution to the observed kinetics (Hsuanyu and Dunford, 1992a).

At low concentrations phenol stimulates and at high concentrations inhibits cyclooxygenase activity (Hemler and Lands, 1980). Studies were extended to a family of phenols (Hsuanyu and Dunford, 1992b). Both enhancement and inhibition of cyclooxygenase activity were increased by phenols which are better electron donors and more hydrophobic. No hydrophobic factors were required to account for Compound II reduction where the results for the family of phenols could be fit solely by the Hammett $\rho\sigma$ equation.

Inhibition of cyclooxygenase activity by phenols can be explained by reduction of Compound I to Compound II, preventing Intermediate II formation, and also by competitive binding in the arachidonic acid site. The latter effect explains the enhanced inhibition by hydrophobic phenols.

Stimulation of cyclooxygenase activity by phenols is readily explained by reduction of Compound II back to native enzyme, ready for another coupled cycle. The lack of a hydrophobicity factor confirms a peroxidase effect, not a direct cyclooxygenase effect. There is no obvious explanation of phenolic stimulation in the branching chain mechanism. “Phenols also protect the enzyme from self-catalyzed inactivation, most likely by removing the free radical of prostaglandin G_2 by reducing it to prostaglandin G_2" (Hsuanyu and Dunford, 1992b, p. 17649).

The kinetics of oxidation of *p*-coumaric acid by PGH synthase-1 was studied using hydrogen peroxide as oxidant. The normal peroxidase cycle was confirmed, and in the presence of arachidonic acid the stimulating effect of *p*-coumaric acid on cyclooxygenase activity was shown (Bakovic and Dunford, 1993). The results are in disagreement with a different peroxidase mechanism proposed when the reducing substrate was TMPD (Kulmacz, 1986).

The stage is now set for the original proposal of the tightly coupled mechanism (Bakovic and Dunford, 1994). Hydrogen peroxide was used as oxidant and ferulic acid as reducing substrate. Peroxidase activity was monitored on a stopped-flow apparatus and cyclooxygenase activity with an oxygen electrode. The results were the following: when equal amounts of ferulic acid and arachidonic acid were present, or when ferulic acid was in excess, cyclooxygenase activity was stimulated. The ratio of ferulic acid to arachidonic acid disappearance rates averaged 2.3 ± 0.2, showing that approximately two moles of ferulic acid react for every one mole of arachidonic acid. When arachidonic acid was in excess enzyme inactivation rapidly occurred and a higher ratio of ferulic acid/arachidonic acid was reacted. According to the branching chain mechanism once cyclooxygenase activity is started PGG_2 formation is independent of the peroxidase cycle. The data therefore disprove the branching chain mechanism. With what is the ferulic acid reacting? Compound II is one obvious choice, and the other is the PGG_2 radical.

This revised tightly coupled mechanism is summarized in Figure 18.4. It shows conversion of Intermediate II back to Compound II in two steps. In the first, the arachidonic free radical is formed and the tyrosyl radical is reconverted to a normal tyrosyl residue. Then upon proton addition to the distal imidazole group in the active site, Compound II is regenerated. Reduction by small reducing substrate

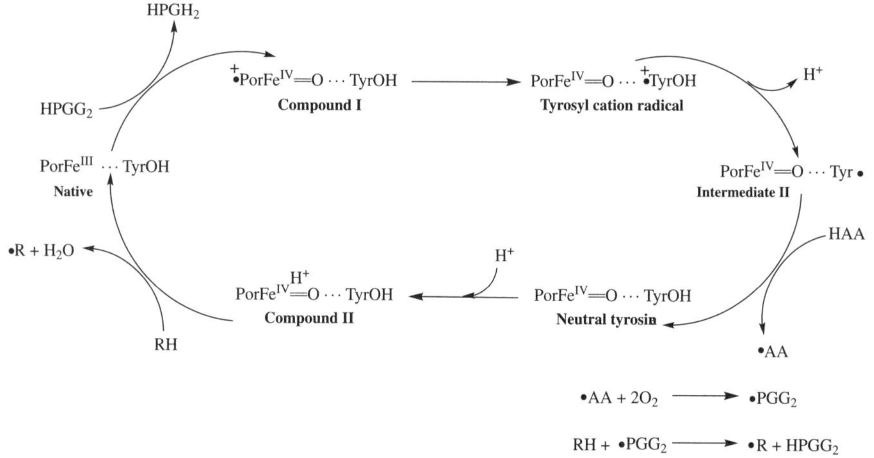

Figure 18.4. The tightly coupled mechanism for the action of prostaglandin H synthase.

regenerates native enzyme, ready for another cycle. The source of this proton for Compound II formation is unknown as is the destination of the proton lost upon Intermediate II formation. The latter proton is common to both mechanisms. In the tightly coupled mechanism the conversion of the $PGG_2^\bullet$ radical to $HPGG_2$ is accomplished by a small free radical scavenger such as a phenol. Thus, the most likely chain terminators in the branching chain mechanism become necessary agents for maintenance of the enzyme cycles in the tightly coupled mechanism.

Work to this point had concentrated on the effect of the peroxidase cycle on the cyclooxygenase reaction. The opposite study, the effect of the cyclooxygenase reaction on the peroxidase cycle, was undertaken (Koshkin and Dunford, 1999a). All measurements were made on a stopped-flow apparatus. PPHP was used as oxidant, *p*-coumaric acid as reducing substrate, and the effect of arachidonic acid addition was examined. It was found that addition of arachidonic acid had no effect on the peroxidase cycle, a result incompatible with the branching chain mechanism.

Reactions of various hydroperoxides with PGH synthase-1 were examined. *t*-Butyl hydroperoxide and hydrogen peroxide are poor oxidants, ethyl hydroperoxide is better, and *m*-chloroperbenzoic acid reacts at about the same fast rate as PGG_2. It was found that peroxides also react with Compound I (Bakovic and Dunford, 1996), a behavior similar to that observed with myeloperoxidase. This previously unobserved reaction appears to compete for the conversion of Compound I to Intermediate II. Also observed for the first time was an apparent reaction of arachidonic acid with Compound I of PGH synthase. This was interpreted as a true reaction with a rate constant of $10^6\ M^{-1}\ s^{-1}$ (Bakovic and Dunford, 1996), but it now appears probable that it is an enhancing effect of arachidonic acid binding on the proximal side on the rate of conversion of Compound I to Intermediate II. This negates a modeling study in which Compound II participated directly in the cyclooxygenase reaction, an example of overinterpretation of kinetic results (Bakovic and Dunford, 1995). The rate constant for the spontaneous conversion of Compound I to Intermediate II was found to be in the range of 5 to 10 s^{-1}, much smaller than previously suggested (Koshkin and Dunford, 1999b), but an enhancing effect of substrates binding in the arachidonic acid channel was found, in accord with the suggestion above of the effect of arachidonic acid binding.

Comparison of the Branching Chain and Tightly Coupled Mechanisms

I shall attempt here to outline (1) what is known and is common to both mechanisms, then (2) where the mechanisms differ, and finally (3) what remains unknown. This leads naturally to an assessment of the current state of knowledge and suggestions of specific problems remaining to be solved.

Mechanistic features in common are the following. The peroxidase cycle is well established, as is the role of Try385 in the cyclooxygenase reaction. Intramolecular conversion of Compound I to Intermediate II is accepted as the route to formation of the tyrosyl radical. A feedback route from the cyclooxygenase to the peroxidase cycle was shown tentatively in the original branching chain mechanism (Dietz et al., 1988), and this route has been used in modeling studies (Wei et al., 1995;

Koshkin and Dunford, 1999a). A hefty rate constant of (0.5–5) × $10^6\ M^{-1}\ s^{-1}$ was assumed for the conversion of Intermediate II (cyclooxygenase active) to Compound II (peroxidase active) by ferulic acid (Wei et al., 1995).

Differences in the coupling step exist in the two mechanisms. According to Dietz et al. (1988) it is Intermediate II that is converted to Compound II. In the tightly coupled mechanism it is not Intermediate II (with tyrosyl radical) but the subsequent species after reaction with arachidonic acid (with neutral tyrosyl residue) that is converted to Compound II. All that is required is proton addition. If the tightly coupled feedback route replaces the one shown by Dietz et al. (1988) then the branching chain mechanism is eliminated, since the chain reaction is broken.

The tightly coupled mechanism leads to the proposal that proton pumping occurs from the proximal to the distal side of the enzyme. A proton is released on the proximal side as the tyrosyl cation radical is converted to a neutral tyrosyl radical, and a proton is added to the imidazole near the $Fe^{IV}{=}O$ group on the distal side when Compound II is formed. The prediction of proton pumping remains to be examined experimentally.

Direct measurement of PGG_2 and PGH_2 formation provided evidence for both the branching chain and tightly coupled mechanisms (Wei et al., 1995). Some PGG_2 always remained in the reaction mixture, supporting the branching chain mechanism, but PGH_2 formation was stimulated by added reducing substrate, particularly phenol, an effect that is a key part of the tightly coupled mechanism and cannot be explained by the branching chain mechanism.

A closer look at the continuing presence of PGG_2 leads to a different suggestion. The concentration of arachidonic acid was 80 μM and the accumulation of PGG_2 never exceeded 2 to 3 μM. The reaction was started by arachidonic acid addition; no hydroperoxide was added. Therefore, an autoxidation product of arachidonic acid was the oxidant converting native PGH synthase to Compound I. This is analogous to the well-studied isobutyraldehyde and indoleacetate reactions with horseradish peroxidase, where autoxidation reactions eliminate the need for added peroxide. PGG_2 has to compete with autoxidation products of arachidonic acid to react with native enzyme. A comprehensive quantitative study of the stability of arachidonic acid, its rate of autoxidation, and the products of autoxidation would answer the question of whether it influences the accumulation of PGG_2.

Another set of required experiments is to examine the kinetics of cyclooxygenase and peroxidase reactions thoroughly with fatty acid peroxides or *m*-chloroperbenzoic acid, the best substrates for native PGH synthase. Hydrogen peroxide is a poor substrate and in the absence of reducing substrates leads to irreproducible results at longer times (Bakovic and Dunford, 1996).

The intramolecular conversion of Compound I to Intermediate II needs to be quantitated and the mechanistic details established. Estimates of the rate vary widely: 65 s^{-1} (Dietz et al., 1988); 350 s^{-1} (Wei et al., 1995); and 5 to 10 s^{-1} with stimulation by binding in the arachidonic acid proximal channel (Koshkin and Dunford, 1999b).

The denaturation pathway of prostaglandin H synthase needs to be quantitated. An initial first step has been taken in a novel analysis of the kinetics of adrenaline oxidation in a peroxidase reaction of PGH synthase, in which Compound I and a Compound I-adrenaline complex are starting points for denaturation (Vrzheshch et al., 2003). The cyclooxygenase pathway with arachidonic acid in excess of small reducing substrate is much faster and less well understood. The corresponding denaturation pathway of horseradish peroxidase starts with Compound I and ends, via several intermediates, as Pigment670 (P670), named after the wavelength of its principal absorbance band. P670 is totally inactive. See the discussion in Chapter 4.

Future models need to address the conversion of Compound I to Compound II by hydroperoxides, a common feature of myeloperoxidase and PGH synthase reactions. Modeling studies will be more constrained by further experimental work, and their predictive power greatly increased.

Finally, all of this work needs to be repeated for COX-2 and COX-3 enzymes. How many more scientific careers are required?

Accuracy of Oxygen Electrode Measurements

It was claimed that the ratio of 2:1 for ferulic acid:arachidonic acid reacted (Bakovic and Dunford, 1994) was grossly in error, and when corrected disproved the tightly coupled mechanism (Wei et al., 1995), an accusation that needs to be addressed. First of all, a fixed ratio, whatever it is, supports the tightly coupled mechanism, and is evidence against the branching chain mechanism, where a variable ratio is expected. Therefore, a fixed ratio indicates tight coupling, and the value of the ratio places a restriction on the details of the tightly coupled mechanism.

In obtaining the initial rate of oxygen consumed, we took tangents to the traces obtained from the oxygen electrode after its dead time had elapsed (Fig. 18.5). There is a lag phase of about 5 s before the electrode responds, because of diffusion across the barrier membrane. Because the time scale was long, we simply ignored

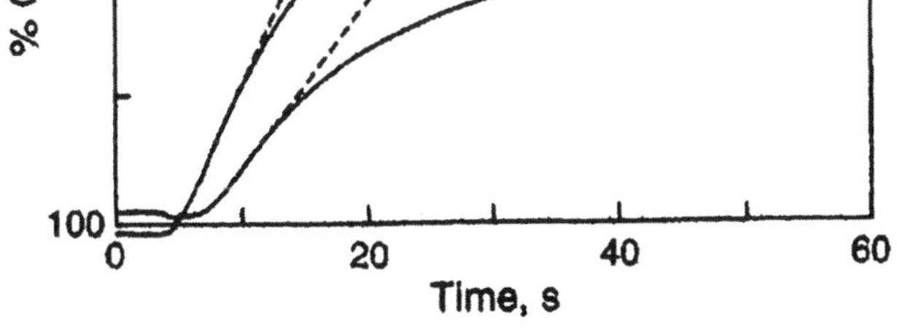

Figure 18.5. Rate of oxygen consumption in the reaction of prostaglandin H synthase with arachidonic and ferulic acids. Presence (upper curve) and absence (lower curve) of ferulic acid. (Reproduced with permission from Bakovic M, Dunford HB. 1994. *Biochemistry* **33**: 6475–6482.)

the initial 5 s interval (Bakovic and Dunford, 1994). I have since taken the oxygen electrode traces in Figure 18.5, fitted the 5 to 30 s intervals to a first-order curve, and extrapolated back 5 s to take into account oxygen reacted in the dead time of the electrode.

The results indicate that the initial rates of oxygen consumption should be increased by ~20%. The difference in the two corrected curves, in the presence and absence of ferulic acid, is negligible. The authors of the criticism (Wei et al., 1995) corrected each data point by an amount proportional to the difference with the data point 5 s later. They included the initial 5 s region, the dead time of the electrode, so it is no wonder they got a different ratio (close to one, without much variation). There is no merit in their claim that the corrected initial velocity of oxygen consumption is higher by a factor of almost two, nor in the claim that the results disprove the tightly coupled mechanism. The section entitled "Comparison of Rates of Ferulic Acid Oxidation and Arachidonic Acid Oxygenation" is incorrect and detracts from an otherwise carefully crafted contribution (Wei et al., 1995).

In summary, the tightly coupled mechanism is the best available at this time, but further refinements will almost certainly be made.

REFERENCES

Bakovic M, Dunford HB. 1993. Kinetics of the oxidation of *p*-coumaric acid by prostaglandin H synthase and hydrogen peroxide. *Biochemistry* **32**: 833–840.

Bakovic M, Dunford HB. 1994. Intimate relation between the cyclooxygenase and peroxidase activities of prostaglandin H synthase: peroxidase reaction of ferulic acid and its influence on the reaction mechanism of arachidonic acid. *Biochemistry* **33**: 6475–6482.

Bakovic M, Dunford HB. 1995. Pre-steady-state kinetics and modeling of the oxygenase and cyclooxygenase reactions of prostaglandin endoperoxide synthase. *Biophys. Chem.* **54**: 237–251.

Bakovic M, Dunford HB. 1996. Reactions of prostaglandin endoperoxide synthase and its Compound I with hydroperoxides. *J. Biol. Chem.* **271**: 2048–2056.

Bambai B, Kulmacz RJ. 2000. Prostaglandin H synthase: effects of peroxidase cosubstrates on cyclooxygenase velocity. *J. Biol. Chem.* **275**: 27608–27614.

Barnett J, Chow J, Ives D, Chiou M, Mackenzie R, Osen E, Nguyen B, Tsing S, Bach C, Freire J, Chan H, Sigal E, Ramesha C. 1994. Purification, characterization and selective inhibition of human prostaglandin G/H synthase 1 and 2 expressed in a baculovirus system. *Biochim. Biophys. Acta* **1209**: 130–139.

Bergström S, Danielsson H, Samuelsson B. 1964. The enzymatic formation of prostaglandin E_2 from arachidonic acid. *Biochim. Biophys. Acta* **90**: 207–210.

Boutard O, Brame CJ, Chaurand P, Li J, Rowlinson SW, Crews BC, Ji C, Marnett LJ, Caprioli RM, Roberts LJ II, Oates JA. 1999. Characterization of the lysyl adducts of prostaglandin H-synthases that are derived from oxygenation of arachidonic acid. *Biochemistry* **40**: 6948–6955.

Boutard O, Brame CJ, Salomon RG, Roberts LJ II, Oates JA. 2001. Characterization of the lysyl adducts formed from prostaglandin H_2 via the levuglandin pathway. *Biochemistry* **40**: 6948–6955.

Callan OH, So O-Y, Swinney DC. 1996. The kinetic factors that determine the affinity and selectivity for slow binding inhibition of human prostaglandin H synthase 1 and 2 by indomethacin and flurbiprofen. *J. Biol. Chem.* **271**: 3548–3554.

Chandrasekharan NV, Dai H, Roos LT, Evanson NK, Tomsik J, Elton TS, Simmons DL. 2002. COX-3, a cyclooxygenase-1 variant inhibited by acetaminophen and other analgesic/antipyretic drugs: cloning, structure, and expression. *Proc. Natl. Acad. Sci. USA* **99**: 13926–13931.

Chen W, Pawelek TR, Kulmacz RJ. 1999. Hydroperoxide dependence and cooperative cyclooxygenase kinetics in prostaglandin H synthase-1 and -2. *J. Biol. Chem.* **274**: 20301–20306.

Curtis JF, Reddy NG, Mason RP, Kalyanaraman B, Eling TE. 1996. Nitric oxide: a prostaglandin H synthase 1 and 2 reducing cosubstrate that does not stimulate cyclooxygenase activity or prostaglandin H synthase expression in murine macrophages. *Arch. Biochem. Biophys.* **335**: 369–3736.

DeGray JA, Lassmann G, Curtis JF, Kennedy TA, Marnett LJ, Eling TE, Mason RP. 1992. Spectral analysis of the protein-derived tyrosyl radicals from prostaglandin H synthase. *J. Biol. Chem.* **267**: 23583–23588.

DeWitt DL, el-Harith EA, Kraemer SA, Andrews MJ, Yao EF, Armstrong RL, Smith WL. 1990. The aspirin and heme-binding sites of ovine and murine prostaglandin endoperoxide synthases. *J. Biol. Chem.* **265**: 5192–5198.

DeWitt DL, Smith WL. 1988. Primary structure of prostaglandin G/H synthase from sheep vesicular gland determined from the complementary DNA sequence. *Proc. Natl. Acad. Sci. USA* **85**: 1412–1416.

Dietz R, Nastainczyk W, Ruf HH. 1988. Higher oxidation states of prostaglandin H synthase: rapid electronic spectroscopy detected two spectral intermediates during the peroxidase reaction with prostaglandin G2. *Eur. J. Biochem.* **171**: 321–328.

Dunford HB. 1999. *Heme Peroxidases*. Wiley-VCH, New York.

Dunford HB. 2005. Kinetics and mechanisms of mammalian heme peroxidase reactions. *Prog. Reaction Kinetics Mech.* **30**: 245–265.

Eling TE, Mason RP, Sivarajah K. 1985. The formation of the aminopyrine cation radical by the peroxidase activity of prostaglandin H synthase and subsequent reactions of the radical. *J. Biol. Chem.* **260**: 1601–1607.

Fletcher BS, Kujubu DA, Perrin DM, Herschman HR. 1992. Structure of the mitogen-inducible TIS10 gene and demonstration that TIS10-encoded protein is a functional prostaglandin G/H synthase. *J. Biol. Chem.* **267**: 4338–4344.

Garg R, Kurup A, Mekapati SB, Hansch C. 2003. Cyclooxygenase (COX) inhibitors: a comparative QSAR study. *Chem. Rev.* **103**: 703–731.

Gaspard S, Chottard G, Mahy J-P, Mansuy D. 1996. Study of the coordination chemistry of prostaglandin G/H synthase by resonance Raman spectroscopy. *Eur. J. Biochem.* **238**: 529–537.

Gierse JK, Koboldt CM, Walker MC, Seibert K, Isakson PC. 1999. Kinetic basis for selective inhibition of cyclo-oxygenases. *Biochem. J.* **339**: 607–614.

Gierse JK, McDonald JJ, Hauser SD, Rangwala SH, Koboldt CM, Seibert K. 1996. A single amino acid difference between cyclooxygenase-1 (COX-1) and -2 (COX-2) reverses the selectivity of COX-2 specific inhibitors. *J. Biol. Chem.* **271**: 15810–15814.

Goodwin DC, Rowlinson SW, Marnett LJ. 2000. Substitution of tyrosine for the proximal histidine ligand to the heme of prostaglandin endoperoxide synthase 2: implications for the mechanism of cyclooxygenase activation and catalysis. *Biochemistry* **39**: 5422–5432.

Gunther MR, His LC, Curtis JF, Gierse JK, Marnett LJ, Eling TE, Mason RP. 1997. Nitric oxide trapping of the tyrosyl radical of prostaglandin H synthase-2 leads to tyrosine iminoxyl radical and nitrotyrosine formation. *J. Biol. Chem.* **272**: 17086–17090.

Graff G. 1982. Preparation of PGG_2 and PGH_2. In: Lands WEM, Smith WL, editors, *Methods in Enzymology, Vol. 86, Prostaglandins and Arachidonate Metabolites*. Academic Press, New York, pp. 376–385.

Hamberg M, Samuelsson B. 1967. On the mechanism of the biosynthesis of prostaglandins E_1 and $F_{1\alpha}$. *J. Biol. Chem.* **242**: 5336–5343.

Hecker M, Ullrich V, Fischer C, Meese CO. 1987. Identification of novel arachidonic acid metabolites formed by prostaglandin H synthase. *Eur. J. Biochem.* **169**: 113–123.

Hemler ME, Crawford CG, Lands WEM. 1978. Lipoxygenase activity of purified prostaglandin-forming cyclooxygenase. *Biochemistry* **17**: 1772–1779.

Hemler ME, Lands WEM. 1980. Evidence for a peroxide-initiated free radical mechanism of prostaglandin biosynthesis. *J. Biol. Chem.* **255**: 6253–6261.

Hemler M, Lands WEM, Smith WL. 1976. Purification of the cycloxygenase that forms prostaglandins. *J. Biol. Chem.* **251**: 5575–5579.

Hla T, Neilson K. 1992. Human cyclooxygenase-2 cDNA. *Proc. Natl. Acad. Sci. USA* **89**: 7384–7388.

Hochegesang GP, Rowlinson SW, Marnett LJ. 2000. Tyrosine-385 is critical for acetylation of cyclooxygenae-2 by aspirin. *J. Am. Chem. Soc.* **122**: 6514–6515.

Hsuanyu Y, Dunford HB. 1992a. Reduction of prostaglandin H synthase Compound II by phenol and hydroquinone, and the effect of indomethacin. *Arch. Biochem. Biophys.* **292**: 213–220.

Hsuanyu Y, Dunford HB. 1992b. Prostaglandin H synthase kinetics: the effect of substituted phenols on cyclooxygenase activity and the substituent effect on phenolic peroxidatic activity. *J. Biol. Chem.* **267**: 17649–17657.

Karthein R, Dietz R, Nastainczyk W, Ruf HH. 1988. Higher oxidation states of prostaglandin H synthase: EPR study of a transient tyrosyl radical in the enzyme during the peroxidase reaction. *Eur. J. Biochem.* **171**: 313–320.

Karthein R, Nastainczyk W, Ruf HH. 1987. EPR study of ferric native prostaglandin H synthase and its ferrous NO derivative. *Eur. J. Biochem.* **166**: 173–180.

Kiefer JR, Pawlitz JL, Moreland KT, Stegeman RA, Hood WF, Gierse JK, Stevens AM, Goodwin DC, Rowlinson SW, Marnett LJ, Stallings WC, Kurumball RG. 2000. Structural insights into the stereochemistry of the cyclooxygenase reaction. *Nature* **405**: 97–101.

Koshkin V, Dunford HB. 1998. Reaction of prostaglandin endoperoxide synthase with *cis,cis*-eicosa-11,14-dienoic acid. *J. Biol. Chem.* **273**: 6046–6049.

Koshkin V, Dunford HB. 1999a. Coupling of the peroxidase and cyclooxygenase reactions of prostaglandin H synthase. *Biochim. Biophys. Acta* **1430**: 341–348.

Koshkin V, Dunford HB. 1999b. Reactions of prostaglandin endoperoxide synthase with hydroperoxide and reducing substrates under single turnover conditions. *Biochim. Biophys. Acta* **1431**: 47–52.

Kozak KR, Prusakiewicz J, Rowlinson SW, Schneider C, Marnett LJ. 2001. Amino acid determinants in cyclooxygenase-2 oxygenation of the endocannabinoid 2-arachidonylglycerol. *J. Biol. Chem.* **276**: 30072–30077.

Kresge N, Simoni RD, Hill RL. 2008. How aspirin interferes with cyclooxygenase activity: the work of William L. Smith. *J. Biol. Chem.* **283**: 12–14.

Kujubu DA, Fletcher BS, Varnum BC, Lim RW, Herschman HR. 1991. TIS10, a phorbolester tumor promoter-inducible mRNA from Swiss 3T3 cells, encodes a novel prostaglandin synthase/cyclooxygenase homologue. *J. Biol. Chem.* **266**: 12866–12872.

Kulmacz RJ. 1986. Prostaglandin H synthase and hydroperoxides: peroxidase reaction and inactivation kinetics. *Arch. Biochem. Biophys.* **249**: 273–285.

Kulmacz RJ. 1987. Attachment of substrate metabolite to prostaglandin H synthase upon reaction with arachidonic acid. *Biochem. Biophys. Res. Commun.* **148**: 539–545.

Kulmacz RJ, Palmer G, Wei C, Tsai A-L. 1994. Reaction and free radical kinetics of prostaglandin H synthase with manganese protoporphyrin IX as the prosthetic group. *Biochemistry* **33**: 5428–5439.

Kulmacz RJ, Tsai A-L, Palmer G. 1987. Heme spin states and peroxide-induced radical species in prostaglandin H synthase. *J. Biol. Chem.* **262**: 10524–10531.

Kulmacz RJ, Wang L-H. 1995. Comparison of hydroperoxide initiator requirements for the cyclooxygenase activities of prostaglandin H synthase-1 and -2. *J. Biol. Chem.* **270**: 24019–24023.

Kurumball RG, Stevens AM, Gierse JK, McDonald JJ, Stegeman RA, Pak JY, Gildehaus D, Miyashiro JM, Penning TD, Seibert K, Isakson PC, Stallings WC. 1996. Structural basis for selective inhibition of cyclooxygenase-2 by anti-inflammatory agents. *Nature* **384**: 644–648.

Lambeir A-M, Markey CM, Dunford HB, Marnett LJ. 1985. Spectral properties of higher oxidation states of prostaglandin H synthase. *J. Biol. Chem.* **260**: 14894–14896.

Landino LM, Crews BC, Timmons MD, Morrow JD, Marnett LJ. 1996. Peroxynitrite, the coupling product of nitric oxide and superoxide, activates prostaglandin biosynthesis. *Proc. Natl. Acad. Sci. USA* **93**: 15069–15074.

Laneuville O, Breuer DK, Xu N, Huang H, Gage DA, Watson JT, Lagarde M, DeWitt DL, Smith WL. 1995. Fatty acid substrate specificities of human prostaglandin-endoperoxide synthase-1 and -2. *J. Biol. Chem.* **270**: 19330–19336.

Lanzo CA, Beechem JM, Talley J, Marnett LJ. 1998. Investigation of the binding of isoform-selective inhibitors to prostaglandin endoperixide synthases using fluorescence spectroscopy. *Biochemistry* **37**: 217–226.

Lou B-S, Snyder JK, Maarshall P, Wang J-S, Wu G, Kulmacz RJ, Tsai A-L, Wang J. 2000. Resonance Raman studies indicate a unique heme active site in prostaglandin H synthase. *Biochemistry* **39**: 12424–12434.

Lu G, Tsai A-L, Van Wart HE, Kulmacz RJ. 1999. Comparison of the peroxidase reaction kinetics of prostaglandin H synthase-1 and -2. *J. Biol. Chem.* **274**: 16162–16167.

Luong C, Miller A, Barnett J, Chow J, Ramesha C, Browner MF. 1996. Flexibility of the NSAID binding site in the structure of human cyclooxygenase-2. *Nature Struct. Biol.* **3**: 927–933.

MacDonald ID, Dunford HB. 1989a. Kinetics and equilibria of cyanide binding to prostaglandin H synthase. *J. Inorg. Biochem.* **37**: 35–44.

MacDonald ID, Dunford HB. 1989b. Similarities in the optical spectra of prostaglandin H sythase during its cyclooxygenase and peroxidase reactions. *Biochem. Cell Biol.* **67**: 301–305.

MacDonald ID, Graff G, Anderson LA, Dunford HB. 1989. Optical spectra and kinetics of reactions of prostaglandin H synthase: effects of the substrates 13-hydroperoxyoctadeca-9,11-dienoic acid, arachidonic acid, *N,N,N′,N′*-tetramethyl–phenylenediamine, and phenol and the nonsteroidal anti-inflammatory drugs aspirin, indomethacin, phenylbutazone and Bromfenac. *Arch. Biochem. Biophys.* **272**: 194–202.

Malkowski MG, Ginell SL, Smith WL, Garavito RM. 2000. The productive conformation of arachidonic acid bound to prostaglandin synthase. *Science* **289**: 1933–1937.

Markey CM, Alward A, Weller PE, Marnett LJ. 1987. Quantitative studies of hydroperoxide reduction by prostaglandin H synthase: reducing substrate specificity and the relationship of peroxidase to cyclooxygenase activities. *J. Biol. Chem.* **262**: 6266–6279.

Marnett LJ, Rowlinson SW, Goodwin DC, Kalgutkar AS, Lanzo CA. 1999. Arachidonic acid oxygenation by COX-1 and COX-2. *J. Biol. Chem.* **274**: 22903–22906.

Masferer JL, Seibert K, Zweifel BS, Needleman P. 1990. Endogenous glucocorticoids regulate an inducible cyclooxygenase enzyme. *Proc. Natl. Acad. Sci. USA* **89**: 3917–3921.

Masferer JL, Zweifel BS, Seibert K, Needleman P. 1992. Selective regulation of cellular cyclooxygenase by dexamethasone and endotoxin in mice. *J. Clin. Invest.* **86**: 1375–1379.

Mason RP, Kalyanaraman B, Tainer BE, Eling TE. 1980. A carbon-centered free radical intermediate in the prostaglandin synthetase oxidation of arachidonic acid. *J. Biol. Chem.* **255**: 5019–5022.

Meade EA, Smith WL, DeWitt DL. 1993. Differential inhibition of prostaglandin endoperoxide synthase (cyclooxygenase isozymes by aspirin and other non-steroidal anti-inflammatory drugs. *J. Biol. Chem.* **268**: 6610–6614.

Mitchell JA, Akarasereenont P, Thiemermann C, Flower RJ, Vane JR. 1994. Selectivity of non-steroidal anti-inflammatory drugs as inhibitors of constitutive and inducible cyclooxygenase. *Proc. Natl. Acad. Sci. USA* **90**: 11693–11697.

Merlie JP, Fagan D, Mudd J, Needleman P. 1988. Isolation and characterization of the complementary DNA for sheep seminal vesicle prostaglandin endoperoxide synthase (cyclooxygenase). *J. Biol. Chem.* **263**: 3550–3553.

Miyamoto T, Ogino N, Yamamoto S, Hayaishi O. 1976. Purification of prostaglandin endoperoxide synthetase from bovine vesicular gland microsomes. *J. Biol. Chem.* **251**: 2629–2636.

Morris JK, Richards JS. 1996. An E-box region within the prostaglandin endoperoxide synthase-2 (PGS-2) promoter is required for transcription in rat ovarian granulose cells. *J. Biol. Chem.* **271**: 16633–16643.

Nastainczyk W, Scuhn D, Ullrich V. 1984. Spectral intermediates of prostaglandin hydroperoxidase. *Eur. J. Biochem.* **144**: 381–385.

Needleman P, Turk J, Jakschik BA, Morrison AR, Lefkowith JB. 1986. Arachidonic acid metabolism. *Annu. Rev. Biochem.* **55**: 69–102.

Odenwaller R, Maddipati KR, Marnett LJ. 1992. Detection of higher oxidation states of manganese-prostaglandin endoperoxide synthase. *J. Biol. Chem.* **267**: 13863–13869.

O'Donnell VB, Coles B, Lewis MJ, Crews BC, Marnett LJ, Freeman BA. 2000. Catalytic consumption of nitric oxide by prostaglandin H synthase-1 regulates platelet formation. *J. Biol. Chem.* **275**: 38239–38244.

Ogino N, Ohki S, Yamamoto S, Hayaishi O. 1978. Prostaglandin endoperoxide synthetase from bovine vesicular gland microsomes: inactivation and activation by heme and other metalloporphyrins. *J. Biol. Chem.* **253**: 5061–5068.

Ohki S, Ogino N, Yamamoto S, Hayaishi O. 1979. Prostaglandin hydroperoxidase, an integral part of prostaglandin endoperoxide synthetase from bovine vesicular gland microsomes. *J. Biol. Chem.* **254**: 829–836.

O'Sullivan MG, Chilton FH, Huggins EM Jr, McCall CE. 1992. Lipopolysaccharide priming of alveolar macrophages for enhanced synthesis of prostanoids involves induction of a novel prostaglandin H synthase. *J. Biol. Chem.* **267**: 14547–14550.

Pagels WR, Sachs RJ, Marnett LJ, DeWitt DL, Day JS, Smith WL. 1983. Immunochemical evidence for the involvement of prostaglandin H synthase in hydroperoxide-dependent oxidations by ram seminal vesicles. *J. Biol. Chem.* **258**: 6517–6523.

Peng S, Okeley NM, Tsai A-L, Wu G, Kulmacz RJ, van der Donk WA. 2001. *J. Am. Chem. Soc.* **123**: 3609–3610.

Petry TW, Eling TE. 1987. The mechanism for the inhibition of prostaglandin H synthase-catalyzed xenoviotic oxidation by methimazole: reaction with free radical oxidation products. *J. Biol. Chem.* **262**: 14112–14118.

Picot D, Loll PJ, Garavito M. 1994. The X-ray crystal structure of the membrane protein prostaglandin H_2 synthasc-1. *Nature* **367**: 243–249.

Plé P, Marnett LJ. 1989. Alkylaryl sulfides as peroxidase reducing substrates for prostaglandin H synthase: probes for the reactivity and environment of the ferryl-oxo complex. *J. Biol. Chem.* **264**: 13983–13993.

Prusakiewicz JJ, Felts AS, Mackenzie BS, Marnett LJ. 2004. Molecular basis of the time-dependent inhibition of cyclooxygenases by indomethacin. *Biochemistry* **43**: 15439–15445.

Rouzer CA, Marnett LJ. 2003. Mechanism of free radical oxygenation of polyunsaturated fatty acids by cyclooxygenases. *Chem. Rev.* **103**: 2239–2304.

Rouzer CA, Marnett LJ. 2008. Non-redundant functions of cyclooxgenases: oxygenation of endocannabinoids. *J. Biol. Chem.* **283**: 8065–8069.

Rowlinson SW, Crews BC, Lanzo CA, Marnett LJ. 1999. The binding of arachidonic acid in the cyclooxygenase active site of mouse prostaglandin endoperoxide synthase-2 (COX-2): a putative L-shaped binding conformation utilizing the top channel region. *J. Biol. Chem.* **274**: 23305–23310.

Ruf HH, Schuhn D, Dietz R, Nastainczyk W, Nielsen M. 1992. Target size analysis of prostaglandin endoperoxide synthase: radiation inactivation of both cyclooxygenase and peroxidase correlated with the monomer of 72 kDa. *Eur. J. Biochem.* **204**: 1069–1073.

Samokyszyn VM, Marnett LJ. 1987. Hydroperoxide-dependent cooxidation of 13-*cis*-retinoic acid by prostaglandin H synthase. *J. Biol. Chem.* **262**: 14119–14133.

Samuelsson B. 1965. On the incorporation of oxygen in the conversion of 8,11,14-eicosatrienoic acid to prostaglandin E_1. *J. Am. Chem. Soc.* **87**: 3011–3013.

Samuelsson B. 1972. Biosynthesis of prostaglandins. *Fed. Proc.* **31**: 1442–1450.

Schneider C, Boegin WE, Prusakiewicz JJ, Rowlinson SW, Marnett LJ, Samel N, Brash AR. 2002. Control of prostaglandin stereochemistry at the 15-carbon by cyclooxygenases-1 and -2. A critical role for serine 530 and valine 349. *J. Biol. Chem.* **277**: 478–485.

Schreiber J, Eling TE, Mason RP. 1986. The oxidation of arachidonic acid by the cyclooxygenase activity of purified prostaglandin H synthase: spin-trapping of a carbon-centered free radical intermediate. *Arch. Biochem. Biophys.* **249**: 126–136.

Seibold SA, Cerda JF, Mulichak AM, Song I, Garavito RM, Arakawa T, Smith WL, Babcock GT. 2000. Peroxidase activity in prostaglandin endoperoxide H synthase-1 occurs with a neutral histidine proximal ligand. *Biochemistry* **39**: 6616–6624.

Selinsky BS, Gupta K, Sharkey CT, Loll PJ. 2001. Structural analysis of NSAID binding by prostaglandin H_2 synthase: time-dependent and time-independent inhibitors elicit identical enzyme conformations. *Biochemistry* **40**: 5172–5180.

Shi W, Hoganson CW, Espe M, Bender CJ, Babcock GT, Palmer G, Kulmacz RJ, Tsai A-L. 2000. Electron paramagnetic resonance and electron nuclear double resonance spectroscopic identification and characterization of the tyrosyl radicals in prostaglandin H synthase 1. *Biochemistry* **39**: 4112–4121.

Shimokawa T, Smith WL. 1992. Prostaglandin endoperoxide synthase: the aspirin acetylation region. *J. Biol. Chem.* **267**: 12387–12392.

Sirois J, Richards JS. 1992. Purification and characterization of a novel, distinct isoform of prostaglandin endoperoxide synthase induced by human chorionic gonadotropin in granulose cells of rat preovulatory follicles. *J. Biol. Chem.* **267**: 6282–6288.

Smith WL, DeWitt DL, Garavito RM. 2000. Cyclooxygenases: structural, cellular, and molecular biology. *Annu. Rev. Biochem.* **69**: 145–182.

Smith WL, Eling TE, Kulmacz RJ, Marnett LJ, Tsai A-L. 1992. Tyrosyl radicals and their role in hydroperoxide-dependent activation and inactivation of prostaglandin endoperoxide synthase. *Biochemistry* **31**: 3–7.

Smith WL, Garavito RM, DeWitt DL. 1996. Prostaglandin endoperoxide synthases (cyclooxygenases)-1 and -2. *J. Biol. Chem.* **271**: 33157–33160.

Smith WL, Lands WEM. 1971. Stimulation and blockade of prostaglandin biosynthesis. *J. Biol. Chem.* **246**: 6700–6702.

Stock BH, Schreiber G, Guenat C, Mason RP, Bend JR, Eling TE. 1986. *J. Biol. Chem.* **261**: 15915–15922.

Strieder S, Schaible K, Scherer H-J, Dietz R, Ruf HH. 1992. Prostaglandin endoperoxide synthase substituted with manganese protporphyrin IX. *J. Biol. Chem.* **267**: 13870–13878.

Swinney DC, Mak AY, Barnett J, Raamesha CS. 1997. Differential allosteric regulation of prostaglandin H synthase 1 and 2 by arachidonic acid. *J. Biol. Chem.* **272**: 12393–12398.

Tang MS, Copeland RA, Penning TM. 1997. Detection of an Fe^{2+}-protophyrin-IX intermediate during aspirin-treated prostaglandin H_2 synthase catalysis of arachidonic acid to 15-HETE. *Biochemistry* **36**: 7527–7534.

Thuresson ED, Lakkides KM, Rieke CJ, Sun Y, Wingerd BA, Micielli R, Mulichak AM, Malkowski MG, Garavito RM, Smith WL. 2001a. Prostaglandin endoperoxide H synthase-1: the functioning of cycloxygenase active site residues in the binding, positioning, and oxygenation of arachidonic acid. *J. Biol. Chem.* **276**: 10347–10357.

Thuresson ED, Malkowski MG, Lakkides KM, Rieke CJ, Mulichak AM, Ginell SL, Garavito RM, Smith WL. 2001b. Mutational and X-ray crystallographic analysis of the interaction of dihomo-γ-linolenic acid with prostaglandin endoperoxide H synthases. *J. Biol. Chem.* **276**: 10538–10365.

Tsai A-L, Kulmacz RJ, Palmer G. 1995. Spectroscopic evidence for reaction of prostaglandin H synthase-1 tyrosyl radical with arachidonic acid. *J. Biol. Chem.* **270**: 10503–10508.

Tsai A-L, Kulmacz RJ, Wang J-S, Wang Y, Van Wart HE, Palmer G. 1993. Heme coordination of prostaglandin H synthase. *J. Biol. Chem.* **268**: 8554–8563.

Tsai A-L, Palmer G, Kulmacz RJ. 1992. Prostaglandin H synthase: kinetics of tyrosyl radical formation and of cyclooxygenase catalysis. *J. Biol. Chem.* **267**: 17753–17759.

Tsai A-L, Palmer G, Wu G, Peng S, Okeley NM, van der Donk WA, Kulmacz RJ. 2002. Structural characterization of arachidonyl radicals formed by aspirin-treated prostaglandin H synthase-2. *J. Biol. Chem.* **277**: 38311–38321.

Tsai A-L, Palmer G, Xiao GS, Swinney DC, Kulmacz RJ. 1998. Structural characterization of arachidonyl radicals formed by prostaglandin H synthase-2 and prostaglandin H synthase-1 reconstituted with mangano protoporphyrin IX. *J. Biol. Chem.* **273**: 3888–3894.

Tsai A-L, Wei C, Baek HK, Kulmacz RJ, Van Wart HE. 1997. Comparison of peroxidase reaction mechanisms of prostaglandin H synthase-1 containing heme and mangano protoporphyrin IX. *J. Biol. Chem.* **272**: 8885–8894.

Upmacis RK, Deeb RS, Hajar DP. 1999. Regulation of prostaglandin H_2 synthase activity by nitrogen oxides. *Biochemistry* **38**: 12505–12513.

Van der Donk WA, Tsai A-L, Kulmacz RJ. 2002. The cyclooxygenase reaction mechanism. *Biochemistry* **41**: 15451–15458.

van der Ouderaa FJ, Buytenhek M, Nugteren DH, van Dorp DA. 1977. Purification and characterization of prostaglandin endoperoxide synthetase from sheep vesicular glands. *Biochim. Biophys. Acta* **487**: 315–331.

van der Ouderaa FJ, Buytenhek M, Slikkerveer FJ, van Dorp DA. 1979. On the haemoprotein character of prostaglandin endoperoxide synthetase. *Biochim. Biophys. Acta* **572**: 29–42.

Van Dorp DA, Beerthuis RK, Nugteren DH, Vonkeman H. 1964. The biosynthesis of prostaglandins. *Biochim. Biophys. Acta* **90**: 204–207.

Vane JR. 1971. Inhibition of prostaglandin synthesis as a mechanism of action for aspirin-like drugs. *Nature New Biol.* **231**: 232–235.

Vane JR. 1994. Towards a better aspirin. *Nature* **367**: 215–216.

Vrzheshch PV, Batanova EA, Mevkh AT, Varfolomeev SD, Gazarian IG, Thorneley RNF. 2003. A novel approach to distinguish between enzyme mechanisms: quasi-steady-state kinetic analysis of the prostaglandin H synthase peroxidase reaction. *Biochem. J.* **372**: 713–724.

Walker MC, Kurumbail RG, Kiefer JR, Moreland KT, Koboldt CM, Isakson PC, Seibert K, Gierse JK. 2001. A three-step kinetic mechanism for selective inhibition of cyclo-oxygenase-2 by diarylheterocyclic inhibitors. *Biochem. J.* **357**: 709–718.

Warner TD, Giuliano F, Vojnovic I, Bukasa A, Mitchell JA, Vane JR. 1999. Nonsteroid drug selectivities for cyclo-oxygenase-1 rather than cyclo-oxygenase-2 are associated with human gastrointestinal toxicity: a full *in vitro* analysis. *Proc. Natl. Acad. Sci. USA* **96**: 7563–7568.

Warner TD, Mitchell JA. 2002. Cyclooxygenase-3 (COX-3): filling in the gaps toward a COX continuum? *Proc. Natl. Acad. Sci. USA* **99**: 13371–13373.

Wei C, Kulmacz RJ, Tsai A-L. 1995. Comparison of branched chain and tightly coupled reaction mechanisms for prostaglandin H synthase. *Biochemistry* **34**: 8499–8512.

Weller PE, Markey CM, Marnett LJ. 1985. Enzymatic reduction of 5-phenyl-4-pentenyl-hydroperoxide: detection of peroxidases and identification of peroxidase reducing substrates. *Arch. Biochem. Biophys.* **243**: 633–643.

Wu G, Kulmacz RJ, Tsai A-L. 2003. Cyclooxygenase inactivation kinetics during reaction of prostaglandin H synthase-1 with peroxide. 2003. *Biochemistry* **42**: 13772–13777.

Wu G, Vuletich JL, Kulmacz RJ, Osawa Y, Tsai A-L. 2001. Peroxidase self-inactivation in prostaglandin H synthase-1 pretreated with cyclooxygenase inhibitors or substituted with mangano protoporphyrin IX. *J. Biol. Chem.* **276**: 19879–19888.

Wu G, Wei C, Kulmacz RJ, Osawa Y, Tsai A-L. 1999. A mechanistic study of self-inactivation of the peroxidase activity in prostaglandin H synthase-1. *J. Biol. Chem.* **274**: 9231–9237.

Xiao G, Tsai A-L, Palmer G, Boyar WC, Marshall PJ, Kulmacz RJ. 1997. Analysis of hydroperoxide-induced tyrosyl radicals and lipoxygenase activity in aspirin-treated human prostaglandin H synthase-2. *Biochemistry* **36**: 1836–1845.

Xie W, Chipman JG, Robertson DL, Erikson RL, Simmons DL. 1991. Expression of a mitogen-responsive gene encoding prostaglandin synthase is regulated by mRNA splicing. *Proc. Natl. Acad. Sci. USA* **88**: 2692–2696.

Xie W, Chipman JG, Robertson DL, Simmons DL. 1992. Mitogen-inducible prostaglandin G/H synthase: a new target for nonsteroidal anti-inflammatory drugs. *Drug Develop. Res.* **25**: 249–265.

Xu X-M, Sansores-Garcia L, Chen X-M, Matijevic-Aleksic N, Du M, Wu KK. 1999. Suppression of inducible cyclooxygenase 2 gene transcription by aspirin and sodium salicylate. *Proc. Natl. Acad. Sci. USA* **96**: 5292–5297.

Yamamoto K, Arakawa T, Ueda N, Yamamoto S. 1995. Transcriptional roles of nuclear factor κB and nuclear factor-interleukin-6 in the tumor necrosis factor α-dependent induction of cycloxygenase-2 in MC3T3-E1 cells. *J. Biol. Chem.* **270**: 31315–31320.

Yeh H-C, Gerfen GJ, Wang J-S, Tsai A-L, Wang L-H. 2009. Characterization of the peroxidase mechanism upon reactionof prostacyclin synthase with peracetic acid: identification of a tyrosyl radical intermediate. *Biochemistry* **48**: 917–928.

Yeh H-C, Hsu P-Y, Tsai A-L, Wang L-H. 2008. Spectroscopic characterization of the oxyferrous complex of prostacyclin synthase in solution and in trapped sol-gel matrix. *FEBS J* **275**: 2305–2314.

Zeldin DC. 2001. Epoxygenase pathways of arachidonic acid metabolism. *J. Biol. Chem.* **276**: 36059–36062.

CHAPTER 19

THYROID PEROXIDASE

INTRODUCTION

A series of six steps describe thyroid function with regard to iodide. First inorganic iodide is taken up by the thyroid gland against an electrochemical gradient. Then iodination of tyrosyl residues in thyroglobulin occurs, followed by a coupling reaction to form the hormones thyroxine, also called tetraiodothyronine (T4), and triiodothyronine (T3). Proteolysis of thyroglobulin releases the hormones into the blood stream. Finally, excess or used organic iodide is reconverted into iodide ion and recycled.

Release of the thyroxine and triiodothyronine from the thyroid is stimulated by signals from the hypothalamus and pituitary glands, other parts of the endocrine gland system. There is a feedback mechanism where the secreted hormones dampen the stimulating signals. T4 and T3 control metabolic rates in the cells of the body. The hormones are transported to other cells in the body by a thyroxine-binding globulin specifically designed for the purpose. Hormone action occurs inside the cell nucleus where regulation of expression of targeted genes occurs. Thyroid malfunctions lead to a host of abnormalities and diseases such as an enlarged thyroid gland (goiter) and autoimmune reactions. Severe iodide deficiency causes cretinism. Neanderthals had heads shaped like those of modern-day cretins, and it has been suggested that a single genetic alteration permitted the more efficient uptake of iodide, a key improvement in human evolution (Dobson, 1998).

The role of thyroid peroxidase in generating free radicals in a medium of cultured Chinese hamster ovary cells has been examined (Moore and Mehlhorn, 1993). The ultimate compilation of data on the physiology and medicine involving the thyroid gland is the volume, periodically updated, edited by Lewis Braverman and Robert Utiger, entitled "*Werner and Ingbar's The Thyroid: A Fundamental and Clinical Text*" (Braverman and Utiger, 2004).

Peroxidases and Catalases: Biochemistry, Biophysics, Biotechnology, and Physiology, Second Edition.
By H. Brian Dunford

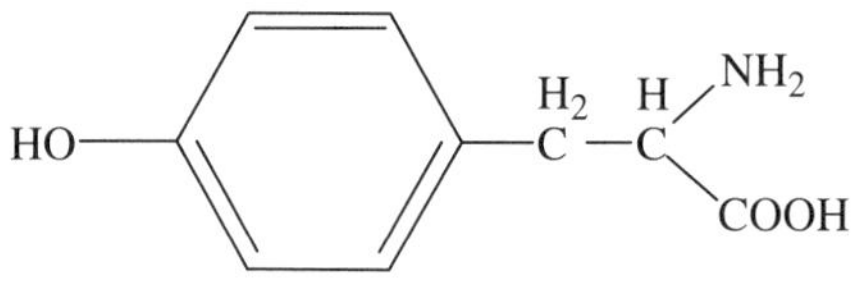

Tyrosine

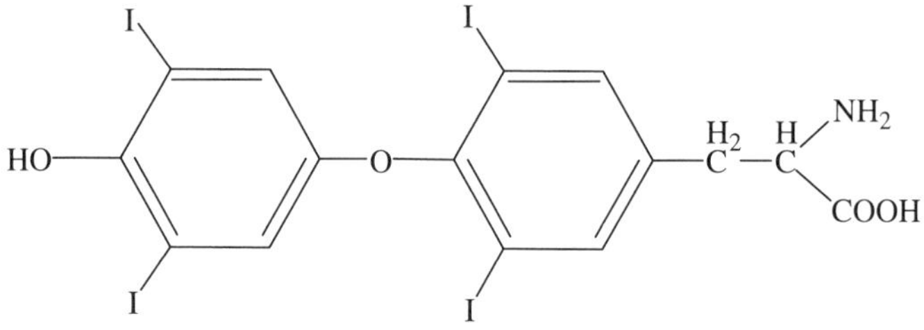

Thyroxine or Tetraiodothyronine, T4

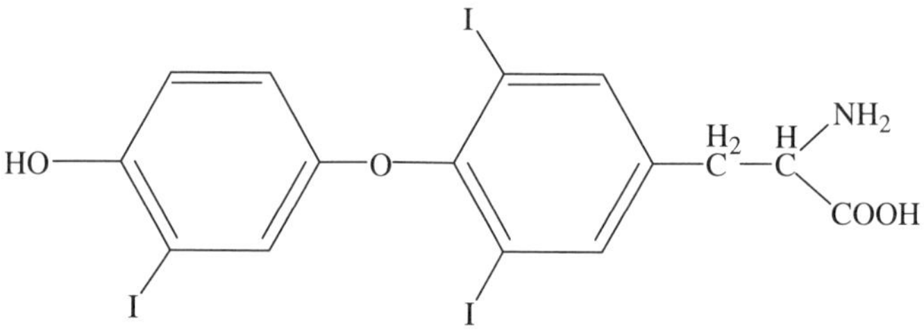

Triiodothyronine, T3

Figure 19.1. Structures of tyrosine, thyroxine or tetraiodothyronine (T4), and triiodothyronine (T3).

This chapter is devoted to a brief history of research progress involving thyroid peroxidase, T4 and T3, and a present-day summary of knowledge. The structures of T4 and T3 are shown in Figure 19.1. Also shown is the structure of tyrosine, an obvious precursor of at least part of the hormone structures.

HORMONE DISCOVERY AND CHEMICAL SYNTHESIS

Thyroxine was first isolated from thyroid glands in 1926 (Harington, 1926) and synthesized the following year (Harington and Barger, 1927). The latter paper has an addendum describing the beneficial effect of synthetic thyroxine administration to patients suffering from myxoedema (Lyon, 1927). Triiodothyronine (T3) was discovered in 1952, and proved to be a more potent hormone (Gross and Pitt-Rivers, 1952, 1954; Gross et al., 1952). Conversion of T4 to T3 was shown also to occur outside the thyroid gland (Braverman et al., 1970).

DETECTION OF THE METHOD OF BIOLOGICAL SYNTHESIS OF THYROXINE

Use of radioactive iodine played a key role in definitive early research. Tracer amounts of ^{131}I were converted into radiothyroxine and radiodiiodotyrosine in slices of thyroid tissue (Morton and Chaikoff, 1943). Monoiodotyrosine was also detected (Taurog et al., 1950). Conversion of ^{131}I from its inorganic form into organic molecules by tissue homogenates was shown by solvent extraction (Weiss, 1953), and by paper chromatography (Fawcett and Kirkwood, 1953). The name of tyrosine iodinase was suggested for the thyroid enzyme (Fawcett and Kirkwood, 1954). As a young postdoctoral fellow at McMaster University (working in a different field), I volunteered to swallow a radioactive iodine cocktail in Sam Kirkwood's lab, and had my neck scanned with a Geiger counter to see if my thyroid gland was normal. (It was).

Further work showed that synthesis of monoiodotyrosine occurs in homogenates of both thyroid and nonthyroid tissue and that hydrogen peroxide is required (Serif and Kirkwood, 1958; Alexander, 1959; Fawcett and McLeod, 1963). The suggested enzyme name became iodide peroxidase.

Progress at isolation and purification was made during the 1960s (Hosoya et al., 1962; Igo et al., 1964; Hosoya and Morrison, 1967). The name thyroid peroxidase became accepted. Iodination of tyrosine by beef thyroid preparations was demonstrated (Klebanoff et al., 1962). One of the better earlier preparations of thyroid peroxidase, though not homogeneous, was obtained by solubilization by trypsin, ammonium sulfate precipitation, chromatography on diethylaminoethyl cellulose, and filtration through Biogel P-100 (Coval and Taurog, 1967). Absorption at 410 nm suggested a heme group. Reactions with hydrogen peroxide (obtained from glucose-glucose oxidase) and iodide, catalyzed by the enzyme, could iodinate protein, free tyrosine, and free monoiodotyrosine. Activity was increased fivefold in a later preparation. Added steps were treatment with deoxycholate and precipitation with acetone in preliminary stages, steps used in an earlier procedure (DeGroot and Davis, 1962), and an additional treatment with DEAE cellulose as the final step (Taurog, 1970). The preparation procedure provided evidence that thyroid peroxidase is bound to particulate matter; it is not a microsomal enzyme. Fifty pounds of hog thyroids were processed in 400 g batches. The Reinheitzahl (RZ) of A_{410}/A_{280} was 0.34, with a value of 0.39 observed in individual fractions. Activity correlated with heme absorption, eliminating the possibility of heme contamination. The heme could be extracted, which may have been facilitated by the use of detergent. Carbohydrate is bound to the enzyme.

In the comprehensive review/research article by Taurog (1970), evidence is summarized that thyroid peroxidase catalyzes formation of mono- and diiodotyrosine, that it can iodinate tyrosyl-containing proteins, that its favorite protein target is thyroglobulin (MW 660,000), and that it plays a role in the coupling reaction of two diiodotyrosine residues to form thyroxine (T4) (Lamas et al., 1972). A later review (Taurog, 2000) contains a pictorial representation of T4 formation as an intramolecular reaction in thyroglobulin, reproduced in Figure 19.2.

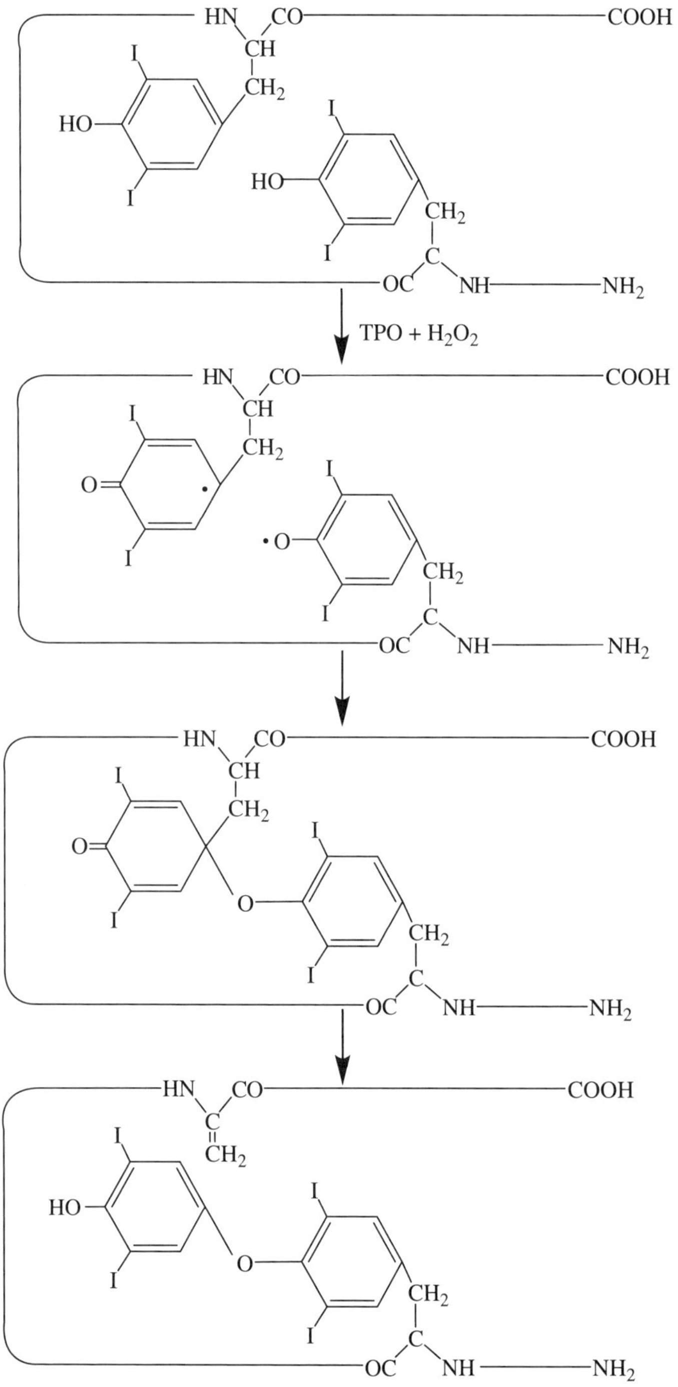

Figure 19.2. Hypothetical coupling scheme for formation of thyroxine (T4) within the thyroglobulin (Tg) molecule, catalyzed by thyroid peroxidase. (Reproduced with permission from

More highly purified hog thyroid peroxidase was obtained in 1979, with an RZ value of 0.54. It contained 10% carbohydrate by weight, had a molecular weight of approximately 90,000 determined by gel filtration, and an isoelectric point of 5.75 (Rawitch et al., 1979). An earlier, remarkably accurate molecular weight of 104,000 was determined, also by gel filtration (Hosoya and Morrison, 1967). See discussion of cloned thyroid peroxidase, below.

Thyroglobulin has design features that favor thyroxine biosynthesis (Lamas et al., 1974; Lamas and Taurog, 1977). Similarities between lactoperoxidase and thyroid peroxidase were pointed out (Taurog et al., 1974). Glycosylation of thyroid peroxidase occurs in the N-terminal third of the polypeptide chain, believed to be extracellular (Rawitch et al., 1992). Most glycosylation occurs to asparagine residues with mannose and glucosamine the dominant sugars. Microheterogeneity accurs in the oligosaccharide units, a common feature of many glycoproteins.

Both lactoperoxidase and thyroid peroxidase undergo a catalatic reaction in which a small amount of iodide causes the decomposition of hydrogen peroxide to oxygen and water (Magnusson et al., 1984a, 1984b). Compound I oxidizes iodide to hypoiodous acid HOI, which in turn oxidizes hydrogen peroxide, as shown in the following scheme:

$$\begin{aligned} \text{Enzyme} + H_2O_2 &\longrightarrow \text{Compound I} + H_2O \\ \text{Compound I} + I^- &\longrightarrow \text{Enzyme-(HOI)} \\ \text{Enzyme-(HOI)} + H_2O_2 &\longrightarrow \text{Enzyme} + O_2 + H_2O + I^- \end{aligned}$$

$$2H_2O_2 = O_2 + 2H_2O$$

Evidence was presented that the hypoiodous acid is enzyme-bound in thyroid peroxidase but not in lactoperoxidase.

Thyroid Peroxidase Inhibition

The mechanisms of action of several antithyroid drugs were examined (Taurog, 1976; Taurog and Dorris, 1989; Taurog et al., 1989; Doerge et al., 1993; Roy and Mugesh, 2005). A study was conducted on the action of the thyroid microsomal antigen that causes production of autoantibodies in patients with autoimmune thyroid disease (Yokoyama et al., 1990). Insulin increases the concentration of a peroxidase inhibitor in cultured thyroid cells. The unidentified inhibitor is hydrophilic with a molecular weight of less than 2000 (Krawiec et al., 2004). No heme binding was found in the thyroid peroxidase obtained from patients with Graves disease (Taurog et al., 1990).

Figure 19.2. (*Continued*) Taurog A. 2000. In: Braverman LE, Utiger RD, editors. Werner and Ingbar's *The Thyroid: A Fundamental and Clinical Text*, 8th edition. Lippincott, Williams & Wilkins, Philadelphia, PA, pp. 61–84.)

cDNA and Sequencing

Two forms of human thyroid peroxidase cDNAs were isolated from a cDNA library prepared from thyroid tissue (Kimura et al., 1987a). The longer, TPO-1, had a reading frame of 933 amino acids, corresponding to a molecular weight of 103,000. This does not include 10% carbohydrate or the heme group. Five potential asparagine glycosylation sites were detected. The shorter peroxidase, TPO-2, lacks 57 amino acids from the middle of the sequence of TPO-1. It was suggested that the two forms of human thyroid peroxidase might result from an alternative splicing of the same gene located in human chromosome 2 (Kimura et al., 1987b; Endo et al., 1995). The structure of the thyroid peroxidase gene was studied in detail and shown to have great similarity to the gene for myeloperoxidase (Kimura and Ikeda-Saito, 1988; Kimura et al., 1989). The first demonstration that thyroid peroxidase could be cloned in large quantities was obtained using the baculovirus expression system (Kendler et al., 1993). However, the cloned enzyme, although exhibiting a high degree of purity by gel electrophoresis, and action as an autoantigen (McLachlan and Rapoport, 1992), has an RZ of only 0.25 compared to 0.5 to 0.6 for conventionally isolated enzyme, which is much more active.

Distal and proximal histidines were identified in TPO-1 by aligning sequences with those of myeloperoxidase for which the crystal structure was available. They are His239 and His494 (Bikker et al., 1997; Taurog and Wall, 1998).

Kinetic Studies, Intermediate Compounds

Many of the first kinetic studies on thyroid peroxidase reactions were conducted in the laboratory of Isao Yamazaki (Nakamura et al., 1983, 1984, 1986, 1989, 1990; Ohtaki et al., 1985). Formation of Compounds I, II, and III was observed in reactions of hog thyroid peroxidase with hydrogen peroxide (Ohtaki et al., 1981). Conversion of Compound I to II occurred spontaneously. A large excess of hydrogen peroxide was required to form Compound III.

The most accurate optical spectra were obtained for Compounds II and III of thyroid- and lactoperoxidases (Kohler et al., 1988). Measurements were made on a Cary spectrophotometer so the dead time was too long to observe Compound I formation and disappearance. Compound II was formed using excess hydrogen peroxide, and Compound III by using a larger excess. Iodide prevented Compound II formation because it started the catalatic reaction, described above. Both Compounds II and III decayed to native enzyme, rapidly in the case of thyroid peroxidase, with considerable denaturation of enzyme.

An estimate of $2 \times 10^7\ M^{-1}\ s^{-1}$ at pH 7.4 for the rate constant for thyroid peroxidase reacting with hydrogen peroxide to form Compound I was deduced from a steady-state study of iodination of tyrosine and monoiodotyrosine, in which the catalatic reaction catalyzed by iodide was neglected (Nakamura et al., 1985).

pH optima were deduced for several steady-state reactions of thyroid peroxidase (Koostra et al., 1993). It was shown that the pH optima depended on both hydrogen peroxide and reducing substrate concentrations.

Kinetic studies were continued in the laboratory of Alvin Taurog. Steady-state experiments indicated a radical mechanism in peroxidase-catalyzed coupling to form T3 and T4 (Taurog et al., 1994). Thyroid peroxidase can simultaneously catalyze both iodination and coupling. It was proposed that both reactions are catalyzed by the porphyrin π-cation radical form of Compound I (Taurog et al., 1996); and in fact there appears to be no other supporting evidence for a proposed protein radical form of thyroid peroxidase Compound I (Virion et al., 1985). In a recent review article properties of the most active highly purified thyroid peroxidase are listed (Taurog, 2000). These are MW of 90,000, A412/A280 (RZ) of 0.50 to 0.54, carbohydrate content about 10%, and isoelectric pH 5.75.

Role of Sulfenyl Iodides

The role of sulfenyl iodides in throid peroxidase catalyzed iodination (Maloof and Soodak, 1965) would appear to be ripe for intensive research. β-Lactoglobulin possesses only one sulfhydryl group and provided an attractive model system. At acid pH it reacts with iodide to form a sulfenyl iodide protein, -S-I, the formation of which can be followed at 355 nm. The sulfenyl iodide forms disulfides with a variety of antithyroid agents such as thiourea, thiouracil, and other compounds that interfere with in vivo iodination of tyrosine (Cunningham, 1964). Absorption at 355 nm was observed when activated thyroid homogenates were supplemented with iodide (Fawcett, 1966). Formation of sulfenyl iodides was demonstrated during iodination of tyrosine by thyroid homogenates (Fawcett, 1968). Could sulfenyl iodides, apparently neglected since the 1960s, be playing the intermediary role in thyroid peroxidase catalyzed iodination reactions that chloramines, and most probably taurine, play in myeloperoxidase reactions?

CONCLUSIONS

Thyroid peroxidase is now known to be genetically related to the other mammalian peroxidases: myeloperoxidase, eosinophil peroxidase, lactoperoxidase, and salivary peroxidase. Research on thyroid peroxidase has lagged behind that on myeloperoxidase partly because of purification difficulties, and partly because there has been a smaller number of laboratories working over more limited time spans. The crystal structure of thyroid peroxidase has yet to be determined. The contributions to thyroid peroxidase research by the late Alvin Taurog over his entire career have been outstanding.

The time would appear ripe for application of a battery of techniques to thyroid peroxidase. As one example of room for more work, the only reference I found to resonance Raman spectra was a preliminary study (Hashimoto et al., 1989). See also the section on iodothyronine deiodinase in Chapter 22.

REFERENCES

Alexander NM. 1959. Iodide peroxidase in rat thyroid and saliva glands and its inhibition by antithyroid compounds. *J. Biol. Chem.* **234**: 1530–1533.

Bikker H, Baas F, De Vijlder JJM. 1987. Molecular analysis of mutated thyroid peroxidase detected in patients with total iodide organification defects. *J. Clin. Endocrin. Metab.* **82**: 649–653.

Bikker H, Baas F, De Vijlder JJM. 1997. Molecular analysis of mutated thyroid peroxidase detected in patients with total iodide organification defects. *J. Clin. Endocrin. Metab.* **82**: 649–653.

Braverman LE, Ingbar SH, Sterling K. 1970. Conversion of thyroxine (T4) to triiodothyronine (T3) in athyreotic human subjects. *J. Clin. Invest.* **49**: 855–864.

Braverman LE, Utiger RD, editors. 2004.*Werner and Ingbar's The Thyroid: A Fundamental and Clinical Text*, 9th edition. Lippincott, Williams & Wilkins, Philadelphia, PA.

Coval ML, Taurog A. 1967. Purification and iodinating activity of hog thyroid peroxidase. *J. Biol. Chem.* **242**: 5510–5523.

Cunningham LW. 1964. The reaction of β-lactoglobulin sulfenyl iodide with several antithyroid agents. *Biochemistry* **3**: 1629–1634.

DeGroot LJ, Davis AM. 1962. Studies on the biosynthesis of iodotyrosines: a soluble thyroid-iodide peroxidase tyrosine-iodinase system. *Endocrinology* **70**: 492–504.

Dobson JE. 1998. The iodine factor in health and evolution. *Geograph. Rev.* **88**: 1–28.

Doerge DR, Decker CJ, Takazawa RS. 1993. Chemical and enzymatic oxidation of bezimidazoline-2-thiones: a dichotomy in the mechanism of peroxidase inhibition. *Biochemistry* **32**: 58–65.

Endo Y, Onogi S, Umeki K, Yamamoto I, Kotani T, Ohtaki S, Fujita T. 1995. Regional location of the gene for thyroid peroxidase to human chromosome 2p25 and mouse chromosome 12C. *Genomics* **25**: 760–761.

Fawcett DM. 1966. The binding of ^{131}iodide by cell-free thyroid homogenates measured by equilibrium dialysis. *Can. J. Biochem.* **44**: 1669–1683.

Fawcett DM. 1968. The formation of sulfenyl iodides as intermediates during the in vitro iodination of tyrosine by calf thyroid homogenates. *Can. J. Biochem.* **46**: 1433–1441.

Fawcett DM, Kirkwood S. 1953. The synthesis of organically bound iodine by cell-free preparations of thyroid tissue. *J. Biol. Chem.* **205**: 795–802.

Fawcett DM, Kirkwood S. 1954. Tyrosine iodinase. *J. Biol. Chem.* **209**: 249–256.

Fawcett DM, McLeod LE. 1963. Comparison of iodide peroxidase-like activity of thyroid and nonthyroid preparations. *Endocrinology* **73**: 279–284.

Gross J, Pitt-Rivers R. 1954. The identification of 3:5:3′-triiodothyronine in human plasma. *Lancet* **259**: 439–441.

Gross J, Pitt-Rivers R. 1952. Triiodothyronine in relation to thyroid physiology. *Recent Prog. Hormone Res.* **10**: 109–128.

Gross J, Pitt-Rivers R, Trotter WR. 1952. Effect of 3:5:3′-L-triiodothyronine in myxoedema. *Lancet* **259**: 1044–1045.

Harington CR. 1926. Chemistry of thyroxine. I. Isolation of thyroxine from the thyroid gland. *Biochem. J.* **20**: 293–299.

Harington CR, Barger G. 1927. Chemistry of thyroxine. III. Constitution and synthesis of thyroxine. *Biochem. J.* **21**: 169–181.

Hashimoto S, Nakajima R, Kotani T, Ohtaki S, Kitigawa T. 1989. Resonance Raman characterization of hog thyroid peroxidase: an SERRS study. *FEBS Lett.* **248**: 205–209.

Hosoya T, Kondo Y, Ui N. 1962. Peroxidase activity in thyroid gland and partial purification of the enzyme. *J. Biochem. (Tokyo)* **52**: 180–189.

Hosoya T, Morrison M. 1967. The isolation and purification of thyroid peroxidase. *J. Biol. Chem.* **242**: 2828–2836.

Igo RP, Mahoney CP, Mackler B. 1964. Studies on the biosynthesis of thyroxine. I. Purification and properties of a particulate iodide peroxidase from thyroid tissue. *J. Biol. Chem.* **239**: 1893–1897.

Kendler DL, Brennan V, Davies TF, Magnusson RP. 1993. Expression of human thyroid peroxidase in insect cells using recombinant baculovirus. *Mol. Cell. Endocrinol.* **93**: 199–206.

Kimura S, Hong Y-S, Kotani T, Ohtaki S, Kikkawa F. 1987a. Structure of the human thyroid peroxidase gene: comparison and relationship to the human myeloperoxidase gene. *Biochemistry* **28**: 4481–4489.

Kimura S, Hong Y-S, Kotani T, Ohtaki S, Kikkawa F. 1989. Structure of the human thyroid peroxidase gene. *Biochemistry* **28**: 4481–4489.

Kimura S, Ikeda-Saito M. 1988. Human myeloperoxidase and thyroid peroxidase, two enzymes with separate and distinct physiological functions, are evolutionarily related members of the same gene family. *Proteins* **3**: 113–120.

Kimura S, Kotani T, McBride OW, Umeki K, Hirai K, Nakayama T, Ohtaki S. 1987b. Human thyroid peroxidase: complete cDNA and protein sequence, chromosome mapping, and identification of two alternately spliced mRNAs. *Proc. Natl. Acad. Sci. USA* **84**: 5555–5559.

Klebanoff SJ, Yip C, Kessler D. 1962. The iodination of tyrosine by beef thyroid preparations. *Biochim. Biophys. Acta* **58**: 563–574.

Kohler H, Taurog A, Dunford HB. 1988. Spectral studies with lactoperoxidase and thyroid peroxidase: interconversions between native enzyme, Compound II, and Compound III. *Arch. Biochem. Biophys.* **264**: 438–449.

Koostra PR, Wever R, De Vijlder JJM. 1993. Thyroid peroxidase: kinetics, pH optima and substrate dependency. 1993. *Acta Endocrinol.* **129**: 328–331.

Krawiec L, Pizarro RA, Aphalo P, De Cavanagh EM, Pisarev MA, Juvenal GJ, Policastro L, Bocanerva LV. 2004. Role of peroxidase inhibition in the bovine thyroid cell proliferation mechanism. *Eur. J. Biochem.* **271**: 2607–2614.

Lamas L, Dorris ML, Taurog A. 1972. Evidence for a catalytic role for thyroid peroxidase in the conversion of diiodotyrosine to thyroxine. *Endocrinology* **90**: 1417–1426.

Lamas L, Taurog A. 1977. The importance of thyroglobulin structure in thyroid peroxidase-catalyzed conversion of diiodotyrosine to thyroxine. *Endocrinology* **100**: 1129–1136.

Lamas L, Taurog A, Salvatore G, Edelhoch E. 1974. Preferential synthesis of thyroxine from early iodinated tyrosyl residues in thyroglobulin. *J. Biol. Chem.* **249**: 2732–2737.

Lyon DM. 1927. Note on physiological test of synthetic thyroxine. *Biochem. J.* **21**: 181–183.

Magnusson RP, Taurog A, Dorris ML. 1984a. Mechanism of iodide-dependent catalatic activity of thyroid peroxidase and lactoperoxidase. *J. Biol. Chem.* **259**: 197–205.

Magnusson RP, Taurog A, Dorris ML. 1984b. Mechanisms of thyroid peroxidase- and lactoperoxidase-catalyzed reactions involving iodide. *J. Biol. Chem.* **259**: 13783–13790.

Maloof F, Soodak M. 1965. Intermediary metabolism of thyroid tissue and the action of drugs. *Pharmocol. Rev.* **15**: 43–95.

McLachlan SM, Rapoport B. 1992. The molecular biology of thyroid peroxidase: cloning, expression and role as autoantigen in autoimmune thyroid disease. *Endocrine Rev.* **13**: 192–206.

Moore KL, Mehlhorn RJ. 1993. Cytostatic effects of horseradish and thyroid peroxidase derived free radicals. *Free Radical Biol. Med.* **14**: 371–379.

Morton ME, Chaikoff IL. 1943. The formation in vitro of thyroxine and diiodotyrosine by thyroid tissue with radioactive iodine as indicator. *J. Biol. Chem.* **147**: 1–9.

Nakamura M, Yamazaki I, Nakagawa H, Ohtaki S. 1983. Steady-state kinetics and regulation of thyroid peroxidase-catalyzed iodination. *J. Biol. Chem.* **258**: 3837–3842.

Nakamura M, Yamazaki I, Nakagawa H, Ohtaki S. 1984. Iodination and oxidation of thyroglobulin catalyzed by thyroid peroxidase. *J. Biol. Chem.* **259**: 359–364.

Nakamura M, Yamazaki I, Nakagawa H, Ohtaki S. 1985. Thyroid peroxidase selects the mechanism of either 1- or 2-electron oxidation of phenols, depending on their substituents. *J. Biol. Chem.* **260**: 13546–13552.

Nakamura M, Yamazaki I, Ohtaki S, Nakamura S. 1986. Characterization of one- and two-electron oxidations of glutathione coupled with lactoperoxidase and thyroid peroxidase reactions. *J. Biol. Chem.* **261**: 13923–13927.

Nakamura M, Yamazaki I, Kotani T, Ohtaki S. 1989. Thyroglobulin-mediated one- and two-electron oxidations of glutathione and ascorbate in thyroid peroxidase systems. *J. Biol. Chem.* **264**: 12909–12913.

Nakamura M, Yamazaki I, Ohtaki S. 1990. Iodothyronine-induced catalatic activity of thyroid peroxidase. *J. Biochem.* **108**: 804–810.

Ohtaki S, Nakagawa H, Kimura S, Yamazaki I. 1981. Analyses of catalytic intermediates of hog thyroid peroxidase during its iodinating reaction. *J. Biol. Chem.* **256**: 805–810.

Ohtaki S, Nakagawa H, Nakamura S, Nakamura M, Yamazaki I. 1985. Characterization of hog thyroid peroxidase. *J. Biol. Chem.* **260**: 441–448.

Rawitch A, Taurog A, Chernoff S, Dorris M. 1979. Hog thyroid peroxidase: physical, chemical, and catalytic properties of the highly purified enzyme. *Arch. Biochem. Biophys.* **194**: 244–257.

Rawitch AB, Pollock G, Yang S-X, Taurog A. 1992. Thyroid peroxidase glycosylation: the location and nature of the N-linked oligosaccharide units in porcine thyroid peroxidase. *Arch. Biochem. Biophys.* **297**: 321–327.

Roy G, Mugesh G. 2005. Anti-thyroid drugs and thyroid hormone synthesis: effect of methimazole derivatives on peroxidase-catalyzed reactions. *J. Am. Chem. Soc.* **127**: 15207–15217.

Serif GS, Kirkwood S. 1958. Enzyme systems concerned with the synthesis of monoiodotyrosine. *J. Biol. Chem.* **233**: 109–115.

Taurog A. 1970. Thyroid peroxidase and thyroxine biosynthesis. *Recent Prog. Hormone Res.* **26**: 189–247.

Taurog A. 1976. The mechanism of action of the thioureylene antithyroid drugs. *Endocrinology* **98**: 1031–1046.

Taurog A. 2000. Hormone synthesis: thyroid iodine metabolism. In: Braverman LE, Utiger RD, editors. *Werner and Ingbar's The Thyroid: A Fundamental and Clinical Text*, 8th edition. Lippincott, Williams & Wilkins, Philadelphia, PA, pp. 61–84.

Taurog A, Dorris ML. 1989. A reexamination of the proposed inactivation of thyroid peroxidase in the rat thyroid by propylthiouracil. *Endocrinology* **124**: 3038–3042.

Taurog A, Dorris M, Doerge DR. 1994. Evidence for a radical mechanism in peroxidase-catalyzed coupling. I. Steady-state experiments with various peroxidases. *Arch. Biochem. Biophys.* **315**: 82–89.

Taurog A, Dorris M, Doerge DR. 1996. Mechanism of simultaneous iodination and coupling catalyzed by thyroid peroxidase. *Arch. Biochem. Biophys.* **330**: 24–32.

Taurog A, Dorris ML, Guziec FS, Uetrecht JP. 1989. Metabolism of ^{35}S- and ^{14}C-labeled propylthiouracil in a model in vitro system containing thyroid peroxidase. *Endocrinology* **124**: 3030–3037.

Taurog A, Dorris ML, Lamas L. 1974. Comparison of lactoperoxidase and thyroid peroxidase-catalyzed iodination and coupling. *Endocrinology* **94**: 1286–1294.

Taurog A, Dorris ML, Yokoyama N, Slaughter C. 1990. Purification and characterization of a large, tryptic fragment of human thyroid peroxidase with high catalytic activity. *Arch. Biochem. Biophys.* **278**: 333–341.

Taurog A, Tong W, Chaikoff IL. 1950. The monoiodotyrosine content of the thyroid gland. *J. Biol. Chem.* **184**: 83–97.

Taurog A, Wall M. 1998. Proximal and distal histidines in thyroid peroxidase: relation to the alternatively spliced form, TPO-2. *Thyroid* **8**: 185–191.

Virion A, Courtin F, Dème D, Michot JL, Kaniewski J, Pommier J. 1985. Spectral characteristics and catalytic properties of thyroid peroxidase-H_2O_2 compounds in the iodination and coupling reactions. *Arch. Biochem. Biophys.* **242**: 41–47.

Weiss B. 1953. Utilization of radioactive iodide by cell-free preparations of beef thyroid tissue. *J. Biol. Chem.* **201**: 31–43.

Yokoyama N, Taurog A, Dorris ML, Klee GG. 1990. Studies with purified human thyroid peroxidase and thyroid microsomal autoantibodies. *J. Clin. Endocrin. Metab.* **70**: 758–765.

CHAPTER 20

LACTO- AND SALIVARY PEROXIDASES

INTRODUCTION

Lactoperoxidase and salivary peroxidase are derived from the same gene, but may be processed differently (Morrison et al., 1965; Gothefors and Marklund, 1975; Månsson- Rahemtulla et al., 1990; Ihalin et al., 2006). There is an excellent summary of earlier work on lactoperoxidase (Pruitt and Tenovuo, 1985), and later on both lacto- and salivary peroxidases (Thomas et al., 1991; Reiter and Perraudin, 1991; Tenovuo, 1991). Both enzymes have a high tendency to stick to surfaces. They belong to a family, variously called the mammalian peroxidase, the animal peroxidase (Taurog, 1999), or the myeloperoxidase superfamily. The molecular evolution of the myeloperoxidase superfamily has been plotted in an unrooted phylogenetic tree (Daiyasu and Toh, 2000).

PROPERTIES

Crystalline lactoperoxidase from bovine milk was obtained in 1943 and shown to be green in color as is myeloperoxidase (Theorell and Åkeson, 1943; Theorell and Paul, 1944). Isolation procedures were simplified and improved using chromatography (Morrison et al., 1957; Morrison and Hultquist, 1963; Paul et al., 1980; Goff et al., 1985). Heterogeneity was demonstrated. Although partly caused by deamidation of asparagine and loss of carbohydrate groups in the isolation procedure (Carlström, 1965), naturally occurring subfractions were identified (Carlström, 1969a). From ultracentrifugation the molecular weight of the main fraction was calculated to be 78,500. Accurate optical spectra were obtained of various complexes (Carlström, 1969b).

There is one calcium ion bound per subunit (Booth et al., 1989). Both cyanide and chloride bind in the sixth coordination position of the heme iron (Kimura and

Peroxidases and Catalases: Biochemistry, Biophysics, Biotechnology, and Physiology, Second Edition.
By H. Brian Dunford

Yamazaki, 1978). The pH-dependent kinetics of cyanide (Dolman et al., 1968) and fluoride (Segal et al., 1968) binding by lactoperoxidase have been studied. NMR (Behere et al., 1985; Thanabal and La Mar, 1989), resonance Raman (Hu et al., 1993), and EPR spectra (Lukat et al., 1987) of the cyanide complex were obtained. Amino acid sequences were obtained for bovine (Dull et al., 1990; Cals et al., 1991) and human (Ueda et al., 1997) lactoperoxidases, and for human salivary peroxidase (Kiser et al., 1996).

It was established that the heme group is covalently bound to the apoprotein but the exact nature of the linkage was difficult to determine (Hultquist and Morrison, 1963; Morell and Clezy, 1963). An iron-porphyrin thiol structure was claimed (Nichol et al., 1987), but contradicted by structures involving two ester linkages only (Rae and Goff, 1996; DePillis et al., 1997; Kooter et al., 1997). It is now established that the lactoperoxidase prosthetic group is a 1,5-bis(hydroxymethyl) derivative of heme *b*. It has two ester linkages formed by an autocatalytic reaction involving Glu275 and Asp125 (DePillis et al., 1997; Rae and Goff, 1998). The sulfonium ion heme link in myeloperoxidase is absent in the milk enzyme. The techniques used included NMR and Fourier transform infrared spectroscopies, Pronase digestion, site-directed mutagenesis, and MALDI and electrospray mass spectrometries. Heme distortions attributed to the ester linkages have been detected by resonance Raman spectra (Zbylut and Kincaid, 2002).

The sixth coordination position of Fe^{III} in lactoperoxidase has been reported to be occupied by water (Kitagawa et al., 1983; Chang et al., 1993; Zbylut and Kincaid, 2002) and to be vacant (Shiro and Morishima, 1986; Modi et al., 1990), a recurring theme in peroxidase chemistry. On the basis of this theme, it may be fair to assume that native highly purified lactoperoxidase has five-coordinate Fe^{III}, and aging or handling converts it to the six-coordinate form.

A lactoperoxidase assay using the reagent 2,2′-azino-di-(3-ethylbenzthiazoline-6-sulfonic acid) (ABTS) was developed (Childs and Bardsley, 1975), and is now widely used. ABTS is a one-electron donor that forms a stable highly colored free radical. Assays for mammalian peroxidases were critically examined (Pruitt et al., 1990). It was shown that pyrogallol, guaiacol, ABTS, and thiocyanate all could be used for satisfactory assays for lactoperoxidase and myeloperoxidase, but that thiocyanate normally present in saliva needed to be removed before a satisfactory salivary peroxidase assay could be conducted.

THE COMPOUNDS OF LACTOPEROXIDASE AND THEIR REACTIONS

Research led to the conclusion that there are two forms of lactoperoxidase Compound I (Courtin et al., 1984, 1984; Ohtaki et al., 1982; Kimura and Yamazaki, 1979; Kohler and Jenzer, 1989; Ghibaudi and Laurenti, 2003). Definitive work showed that after formation of a traditional $Fe^{IV}{=}O$ porphyrin π-cation radical a protein radical is generated which was spin-trapped (Lardinois et al., 1999). A second radical site leads to dimer formation withTyr289 cross-links. The rate constant for Compound I formation using hydrogen peroxide was found to be $(9.2 \pm 0.9) \times 10^6\ M^{-1}\ s^{-1}$ at 25°C over the pH range 3 to 11 (Maguire and Dunford, 1971), and $(1.1 \pm 0.1) \times 10^7\ M^{-1}\ s^{-1}$ at pH 7 and 15°C. With hypochlorous acid the rate of Compound I formation is $(3.2 \pm 0.1) \times 10^7\ M^{-1}\ s^{-1}$ at pH 7 and 15°C (Furtmüller et al., 2002) and this rate constant is for formation of the Fe^{IV}=O porphyrin π-cation radical. The problem in sorting out the rate of interconversion of the lactoperoxidase Compound I species and their reactivities is that the optical spectra of Compound II and the protein radical form of Compound I would be expected to be similar if not identical, as in the case of yeast cytochrome *c* peroxidase. The porphyrin π-cation radical form of Compound I decays with a rate constant of $(2.0 \pm 0.5)\ s^{-1}$ at pH 7 and 15°C and this would appear to be conversion to the protein free radical form of Compound I, stable for about 20 s (Furtmüller et al., 2002).

In early transient-state kinetic studies a Compound II-like species was formed by the addition of only one molar equivalent of hydrogen peroxide to lactoperoxidase. Reactions with iodide were similar to those of horseradish peroxidase Compound II but were an order of magnitude faster (Maguire and Dunford, 1972). Reactions with *p*-cresol were of a similar rate to those of HRP Compound II (Maguire and Dunford, 1973). These reactions of lactoperoxidase attributed to Compound II may be reactions with the protein free radical form of lactoperoxidase Compound I. Compound II of lactoperoxidase reacts with sulfide and sulhydryl compounds to form a typical sulfheme protein (Nakamura et al., 1984).

A steady-state study of the oxidation of phenolic compounds by lactoperoxidase provided evidence that the iron-oxo group of Compound I reacted at low pH and the protein free radical site at high pH. Compound I is readily converted into Compound III by reaction with hydrogen peroxide (Monzani et al., 1997). Conversion of lactoperoxidase Compound II to Compound III by excess hydrogen peroxide and ferrous enzyme plus dioxygen to Compound III was demonstrated using resonance Raman spectroscopy (Hu and Kincaid, 1991). Catalase activity of lactoperoxidase through Compound III formation was demonstrated (Galijasevic et al., 2004). Slow inactivation of lactoperoxidase by hydrogen peroxide was demonstrated (Huwiler et al., 1986).

Reactions of lactoperoxidase Compounds I, II, and III with nitrite were investigated by both steady-state and transient-state kinetics (Brück et al., 2001). Nitrite appears to react with Compound I via a two-electron reaction, unlike its horseradish peroxidase counterpart. Increased hydrogen peroxide concentration favors Compound III formation. Nitrite reduces Compound III to Compound II and Compound II to native enzyme, thus helping to maintain a steady-state cycle. Compounds I and II of lactoperoxidase oxidize benzylpenicillin G (Metodiewa and Dunford, 1990). 3-Aminotriazole is a substrate for Compounds I, II and III of lactoperoxidase (Metodiewa and Dunford, 1991a).

Optical spectra were obtained for native lactoperoxidase, Compounds I and II (Furtmüller et al., 2002), and for native lactoperoxidase, Compounds II and III (Kohler et al., 1988). Details of the native spectrum in the visible region are sharper in the earlier data. The earlier data were measured directly using a small excess of hydrogen peroxide to generate Compound II and a larger excess to obtain Compound III. In the later data lactoperoxidase was reacted with hypochlorous acid, and the spectra obtained by computer deconvolution. The Compound II spectral data are in agreement. Reaction of hypochlorous acid with lactoperoxidase is analogous to its reaction with myeloperoxidase.

The reduction potential for the ferric-ferrous couple of lactoperoxidase is -188 ± 1 mV at pH 7 (Ohlsson and Paul, 1983). Resonance Raman spectra indicated that in the cyanoferrous and nitric oxide complexes the heme iron is displaced from the heme plane and away from the proximal histidine ligand (Manthey et al., 1986). A nine-line hyperfine splitting of the ferrous-NO EPR spectrum was observed. A nitrogen fifth ligand was confirmed (Lukat et al., 1987).

In a continuation of autocatalytic covalent bond formation, it was shown that the heme of horseradish peroxidase but not lactoperoxidase was modified by thiocyanate oxidation products. This demonstrated that the autocatalytic covalent cross-links to the heme of lactoperoxidase have a protective effect (Wojciechowski et al., 2005). Cross-links between lactoperoxidase and myoglobin were formed by reaction with hydrogen peroxide (Lardinois and Ortiz de Montellano, 2001). Irreversible inactivation of lactoperoxidase by excess hydrogen peroxide was investigated by a variety of techniques (Jenzer et al., 1986, 1987b), as was inactivation during iodide oxidation (Jenzer et al., 1987a).

Iodide causes pseudo-catalase activity in thyroid and lactoperoxidase (Magnusson and Taurog, 1983; Magnusson et al., 1984), as it does in horseradish peroxidase. L-Tyrosine binds preferentially to lactoperoxidase compared to D-tyrosine (Casella et al., 1991). Aromatic donor molecules were reported to bind away from the heme (Modi et al., 1989a), but near a heme edge (Hosoya et al., 1989). Iodination of tyrosine peptides (Bayse et al., 1972) and halogenation of tyrosine by lactoperoxidase was investigated (Huwiler et al., 1985; Taurog and Dorris, 1991). Lactoperoxidase catalyzes iodination of a serine proteinase, the first component of complement, and abolishes its calcium ion-dependent dimer formation (Illy et al., 1991).

Enantiomeric sulfoxidation of thioethers occurs (Doerge et al., 1991; Colonna et al., 1995; Tuyman et al., 1998). An oxygen rebound mechanism occurs in which the ferryl oxygen is transfered to a sulfur cation radical.

Lactoperoxidase oxidizes NADH (Hogg and Jago, 1970). Sustained oscillations occur in the lactoperoxidase, NADPH, and O_2 system (Nakamura et al., 1969). Substituted catechols react with lactoperoxidase (Metodiewa et al., 1989a, 1989b), as does serotonin (5-hydroxy-tryptamine) and 5,7-dihydroxytrptamine (Metodiewa and Dunford, 1991b, 1992). Lactoperoxidase reacts with the water-soluble derivative of vitamin E, Trolox C (Metodiewa and Dunford, 1991c) and activates carcinogenic amines (Gorlewska-Roberts et al., 2004). The chromophores of the anticancer anthracyclines doxorubicin and daunorubicin are destroyed in lactoperoxidase/hydrogen peroxide and lactoperoxidase/hydrogen peroxide/nitrite systems (Reszka et al., 2001). Lactoperoxidase catalyzes DNA damage using lipid hydroperoxides (Adam et al., 2000). Singlet oxygen is generated from the decomposition of α-linolenic acid hydroperoxide catalyzed by lactoperoxidase (Sun et al., 2007).

Bactericidal Activity

The bactericidal activity of salivary and lactoperoxidase appears to be due to the formation of hypothiocyanite $OSCN^-$ from thiocyanate SCN^- (Wever et al., 1982; Pollock and Goff, 1992; Nagy et al., 2006). Thiocyanate is called a pseudo-halide because of similarities in its chemistry. It appears to bind to lactoperoxidase at the same site as iodide (Modi et al., 1989b). The conjugate acid, hypothiocyanous acid, HOSCN, has a pK_a value of 5.3. Addition of thiocyanate and hydrogen peroxide is used to help preserve milk in tropical climates.

Formation of $OSCN^-$ or its conjugate acid may occur via the one-electron oxidation of SCN^- as follows.

$$2SCN^- \longrightarrow 2{\cdot}SCN$$
$$2{\cdot}SCN \longrightarrow (SCN)_2$$
$$(SCN)_2 + H_2O \longrightarrow HOCN + SCN^- + H^+$$

Alternatively, it could occur by direct oxygen addition, which by analogy with halide reactions, would be expected if reaction occurs with the $Fe^{IV}{=}O$ porphyrin π-cation radical form of Compound I.

$$SCN^- + H_2O_2 \longrightarrow OSCN^- + H_2O$$

REFERENCES

Adam W, Kurz A, Saha-Möller CR. 2000. Peroxidase-catalyzed oxidative damage of DNA and 2′-deoxyguanosine by model compounds of lipid hydroperoxides: involvement of peroxyl radicals. *Chem. Res. Toxicol.* **13**: 1199–1207.

Bayse GS, Michaels AW, Morrison M. 1972. Lactoperoxidase-catalyzed iodination of tyrosine peptides. *Biochim. Biophys. Acta* **284**: 30–33.

Behere DV, Gonzalez-Vergara E, Goff HM. 1985. Unique cyanide nitrogen-15 nuclear magnetic resonance chemical shift values for cyanide-peroxidase complexes: relevance to the heme active-site structure and mechanism of peroxidase activation. *Biochim. Biophys. Acta* **832**: 319–325.

Booth KS, Kimura S, Lee HC, Ikeda-Saito M, Caughey WS. 1989. Bovine myeloperoxidase and lactoperoxidase each contain a high affinity site for calcium. *Biochem. Biophys. Res. Commun.* **160**: 897–902.

Brück TB, Fielding RJ, Symons MCR, Harvey PJ. 2001. Mechanism of nitrite-stimulated catalysis by lactoperoxidase. *Eur. J. Biochem.* **268**: 3214–3222.

Cals M-M, Mailliart P, Brignon G, Anglade P, Dumas BR. 1991. Primary structure of lactoperoxidase, a fourth member of a mammalian peroxidase family. *Eur. J. Biochem.* **198**: 733–739.

Carlström A. 1965. The heterogeneity of lactoperoxidase. *Acta Chem. Scand.* **19**: 2387–2394.

Carlström A. 1969a. Physical and compositional investigations of the subfractions of lactoperoxidase. *Acta Chem. Scand.* **23**: 185–202.

Carlström A. 1969b. Lactoperoxidase: some spectral properties of a haemoprotein with a prosthetic group of unknown structure. *Acta Chem. Scand.* **23**: 203–212.

Casella L, Gullotti M, Poli S, Bonfà M, Ferrari RP. 1991. Spectroscopic and binding studies on the stereoselective interaction of tyrosine with horseradish peroxidase and lactoperoxidase. *Biochem. J.* **279**: 245–250.

Chang C-S, Sinclair R, Khalid S, Yamazaki I, Nakamura S, Powers L. 1993. An extended X-ray absorption fine structure investigation of the structure of the active site of lactoperoxidase. *Biochemistry* **32**: 2780–2786.

Childs RE, Bardsley WG. 1975. The steady-state kinetics of peroxidase with 2,2′-azino-di-(3-ethylbenzthiazoline-6-sulfonic acid) as chromogen. *Biochem. J.* **145**: 93–103.

Colonna S, Gaggero N, Richelmi C, Carrea G, Pasta P. 1995. Enantioselective sulphoxidations catalyzed by LPO. *Gaz. Chim. Ital.* **125**: 479–482.

Courtin F, Deme D, Virion A, Michot JL, Pommier J, Nunez J. 1982. The role of lactoperoxidase compounds in the catalysis of throglobulin iodination and thyroid hormone synthesis. *Eur. J. Biochem.* **124**: 603–609.

Courtin F, Michot J-L, Virion A, Pommier J, Nunez J. 1984. Reduction of lactoperoxidase-H_2O_2 compounds by ferrocyanide: indirect evidence of an apoprotein site for one of the two oxidizing equivalents. *Biochem Biophys. Res. Commun.* **121**: 463–470.

Daiyasu H, Toh H. 2000. Molecular evolution of the myeloperoxidase family. *J. Mol. Evol.* **51**: 433–445.

DePillis GD, Ozaki S, Kuo JM, Maltby DA, Ortiz de Montellano PR. 1997. Autocatalytic processing of heme by lactoperoxidase produces the native protein-bound prosthetic group. *J. Biol. Chem.* **272**: 8857–8860.

Doerge DR, Cooray NM, Brewster ME. 1991. Peroxidase-catalyzed S-oxygenation: mechanism of oxygen transfer for lactoperoxidase. *Biochemistry* **30**: 8960–8964.

Dolman D, Dunford HB, Chowdhury DM, Morrison M. 1968. The kinetics of cyanide binding by lactoperoxidase. *Biochemistry* **7**: 3991–3996.

Dull TJ, Uyeda C, Strosberg AD, Nedwin G, Seilhamer JJ. 1990. Molecular cloning of cDNAs encoding bovine and human lactoperoxidase. *DNA Cell Biol.* **9**: 499–509.

Furtmüller PG, Jantschko W, Regelsberger G, Jakopitsch, Arnhold J, Obinger C. 2002. Reaction of lactoperoxidase Compound I with halides and thiocyanate. *Biochemistry* **41**: 11895–11900.

Galijasevic S, Saed GM, Diamond MP, Abu-Soud HM. 2004. High dissociation rate constant of ferrous-dioxy complex linked to the catalase-like activity in lactoperoxidase. *J. Biol. Chem.* **279**: 39465–39470.

Ghibaudi E, Laurenti E. 2003. Unravelling the catalytic mechanism of lactoperoxidase and myeloperoxidase. *Eur. J. Biochem.* **270**: 4403–4412.

Goff HM, Gonzalez-Vergara E, Ales DC. 1985. High resolution proton nuclear magnetic resonance spectroscopy of lactoperoxidase. *Biochem. Biophys. Res. Commun.* **133**: 794–799.

Gorlewska-Roberts KM, Teitel CH, Lay JO Jr, Roberts DW, Kadlubar FF. 2004. Lactoperoxidase-catalyzed activation of carcinogenic aromatic and heterocyclic amines. *Chem. Res. Toxicol.* **17**: 1659–1666.

Gothefors L, Marklund S. 1975. Lactoperoxidase activity in human milk and in saliva of newborn infants. *Infection Immunity* **11**: 1210–1215.

Hogg DM, Jago GR. 1970. The oxidation of reduced nicotinamide nucleotides by hydrogen peroxide in the presence of lactoperoxidase and thiocyanate, iodide or bromide. **117**: 791–797.

Hosoya T, Saakurada J, Kurokawa C, Toyoda R, Nakamura S. 1989. Interaction of aromatic donor molecules with lactoperoxidase probed by optical difference spectra. *Biochemistry* **28**: 2639–2644.

Hu S, Kincaid JR. 1991. Resonance Raman structural characterization and the mechanism of formation of lactoperoxidase Compound III. *J. Am. Chem. Soc.* **113**: 7189–7194.

Hu S, Treat RW, Kincaid JR. 1993. Distinct heme active-site structure in lactoperoxidase revealed by resonance Raman spectroscopy. *Biochemistry* **32**: 10125–10130.

Hultquist DE, Morrison M. 1963. Lactoperoxidase. I. The prosthetic group of lactoperoxidase. *J. Biol. Chem.* **238**: 2843–2846.

Huwiler M, Bürgi U, Kohler H. 1985. Mechanism of enzymatic and non-enzymatic tyrosine iodination. *Eur. J. Biochem.* **147**: 469–476.

Huwiler M, Jenzer H, Kohler H. 1986. The role of Compound III in reversible and irreversible inactivation of lactoperoxidase. *Eur. J. Biochem.* **158**: 609–614.

Ihalin R, Loimaranta V, Tenovuo J. 2006. Origin, structure, and biological activities of peroxidases in human saliva. *Arch. Biochem. Biophys.* **445**: 261–268.

Illy C, Thielens NM, Gagnon J, Arlaud GJ. 1991. Effect of lactoperoxidase-catalyzed iodination on the Ca^{2+}-dependent interactions of human C1s: location of the iodination sites. *Biochemistry* **30**: 7135–7141.

Jenzer H, Jones W, Kohler H. 1986. On the molecular mechanism of lactoperoxidase-cataalyzed H_2O_2 metabolism and irreversible enzyme inactivation. *J. Biol. Chem.* **261**: 15550–15556.

Jenzer H, Bürgi U, Kohler H. 1987a. Irreversible inactivation of lactoperoxidase in the course of iodide oxidation. *Biochem. Biophys. Res. Commun.* **142**: 552–558.

Jenzer H, Kohler H, Broger C. 1987b. The role of hydroxyl radicals in irreversible inactivation of lactoperoxidase by excess H_2O_2. *Arch. Biochem. Biophys.* **258**: 381–390.

Kimura S, Yamazaki I. 1978. Heme-linked ionization and chloride binding in intestinal peroxidase and lactoperoxidase. *Arch. Biochem. Biophys.* **189**: 14–19.

Kimura S, Yamazaki I. 1979. Comparisons between hog intestinal peroxidase and bovine lactoperoxidase: Compound I formation and inhibition by benzhydroxamic acid. *Arch. Biochem. Biophys.* **198**: 580–588.

Kiser C, Caterina J, Engler JA, Rahemtulla B, Rahemtulla F. 1996. Cloning and sequence analysis of the human salivary peroxidase-encoding cDNA. *Gene* **173**: 261–264.

Kitagawa T, Hashimoto S, Teraoka J, Nakamura S, Yajima H, Hosoya T. 1983. Distinct heme-substrate interactions of lactoperoxidase probed by resonance Raman spectroscopy: difference between animal and plant peroxidases. *Biochemistry* **22**: 2788–2792.

Kohler H, Jenzer H. 1989. Interaction of lactoperoxidase with hydrogen peroxide: formation of enzyme intermediates and generation of free radicals. *Free Radical Biol. Med.* **6**: 323–339.

Kohler H, Taurog A, Dunford HB. 1988. Spectral studies with lactoperoxidase and thyroid peroxidase: interconversions between native enzyme, Compound II, and Compound III. *Arch. Biochem. Biophys.* **264**: 438–449.

Kooter IM, Pierik AJ, Merkx M, Averill BA, Moguilevsky N, Bollen A, Wever R. 1997. Difference Fourier transform infrared evidence for ester bonds linking the heme group in myeloperoxidase, lactoperoxidase, and eosinophil peroxidase. *J. Am. Chem. Soc.* **119**: 11542–11543.

Lardinois OM, Medzihradszky KF, Ortiz de Montellano PR. 1999. Spin trapping and protein cross-linking of the lactoperoxidase protein radical. *J. Biol. Chem.* **274**: 35441–35448.

Lardinois OM, Ortiz de Montellano PR. 2001. H_2O_2-mediated cross-linking between lactoperoxidase and myoglobin. *J. Biol. Chem.* **276**: 23186–23191.

Lukat GS, Rodgers KR, Goff HM. 1987. Electron paramagnetic resonance spectroscopy of lactoperoxidase complexes: clarification of hyperfine splitting of the NO adduct of lactoperoxidase. *Biochemistry* **26**: 6927–6932.

Magnusson RP, Taurog A. 1983. Iodide-dependent activity of thyroid peroxidase and lactoperoxidase. *Biochem. Biophys. Res. Commun.* **112**: 475–481.

Magnusson RP, Taurog A, Dorris ML. 1984. Mechanism of iodide-dependent catalatic activity of thyroid peroxidase and lactoperoxidase. *J. Biol. Chem.* **259**: 197–205.

Maguire RJ, Dunford HB. 1971. The kinetics of the formation of the primary lactoperoxidase: hydrogen peroxide compound. *Can. J. Biochem.* **49**: 1165–1171.

Maguire RJ, Dunford HB. 1972. Kinetics of the oxidation of iodide ion by lactoperoxidase Compound II. *Biochemistry* **11**: 937–941.

Maguire RJ, Dunford HB. 1973. Studies on lactoperoxidase. VI. Kinetics of the oxidation of *p*-cresol by Compound II. *Can. J. Biochem.* **51**: 1721–1723.

Månsson-Rahemtulla B, Rahemtulla F, Humphreys-Beher MG. 1990. Human salivary peroxidase and bovine lactoperoxidase are cross-reactive. *J. Dent. Res.* **69**: 1839–1846.

Manthey JA, Boldt NJ, Bocian DF, Chan SI. 1986. Resonance Raman studies of lactoperoxidase. *J. Biol. Chem.* **261**: 6734–6741.

Metodiewa D, Dunford HB. 1990. Evidence for one-electron oxidation of benzylpenicillin G by lactoperoxidase Compounds I and II. *Biochem. Biophys. Res. Commun.* **169**: 1211–1216.

Metodiewa D, Dunford HB. 1991a. 3-Aminotriazole is a substrate for lactoperoxidase but not for catalase. *Biochem. Biophys. Res. Commun.* **180**: 585–590.

Metodiewa D, Dunford HB. 1991b. Involvement of lactoperoxidase in the peroxidative degradation of serotonin: a potential pathway for indolaminergic melanin formation. *Biochem. Int.* **23**: 183–191.

Metodiewa D, Dunford HB. 1991c. On the ability of lactoperoxidase to catalyze the peroxidase-oxidase oxidation of the vitamin E water soluble derivative (Trolox C). *Biochem. Int.* **25**: 895–904.

Metodiewa D, Dunford HB. 1992. Peroxidase-promoted oxidation and peroxidation of the serotonergic neurotoxin 5,7-dihydroxytryptamine: a new pathway for its metabolic degradation. *Mol. Cell Biochem.* **112**: 35–44.

Metodiewa D, Reszka K, Dunford HB. 1989a. Evidence for a peroxidatic oxidation of norepinephrine, a catecholamine by lactoperoxidase. *Biochem. Biophys. Res. Commun.* **160**: 1183–1188.

Metodiewa D, Reszka K, Dunford HB. 1989b. Oxidation of the substituted catechols dihydroxyphenylalanine methyl ester and trihydroxyphenylalanine by lactoperoxidase and its compounds. *Arch. Biochem. Biophys.* **274**: 601–608.

Modi S, Behere DV, Mitra S. 1989a. Binding of aromatic donor molecules to lactoperoxidase: proton NMR and optical difference spectroscopic studies. *Biochem. Biophys. Acta* **996**: 214–225.

Modi S, Behere DV, Mitra S. 1989b. Binding of thiocyanate to lactoperoxidase: ^{1}H and ^{15}N nuclear magnetic resonance studies. *Biochemistry* **28**: 4689–4694.

Modi S, Behere DV, Mitra S. 1990. Coordination geometry of heme in lactoperoxidase: pH-dependent ^{1}H relaxivity and optical spectral studies. *J. Inorg. Biochem.* **38**: 17–25.

Monzani E, Gatti AL, Profumo A, Casella L, Gulloti M. 1997. Oxidation of phenolic compounds by lactoperoxidase: evidence for the presence of a low-potential Compound II during catalytic turnover. *Biochemistry* **36**: 1918–1926.

Morell DB, Clezy PS. 1963. The haematin prosthetic groups of some animal peroxidases. I. The preparation and properties of an ether-soluble haematin from milk peroxidase. *Biochim. Biophys. Acta* **71**: 157–164.

Morrison M, Allen PZ, Bright J, Jayasinghe W. 1965. Lactoperoxidase. V. Identification and isolation of lactoperoxidase from salivary gland. *Arch. Biochem. Biophys.* **111**: 126–133.

Morrison M, Hamilton HB, Stotz E. 1957. The isolation and purification of lactoperoxidase by ion exchange chromatography. *J. Biol. Chem.* **228**: 767–776.

Morrison M, Hultquist DE. 1963. Lactoperoxidase. II. Isolation. *J. Biol. Chem.* **238**: 2847–2849.

Nagy P, Alguindigue SS, Ashby MT. 2006. Lactoperoxidase-catalyzed oxidation of thiocyanate by hydrogen peroxide: a reinvestigation of hypothiocyanite by nuclear magnetic resonance and optical spectroscopy. *Biochemistry* **45**: 12610–12616.

Nakamura S, Nakamura M, Yamazaki I, Morrison M. 1984. Reactions of ferryl lactoperoxidase (Compound II) with sulfide and sulfhydryl compounds. *J. Biol. Chem.* **259**: 7080–7085.

Nakamura S, Yokota K, Yamazaki I. 1969. Sustained oscillations in a lactoperoxidase, NADPH and O_2 system. *Nature* **222**: 794.

Nichol AW, Angel LA, Moon T, Clezy PS. 1987. Lactoperoxidase haem, an iron-porphyrin thiol. **247**: 147–150.

Ohtaki S, Nakagawa H, Nakamura M, Yamazaki I. 1982. Reactions of purified hog thyroid peroxidase with H_2O_2, tyrosine, and methyl mercaptoimidazole (goitrogen) in comparison with bovine lactoperoxidase. *J. Biol. Chem.* **257**: 761–766.

Ohlsson P-I, Paul K-G. 1983. The reduction potential of lactoperoxidase. *Acta Chem. Scand.* **B37**: 917–921.

Paul KG, Ohlsson PI, Henriksson A. 1980. The isolation and some liganding properties of lactoperoxidase. *FEBS Lett.* **110**: 200–204.

Pollock JR, Goff HM. 1992. Lactoperoxidase-catalyzed oxidation of thiocyanate ion: a carbon-13 nuclear magnetic resonance study of the oxidation products. *Biochim. Biophys. Acta* **1159**: 279–285.

Pruitt KM, Kamau DN, Miller K, Månsson-Rahemtulla B, Rahemtulla F. 1990. Quantitative, standardized assays for determining the concentrations of bovine lactoperoxidase, human salivary peroxidase, and human myeloperoxidase. *Anal. Biochem.* **191**: 278–286.

Pruitt KM, Tenovuo JO, editors. 1985. *The Lactoperoxidase System, Chemistry and Biological Significance*. Marcel Dekker, New York.

Rae TD, Goff HM. 1996. Lactoperoxidase heme structure characterized by paramagnetic proton NMR spectroscopy. *J. Am. Chem. Soc.* **118**: 2103–2104.

Rae TD, Goff HM. 1998. The heme prosthetic group of lactoperoxidase. *J. Biol. Chem.* **273**: 27968–27977.

Reiter B, Perraudin J-P. 1991. Lactoperoxidase: biological function. In: Everse J, Everse KE, Grisham MB, editors. *Peroxidases in Chemistry and Biology*, Vol. 1. CRC Press, Boca Raton, FL, pp. 143–180.

Reszka KJ, McCormick ML, Britigan BE. 2001. Peroxidase- and nitrite-dependent metabolism of the anthracyclin anticancer agents daunorubicin and doxorubicin. *Biochemistry* **40**: 15349–15361.

Segal R, Dunford HB, Morrison M. 1968. The kinetics of fluoride binding by lactoperoxidase. *Can. J. Biochem.* **46**: 1471–1474.

Shiro Y, Morishima I. 1986. Structural characterization of lactoperoxidase in the heme environment by proton NMR spectroscopy. *Biochemistry* **25**: 5844–5849.

Sun S, Bao Z, Ma H, Zhang D, Zheng X. 2007. Singlet oxygen generation from the decomposition of α-liolenic acid hydroperoxide by cytochrome *c* and lactoperoxidase. *Biochemistry* **46**: 6668–6673.

Taurog A. 1999. Molecular evolution of thyroid peroxidase. *Biochimie* **81**: 557–562.

Taurog A, Dorris ML. 1991. Peroxidase-catalyzed bromination of tyrosine, thyroglobulin, and bovine serum albumin: comparison of thyroid peroxidase and lactoperoxidase. *Arch. Biochem. Biophys.* **287**: 288–296.

Tenovuo JO. 1991. Salivary peroxidase. In: Everse J, Everse KE, Grisham MB, editors. *Peroxidases in Chemistry and Biology*, Vol. 1. CRC Press, Boca Raton, Fl, pp. 181–198.

Thanabal V, La Mar GN. 1989. A nuclear Overhauser effect investigation of the molecular and electronic structure of the heme crevice in lactoperoxidase. *Biochemistry* **28**: 7038–7044.

Theorell H, Åkeson Å. 1943. Highly purified milk peroxidase. *Arkiv Kemi Mineral Geol.* **17B**(7): 1–6.

Theorell H, Paul KG. 1944. Dissociation constants and their relations to the activity in peroxidases. *Arkiv Kemi Mineral Geol.* **18A**(12): 1–23.

Thomas EL, Bozeman PM, Learn DB. 1991. Lactoperoxidase: structure and catalytic properties. In: Everse J, Everse KE, Grisham MB, editors. *Peroxidases in Chemistry and Biology*, Vol. 1. CRC Press, Boca Raton, FL, pp. 123–142.

Tuyman A, Vink MKS, Dekker HL, Schoemaker HE, Wever R. 1998. The sulphoxidation of thioanisole catalyzed by lactoperoxidase and *Coprinus cinerus* peroxidase: evidence for an oxygen rebound mechanism. *Eur. J. Biochem.* **258**: 906–913.

Ueda T, Sakamaki K, Kuroki T, Yano I, Nagata S. 1997. Molecular cloning and characterization of the chromosomal gene for human lactoperoxidase. *Eur. J. Biochem.* **243**: 32–41.

Wever R, Kast WM, Kasinoedin J, Boelens R. 1982. The peroxidation of thiocyanate catalyzed by myeloperoxidase and lactoperoxidase. *Biochim. Biophys. Acta* **709**: 212–219.

Wojciechowski G, Huang L, Ortiz de Montellano PR. 2005. Autocatalytic modification of the prosthetic heme of horseradish but not lactoperoxidase by thiocyanate oxidation products: a role for heme-protein covalent cross-linking. *J. Am. Chem. Soc.* **127**: 15871–15879.

Zbylut SD, Kincaid JR. 2002. Resonance Raman evidence for protein-induced out-of-plane distortion of the heme prosthetic group of mammalian lactoperoxidase. *J. Am. Chem. Soc.* **124**: 6751–6758.

CHAPTER 21

CHLOROPEROXIDASE FROM *C. FUMAGO*

INTRODUCTION

Heme-containing chloroperoxidase from *Caldariomyces fumago* is a unique member of the peroxidase superfamily of enzymes. As its name implies it can catalyze chlorination reactions, and it has recently been shown to be a dehalogenase (Osborne et al., 2006, 2007). It can catalyze enantioselective rections, and it also has considerable catalase-like (catalatic) activity (Manoj and Hager, 2008). For a summary of its reactions see Table 25.1.

Chloroperoxidase is not related to the Welinder Classes I to III of the plant, fungal, and bacterial peroxidases. It shares, in common with the cytochromes P450, a thiolate proximal iron ligand. The common abbreviation for chloroperoxidase is CPO. I once tried to change this to ClP, with limited success (Dunford and Stillman, 1976; Samokyszyn and Ortiz de Montellano, 1991), since otherwise the "O" ending is reserved for mammalian peroxidases such as LPO, MPO, and TPO.

HISTORY

The story of chloroperoxidase begins with a single report (Clutterbuck et al., 1940), a survey of chlorine metabolism by molds, which showed the unusual activity of an enzyme in *Caldariomyces fumago*. Nineteen years later an intensive investigation was started in the laboratory of Lowell Hager, whose name is now universally associated with work on chloroperoxidase. He initially called the enzyme system β-ketoadipate chlorinase (Shaw and Hager, 1959a, 1959b). This multienzyme system contained glucose oxidase which, along with glucose, provided the hydrogen peroxide necessary

Peroxidases and Catalases: Biochemistry, Biophysics, Biotechnology, and Physiology, Second Edition.
By H. Brian Dunford

for the reaction of chloroperoxidase (Shaw and Hager, 1961). Chloroperoxidase is a monomeric glycoprotein of molecular weight ~42,000 that contains ferriprotoporphyrin IX (heme b) as its prosthetic group. It was obtained in pure crystalline form (Morris and Hager, 1966), but 29 more years elapsed before the crystal structure was published (Sundaramoorthy et al., 1995).

Important contributions to the preparation and properties of chloroperoxidase have also been made by the group of Michael Pickard. Replacement of a glucose-malt growth medium for *Caldariomyces fumago* with a fructose medium led to faster production with less pigmentation in the medium (Pickard, 1981; Pickard and Hashimoto, 1982, 1987; Hashimoto and Pickard, 1984). Multiple forms exist differing only in carbohydrate composition and orientation. Batch to continuous production was developed (Carmichael et al., 1986; Carmichael and Pickard, 1989; Blanke et al., 1989). A colorimetric assay for immobilized chloroperoxidase was developed using *N,N,N′,N′*-tetramethyl-p-phenylenediamine (TMPD) as reducing substrate (Kadima and Pickard, 1989).

It was shown that chloroperoxidase can oxidize chloride, bromide, and iodide, and halogenate β-keto acids, cyclic β-diketones, and substituted phenols (Hager et al., 1966). An assay procedure was developed based on the conversion of monochlorodimedone to dichlorodimedone (Fig. 21.1). It was shown that chlorination of anisole by chloroperoxidase was a two-electron reaction, producing chlorine in a +1 oxidation state (Brown and Hager, 1967).

Reactions of chloroperoxidase were classified into four categories. First, chlorination reactions had acid pH optima, and utilized three substrates: hydrogen peroxide or other hydroperoxides, chloride ion, and a wide range of compounds subject to chlorination. Second, some substrates such as uric acid also require hydrogen

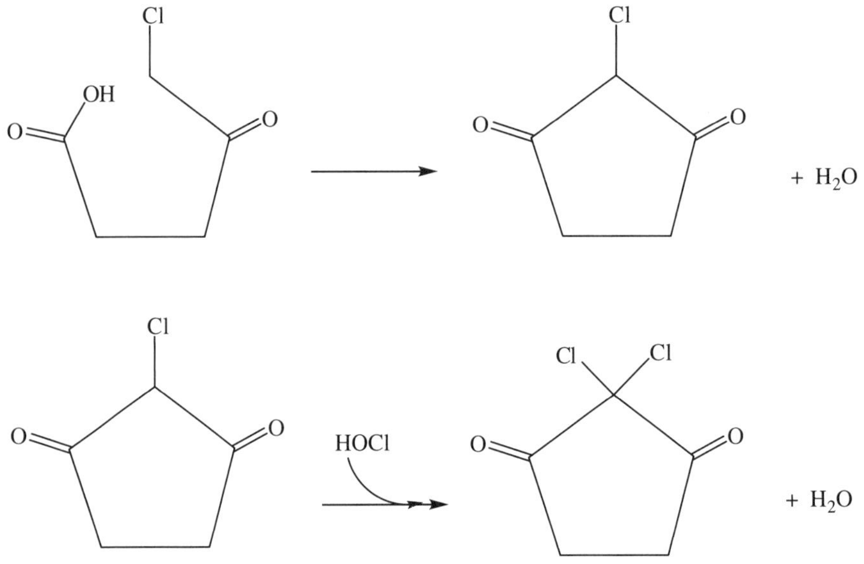

Figure 21.1. Conversion of δ-chlorolevulinic acuid to monochlorodimedone (top). Conversion of monochlorodimedone to dichlorodimedone (bottom).

peroxide and chloride ion, but the final product is oxidized, not chlorinated. Such substrates also have an acid pH optimum. Third, in the absence of chloride, classic peroxidase reactions were observed, with pH optima in the 4 to 7 range, depending on the organic substrate. Fourth, in the absence of chloride and organic substrate, the disproportionation of hydrogen peroxide was observed to produce water and dioxygen, the catalatic reaction (Thomas et al., 1970a). Chloroperoxidase was shown to possess one of the properties of cytochrome P450, the 450 nm peak in the spectrum of the ferrous-CO complex (Makino et al., 1976). Redox potentials for ferric and ferrous chloroperoxidase were measured.

OPTICAL SPECTRA

Chloroperoxidase forms conventional Compounds I and II with spectra of similar form as those for HRP (Thomas et al., 1970b). Later, the spectra of native enzyme and Compound I were optimized and it was shown that the Compound I spectrum is unique for peroxidases (Palcic et al., 1980). The native enzyme has a broad Soret peak with a maximum at 408 nm. A molar absorptivity of 86 $mM^{-1} cm^{-1}$ for native chloroperoxidase was used as a reference for the Compound I spectrum. Visible region peaks of native enzyme are at 515, 544, 585, and 652 nm. The RZ value (A400/A280 nm) for pure chloroperoxidase is 1.44. Upon Compound I formation, using peracetic acid to avoid the catalatic reaction (see below), the Soret maximum is shifted dramatically to 367 nm, with the molar absorptivity decreasing to 49 $mM^{-1} cm^{-1}$. The dominant visible region peak for Compound I occurs at 688 nm, with a molar absorptivity of 11.7 $mM^{-1} cm^{-1}$. All of the above spectral data were obtained at pH 3.5. In later work the Soret maximum for native chloroperoxidase at pH 3.0 was found at 399 nm, with a molar absorptivity of 91 $mM^{-1} cm^{-1}$. Visible region peaks at 515 and 650 nm were reported (Hollenberg and Hager, 1973). Further accurate work to determine molar absorptivities of chloroperoxidase and its compounds as a function of pH would be welcome.

The spectrum of Compound II shows peaks at 438, 542, and 571 nm, and of Compound III at 432, 555, and 586 nm. Compounds II and III were prepared by successive additions of hydrogen peroxide to native chloroperoxidase (Nakajima et al., 1985). Thus twin peaks in the visible region are shifted upon going from Compound II to III, as in other peroxidases.

Compound III was also prepared by oxygen addition to dithionite-reduced chloroperoxidase. The ferrous enzyme has peaks at 408 and 550 nm (Sono et al., 1985; Lambeir and Dunford, 1985). The spectral results for Compound III from three different laboratories, different preparation methods, and different temperatures, are in close agreement. Compound III has peaks at 430 ± 2, 555 ± 1, and 587 ± 1 nm. Molar absorptivities for the three peaks are 94, 16.5, and 12.5 $M^{-1} cm^{-1}$, obtained at $-30°C$ (Sono et al., 1985). A peak at 354 nm was also reported in the latter study.

Chloroperoxidase undergoes complicated reactions as the pH is raised. In the region of pH 7 small increases in pH lead to dramatic changes involving four or more ionizable groups, and the enzyme is inactivated. A further transition occurs at pH 11 (Lambeir and Dunford, 1983a). These complicated alkaline transitions were further investigated for both the ferric and ferrous enzymes using electronic absorption, magnetic circular dichroism, and electron paramagnetic resonance spectroscopy (Blanke et al., 1996).

The cytochrome P450 enzymes got their name from the unusual Soret region spectra of their ferrous-carbon monoxide complexes with maxima at ~450 nm. The same maximum for the chloroperoxidase ferrous-CO complex immediately showed that the enzyme was different from other peroxidases, where the corresponding maxima are found at ~420 nm (Hollenberg and Hager, 1973). It also showed that there might be a structural relationship between chloroperoxidase and the P450s, which gradually led to the conclusion that the iron fifth coordination position is occupied by the same axial thiolate ligand (Dawson et al., 1976; Dawson and Sono, 1987; Sono et al., 1984, 1985, 1986; Blanke and Hager, 1988). By the time that the crystal structure of chloroperoxidase showed that the sulfur of Cys29 is coordinated to iron (Sundaramoorthy et al., 1995), the existence of the thiolate ligand had been proven many times over. The activation volume $\Delta V^{\ddagger}$ for CO binding is unusually small when the axial ligand is Cys (Lange et al., 1994).

Ferrous peroxidase at room temperature is high spin. The one-electron reduction of chloroperoxidase in frozen glasses at 77 K by electrons generated by ^{60}Co γ rays is a mixture of low- and high-spin iron(II) (Metodiewa and Dunford, 1990).

ESR, ENDOR, MÖSSBAUER, EXAFS, AND RESONANCE RAMAN SPECTRA

Ferric and ferrous forms of chloroperoxidase were investigated using Mössbauer spectroscopy (Champion et al., 1973). Similarities between chloroperoxidase and cytochrome P450 were noted, both in Mössbauer and resonance Raman spectra (Champion et al., 1975; Remba et al., 1979). By use of ^{34}S and ^{13}CO labeling the same thiolate fifth ligand as in P450 was shown by resonance Raman results to occur in chloroperoxidase (Bangcharoenpaurpong et al., 1986).

Spin states of chloroperoxidase in the presence and absence of several ligands were determined from ESR results (Hollenberg et al., 1980). An ENDOR study showed water occupies the sixth coordination position of the heme iron at low temperature (Fann et al., 1994). Chloroperoxidase will catalyze epoxidation reactions (Manoj et al., 1999). With allylbenzene as substrate the enzyme is inactivated by the epoxidation. ENDOR studies showed the allylbenzene becomes covalently bound to the heme (Lee et al., 1997). Fe-N and Fe-S distances in ferric chloroperoxidase were determined by EXAFS spectroscopy (Cramer et al., 1978).

INVESTIGATIONS OF COMPOUNDS I AND II

Chloroperoxidase Compound I has an ESR signal similar to that in horseradish peroxidase but stronger porphyrin π-cation radical coupling to the heme iron is indicated (Rutter and Hager, 1982). These studies were extended using Mössbauer spectroscopy

(Rutter et al., 1984). An intensive study of optical and resonance Raman spectra of Compounds I and II showed that they can be categorized as hyperporphyrins (Egawa et al., 2001).

By use of x-ray absorption and resonance Raman spectroscopies it has been shown that the Fe–O bond length in Compound II is exceptionally long (Green et al., 2004), indicating that the structure is Fe–O–H (Stone et al., 2005, 2006), a conclusion that is widely accepted. The same is not true of claims of an Fe^{IV}—OH structure in horseradish peroxidase Compound II. See the discussion in Chapter 8 on resonance Raman spectra.

STRUCTURE OF COMPOUND I AND THE CATALATIC REACTION

Reaction of chloroperoxidase with *m*-chloroperbenzoic acid showed that one atom of oxygen is incorporated into Compound I. By analogy with other peroxidases the Compound I structure must contain $Fe^{IV}{=}O$ and a porphyrin π-cation radical. By use of double labeling ^{18}O experiments with both hydrogen peroxide and *m*-chloroperbenzoic acid, the catalatic reaction was shown to occur by a nonscrambling mechanism (Hager et al., 1972). Using three different hydroperoxides ROOH, the same Compound I was obtained.

At pH 4.7 the rate constants for Compound I formation were $2.7 \times 10^6\ M^{-1}\,s^{-1}$ for hydrogen peroxide, $6.0 \times 10^5\ M^{-1}\,s^{-1}$ for methyl hydroperoxide, and $3.8 \times 10^6\ M^{-1}\,s^{-1}$ for peracetic acid. Rates of reaction with Compound I to form O_2 and ROH were $3.4 \times 10^5\ M^{-1}\,s^{-1}$ for hydrogen peroxide, $2.9 \times 10^5\ M^{-1}\,s^{-1}$ for methyl hydroperoxide, and 0 for peracetic acid. Thus peracetic acid can be used to obtain a pure Compound I spectrum (Araiso et al., 1981).

Accurate bond lengths in Compound I were determined by EXAFS (Stone et al., 2005). The Fe=O bond length is 1.65 Å, indicative of a true ferryl species. The Fe—S proximal bond length is an exceptionally long 2.48 Å. These measurements are relevant to the elusive Compound I of cytochrome P450. Compound I formation of chloroperoxidase has been examined in a QM/MM theoretical study (Chen et al., 2009). The results confirmed a Compound 0 intermediate and the heterolytic cleavage of the hydrogen peroxide O—O bond. An activation energy of 12.4 kcal/mol was calculated for Compound 0 formation. Thus, the overall activation energy barrier for Compound I formation must be at least this high. Experimental results on horseradish peroxidase Compound I formation indicate a much lower activation energy.

At pH 6.2 and 25°C, saturation kinetics were observed with hydrogen peroxide as the only substrate, and the kinetic data provided quantitative evidence for a Compound I-hydrogen peroxide complex.

$$\text{CPO} + H_2O_2 \xrightarrow{k_1} \text{CPO-I} + H_2O$$

$$\text{CPO-I} + H_2O_2 \xrightarrow{k_2} \text{CPO-I-}H_2O_2$$

$$\text{CPO-I-}H_2O_2 \xrightarrow{k_3} \text{CPO} + O_2 + H_2O$$

Values of the rate constants, obtained are $k_1 = (2.3 \pm 0.1) \times 10^6\,M^{-1}\,s^{-1}$, $k_2 = (3.4 \pm 0.1) \times 10^5\,M^{-1}\,s^{-1}$, and $k_3 = (8.0 \pm 0.2) \times 10^2\,M^{-1}\,s^{-1}$. Depending on reaction conditions the percentage of the total enzyme in the form of the Compound I—H_2O_2 complex ranged from 13% to 89%. In the presence of *N,N,N′,N′*-tetramethyl-*p*-phenylenediamine (TMPD) and hydrogen peroxide, the catalatic reaction persists and cannot be eliminated, a result expected to be applicable to reactions of all conventional reducing substrates. Because of the catalatic reaction, pure Compound I of CPO cannot be obtained using hydrogen peroxide (Sun et al., 1994).

In the presence of both hydrogen peroxide and chloride at pH 2.85 some singlet oxygen is formed, as detected by the $^1\Delta_g \rightarrow {}^3\Sigma_g^-$ emission at 1.30 μm (Khan et al., 1983). Ferric chloroperoxidase reacts with superoxide to form Compound III which decays to Compound II. Thus chloroperoxidase acts as a superoxide scavenger, but not as efficiently as superoxide dismutase (Metodiewa et al., 1989).

LIGAND BINDING

Optical spectral data are available for a large number of ligands bound to chloroperoxidase (Morris and Hager, 1966; Sono et al., 1986), Cyanide binds reversibly to a basic form of chloroperoxidase with an "on" rate constant of $(5.2 \pm 0.5) \times 10^4$ $M^{-1}\,s^{-1}$ and an "off" constant of $(5.0 \pm 1.4)\,s^{-1}$. Nitrate and cyanide compete for the same binding site on iron (Lambeir et al., 1983a). A pK_a of the enzyme active site acid group of ~2.0 was indicated by data in the absence of nitrate. Transition state $\Delta V^{\ddagger}$, $\Delta H^{\ddagger}$, and $\Delta S^{\ddagger}$ parameters were obtained by studying cyanide binding as a function of pressure and temperature (Lambeir et al., 1983b).

The binding of over 20 different ligands to chloroperoxidase was studied from pH 2 to 9. Competitive experiments with cyanide showed they bind to iron. Weakly acidic ligands bind in their protonated form, whereas strongly acidic ligands such as nitric acid bind in their dissociated form. An active site pK_a value of <2 is indicated. The results showed that chloroperoxidase, although it has some properties in common with cytochrome P450, clearly belongs in the peroxidase family (Sono et al., 1986).

The binding of azide to chloroperoxidase is too fast to be measured on a stopped-flow apparatus. Data at 9.5°C and ionic strength 0.4 M were obtained using an iodine laser temperature jump apparatus. The results are shown in Figure 21.2.

The pH dependence of the "on" rate constant provides an accurate measure of the dissociation constant of hydrazoic acid, 2.5×10^{-5} M. The "off" rate is pH independent. The mechanism is binding of HN_3 by chloroperoxidase with the proton binding to a distal basic group and N_3^- to Fe^{III}, with recombination of H^+ and N_3^- for dissociation. The alternative binding mechanism involving the protonated enzyme combining with N_3^- is excluded since the rate constant exceeds the diffusion controlled limit. Measurements in H_2O/D_2O showed the "on" rate constant ratio k_H/k_D was 1.16 and "off" 1.7. Therefore, proton transfer is partially rate controlling, at least for the dissociation (Holzwarth et al., 1988).

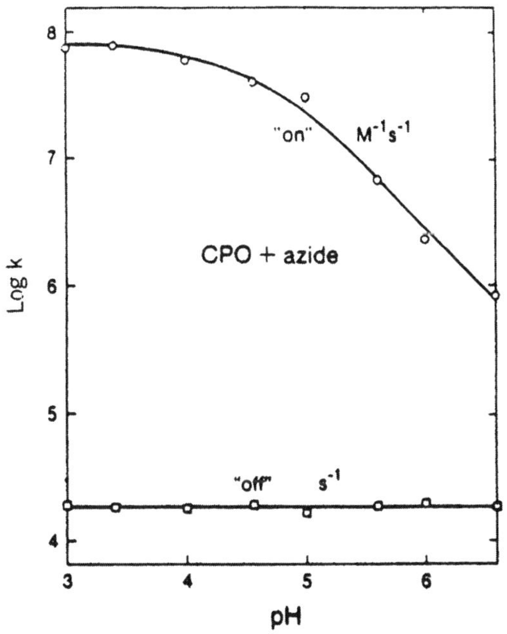

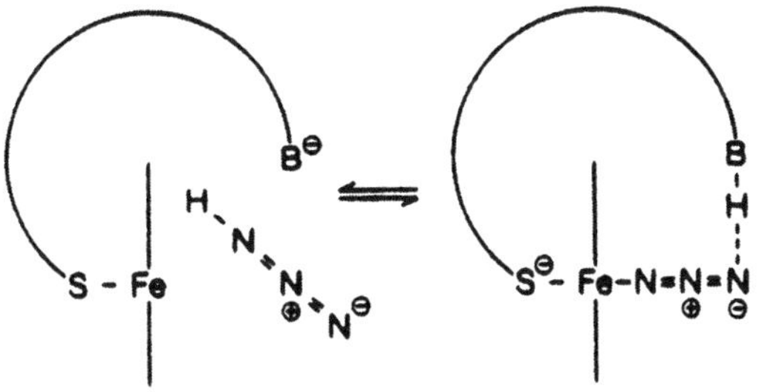

Figure 21.2. pH rate profiles for the binding (on) and dissociation (off) of azide with chloroperoxidase. (Reproduced with permission from Holzwarth et al., 1988. *Biochemistry* **27**: 6628–6633.)

KINETICS AND MECHANISMS OF CHLORINATION AND OXIDATION

The active site of chloroperoxidase was probed by studying the oxidations of phenylhydrazine and azide (Samokyszyn and Ortiz de Montellano, 1991). An iron-phenyl complex is formed from phenylhydrazine which can change to *N*-phenyl heme isomers. Thus the heme pocket is much more accessible than in horseradish peroxidase, in accord with the fact that oxygen atom transfers (monooxygenase reactions) are more common for chloroperoxidase (Osborne et al., 2007). Azide is oxidized to the azidyl radical. Loss of heme absorbance occurs and δ-mesoazidoheme is formed.

The kinetics of the chlorination of monochlorodimedone to the dichloro compound has been studied exhaustively (Lambeir and Dunford, 1983b; Dunford et al., 1987). Peracetic acid was used to generate Compound I so the catalatic reaction was eliminated. A transient chlorinating intermediate, formed upon reaction with Compound I, is responsible for the chlorination reaction. This was shown as an Fe=O—Cl species, but because of its transient nature, its structure will be difficult to prove. Its nature is also of possible relevance to the chlorinating mechanism of myeloperoxidase. A mechanism involving free hypochlorous acid as the chlorinating agent of monochlorodimedone was eliminated by the kinetic data.

Chloride binds reversibly to the Fe^{III} of native enzyme, removing some enzyme from the chlorinating cycle. The relevant rate constants at pH 3.2 are as follows: k_1 for Compound I formation from chloroperoxidase and peracetic acid, 2.3×10^6 $M^{-1} s^{-1}$; k_2 for chloride addition to Compound I, 9×10^4 $M^{-1} s^{-1}$; and k_3 for the chlorination reaction 2.4×10^7 $M^{-1} s^{-1}$. Therefore, the rate constant for the reaction of the chlorinating intermediate with monochlorodimedone is more than two orders of magnitude faster than for its formation. The results indicate monochlorodimedone binding adjacent to the chlorinating active site. If a poorly oriented or nonbinding substrate is used, then release of free hypochlorous acid would be expected to occur.

In summary, an enzyme-assisted chlorination occurs if the substrate binds appropriately (Libby et al., 1982, 1996). Otherwise, HOCl dissociates from the enzyme and becomes the chlorinating reagent. There appears little possibility that Cl_2 plays a role.

N-Dealkylation reactions are faster with chloroperoxidase compared to cytochrome P450. Aminium radical intermediates are formed in a one-electron reaction (Okazaki and Guengerich, 1993; Guengerich et al., 1995). With *N,N*-dimethylaniline the products are the same as in the horseradish peroxidase reaction: *N*-methylaniline and formaldehyde (Kedderis et al., 1980).

The breadth of chloroperoxidase substrates is illustrated in the following examples. Enantioselective reactions with sulfides were observed (Collonna et al., 1992; Fu et al., 1992). Disubstituted alkenes are epoxidized enantioselectively (Alain et al., 1993; Lakner et al., 1997). With *N*-acetyltyrosine as substrate only the L isomer is oxidized (Casella et al., 1994).

Halogenation reactions appear not to be stereoselective (Ramakrishnan et al., 1983). 4-Chloroaniline is oxidized to 4-chloronitrosobenzene by chloroperoxidase. In the presence of chloride or bromide both oxidation and ring halogenation occur, with the latter dominant below pH 3.5 (Corbett et al., 1978, 1979, 1980). Indole is oxidized to oxindole (Corbett and Chipko, 1979). Steroids, alkynes, and cyclopropanes can be halogenated by chloroperoxidase (Neidleman et al., 1966; Geigert et al., 1983). Antipyrene is both oxidized and chlorinated (Ashley and Griffin, 1981). Thiols and disulfides are oxidized by chloroperoxidase with sulfenyl chlorides as the final products (Silverstein and Hager, 1974). Phenol oxidation by chloroperoxidase was studied as a possible method of wastewater purification (Carmichael et al., 1985). A large number of phenols were studied as substrates. Those that react bind in the heme pocket (Casella et al., 1994). Oxidation reactions with substrates such as phenols proceed in one-electron steps, as for horseradish peroxidase. Chloroperoxidase appears to prefer the dimeric form of 4,6-dimethyldibenzothiphene as a substrate (Torres and Aburto, 2005).

Chloroperoxidase also catalyzes oxidative dehalogenation of chlorophenols (Osborne et al., 2006, 2007). Both Compounds I and II undergo one-electron reactions to form phenoxy radicals as initiation steps. Peroxidatic reactions appear to occur at the heme edge, whereas monoxygenase reactions require that substrate binds close to the Fe=O group (Manoj and Hager, 2008).

AMINO ACID SEQUENCE AND CRYSTAL STRUCTURE

The sequencing of the cDNA of chloroperoxidase (Fang et al., 1986; Kenigsberg et al., 1987; Nuell et al., 1988), followed by determination of the crystal structure, were major triumphs (Sundaramoorthy et al., 1995). Chloroperoxidase has two isomeric forms (Sae and Cunningham, 1979; Pickard and Hashimoto, 1982), one more heavily glycosylated than the other. The crystal structure was determined for the more heavily glycosylated isoenzyme. Cys29 is the proximal ligand. One of the biggest surprises was that the side chain of Glu183 is close to the iron on the distal side, and must replace the acid-base catalytic role of a distal imidazole side chain found in most heme peroxidases. The arrangement of helices in chloroperoxidase is totally different. Another surprise was a heavy metal ion binding site, believed to be for manganese. Two-dimensional NMR spectra of chloroperoxidase supplement crystallographic information (Wang et al., 2003).

Chloroperoxidase from *Caldariomyces fumago* has been expressed in *Aspergillus niger* (Conesa et al., 2001). The proximal thiolate ligand of chloroperoxidase has been replaced with a histidine residue. Resultant characterization led to the conclusion that thiolate is not the dominant factor in determining the unusual properties of chloroperoxidase (Yi et al., 1999). A non-heme chloroperoxidase obtained from *Curvularia inequalis* was isolated and characterized (Liu et al., 1987).

REFERENCES

Alain EJ, Hager LP, Deng L, Jacobsen EN. 1993. Highly enantioselective epoxidation of disubstituted alkenes with hydrogen peroxide catalyzed by chloroperoxidase. *J. Am. Chem. Soc.* **115**: 4415–4416.

Araiso T, Rutter R, Palcic MM, Hager LP, Dunford HB. 1981. Kinetic analysis of Compound I formation and the catalatic activity of chloroperoxidase. *Can. J. Biochem.* **59**: 233–236.

Ashley PL, Griffin BW. 1981. Chloroperoxidase-catalyzed halogenation of antipyrene, a drug substrate of liver microsomal cytochrome P-450. *Arch. Biochem. Biophys.* **210**: 167–178.

Bangcharoenpaurpong O, Champion PM, Hall KS, Hager LP. 1986. Resonance Raman studies of isotopically labeled chloroperoxidase. *Biochemistry* **25**: 2374–2378.

Blanke SR, Hager LP. 1988. Identification of the fifth axial heme ligand of chloroperoxidase. *J. Biol. Chem.* **263**: 18739–18743.

Blanke SR, Martinis SA, Sligar SG, Hager LP. 1996. Probing the heme iron coordination structure of alkaline chloroperoxidase. *Biochemistry* **35**: 37–43.

Blanke SR, Yi S, Hager LP. 1989. Development of semi-continuous and continuous flow bioreactors for the high level production of chloroperoxidase. *Biotech. Lett.* **11**: 769–774.

Brown FS, Hager LP. 1967. Chloroperoxidase. IV. Evidence for an ionic electrophilic substitution mechanism. *J. Am. Chem. Soc.* **89**: 719–720.

Carmichael RD, Fedorak PM, Pickard MA. 1985. Oxidation of phenols by chloroperoxidase. *Biotech. Lett.* **7**: 289–294.

Carmichael RD, Jones A, Pickard MA. 1986. Semicontinuous and continuous production of chloroperoxidase by *Caldariomyces fumago* immobilized in k-carrageenan. *Appl. Environ. Microbiol.* **51**: 276–280.

Carmichael RD, Pickard MA. 1989. Continuous and batch production of chloroperoxidase by mycelial pellets of *Caldariomyces fumago* in an airlift fermentor. *Appl. Environ. Microbiol.* **55**: 17–20.

Casella L, Poli S, Gullotti M, Selaggini C, Beringhelli T, Marchesini A. 1994. The chloroperoxidase-catalyzed oxidation of phenols: mechanism, selectivity, and characterization of enzyme-substrate complexes. *Biochemistry* **33**: 6377–6386.

Champion PM, Chiang R, Münck E, Debrunner PG, Hager LP. 1975. Mössbauer investigations of high-spin ferrous heme proteins. II. Chloroperoxidase, horseradish peroxidase, and hemoglobin. *Biochemistry* **14**: 4159–4166.

Champion PM, Münck E, Debrunner PG, Hollenberg PF, Hager LP. 1973. Mössbauer investigations of chloroperoxidase and its halide complexes. *Biochemistry* **12**: 426–435.

Champion PM, Remba RD, Chiang R, Fitchen DB, Hager LP. 1976. Resonance Raman spectra of chloroperoxidase. *Biochim. Biophys. Acta* **446**: 486–492.

Chen H, Hirao H, Derat E, Schlichting I, Shaik S. 2009. Quantum mechanical/molecular mechanical study on the mechanisms of Compound I formation in the catalytic cycle of chloroperoxidase: an overview of the heme enzymes. *J. Phys. Chem. B.* **112**: 9490–9500.

Clutterbuck PW, Mukhpadhyay SL, Oxford AE, Raistrick H. 1940. Studies in the biochemistry of microorganisms. (A) A survey of chlorine metabolism by molds. (B) Caldariomycin, $C_5H_8O_2Cl_2$, a metabolic product of Caldariomyces fumago woronichin. *Biochem. J.* **34**: 664–677.

Collonna S, Gaggero N, Casella L, Carrea G, Pasta P. 1992. Chloroperoxidase and hydrogen peroxide: an efficient system for enzymatic enantioselective sulfoxidations. *Tetrahedron Asymmetry* **3**: 95–106.

Conesa A, van de Velde F, van Rantwijk F, Sheldon RA, van den Hondel CAMJJ, Punt PJ. 2001. Expression of the *Caldariomyces fumago* chloroperoxidase in *Aspergillus niger* and characterization of the recombinant enzyme. *J. Biol. Chem.* **276**: 17635–17640.

Corbett MD, Baden DG, Chipko BR. 1979. Arylamine oxidations by chloroperoxidase. *Bioorg. Chem.* **8**: 91–95.

Corbett MD, Chipko BR. 1979. Peroxidase oxidation of indole to oxindole by chloroperoxidase catalysis. *Biochem. J.* **183**: 269–276.

Corbett MD, Chipko BR, Baden DG. 1978. Chloroperoxidase-catalysed oxidation of 4-chloroaniline to 4-chloronitrosobenzene. *Biochem. J.* **175**: 353–360.

Corbett MD, Chipko BR, Batchelor AO. 1980. The action of chloride peroxidase on 4-chloroaniline: N-oxidation and ring halogenation. *Biochem. J.* **187**: 893–903.

Cramer SP, Dawson JH, Hodgson KO, Hager LP. 1978. Studies on the ferric forms of cytochrome P-450 and chloroperoxidase by extended X-ray absorption fine structure: characterization of the Fe-N and Fe-S distances. *J. Am. Chem. Soc.* **100**: 7282–7290.

Dawson JH, Sono M. 1987. Cytochrome P-450 and chloroperoxidase: thiolate-ligated heme enzymes. Spectroscopic determination of the active site structures and mechanistic implications of thiolate ligation. *Chem. Rev.* **87**: 1255–1276.

Dawson JH, Trudell JR, Barth G, Linder RE, Bennenberg E, Djerassi C, Chiang R, Hager LP. 1976. Chloroperoxidase: evidence for P-450 type heme environment from magnetic circular dichroism spectroscopy. *J. Am. Chem. Soc.* **98**: 3709–3710.

Dunford HB, Lambeir A-M, Kashem MA, Pickard MA. 1987. On the mechanism of chlorination by chloroperoxidase. *Arch. Biochem. Biophys.* **252**: 292–302.

Dunford HB, Stillman JS. 1976. On the function and mechanism of action of peroxidases. *Coord. Chem. Rev.* **19**: 187–251.

Egawa T, Proshlyakov DA, Miki H, Makino R, Ogura T, Kitagawa T, Ishimura Y. 2001. Effects of thiolate axial ligand on the π-π^* electronic states of oxyferryl porphyrins: a study of the optical and resonance Raman spectra of Compounds I and II of chloroperoxidase. *J. Biol. Inorg. Chem.* **6**: 46–54.

Fang G-H, Kenibsberg P, Axley MJ, Nuell M, Hager LP. 1986. Cloning and sequencing of chloroperoxidase cDNA. *Nucleic Acids Res.* **14**: 8061–8071.

Fann Y-C, Gerber NC, Osmulski PA, Hager LP, Sligar SG, Hoffman BM. 1994. ENDOR determination of heme ligation in chloroperoxidase and comparison with cytochrome P-450_{cam}. *J. Am. Chem. Soc.* **116**: 5989–5990.

Fu H, Kondo H, Ichikawa Y, Look GC, Wong C-H. 1992. Chloroperoxidase-catalyzed asymmetric synthesis: enantioselective reactions of chiral hydroperoxides with sulfides and bromohydration of glycals. *J. Org. Chem.* **57**: 7265–7270.

Geigert J, Neidleman SL, Dalietos DJ. 1983. Novel haloperoxidase substrates: alkynes and cyclopropanes. *J. Biol. Chem.* **258**: 2273–2277.

Green MT, Dawson JH, Gray HB. 2004. Oxoiron(IV) in chloroperoxidase is basic: implications for P450 chemistry. *Science* **304**: 1653–1656.

Guengerich FP, Okazaki O, Seto Y, MacDonald TL. 1995. Radical cation intermediates in N-dealkylation reactions. *Xenobiotica* **25**: 689–709.

Hager LP, Doubek DL, Silverstein RM, Hargis JH, Martin JC. 1972. Chloroperoxidase. IX. The structure of Compound I. *J. Am. Chem. Soc.* **94**: 4364–4366.

Hager LP, Morris DR, Brown FS, Eberwein H. 1966. Chloroperoxidase. II. Utilization of halogen anions. *J. Biol. Chem.* **241**: 1769–1777.

Hashimoto A, Pickard MA. 1984. Chloroperoxidases from Caldariomyces (= Leptoxyphium) cultures: glycoproteins with variable carbohydrate content and isoenzymic forms. *J. Gen. Microbiol.* **130**: 2051–2058.

Hollenberg PF, Hager LP. 1973. The P-450 nature of the carbon monoxide complex of ferrous chloroperoxidase. *J. Biol. Chem.* **248**: 2630–2633.

Hollenberg PF, Hager LP, Blumberg WE, Peisach J. 1980. An electron paramagnetic resonance study of the high and low spin forms of chloroperoxidase. *J. Biol. Chem.* **255**: 4801–4807.

Holzwarth JF, Meyer F, Pickard MA, Dunford HB. 1988. Mechanism of azide binding to chloroperoxidase and horseradish peroxidase using an iodine laser temperature-jump apparatus. *Biochemistry* **27**: 6628–6633.

Kadima TA, Pickard MA. 1989. A colorimetric assay for immobilized chloroperoxidase. *Can. J. Microbiol.* **36**: 302–304.

Kedderis GL, Koop DR, Hollenberg PF. 1980. N-Demethylation reactions catalayzed by chloroperoxidase. *J. Biol. Chem.* **255**: 10174–10182.

Kenigsberg P, Fang G-H, Hager LP. 1987. Post-translational modifications of chloroperoxidase from *Caldariomyces fumago*. *Arch. Biochem. Biophys.* **254**: 409–415.

Khan AU, Gebauer P, Hager LP. 1983. Chloroperoxidase generation of singlet Δ molecular oxygen observed directly by spectroscopy in the 1–1.6-μm region. *Proc. Natl. Acad. Sci. USA* **80**: 5195–5197.

Lakner FJ, Cain KP, Hager LP. 1997. Enantioselective epoxidation of ω-bromo-2-methyl-1-alkenes catalyzed by chloroperoxidase: effect of chain length on selectivity and efficiency. *J. Am. Chem. Soc.* **119**: 443–444.

Lambeir A-M, Dunford HB. 1983a. A kinetic and spectral study of the alkaline transitions of chloroperoxidase. *Arch. Biochem. Biophys.* **220**: 549–556.

Lambeir A-M, Dunford HB. 1983b. A steady state kinetic analysis of the reaction of chloroperoxidase with peracetic acid, chloride, and 2-chlorodimedone. *J. Biol. Chem.* **258**: 13558–13563.

Lambeir A-M, Dunford HB. 1985. Oxygen binding to dithionite-reduced chloroperoxidase. *Eur. J. Biochem.* **147**: 93–96.

Lambeir A-M, Dunford HB, Pickard MA. 1983a. Kinetics of cyanide binding to chloroperoxidase in the presence of nitrate: detection of the influence of a heme-linked acid group by shift in the apparent pK_a value. *J. Inorg. Biochem.* **19**: 291–300.

Lambeir A-M, Heremans K, Dunford HB. 1983b. High-pressure effect on the equilibrium and kinetics of cyanide binding to chloroperoxidase. *Biophys. Chem.* **18**: 195–201.

Lange R, Heiber-Langer I, Bonfils C, Fabre I, Negishi M, Balny C. 1994. Activation volume and energetic properties of the binding of CO to hemoproteins. *Biophys. J.* **66**: 89–98.

Lee H-I, Dexter AF, Fann Y-C, Lakner FJ, Hager LP, Hoffman BM. 1997. Structure of the modified heme in allylbenzene-inactivated chloroperoxidase determined by Q-band CW and pulsed ENDOR. *J. Am. Chem. Soc.* **119**: 4059–4069.

Libby RD, Beachy TM, Phipps AK. 1996. Quantitating direct chlorine transfer from enzyme to substrate in chloroperoxidase-catalyzed reactions. *J. Biol. Chem.* **271**: 21820–21827.

Libby RD, Thomas JA, Kaiser LW, Hager LP. 1982. Chloroperoxidase halogenation reactions: chemical versus enzymic halogenating intermediates. *J. Biol. Chem.* **257**: 5030–5037.

Liu TE, M'Timkulu, Geigert J, Wolf B, Neidleman SL, Silva D, Hunter-Cevera JC. 1987. Isolation and characterization of a novel nonheme chloroperoxidase. *Biochem. Biophys. Res. Commun.* **142**: 329–333.

Makino R, Chiang R, Hager LP. 1976. Oxidation-reduction potential measurements on chloroperoxidase and its complexes. *Biochemistry* **15**: 4748–4754.

Manoj KM, Hager LP. 2008. Chloroperoxidase, a Janus enzyme. *Biochemistry* **47**: 2997-3003.

Manoj KM, Yi X, Rai GP, Hager LP. 1999. A kinetic epoxidation assay for chloroperoxidase. *Biochem. Biophys. Res. Commun.* **206**: 301–303.

Metodiewa D, Dunford HB. 1990. One-electron reduction of chloroperoxidase by radiolytically generated electrons. *Biochem. Biophys. Res. Commun.* **168**: 1311–1317.

Metodiewa D, Pickard M, Dunford HB. 1989. The reactions of chloroperoxidase in the presence of xanthine/xanthine oxidase. *Biochem. Biophys. Res. Commun.* **159**: 1086–1092.

Morris DR, Hager LP. 1966. Chloroperoxidase. I. Isolation and properties of the crystalline glycoprotein. *J. Biol. Chem.* **241**: 1763–1768.

Nakajima R, Yamazaki I, Griffin BW. 1985. Spectra of chloroperoxidase Compounds II and III. *Biochem. Biophys. Res. Commun.* **128**. 1–6.

Neidleman SL, Diassi PA, Junta B, Palmere RM, Pan SC. 1966. The enzymatic halogenation of steroids. *Tetrahedron Lett.* **44**: 5337–5342.

Nuell MJ, Fang G-H, Axley MJ, Kenigssberg P, Hager LP. 1988. Isolation and nucleotide sequence of the chloroperoxidase gene from *Caldariomyces fumago*. *J. Bacteriol.* **170**: 1007–1011.

Okazaki O, Guengerich FP. 1993. Evidence for specific base catalysis in N-dealkylation reactions catalyzed by cytochrome P450 and chloroperoxidase. *J. Biol. Chem.* **268**: 1546–1552.

Osborne RL, Coggins MK, Terner J, Dawson JH. 2007. *Caldariomyces fumago* chloroperoxidase catalyzes the oxidative dehalogenation of chlorophenols by a mechanism involving two one-electron steps. *J. Am. Chem. Soc.* **129**: 14838–14839.

Osborne RL, Raner GM, Hager LP, Dawson JH. 2006. *C. fumago* chloroperoxidase is also a dehaloperoxidase: oxidative dehalogenation of halophenols. *J. Am. Chem. Soc.* **128**: 1036–1037.

Palcic MM, Rutter R, Araiso T, Hager LP, Dunford HB. 1980. Spectrum of chloroperoxidase Compound I. *Biochem. Biophys. Res. Commun.* **94**: 1123–1127.

Pickard MA. 1981. A defined growth medium for the production of chloroperoxidase from *Caldariomyces fumago*. *Can. J. Microbiol.* **27**: 1298–1305.

Pickard MA, Hashimoto A. 1982. Isoenzymes of chloroperoxidase from *Caldariomyces fumago*. *Can. J. Microbiol.* **28**: 1382–1388.

Pickard MA, Hashimoto A. 1987. Stability and carbohydrate composition of chloroperoxidase from *Caldariomyces fumago* grown in a fructose-salts medium. *Can. J. Microbiol.* **34**: 998–1002.

Ramakrishnan K, Oppenhuizen ME, Saunder S, Fisher J. 1983. Stereoselectivity of chloroperoxidase-dependent halogenation. *Biochemistry* **22**: 3271–3277.

Remba RD, Champion PM, Fitchen DB, Chiang R, Hager LP. 1979. Resonance Raman investigations of chloroperoxidase, horseradish peroxidase, and cytochrome *c* using Soret band laser excitation. *Biochemistry* **18**: 2280–2290.

Rutter R, Hager LP. 1982. The detection of two electron paramagnetic resonance Raman signals associated with chloroperoxidase Compound I. *J. Biol. Chem.* **257**: 7958–7961.

Rutter R, Hager LP, Dhonau H, Hendrich M, Valentine M, Debrunner P. 1984. Chloroperoxidase Compound I: electron paramagnetic resonance and Mössbauer studies. *Biochemistry* **23**: 6809–6815.

Sae ASW, Cunningham BA. 1979. Isolation and properties of chloroperoxidase isozymes. *Phytochemistry* **18**: 1785–1787.

Samokyszyn VM, Ortiz de Montellano PR. 1991. Topology of the chloroperoxidase active site: regiospecificity of heme modification by phenyl hydrazine and sodium azide. *Biochemistry* **30**: 11646–11653.

Shaw PD, Hager LP. 1959a. Biological chlorination. III. β-Ketoadipate chlorinase: a soluble enzyme system. *J. Biol. Chem.* **234**: 2565–2569.

Shaw PD, Hager LP. 1959b. Biological chlorination. IV. Peroxidative nature of enzymatic chlorination. *J. Am. Chem. Soc.* **81**: 6527–6528.

Shaw PD, Hager LP. 1961. Biological chlorination. VI. Chloroperoxidase: a component of the β-ketoadipate chlorinase system. *J. Biol. Chem.* **236**: 1626–1630.

Silverstein RM, Hager LP. 1974. The chloroperoxidase-catalyzed oxidation of thiols and disulfides to sulfenyl chlorides. *Biochemistry* **13**: 5069–5073.

Sono M, Dawson JH, Hager LP. 1984. The generation of a hyperporphyrin spectrum upon thiol binding to ferric chloroperoxidase: further evidence of endogenous thiolate ligation to the ferric enzyme. *J. Biol. Chem.* **259**: 13209–13216.

Sono M, Dawson JH, Hager LP. 1985. Phosphine binding as a structural probe of the chloroperoxidase active site: spectroscopic evidence for endogenous thiolate ligation to the heme iron. *Inorg. Chem.* **24**: 4339–4343.

Sono M, Dawson JH, Hall K, Hager LP. 1986. Ligand and halide binding properties of chloroperoxidase: peroxidase-type active site heme environment with cytochrome P-450 type endogenous axial ligand and spectroscopic properties. *Biochemistry* **25**: 347–356.

Sono M, Eble KM, Dawson JH, Hager LP. 1985. Preparation and properties of ferrous chloroperoxidase complexes with dioxygen, nitric oxide, and an alkyl cyanide. *J. Biol. Chem.* **260**: 15530–15535.

Stone KL, Behan RK, Green MT. 2005. X-ray absorption spectroscopy of chloroperoxidase compound I: insight into the reactive intermediate of P450 chemistry. *Proc. Natl. Acad. Sci. USA* **102**: 16563–16565.

Stone KL, Behan RK, Green MT. 2006. Resonance Raman spectroscopy of chloroperoxidase compound II provides direct evidence for the existence of an iron(IV)-hydroxide. *Proc. Natl. Acad. Sci. USA* **103**: 12307–12310.

Sun W, Kadima TA, Pickard MA, Dunford HB. 1994. Catalase activity of chloroperoxidase and its interaction with peroxidase activity. *Biochem. Cell Biol.* **72**: 321–331.

Sundaramoorthy M, Terner J, Poulos TL. 1995. The crystal structure of chloroperoxidase: a heme peroxidase-cytochrome P450 functional hybrid. *Structure* **3**: 1367–1377.

Thomas JA, Morris DR, Hager LP. 1970a. Chloroperoxidase. VII. Classical peroxidatic, catalatic, and halogenating forms of the enzyme. *J. Biol. Chem.* **245**: 3129–3134.

Thomas JA, Morris DR, Hager LP. 1970b. Chloroperoxidase. VIII. Formation of peroxide and halide complexes and their relation to the mechanism of the halogenation reaction. *J. Biol. Chem.* **245**: 3135–3142.

Torres E, Aburto J. 2005. Chloroperoxidase-catalyzed oxidation of 4,6-dimethyldibenzothiophene as dimer complexes: evidence for kinetic cooperativity. *Arch. Biochem. Biophys.* **437**: 224–232.

Wang X, Tachikawa H, Yi X, Manoj KM, Hager LP. 2003. Two-dimensional NMR study of the heme active site structure of chloroperoxidase. *J. Biol. Chem.* **278**: 7765–7774.

Yi X, Mroczko M, Manoj KM, Wang X, Hager LP. 1999. Replacement of the proximal heme thiolate ligand in chloroperoxidase with a histidine residue. *Proc. Natl. Acad. Sci. USA* **96**: 12412–12417.

CHAPTER 22

SELENIUM-CONTAINING ENZYMES: GLUTATHIONE PEROXIDASE AND IODOTHYRONINE DEIODINASE

INTRODUCTION

Selenocysteine, SeCys, is the 21st amino acid (Böck et al., 1991). Its codon for incorporation into glutathione peroxidase is TGA (Chambers et al., 1986), and into iodothyronine deiodinase UGA (Berry et al., 1991). Selenocysteine-containing proteins are widespread (Margis et al., 2008). Thioredoxin reductase is a selenium-containing enzyme (Lacey et al., 2008; Cheng et al., 2009). One unusual example is a citrus salt-stress-associated protein similar to mammalian glutathione peroxidase (Holland et al., 1993, 1994). Selenoproteins have been reviewed (Stadtman, 1991, 1996; Lu and Holmgren, 2009).

Analogs of glutathione peroxidases exist in *Trypanosoma brucei*, the cause of African sleeping sickness. These enzymes have cysteine analogs of selenocysteine (Hillebrand et al., 2003). The remainder of this chapter is restricted to mammalian glutathione peroxidase and iodothyronine deiodinase.

GLUTATHIONE PEROXIDASE

Glutathione is a tripeptide, γ-L-glutamyl-L-cysteinylglycine, present in millimolar concentrations in cells. The cysteinyl–SH side chain can act as a reductant of hydrogen peroxide when catalyzed by the enzyme glutathione peroxidase:

$$\underset{\text{Glutathinone}}{2\,\text{GSH}} + \text{H}_2\text{O}_2 \longrightarrow \underset{\substack{\text{Oxidized}\\\text{glutathione}}}{\text{GSSG}} + 2\,\text{H}_2\text{O}$$

Peroxidases and Catalases: Biochemistry, Biophysics, Biotechnology, and Physiology, Second Edition.
By H. Brian Dunford

Thus the combination of catalase and glutathione peroxidase provides excellent protection of hemoglobin in red blood cells from oxidative damage. Oxyhemoglobin is an analog of heme peroxidase Compound III, which can be a source of hydrogen peroxide, for example by reaction with ascorbate.

Glutathione peroxidases are widely distributed (Takahashi et al., 1990; Yang and Zhou, 2006). Four glutathione peroxidases, obtained from four different genes, have been identified (Takahashi and Cohen, 1986). Similar if not identical glutathione peroxidases are found in erythrocytes, leukocytes, and platelets. The glutathione peroxidase in plasma is different, as is a phospholipid glutathione peroxidase which converts harmful hydroperoxy lipids and oxidized 5-lipoxygenase to corresponding alcohols (Schnurr et al., 1996; Mauri et al., 2003). In addition, a unique glutathione peroxidase has been identified in human hepatoma cells (Chu et al., 1993).

The erythrocyte enzyme was isolated and purified (Mills, 1958) and its crystal structure has been obtained (Ladenstein et al., 1979; Epp et al., 1983). The enzyme is a tetramer composed of identical subunits. The active site in each subunit is centered on SeCys35 (Forstrom et al., 1978). The crystal structure of the erythrocyte glutathione peroxidase shows the presence of ~80 water molecules per monomer. Treatment with excess hydrogen peroxide yielded a seleninic acid:

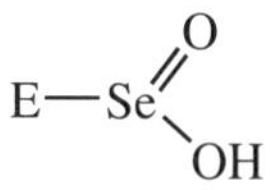

However, this is not believed to be the active oxidized form. The resting enzyme is also the starting point for the enzymatic cycle. The pK_a for dissociation of the —SeH group of selenocysteine is ~5.2; it is more acidic than the corresponding —SH group of cysteine. The ionized form $-Se^-$ may be the starting point of the glutathione peroxidase cycle, but for simplicity in balancing equations the —SeH form is shown below. The glutathione peroxidase cycle fits ping pong kinetics (Flohé et al., 1972; Chaudiere and Tappel, 1984).

A plausible reaction scheme is the following:

$$\begin{aligned} E{-}SeH + H_2O_2 &\longrightarrow ESeOH + H_2O \\ ESeOH + HSG &\longrightarrow ESe^{\bullet} + {}^{\bullet}SG + H_2O \\ ESe^{\bullet} + HSG &\longrightarrow ESeH + {}^{\bullet}SG \\ GS^{\bullet} + {}^{\bullet}SG &\longrightarrow GSSG \end{aligned}$$

By analogy with heme peroxidase cycles, a nearby amino acid side chain would play a key role in proton transfers. There are several water molecules in the active site that could be displaced (Epp et al., 1983), perhaps by the first GSH molecule to react in the cycle, providing an efficient method for GSSG formation.

As might be expected glutathione peroxidase reduces thyroid peroxidase activity, since they are competitors for hydrogen peroxide (Ekholm and Björkman, 1997).

IODOTHYRONINE DEIODINASE

This section provides a complement to Chapter 19 on thyroid peroxidase. Iodothyronine deiodinase is a selenium-containing enzyme, proven by expression cloning in the *Xenopus* oocyte. The mRNA for this enzyme contains the UGA codon for selenocysteine, necessary for maximal enzyme activity (Berry et al., 1991). The enzyme is found in the thyroid, but mainly in liver and kidney (Stadtman, 1991). It reacts in the following manner:

$$\text{SeH} + \text{IR} \longrightarrow \text{SeI} + \text{HR}$$

where SeH is iodothyronine deiodinase, and IR is amino acid-bound iodide.

By such a reaction the 5′ iodine of thyroxine is removed to form the more active hormone triiodothyronine. Reaction with reduced glutathione regenerates the reduced enzyme. The iodothyronine deiodinase has no significant homology with glutathione peroxidase (Berry et al., 1991).

Thus the amino acids cysteine and selenocysteine play key roles in iodine metabolism. Iodination of cysteine residues in thyroglobulin may be key intermediates in thyroxine synthesis and a selenocysteine residue is the key to the deiodinase reaction. Reduced glutathione is required to regenerate the resting deiodinase.

The pattern of small molecules acting as transfer agents for peroxidases appears to continue for the mammalian peroxidases. Could glutathione act as the iodine transfer reagent for thyroid peroxidase, as taurine appears to be for chlorine transfer from myeloperoxidase?

REFERENCES

Berry MJ, Banu L, Larsen PR. 1991. Type I iodothyronine deiodinase is a selenocysteine-containing enzyme. *Nature* **349**: 438–440.

Böck A, Forchhammer K, Heider J, Baron C. 1991. Selenoprotein synthesis: an expansion of the genetic code. *Trends Biochem. Sci.* **16**: 463–467.

Chambers I, Frampton J, Goldfarb P, Affara N, McBain W, Harrison PR. 1986. The structure of the mouse glutathione peroxidase gene: the selenocysteine in the active site is encoded by the "termination" codon, TGA. *EMBO J* **5**: 1221–1227.

Chaudiere J, Tappel AL. 1984. Interaction of gold(I) with the active site of selenium-glutathione peroxidase. *J. Inorg. Biochem.* **20**: 313–325.

Cheng Q, Sanadalova T, Lindqvist Y, Arnér ESJ. 2009. Crystal structure and catalysis of the selenoprotein thioredoxin reductase 1. *J. Biol. Chem.* **284**: 3998–4008.

Chu F-F, Doroshow JH, Esworthy RS. 1993. Expression, characterization and tissue distribution of a new cellular selenium-dependent glutathione peroxidase, GSHPx-GI. *J. Biol. Chem.* **268**: 2571–2576.

Ekholm R, Björkman. 1997. Glutathione peroxidase degrades intracellular hydrogen peroxide and thereby inhibits intracellular protein iodination in thyroid epithelium. *Endocrinology* **138**: 2871–2878.

Epp O, Ladenstein R, Wendel A. 1983. The refined structure of the selenoenzyme glutathione peroxidase at 0.2-nm resolution. *Eur. J. Biochem.* **133**: 51–69.

Flohé L, Loschen G, Günzler WA, Eichele E. 1972. Glutathione peroxidase. V. The kinetic mechanism. *Hoppe-Seyler's Z. Physiol. Chem.* **353**: 987–999.

Forstrom JW, Zakowski JJ, Tappel AL. 1978. Identification of the catalytic site of rat liver glutathione peroxidase as selenocysteine. *Biochemistry* **17**: 2639–2644.

Hillebrand H, Schmidt A, Krauth-Siegel RL. 2003. A second class of peroxidases linked to trypanothione metabolism. *J. Biol. Chem.* **278**: 686-09-6815.

Holland D, Ben-Hayyim G, Faltin Z, Camoin L, Strosberg AD, Eshdat Y. 1993. Molecular characterization of salt-stress-associated protein in citrus: protein and cDNA sequence homology to mammalian glutathione peroxidases. *Plant Mol. Biol.* **21**: 923–927.

Holland D, Faltin Z, Perl A, Ben-Hayyim G, Eshdat Y. 1994. A novel plant glutathione peroxidase-like protein provides tolerance to oxygen radicals generated by paraquat in *Escherichia coli*. *Plant Mol. Biol.* **21**: 923–927.

Lacey BM, Eckenroth BE, Flemer S Jr, Hondal RJ. 2008. Selenium in thioredoxin reductase: a mechanistic perspective. *Biochemistry* **47**: 12810–12821.

Ladenstein R, Epp O, Bartels K, Jones A, Huber R, Wendel A. 1979. Structure analysis and molecular model of the selenoenzyme glutathione peroxidase at 2.8 Å resolution. *J. Mol. Biol.* **134**: 199–218.

Lu J, Holmgren A. 2009. Selenoproteins. *J. Biol. Chem.* **284**: 723–727.

Margis R, Dunand C, Teixeira FK, Margis-Pinheiro M. 2008. Glutathione peroxidase family: an evolutionary overview. *FEBS J.* **275**: 3959–3970.

Mauri P, Benazzi L, Flohé L, Maiorino M, Pietta PG, Pilawa S, Roveri A, Ursini F. 2003. Versatility of selenium catalysis in PHGPx unraveled by LC/ESI-MS/MS. *Biol. Chem.* **384**: 575–588.

Mills GC. 1958. The purification and properties of glutathione peroxidase of erythrocytes. *J. Biol. Chem.* **234**: 502–506.

Schnurr K, Belkner J, Ursini F, Schewe T, Kühn H. 1996. The selenoenzyme phospholipid hydroperoxide glutathione peroxidase controls the activity of the 5-lipoxygenase with complex substrates and preserves specificity of the oxygenation products. *J. Biol. Chem.* **271**: 4653–4658.

Stadtman TC. 1991. Biosynthesis and function of selenocysteine-containing enzymes. *J. Biol. Chem.* **266**: 16257–16260.

Stadtman TC. 1996. Selenocysteine. *Annu. Rev. Biochem.* **65**: 83–100.

Takahashi K, Akasaka M, Yamamoto Y, Kobayashi C, Mizoguchi J, Koyaxna J. 1990. Primary structure of human plasma glutathione peroxidase deduced from eDNA sequences. *J. Biochem. (Tokyo)* **108**: 145–148.

Takahashi K, Cohen HJ. 1986. Selenium-dependent glutathione peroxidase protein and activity: immunological investigations on cellular and plasma enzymes. *Blood* **68**: 640–645.

Yang Z, Zhou C-Z. 2006. Purification, crystallization and preliminary X-ray analysis of glutathione peroxidase Gpx3 from *Saccharomyces cerevisiae*. *Acta Crystal.* **F62**: 593–596.

CHAPTER 23

STRUCTURE AND FUNCTION OF VANADIUM HALOPEROXIDASES

RON WEVER and ROKUS RENIRIE

SUMMARY

The structure and function of vanadium haloperoxidases are discussed in this chapter. These enzymes are very different from the heme peroxidases. They contain a bare metal oxide (vanadate) as a prosthetic group, which makes them very resistant towards oxidative inactivation. These enzymes have a narrow substrate specificity; they are only able to oxidize halides in the presence of hydrogen peroxide to hypohalous acids and catalyze sulfoxidation reactions though this catalytic activity is low.

Iodo- and bromoperoxidases found in the marine environment in several seaweeds are involved in the production of the huge amounts of brominated and iodinated compounds that have an impact on atmospheric chemistry. Vanadium chloroperoxidases (VCPO) occur in a group of common terrestrial fungi. They are involved in the decay of lignocellulose and plant debris and these enzymes have roles in the natural production of high molecular weight chloro-aromatics and in lignin breakdown. Several x-ray structures are available now highlighting the nature of the active site. In the molecular mechanism hydrogen peroxide is bound first in a side-on fashion to the vanadium and in a second reaction a nucleophilic attack occurs by an incoming halide. Interestingly the active site residues are conserved and identical to those found in acid phosphatases that hydrolyze phosphate monoesters, pointing to an evolutionary relationship between these groups of phosphatases and vanadium haloperoxidases. Indeed when the vanadate cofactor is removed from VCPO the enzyme low phosphatase activity is observed. Conversely when vanadate is substituted in acid phosphatase this artificial enzyme shows haloperoxidase activity.

Peroxidases and Catalases: Biochemistry, Biophysics, Biotechnology, and Physiology, Second Edition.
By H. Brian Dunford

ABBREVIATIONS

EPR	Electron paramagnetic resonance
EXAFS	Extended x-ray fine structure analysis
ESEEM	Electron spin echo envelope spectroscopy
VCPO	Vanadium chloroperoxidase
VBPO	Vanadium bromoperoxidase

INTRODUCTION

Vanadium is universally distributed in the soil and is the second most abundant metal in seawater. It is a structural and electronic analog of phosphate. The metal is essential for the functioning of the bacterial nitrogen fixation system, and the metal oxide is the prosthetic group in the vanadium haloperoxidases from seaweeds and in a group of fungi, the dematiaceous hyphomycetes. The first evidence for the involvement of vanadium in the vanadium-dependent haloperoxidase enzymes was obtained in 1983 (Vilter, 1984). He showed that the bromoperoxidase isolated from *Ascophyllum nodosum* was inactivated by dialysis at pH 3.8 in a citrate-phosphate buffer containing EDTA and that the enzyme was reactivated by vanadate in suitable buffers. A few years later it was demonstrated that vanadate was the prosthetic group in the bromoperoxidase and that in the low pH procedure vanadate was removed from the enzyme (de Boer et al., 1986a, 1986b). The vanadium chloroperoxidases with very similar properties as the bromoperoxidases were later detected in a family of terrestrial fungi (Van Schijndel et al., 1993). In the native form of the haloperoxidases the oxidation state of the metal is V and it can be reduced to the catalytically inactive IV state by reducing agents. As a d1 metal ion, vanadium IV has a single electron that is strongly coupled to the ^{51}V nucleus ($I = 7/2$).

In EPR spectra this gives rise to a signal of either eight or two sets of eight overlapping lines. This typical signal is easily detectable and it is possible to observe relatively low concentrations of these enzymes by EPR. Unfortunately the enzymes in their reduced state are inactive. In contrast to the heme peroxidases (see Fig. 23.1) and catalases that have a distinct and intense Soret absorption around 400 to 430 nm, the UV-VIS spectra of these vanadium enzymes show a modest absorption in the optical spectra around 315 nm due to the bound cofactor. The 315 nm shoulder is sensitive to additions of substrates and has been used as probe to monitor binding and kinetics of substrates and binding of vanadate to active-site mutants (Renirie et al., 2000a, 2000b).

Haloperoxidases catalyze a two-electron oxidation of a halide (Cl^-, Br^-, I^-) by H_2O_2 to hypohalous acid:

$$H_2O_2 + H^+ + X^- \longrightarrow HOX + H_2O \qquad (23.1)$$

This reaction may also be seen as an oxygen-transfer reaction from the peroxide to the halide ion. During catalysis a reactive and halogenating species is formed that is

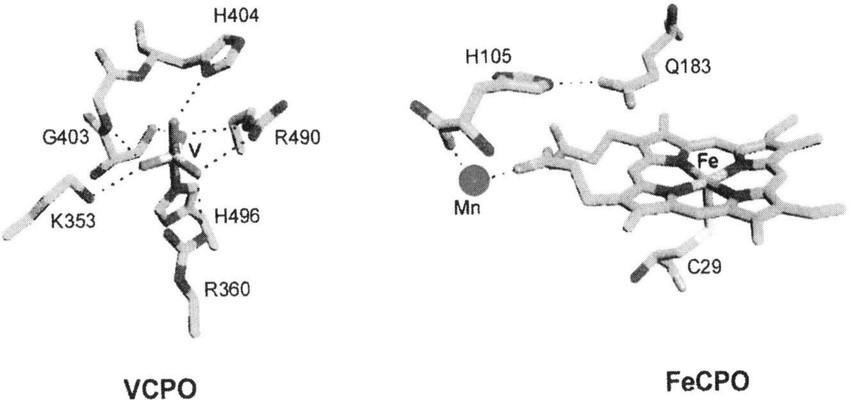

Figure 23.1. Comparison between the active sites of vanadate- and heme-containing chloroperoxidases (CPOs). VCPO from *Curvularia inaequalis* (pdb: 1idq) and the well studied FeCPO from *Caldariomyces fumago* (pdb: 1CPO).

released from the active site and may react specifically or nonspecifically with a variety of organic compounds (RH) resulting in the production of a diversity of halogenated compounds (RX; Eq. 23.2):

$$HOX + RH \longrightarrow RX + H_2O \qquad (23.2)$$

The historical nomenclature convention of the haloperoxidases is based on the most electronegative halide that these enzymes can oxidize. Thus, the chloroperoxidases can oxidize Cl^-, Br^-, and I^-; bromoperoxidases can oxidize Br^- and I^-; and iodoperoxidases oxidize only I^-. However, the distinction between the haloperoxidases is somewhat arbitrary since a bromoperoxidase may also oxidize chloride albeit with a low specificity constant.

OCCURRENCE AND BIOLOGICAL FUNCTION OF VANADIUM IODO- AND BROMOPEROXIDASES

A striking array of iodinated and brominated products produced by natural sources is found in the biosphere and in particular the marine environment is a rich source. A majority of these products are probably formed by the enzymatic activity of the enzyme family of vanadium bromo- and iodoperoxidases. Iodo- and bromoperoxidases have been found in red and brown macro-algae from all over the world (Wever and Hemrika, 2001). In the green macro-alga *Ulvella lens* a vanadium-dependent enzyme is probably also present (Ohshiro et al., 1999). Vanadium bromoperoxidases are responsible for the formation of huge amounts of bromoform and other volatile brominated compounds by macroalgae and phytoplankton (Wever et al., 1991). Oxidative stress as well as UV light result in the release of these compounds by seaweeds. Estimates (Carpenter and Liss, 2000) give a global source strength of $2.2\ 10^{11}$ g $CHBr_3$ by macro-algae. The brominated compounds are ventilated to the atmosphere and may reach the lower region of the stratosphere in which they may play a natural role in the regulation of the ozone concentration (Simpson et al., 2007 and references therein). Thus, macro-algae have a dominant role in producing bromoform that effects the environment. In some seaweed species the iodo- and bromoperoxidases are extracellularly located and when these seaweeds are triggered by changes in the environment they produce HOBr. This reactive compound has a direct antimicobial effect (Hansen et al., 2003; Renirie et al., 2008) and as shown (Borchardt et al., 2001) the oxidized halogen species inactivate homoserine lactones. Homoserine lactones play an important role as communication signals between bacteria and as a result of inactivation of these lactones biofilm formation and fouling by bacteria and fungi of the surface of the seaweeds is prevented.

The brown seaweeds of the genus *Laminaria* (kelp) are the most efficient iodine accumulators among all living species. In coastal regions Laminariales are the major contributor to the iodine flux from the oceans through the production of volatile iodocarbons, and the iodoperoxidases that are found in these species play a prime role in the production of iodine and iodocarbons (Leblanc et al., 2006). Again oxidative stress as well as UV light induce a release of volatile iodocarbons and molecular iodine by *Laminari digitata* (Palmer et al., 2005; Laturnus et al., 2004). Interestingly, iodocarbons and iodine-containing particles produced in coastal areas provide condensation nuclei for coastal cloud formation (O'Dowd et al., 2002; McFiggans et al., 2004). Why these seaweeds produce these iodinated compounds is not clear. There is considerable debate as to the role of the iodoperoxidases in the *Laminaria* species and the involvement in iodide accumulation. Recently it was proposed that iodide accumulation in kelp constitutes an extracellular protection against oxidative stress (Kupper et al., 2008).

OCCURRENCE AND BIOLOGICAL FUNCTION OF VANADIUM CHLOROPEROXIDASES

During the course of screening for haloperoxidases with tolerance to high pH environments and high temperatures it was observed that dematiaceous hyphomycetes isolated from plant material and soils collected in Death Valley possessed haloperoxidase activity. Some years later it was shown (Van Schijndel et al., 1993) that the nonheme chloroperoxidase produced by such a hyphomycete, the fungus *Curvularia inaequealis*, also contains vanadate as the prosthetic group in the enzyme. The vanadium enzyme is produced in the secondary growth phase, when nutrients become limiting. In the group of the dematiaceous hyphomycetes several species have been detected that produce vanadium haloperoxidases (e.g. *Drechslera biseptata, D. subpapendorfii*, *Embellisiadidymospora*, and *Ulocladium chartarum*; Vollenbroek et al., 1995). Many of these hyphomycetes are phytopathogenic and/or saprophytes and this raises the question as to the physiological function of the vanadium chloroperoxidases. In 1997 it was proposed that the generated HOCl is used by the fungus as an attack mechanism to oxidize the lignocellulose in the cell walls of the plant in order to facilitate penetration of the fungal hyphen into the host. It may well be that HOCl which is generated by fungi in soils also reacts with organic matter resulting in the formation of the organohalogens. Since chloroperoxidase-producing fungi are ubiquitous in decaying lignocellulose and plant debris and occur widely in nature these enzymes have roles in the natural production of high molecular weight chloroaromatics and in lignin breakdown. Chlorolignin produced by this route and humus derived from it are probably significant components of the global chlorine cycle (Ortiz-Bermudes et al., 2007) because lignin is the earth's most abundant aromatic substance. The view that haloperoxidases were the full story changed when researchers recently identified another group of enzymes the iron (II)- and α-oxoglutarate-dependent halogenases (Fujimori and Walsh, 2007). These halogenases have a mechanism for chlorination in which a key intermediate is a Cl-Fe(IV)- oxo species, substrate chlorination can then proceed through "rebound" of a chloride radical, analogous to the hydroxyl radical rebound postulated for both heme and nonheme hydroxylases. This family of enzymes appears to be involved in more specific halogenation of more complex natural products.

CATALYTIC PROPERTIES OF BROMOPEROXIDASE

The vanadium peroxidases differ from the heme-containing peroxidases in that only halides and sulfides are used as substrates. Classical heme peroxidase substrates such as guaiacol, *o*-dianisidine, and benzidine are not oxidized directly by these enzymes. The activity of these enzymes is routinely determined using scavengers that react with HOCl, HOBr, or HOI. A quantitative assay to determine the halogenating activity of these enzymes is the bromination or chlorination of the cyclic diketone monochlorodimedone to the dihalogenated compound. At more alkaline pH values and in particular at high concentrations of H_2O_2 a competing reaction (Eq. 23.3) between H_2O_2 and HOX occurs. This results in singlet oxygen formation and a decrease in the rate of bromination or chlorination of monochlorodimedon.

$$H_2O_2 + HOX \longrightarrow O_2 + H_2O + H^+ + X^- \qquad (23.3)$$

In fact the vanadium chloroperoxidase that will be discussed later has been used as a highly efficient catalyst for the production of singlet oxygen and is 10^3 to 10^4 times faster than the molybdate-peroxide system (Renirie et al., 2003). A convenient qualitative method to assay the haloperoxidase activity is based on bromination of phenol red to bromophenol blue. The bromination of the dye results in a marked color change which allows screening of haloperoxidase activity of large numbers of samples by simple visual inspection. This assay has been used in high throughput screening of mutant libraries of vanadium chloroperoxidase (Hasan et al., 2006). Extensive steady-state kinetic studies on vanadium bromoperoxidases have been carried out, which show that the substrates hydrogen peroxide and bromide react with the enzyme in a sequential order and the kinetic mechanisms for the bromoperoxidases from various sources appear to be very similar (De Boer and Wever, 1988; Soedjak et al., 1995). Vanadium bromoperoxidases are very resistant towards oxidative inactivation and only a slight inhibition of these enzymes by high concentrations of H_2O_2 (120 to 400 mM) with stronger inhibition at higher pH is seen (Soedjak et al., 1995).

Interestingly, the log K_m value for H_2O_2 decreases linearly with increasing pH concentration but levels off above pH 6. This indicates that hydrogen peroxide reacts preferentially with a deprotonated group in the enzyme with a pK_a of 5.7 to 6.7 that upon protonation does not bind peroxide. The second-order rate constant that was derived for the binding of hydrogen peroxide to the bromoperoxidase is 2.5 × 106 M^{-1} s^{-1} at pH > 6. In contrast the K_m for bromide is hardly affected by pH. Table 23.1 gives a summary of some of the kinetic constants of the bromoperoxidase and of the chloroperoxidase that will be discussed later.

On the basis of these early kinetic studies the presence of a peroxo-intermediate was postulated. The primary reaction products of the enzyme-mediated peroxidation of bromide appear to be Br_3, Br_2, or HOBr. Phenylperacetic acid, *m*-chloroperoxybenzoic acid, and *p*-nitroperoxybenzoic acid are also used as oxidants, but alkyl peroxides are not substrates (De Boer and Wever, 1988). Specificity with regard to bromination of various organic nucleophilic acceptors could not be demonstrated for the enzyme from the brown seaweed *Ascophyllum nodosum*. This suggests a

TABLE 23.1. Kinetic Parameters of the Oxidation of Halides Catalyzed by VCPO and VBPO

		X^-			H_2O_2		
Enzyme	pH	K_m (μM)	k_{cat} (s^{-1})	k_{cat}/K_m ($M^{-1}s^{-1}$)	K_m (μM)	k_{cat} (s^{-1})	k_{cat}/K_m ($M^{-1}s^{-1}$)
VBPO bromide oxidation[a]	4.5	3400	278	8.2×10^4	1100	278	2.5×10^5
	5.2	5100	166	3.2×10^4	162	166	1.0×10^6
	6.6	9300	113	1.2×10^4	42	113	2.7×10^6
	7.9	18000	50	2.8×10^3	22	50	2.3×10^6
VCPO bromide oxidation[b,c]	4.2	<5	253	$>5.1 \times 10^7$	90	250	2.6×10^6
	5.2	9	248	2.8×10^7	35	203	5.8×10^6
	6.3	7	37	5.3×10^6	<5	33	6.6×10^6
	8.0	120	1	8.3×10^3	<5	1	$>2 \times 10^5$
VCPO chloride oxidation[d,e]	5.0	1100	23	2.0×10^4	49	183	3.7×10^6
	6.3	10000	7.3	7.1×10^2	2.6	152	5.8×10^6
	8.0	110000	n.d.	n.d.	n.d.	n.d.	n.d.
VCPO alkaline mutant bromide oxidation[c]	5.0	n.d.	575	n.d.	n.d.	n.d.	n.d.
	8.0	3100	100	3.2×10^4	16	100	6.2×10^6
VCPO alkaline mutant chloride oxidation[c]	5.0	n.d.	36	n.d.	n.d.	n.d.	n.d.

[a]De Boer and Wever, 1988.
[b]Hemrika et al., 1999.
[c]Hasan et al., 2006.
[d]Tanaka et al., 2003.
[e]Van Schijndel et al., 1993.

mechanism in which the enzyme releases oxidized bromine species into solution. This would be in line with the production of simple halohydrocarbons such as CH_2Br_2 and $CHBr_3$ by this brown seaweed, which is probably the result of enzymatically generated HOBr with organic matter in seawater. Recently the diffusion behavior of these bromonium species produced by a vanadium peroxidase from the species *Curvularia verruculosa* was studied (Martinez et al., 2008) by in situ observation of single oxidation/halogenation events using single molecule fluorescence microscopy. It was found that in steady-state conditions hypobromite is able to diffuse over 800 nm in the bulk solution before it reacts with organic substrates. Figure 23.2 gives a simple model for the reaction mechanism of the enzyme.

Some seaweed species contain more elaborate and complex brominated compounds such as halogenated sequiterpenes and there appears to be a more specific interaction between organic substrate and an enzyme bound halogenating intermediate interaction in these seaweeds. As shown (Carter-Franklin and Butler, 2004; Butler and Carter-Franklin, 2004), vanadium bromoperoxidases isolated from marine red algae (species of *Laurencia*, *Plocamium*, and *Corallina*) catalyze the asymmetric bromination and cyclization of a terpene, producing products that are also found in red algae. These vanadium bromoperoxidases likely function by catalyzing the two-electron oxidation of bromide ion by hydrogen peroxide producing a bromonium ion or equivalent in the active site that brominates one face of the terminal olefin that docks within the active site cavity in a specific orientation. Interestingly this cyclic terpene is only formed in mixtures of water and organic media. Thus, these vanadium enzymes may also have a role in the biosynthesis of brominated cyclic sesquiterpenes that are

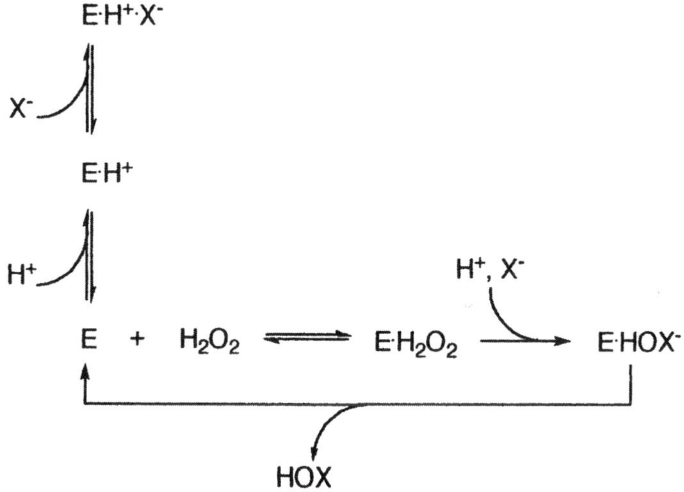

Figure 23.2. General and simplified reaction mechanism of the vanadium peroxidase. The enzyme species in EH^+ and $EH^{+\bullet}X^-$ represent the inhibited forms of the enzymes. The K_m for chloride is linearly dependent on pH, whereas in the reaction of bromide with the bromoperoxidase peroxo complex ($E^\bullet$ HOX) no protons are involved.

found in marine red algae. As will be discussed later the haloperoxidase from *Ascophyllum nodosum* catalyzes enantioselective sulfoxidation of aromatic sulfides and thus there is also specific binding of organic molecules possibly close to or at the active site of this enzyme.

PROPERTIES OF THE PROSTHETIC GROUP IN BROMOPEROXIDASE

Vanadium bromoperoxidases rapidly lose their enzymatic activity at low pH in buffers containing phosphate and EDTA. Also during purification procedures the enzyme slowly inactivates if phosphate buffers are used. However, in Tris buffers the enzyme retains its activity for months. Incubation of the apoenzyme at neutral pH with orthovanadate (HVO_4^{2-}) results in complete recovery of the enzymatic activity (Vilter, 1984) Reconstitution is inhibited competitively by structural analogs such as phosphate (PO_4^{3-}), arsenate (AsO_4^{3-}), molybdate (MoO_4^{2-}), fluoroberyllate, and fluoroaluminate. The reconstitution of the apoenzyme by vanadate was taken as early evidence that vanadate is present in the active site. The x-ray structures confirm that vanadate is indeed the prosthetic group in these enzymes (Weyand et al. 1999; Isupov et al., 2000). The affinity of the apobromoperoxidase for vanadate is quite high. Values of 35–55 nM have been reported which are close to the concentration of vanadate in seawater of 35 nM (Butler, 1998). Thus, for incorporation of vanadate into the apoenzyme, in vivo, no additional enzyme system is required. It is possible to incorporate also molybdate in the active site of the bromoperoxidase. However, the enzyme reconstituted with molybdate is inactive (De Boer et al., 1988a).

Only weak bands are detected in the region 300 to 330 nm of the optical absorption spectrum of the bromoperoxidases (Tromp et al., 1990) and therefore it is difficult to gain information by optical spectroscopy. However, when the enzyme is allowed to react the bands in the region diminish in intensity and they are restored by addition of bromide. This observation supported the steady-state kinetic data that pointed to the existence of an enzyme peroxide intermediate. In the presence of substrates or during turnover the vanadium (IV) or (III) states are not observed by EPR (De Boer et al., 1988a) or by K edge x-ray absorption studies (Arber et al., 1989). Further, the reduced enzyme is inactive and apparently the redox state of the metal remains (V) and does not change during the catalytic turnover. Hence it has been proposed that vanadium due to its low radius/charge ratio plays the role as a Lewis acid with respect to H_2O_2 (De Boer et al., 1988a). In this mechanism the metal effectively withdraws electron density from the bound peroxide in such a way that the nucleophilic halide is able to react with this activated peroxide intermediate of the enzyme to yield hypohalous acid. In this mechanism, no redox changes occur in the vanadium metal center. These spectroscopic methods have not only yielded insight into the mechanism but also valuable structural data have also been obtained. Based on the EXAFS studies it was concluded (Arber et al., 1989; Carrano et al., 1994) that the geometry of the metal oxide bound in the active site was five-coordinate. Based on the multiple scattering effects from outer atoms of a group it was concluded that histidine is ligated to the vanadate. An ESEEM study (De Boer et al., 1988b) of the reduced enzyme demonstrated clearly that nitrogen atoms are in the direct coordination sphere of the metal oxide site. The x-ray structure of the vanadium bromoperoxidase from *Ascophyllum nodosum* (Weyand et al., 1999) confirms that vanadate is forming a bond with a nitrogen atom of a histidine residue.

KINETIC AND OPTICAL PROPERTIES OF VANADIUM CHLOROPEROXIDASES

The expression of active recombinant bromoperoxidases has not been very successful up to now (Oshiro et al., 2002; Coupe et al., 2007). In contrast cloning and expression of an active vanadium chloroperoxidase has been straightforward. The enzyme has been expressed in both the yeast *Saccharomyces cerevisiae* (Hemrika et al., 1997) and in *Escherichia coli* (Hasan et al., 2006). Activation of the apoform of the enzyme secreted either by the native fungus or of the recombinant enzyme is easily achieved by addition of vanadate to either the growth medium or to the isolated enzymes. Most soils contain a high concentration (100 ppm) of vanadium. Because the affinity of the apochloroperoxidase for vanadate is high the amount of vanadium is not a limiting factor in the conversion of the apoenzyme into the holo-enzyme. Additionally, the presence of H_2O_2 increases the affinity of the apoenzyme for vanadate at least 200-fold. This is due to the formation of pervanadate that is bound much more strongly to the active site than vanadate (Renirie et al., 2000a). The steady-state kinetics of the enzyme resembles those of the vanadium bromoperoxidases (Van Schijndel et al., 1994; Hemrika et al., 1999; Tanaka et al., 2003). The chloroperoxidase exhibits a pH profile similar to vanadium bromoperoxidases although the optimal pH of 4.5 to 5.0 is at a lower pH value. The enzyme is inhibited at low pH by chloride in a competitive way, whereas at higher pH values a normal Michaelis–Menten type of behavior is observed. The log K_m for chloride increases linearly with pH, whereas that for hydrogen peroxide decreases with increasing pH, demonstrating that protons are involved in the catalytic mechanism (Tanaka et al., 2003). These observations suggest a ping-pong type of mechanism for the chloroperoxidase as shown in Figure 23.2. The enzyme reacts first with peroxide to form a peroxo-intermediate after which a chloride ion and a proton react resulting in the formation of an enzyme-HOCl intermediate. This intermediate decays in a rate-determining step to enzyme and free HOCl. The linear dependency of the log K_m for hydrogen peroxide on pH suggests that an ionizable group is involved in the binding of hydrogen peroxide. When this group is protonated hydrogen peroxide is unable to bind. This is similar to the observations in vanadium bromoperoxidases. The linear dependence of the log K_m for chloride on pH shows that for binding of chloride to occur, protonation of a group is essential. As shown in Figure 23.3 vanadium chloroperoxidase also exhibits a weak absorption band ($\varepsilon = 2.8\ mM^{-1}\ cm^{-1}$) in the near UV which is not found in the apoenzyme.

The intensity of the band decreases upon addition of hydrogen peroxide and chloride restores the original intensity of the band. Thus the band can be used to monitor binding of peroxide to the enzyme and also to obtain information about halide oxidation by the peroxo-intermediate.

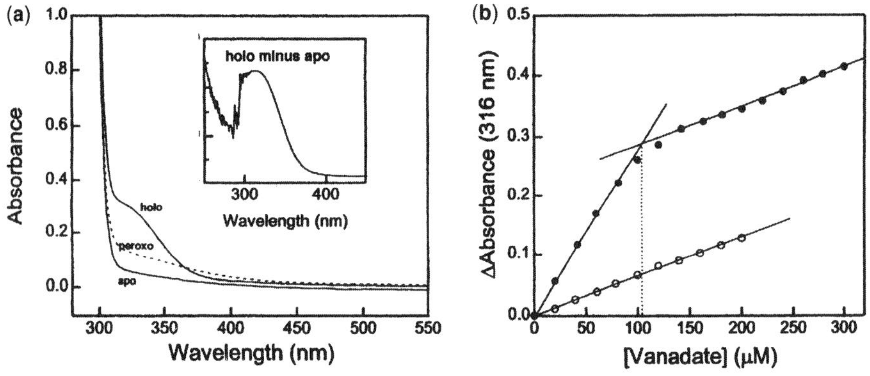

Figure 23.3. Panel A: UV-VIS absorption spectra of 100 μM vanadium chloroperoxidase at pH 8.3. The lower solid line shows the apoenzyme, the upper solid line the holoenzyme after addition of 100 μM vanadate. The dotted line is the peroxo-form of the enzyme after addition of 100 μM H_2O_2. Panel B: Titration of apo-CPO with vanadate: changes in the optical absorbance at 316 nm (●). Absorbance of free vanadate at the same buffer conditions (○). (From Renirie et al., 2000a. *Biochemistry* **39**: 1133–1141. Reprinted with permission of the American Chemical Society.)

Figure 23.4 illustrates the titration of the holo-enzyme with the substrate H_2O_2, showing the high affinity of the enzyme for this substrate. These UV-VIS studies require very pure recombinant material since the wild-type chloroperoxidase contains dyes that are very difficult to remove.

Stopped-flow studies confirmed the catalytic sequence of events (Renirie et al., 2000a). Only at pH 5 was it possible to obtain a first-order rate constant for the binding

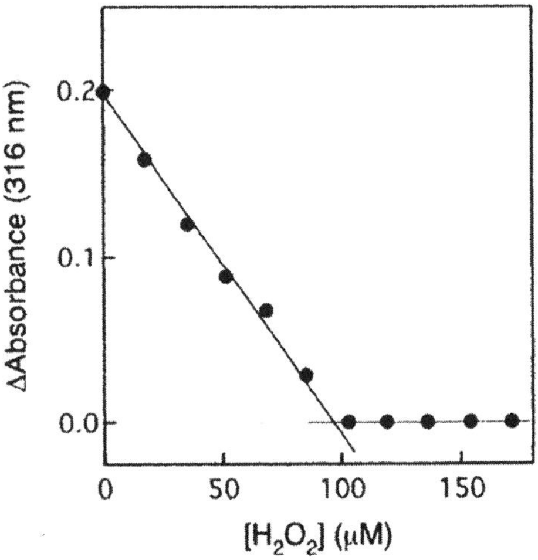

Figure 23.4. Titration of 100 μM vanadium chloroperoxidase with H_2O_2 at pH 8.3. (From Renirie et al., 2000a. *Biochemistry* **39**: 1133–1141. Reprinted with permission of the American Chemical Society.)

of hydrogen peroxide to the enzyme ($8 \times 10^4\ M^{-1}\ s^{-1}$). At higher pH values the reaction between enzyme and peroxide is too fast to be measured. It was also possible to monitor the rate of chloride oxidation by the peroxo-intermediate using a sequential stopped flow. The first-order rate constant of $7.0\ s^{-1}$ is similar to that obtained from steady-state kinetics at the same temperature.

The UV-VIS spectra of several mutants of this enzyme together with the known x-ray data confirmed the importance of several structural elements. A His496Ala mutant showed no optical spectrum (Renirie et al., 2000b), whereas the crystal structure of this mutant showed the presence of tetrahedral vanadate in the active site (Macedo-Ribeiro et al. 1999), indicating that the V—N bond is crucial for the observed spectrum. TD-DFT and quantum mechanical calculations (Borowski et al., 2004) resulted in UV-VIS absorption spectra that were in line with the optical absorption spectra, including the effect of peroxide addition, change of pH, and mutation of active-site residues. In the native enzyme a His496 $\pi \rightarrow$ V 3d transition dominates, whereas in the peroxide form a peroxide $\pi^* \rightarrow$ V 3d CT transition is most intensive, as was previously suggested (Renirie et al., 2000a) on the basis of analogy to well-studied inorganic imidazoleperoxovanadate complexes. The results of theoretical calculations for reliable models of active sites in chloroperoxidase allow the interpretation of the observed spectral features.

The bromoperoxidase activity of chloroperoxidase has also been studied (Hemrika et al., 1999). Compared to the bromoperoxidase which has a K_m for bromide at pH 5.2 of 5.1 mM the chloroperoxidase has a very high affinity for bromide (K_m about 10 μM). See Table 23.1 for details. The k_{cat} at pH 5 is about $250\ s^{-1}$ which is similar to that of the bromoperoxidase from *A. nodosum*. The enzyme is strongly inhibited by excess bromide and hardly any activity is observed at 100 mM bromide. An interesting difference is seen between the pH dependence of K_m values for the halides. The K_m for chloride is strongly pH dependent, whereas the K_m for bromide is hardly affected by pH (Hemrika et al., 1999; Tanaka et al., 2003). Also the K_m for bromide of bromoperoxidase from *Ascophyllum nodosum* (De Boer and Wever, 1988) is hardly pH dependent (see Table 23.1). It is interesting to note that as shown in Table 23.1 the specificity constant for the reaction of hydrogen peroxide with the peroxidases and mutants is nearly the same in all cases and not affected by pH. In contrast the specificity constant for the reaction with the halides varies 1000-fold.

SULFOXIDATION REACTIONS

The vanadium haloperoxidases are also capable of catalyzing the sulfoxidation of organic sulfides in the absence of halides. For the bromoperoxidases this occurs enantioselectively (Andersson et al., 1997; Ten Brink et al., 1998). The bromoperoxidase from the brown seaweed *A. nodosum* produces the R-enantiomer of methyl phenyl sulfoxide (96% e.e.) under slightly acidic conditions, while the bromoperoxidase from the red seaweed *C. pilulifera* produces the S-enantiomer of this sulfoxide (55% e.e.). The bromoperoxidase from the seaweed *C. officinalis* was also observed (Andersson et al., 1997) to convert organic sulfides that structurally resemble indenes to the S-enantiomer of the corresponding sulfoxides. Also small sulfides, possessing a cis-positioned carboxyl group with respect to the sulfur atom, are rapidly converted to the corresponding sulfoxides with selectivities exceeding 95 % e.e. Studies using labeled hydrogen peroxide showed that the vanadium bromoperoxidases mediate the direct transfer of the peroxide oxygen to the sulfide in a selective manner, strongly suggesting that the aromatic sulfide binds near or at the active site with a relatively high K_m value (Ten Brink et al., 2001). The k_{cat} for the sulfoxidation reactions (1 min^{-1}) is low compared to the brominating activity of the bromoperoxidase from *A. nodosum* (166 s^{-1}). Vanadium chloroperoxidase in contrast to vanadium bromoperoxidase does not mediate direct and selective transfer of peroxide oxygen to methyl phenyl sulfide (Ten Brink et al., 2001). It was suggested that due to the high reactivity of the chloroperoxidase an electron is abstracted from the substrate forming a positively charged sulfur radical, which migrates from the enzyme and is subsequently converted into a sulfoxide.

STABILITY OF BROMO- AND CHLOROPEROXIDASES

An extraordinary feature of these enzymes is their remarkable stability. The vanadium bromoperoxidases and in particular the chloroperoxidase are highly resistant towards elevated temperatures; the midpoint temperatures determined from thermal denaturation curves of the chloroperoxidases range from 82°C to 90°C (Van Schijndel et al., 1994). The haloperoxidases are very resistant towards detergents like sodium dodecyl sulfate (SDS) and are unaffected in their activity. It was also shown that the vanadium chloroperoxidases remained fully active and are stable in microemulsions containing 10% nonionic surfactant (Renirie et al., 2009). Pronounced chemostability and extraordinary resistance towards oxidative agents such as H_2O_2, HOCl, and singlet oxygen have been reported (Renirie et al., 2003).

Heme-containing peroxidases with their electron-rich aromatic ring system are much more easily oxidized and inactivated by oxidative agents. In this respect the vanadium haloperoxidases are unique and the ability of the enzymes to handle the oxidative substrate and product are the consequences of the nature of the active site. These fascinating features are of considerable interest for the potential application of the vanadium peroxidases as (industrial) biocatalyst in organic synthesis, their use in disinfection formulations (Renirie et al., 2008), and as bleaching enzymes.

X-RAY STRUCTURES OF VANADIUM BROMOPEROXIDASES

Three x-ray structures of bromoperoxidases are presently known, the homodimeric structure of the enzyme from the brown seaweed *A. nodosum* and those of the red seaweeds *Corallina pilulifera* and *C. officinalis*. The enzymes from the red seaweeds crystallize as dodecamers and the structures are made up of six homodimers. The subunit in the bromoperoxidases from *Cor. officinalis* (Littlechild et al., 2002) measures approximately 85 × 56 × 55 Å in size (Fig. 23.5). It is folded into a single α–β type

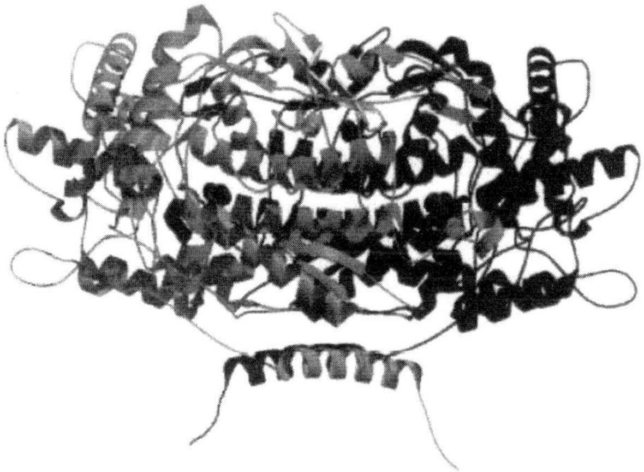

Figure 23.5. Ribbon diagram representation of the *Corallina* bromoperoxidase dimer viewed normal to the molecular dyad. In the structure the two inorganic phosphate ions that co-crystallizes in the active site are visible. (From Littlechild et al., 2002. *J. Mol. Recogn.* **15**: 291–296. Reproduced by permission of John Wiley & Sons, Ltd.)

domain of 595 amino acids. The first 10 N-terminal amino acids point away from the body of the domain that is composed of α-helices and β-strands. The β-strands are mainly involved in β-hairpins. The two subunits form a homodimer. The residues 123–124 from one subunit and 368–369 from the other subunit in the dimer form a parallel two-stranded β-sheet. The β-hairpins 270–273 and 276–279 from each subunit in the dimer form an antiparallel fourstranded β-sheet. The *Cor. officinalis* bromoperoxidase dodecamer measures 150 Å in diameter. Twelve subunits are arranged within a 23 cubic point group symmetry. The vanadate-binding site is located at the bottom of the active-site cleft, which is about 20 Å deep and 14 Å wide. The active-site cleft is formed predominantly from residues of two different subunits in the dimer. The intertwined structure suggests that the subunit by itself will show no activity and that reconstitution of activity is a difficult process. The secondary structure of the vanadium chloroperoxidase from the fungus *Curvularia inaequalis* that will be discussed later can be superimposed on the *Corallina* bromoperoxidase dimer. Many of the α-helices of each chloroperoxidase domain are structurally equivalent to the α-helices in the *Corallina* bromoperoxidase dimer.

ACTIVE SITE OF VANADIUM BROMOPEROXIDASE FROM *A. NODOSUM*

The residues contributing to the active site are conserved in the bromoperoxidases. Figure 23.6 illustrates the active site in the bromoperoxidase from *A. nodosum* (Weyand et al., 1999).

Vanadium is bound as orthovanadate on the N-terminal side of a four-helix bundle and the vanadium metal is the center of a trigonal bipyramid with three oxygen atoms

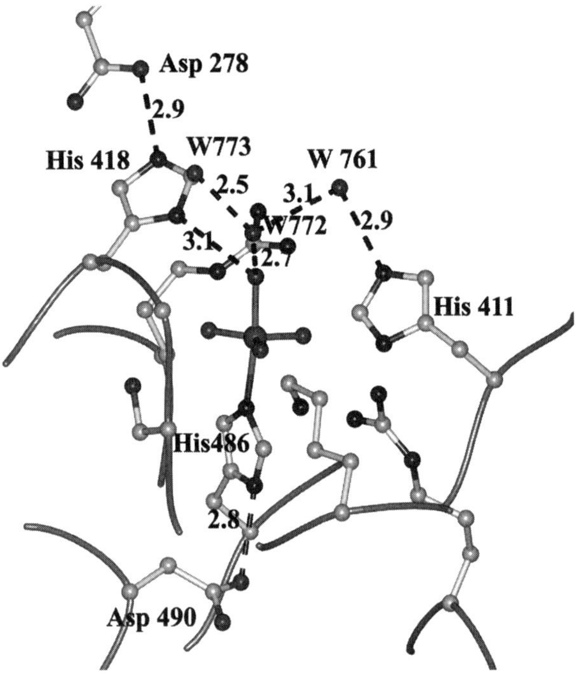

Figure 23.6. The active site residues around the vanadate center of vanadium bromoperoxidase from *A. nodosum*. There are also three water molecules (W772, W773, and W761) in the active site. (From Weynand et al., 1999. *J. Mol. Biol.* **293**: 595–611. Reproduced with permission of Academic Press.)

coordinated in the equatorial plane, one oxygen in the axial position (Fig. 23.6) and a covalent linkage to the Nε2 of His486. The negative charge of the oxygen atoms in the equatorial plane is compensated by hydrogen bonds donated by several hydrophilic and positively charged amino acids in the vicinity, including Lys341, Arg349, Ser416, and Arg480. The axial oxygen atom forms a hydrogen bond with the Nε1 of the distal His418, which is a conserved catalytic residue in vanadium peroxidases. Both hydrophilic and hydrophobic residues cover the entrance channel, with hydrophobic residues dominating the lower part near the vanadate-binding site. As in the enzyme from *Cor. officinalis* amino acid residues from both monomers contribute to the substrate access channel of each active center. The channel to the active site of each monomer, formed by both hydrophilic and hydrophobic residues, is 15 Å deep, with a diameter of 12 Å at the entrance, decreasing to 8 Å near the axial oxygen of the vanadate prosthetic group, where the channel exhibits a more hydrophobic character.

X-RAY STRUCTURES OF THE VANADIUM CHLOROPEROXIDASE AND DETAILS OF THE ACTIVE SITE

Both the 2.1 Å crystal structure of the native chloroperoxidase and the structure at recombinant chloroperoxidase at 1.66 Å resolution have been reported (Messerschmidt and Wever, 1996; Maceido-Ribeiro et al., 1996). The enzyme molecule crystallizes as a monomer and has an overall cylindrical shape (not shown) with a length of about 80 Å and a diameter of 55 Å. The protein fold is mainly α-helical with two four-helix bundles as the main structural motifs. There are no disulfide bridges, which is in contrast to the vanadium bromoperoxidase from *A. nodosum*. The vanadium-binding center is located on top of the second four-helix bundle.

The x-ray structure analysis of the active site (Fig. 23.7a) shows that vanadium is bound as orthovanadate in the same way as in the bromoperoxidases. Three oxygen atoms form a plane (bond length about 1.65 Å), the fourth oxygen is found at the apex (bond length 1.93 Å), which is in the range of a vanadium hydroxyl bond. It is hydrogen bonded to the nitrogen Nε1 of His404. The trigonal bipyrimidal coordination is completed by an Nε2 atom from His496 ligating directly to the metal. The negative charge of the vanadate group is compensated by hydrogen bonds to three positively charged residues (Arg360, Arg490, and Lys353). Further, the vanadate forms hydrogen bonds with a glycine (Gly403) and a serine residue (Ser402). There is also a histidine residue (His404) close to the apical hydroxyl moiety of the vanadate that may act as an acid/base group in catalysis. These residues are the same as in bromoperoxidase and the active sites are structurally nearly superimposable (Weyand et al., 1999). The pocket where vanadate is bound is at the end of a channel that is mainly hydrophobic. The other half of the channel is predominantly polar, with several carbonyl oxygens and the ion pair Arg490-Asp292. A prominent difference in the active site architecture of the bromo- and chloroperoxidase is the presence of a

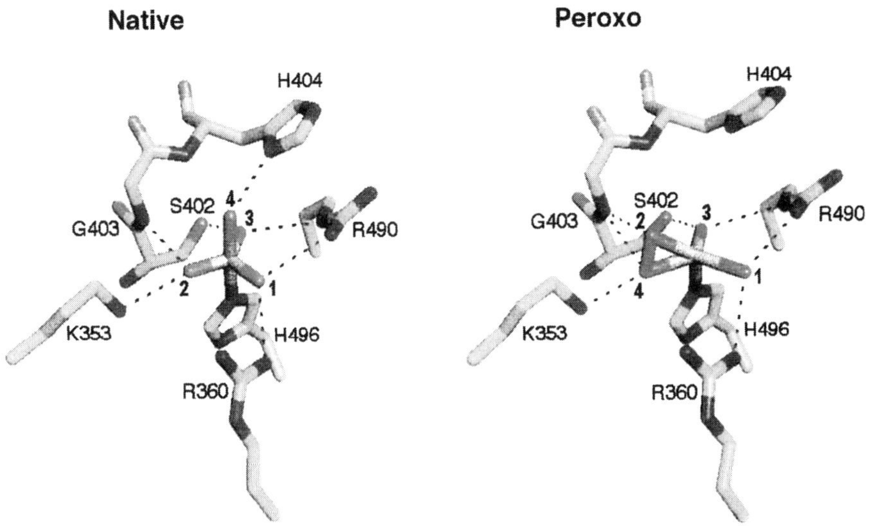

Figure 23.7. The active site of native chloroperoxidase (left) from *C. inequalis* and after the addition of H_2O_2. The vanadium center is shown in grey. The numbering of the oxygen atoms is that of Messerschmidt et al. (1997). All the amino acid residues shown are conserved in all known sequences and crystal structures of vanadium haloperoxidases. (Figure constructed using PyMOL, http://www.pymol.org.)

second histidine in the bromoperoxidase, a residue substituted by a phenylalanine in chloroperoxidase.

X-RAY STRUCTURE OF THE PEROXO-INTERMEDIATE OF VANADIUM CHLOROPEROXIDASE AND DIFFERENCE IN REACTIVITY BETWEEN CHLORO- AND BROMOPEROXIDASES

The x-ray structure of the important catalytic peroxo-intermediate has also been obtained by incubating native crystals that were shock frozen in cryobuffer in mother liquor containing peroxide. The peroxide is bound side-on and the resulting coordination geometry (Fig. 23.7b) is that of a distorted tetragonal pyramid with only four oxygen atoms and one nitrogen atom. The apical ligand is oxygen VO3 with a bond length of about 1.6 Å, identifying this as a V═O bond. The two peroxide oxygen atoms are at a distance of about 1.87 Å and the distance between the two peroxide oxygens OV2 and OV4 is 1.47 Å. There is a longer VO bond at about 1.93 Å. There are several changes in the hydrogen bonding network. His404 is no longer bonded to any of the vanadate oxygens and a hydrogen bond is formed to Lys353 that has important implications, as discussed below. The molecular mechanism of vanadium chloroperoxidase and difference in reactivity between chloro- and bromoperoxidases.

As shown in Figure 23.7b peroxide is bound side-on in the chloroperoxidase. Based on the effects of mutations of the active-site resides the following mechanism has been proposed (Hemrika et al., 1999): Electron density is withdrawn from the bound peroxide by the Lewis acid properties of the two positively charged arginine residues (Arg490 and Arg360) and the metal ion itself. The bound peroxide is further polarized and activated by the conserved Lys353 that forms a hydrogen bond to one of the oxygen atoms. It has been proposed that protonation of one of the oxygen atoms of the peroxide is required to allow the more difficult oxidation of the chloride ion. After formation of the peroxide intermediate the next step in catalysis is the nucleophilic attack by the halide on the electrophilic hydrogen peroxide and breaking the peroxide bond, resulting in formation of HOX or XO^-. All mutations of the active-site residues result in enzymes that are dead or kinetically crippled (Tanaka et al., 2003; Hemrika et al., 1999; Renirie et al., 2000a, 2000b). Most or all of the chloroperoxidase activity is lost; however, some mutants retain some bromoperoxidase activity. Chloride is more difficult to oxidize than bromide and since detailed structural data are available it should be possible to explain the difference in reactivity between the vanadium bromo- and chloroperoxidases. As illustrated in Figure 23.8 the only residue that differs in the two enzymes is Phe397 in chloroperoxidase that is a His in bromoperoxidase. It has been suggested (Renirie et al., 2000b; Hasan et al., 2006) that this histidine accepts the proton from the bound peroxide and in this way reduces the oxidative strength of the bound peroxide and the reactivity of the peroxidase.

Mutagenesis studies on both bromoperoxidase and chloroperoxidase of the only residue that differs in the two enzymes (Phe in chloroperoxidase versus His in bromoperoxidase) have given no clear picture whether this proposal is correct. By

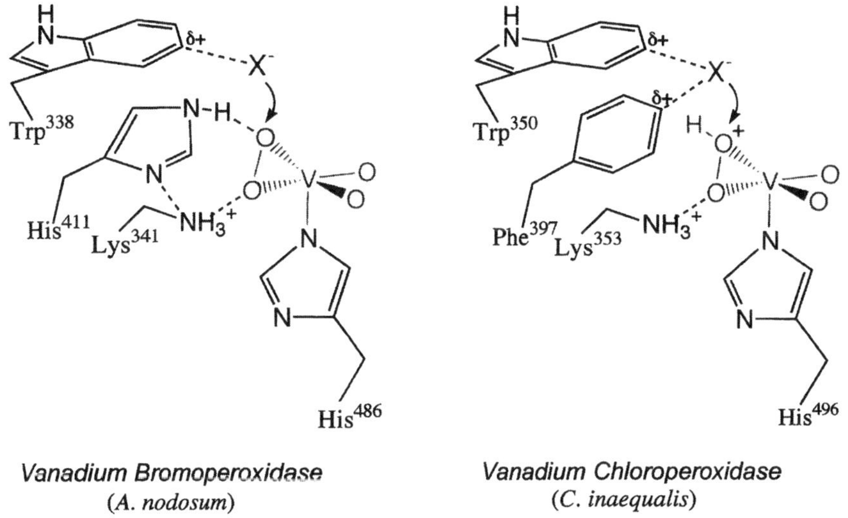

Figure 23.8. A possible explanation for the difference in reactivity between vanadium chloroperoxidase and bromoperoxidase. (From R. Renirie et al., 2000b. *J. Biol. Chem.* **275**: 11650–11657. Reproduced with permission of the American Chemical Society.)

site-directed mutagenesis the Phe397 was replaced by a histidine. Indeed it was observed (Tanaka et al., 2003) that the specificity constant for the oxidation of chloride decreases 100-fold. However, the specificity constant for bromide oxidation also decreased 10-fold and further the mutant rapidly inactivated during turnover. An attempt to induce substantial chlorinating activity in the bromoperoxidase from *C. pulilifera* met limited success (Ohshiro et al., 2004). When the Arg 397 in this bromoperoxidase which a Trp in chloroperoxidase was replaced by a Phe or a Trp substantial chlorinating activity was found. However, the reported K_m values for chloride of about 700 mM for the two mutants were very high. Satisfactory explanations have not been offered for these observations and the factors that are important in tuning the reactivity of the haloperoxidases are not fully understood. As will be discussed later there are also bound water molecules in the active site and these may play important roles in tuning the reactivity of these enzymes.

Directed evolution was used to increase the brominating activity of the vanadium chloroperoxidase from *C. inaequalis* at mildly alkaline pH and activities were created that were the highest ever measured for vanadium haloperoxidases (Hasan et al., 2006). After two rounds of random mutagenesis, saturation mutagenesis of a hot-spot, and rational recombination of results a triple mutant was created, showing synergistic mutations in the first and second coordination sphere around the vanadate cofactor. A 100-fold increase in activity at pH 8 (100 s^{-1}), a 5-fold higher brominating activity at pH 5 (575 s^{-1}) and a 1.8 times higher chlorinating activity at pH 5 (36 s^{-1}) were observed. Some increase of K_m was observed but overall the specificity constants k_{cat}/K_m for bromide and chloride were increased (Table 23.1). The mutants are in the first and second coordination sphere of the vanadate cofactor and the catalytic effects suggest that fine tuning of residues Lys353 and Phe397 along with the addition of negative charge or removal of positive charge near one of the vanadate oxygens is very important.

NATURE OF THE VANADATE COFACTOR

A detailed molecular picture of the catalytic mechanism of VCPO requires understanding of the structure of the vanadate cofactor, the peroxo–intermediate, and the relevant interactions between vanadate and the protein cavity. This cannot be determined unequivocally from the crystal structures and several theoretical investigations have now been performed on small vanadium complexes that model the native VCPO and on the enzyme itself. DFT calculations (Zampella et al., 2006) resulted in a doubly protonated vanadate with an axial hydroxo ligand in line with the crystal structure data with one hydroxo group in the equatorial plane, creating an interesting asymmetry. A time-dependent DFT study (Bangesh and Plass, 2005) led to the same conclusion, as well as a QM/MM study (Raugei and Carloni, 2006). This picture was also concluded by a solid-state 51V-NMR study on VCPO in combination with DFT calculations of the NMR observables using a series of small models (Pooransing et al., 2006). The importance of including the protein environment into the QM models was also shown (Kravitz et al., 2005), which again led to a minimum with two protons, but now in the form of an axially bound water ligand. The 51V-NMR chemical shifts (Pooransingh et al., 2006) were remodeled (Waller et al., 2007) using QM/MM-optimized models of the complete protein and systematically increasing the size of the QM region. They concluded that the resting state of the cofactor is either the proposed doubly protonated vanadate or a triply protonated state in which there is an axially bound water and an equatorial hydroxide. Finally, a comprehensive analysis of the most likely ground state configuration based on quantum mechanics/molecular mechanics evaluations of ground state properties, UV-VIS spectra, and NMR chemical shifts was given (Zhang and Gascón, 2008). They concluded that the water molecules that are present in the crystallographic structure are very important in explaining the spectral and NMR data. Further the most likely protonation state is vanadate that has an axially bound water molecule that is hydrogen bonded to His404. This apical water is stabilized by this hydrogen bond. Mutation experiments already showed that the mutant H404A was strongly impaired in its activity and ability to bind vanadate.

These computational studies and the studies of the effect of mutations on the enzymatic activity and structure show clearly that the enzyme has a very complex hydrogen bonding network around the cofactor. Two water molecules also participate in this but their possible role in tuning the activity is not well understood. Several groups (Stukey and Carman, 1997; Neuwald, 1997; Hemrika et al., 1997) showed that three regions providing the metal anion binding site in the vanadium enzymes showed a high similarity to stretches that were present in a large group of acid phosphatases that were previously considered unrelated. Based on the sequence similarity and the fact that vanadate is a structural and electronic analog of phosphate, it was concluded that the architecture of the active sites in the two classes of enzymes is very similar. The x-ray structure of the acid phosphatase (Ishikawa et al., 2000) from *Escherichia blattae* and that (Makde et al., 2007) of *Salmonella enterica* confirmed the remarkable similarity of the residues binding the vanadate in the vanadium enzymes (Fig. 23.9a) and phosphate (Fig. 23.9b) in the phosphatases.

These phosphatases are not only found in bacteria but are widespread in nature and include, for example, the membrane-bound glucose-6-phosphatase, an enzyme involved in regulation of glucose levels in the blood of mammals. Also the large family of lipid phosphohydrolases has the same architecture for the active site as the vanadium haloperoxidases.

Since these two different types of enzymes share the same supramolecular environment at their active sites, it is reasonable to assume that both of them may exhibit the dual activities. Indeed, it was shown (Hemrika et al., 1997; Renirie et al., 2000b) that the apoprotein of the vanadium chloroperoxidase isolated from *C. inaequalis* has phosphatase activity. However, since the phosphatase activity is low this active site of the enzyme is apparently not optimized for phosphatase activity and this allowed pre-steady-state studies on the phosphatase activity of both on native chloroperoxidase and mutants (Renirie et al., 2000b). Hydrolysis of a phospho-intemediate is the rate-determining step in the phosphatase activity of the apochloroperoxidase and in the mechanism proposed (Renirie et al., 2000b) the His496 that binds the vanadate cofactor in holo-chloroperoxidase acts as a nucleophile forming a phosphoryl intermediate. By incubating apochloroperoxidase crystals with *p*-nitrophenylphosphate and subsequent flash-cooling of the crystals it was possible to trap this intermediate and obtain a high resolution x-ray structure (Macedo-Ribeiro et al., 2008). The

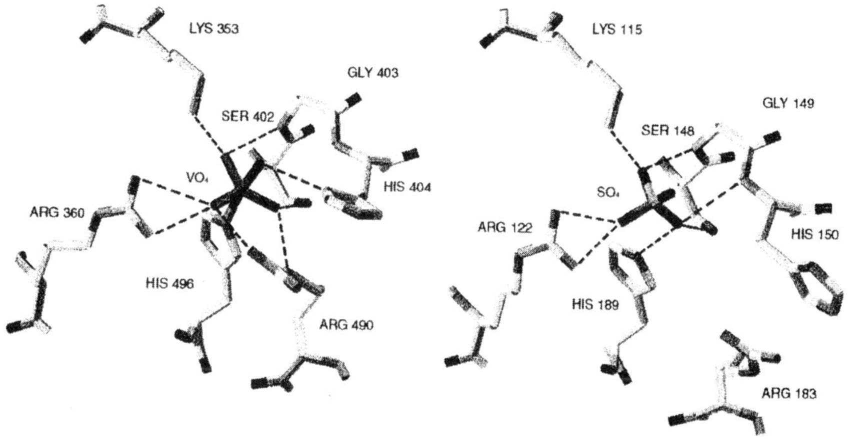

Figure 23.9. The active site of vanadium chloroperoxidase from *C. inequalis* (left) and that of the acid phosphatase from *E. blattea* (right) (Ishikawa et al., 2000). The acid phosphatase cocrystallized with sulfate that was present in the crystallization buffer. (From Wever et al., 2006. In Scott RA, Section editor. *Encyclopedia Inorg. Chem. Section Bioinorganic Chemistry*, www.wiley.com/eic2. Reproduced with permission of John Wiley & Sons, Ltd.)

intermediate consists of a metaphosphate anion PO_3^- covalently bound via its phosphorous atom to the Nε2 atom of the His496 with a water molecule in the position for a nucleophilic attack on the phosphorus. This is the first structural characterization of a real reaction intermediate of the phosphate group release in a dephosphorylation reaction. Conversely it was shown that vanadate-substituted acid phosphatase (Tanaka et al., 2002) exhibits bromoperoxidase activity and enantioselective sulfoxidation activity. The turnover values are, however, much lower (3 to 30 min^{-1}) than for native bromoperoxidase from *A. nodosum.* Thus, the analogy between these classes of enzymes includes both structural and catalytic aspects. When vanadate is substituted in the phosphatase phytase (van de Velde et al., 2000) sulfoxidation and bromoperoxidase activity is observed also. However, phytase does not belong to the class of acid phosphatases and the active site of phytase is structurally not related to that of the acid phosphatases. It is obvious that the phosphatases described have very different physiological functions and are found in species varying from *E. coli* to humans. However, vanadium haloperoxidases have thus far only been found in eukaryotes. Since phosphate and phosphate metabolizing enzymes entered evolution at an early stage, in which oxygen and hydrogen peroxide were not yet present, it seems likely that vanadate was coined by nature to become the prosthetic group in the vanadium enzymes in a more recent stage and thus these enzymes have evolved from the phosphatases.

REFERENCES

Andersson M, Willetts A, Allenmark S. 1997. Asymmetric sulfoxidation catalyzed by a vanadium-containing bromoperoxidase. *J. Org. Chem.* **62**: 8455–8458.

Arber JM, De Boer E, Garner CD, Hasnain SS, Wever R. 1989. Vanadium K-Edge X-ray absorption spectroscopy of bromoperoxidase from *Ascophyllum nodosum. Biochemistry* **28**: 7968–7973.

Bangesh M, Plass W. 2005. TD-DFT studies on the electronic structure of imidazole bound vanadate in vanadium containing haloperoxidases (VHPO). *J. Mol. Struct. Theochem.* **725**: 163–175.

Barnett P, Kruitbosch DL, Hemrika W, Dekker HL, Wever R. 1997. The regulation of the vanadium chloroperoxidase from *Curvularia inaequalis. Biochim. Biophys. Acta* **1352**: 73–84.

Borchardt SA, Allain EJ, Michels JJ, Stearns GW, Kelly RF, Mccoy WF. 2001. Reaction of acetylated homoserine lactone signaling molecules with oxidized halogen antimicrobials. *Appl. Environ. Microbiol.* **67**: 3174–3179.

Borowski T, Szczepanik W, Chruszcz M, Broclawik E. 2004. First-principle calculations for the active centers in vanadium-containing chloroperoxidase and its functional models: geometrical and spectral properties. *Int. J. Quantum Chem.* **99**: 864–875.

Butler A. 1998. Acquisition and utilization of transition metal ions by marine organisms. *Science* **281**: 207–210.

Butler A, Carter-Franklin JN. 2004. The role of vanadium bromoperoxidase in the biosynthesis of halogenated marine natural products. *Nat. Prod. Rep.* **21**: 180–188.

Carpenter LJ, Liss PS. 2000. On temperate sources of bromoform and other reactive organic bromine gases. *J. Geophys. Res. Atm.* **105**: 20539–20547.

Carrano CJ, Mohan M, Holmes SM, Delarosa R, Butler A, Charnock JM, Garner CD. 1994. Oxovanadium (V) alkoxy chlorocomplexes of the hydridotripyrozolylborates as models for the binding site in bromoperoxidase. *Inorg. Chem.* **33**: 646–655.

Carter-Franklin JN, Butler A. 2004. Vanadium bromoperoxidase-catalyzed biosynthesis of halogenated marine natural products. *J. Am. Chem. Soc.* **126**: 15060–15066.

Coupe EE, Smyth MG, Fosberry A, Hall RM, Littlechild JA. 2007. The dodecameric vanadium-dependent haloperoxidase from the marine algae *Corallina officinalis*: cloning, expression, and refolding of the recombinant enzyme. *Prot. Expres. Purification* **52**: 265–272.

De Boer E, Boon K, Wever R. 1988a. Electron paramagnetic resonance studies on conformational states and metal ion exchange properties of vanadium bromoperoxidase. *Biochemistry* **27**: 1629–1635.

De Boer E, Keijzers CP, Klaasen AAK, Reijerse EJ, Collison D, Garner CD, Wever R. 1988b. N-14-coordination to VO^{2+} in reduced vanadium bromoperoxidase: an electron spin echo study. *FEBS Lett.* **235**: 93–97.

De Boer E, Plat H, Tromp MGM, Wever R, Franssen MCR, van der Plas HC, Meijer EM, Schoemaker HE. 1987. Vanadium containing bromoperoxidase: an example of an oxidoreductase with high operational stability in aqeous and organic media. *Biotechnol. Bioeng.* **30**: 607–610.

De Boer E, Tromp MGM, Plat H, Krenn GE, Wever R. 1986a. Vanadium(V) as an essential element for haloperoxidase activity in marine brown algae: purification and characterization of a vanadium(V)-containing bromoperoxidase from *Laminaria saccharina*. *Biochim. Biophys. Acta* **872**: 104–115.

De Boer E, van Kooy Y, Tromp MGM, Plat H, Wever R. 1986b. Bromoperoxidase from *Ascophyllum nodosum*: a novel class of enzymes containing vanadium as a prosthetic group. *Biochim. Biophys. Acta* **869**: 48–53.

De Boer E, Wever R. 1988. The reaction mechanism of the novel vanadium bromoperoxidase, a steady-state kinetic analysis. *J. Biol. Chem.* **263**: 12326–12332.

Fujimori DG, Walsh CT. 2007. What's new in enzymatic halogenations. *Current Opin. Chem. Biol.* **11**: 553–560.

Hansen EH, Albertsen L, Schafer T, Johansen C, Frisvad JC, Molin S, Gram L. 2003. *Curvularia* haloperoxidase: antimicrobial activity and potential application as a surface disinfectant. *Appl. Environ. Microbiol.* **69**: 4611–4617.

Hasan Z, Renirie R, Kerkman R, Ruijssenaars HJ, Hartog AF, Wever R. 2006. Laboratory-evolved vanadium chloroperoxidase exhibits 100-fold higher halogenating activity at alkaline pH: catalytic effects from first and second coordination sphere mutations. *J. Biol. Chem.* **281**: 9738–9744.

Hemrika W, Renirie R, Dekker HL, Barnett P, Wever R. 1997. From phosphatases to peroxidases: a similar architecture of the active site. *Proc. Natl. Acad. Sci. USA* **94**: 2145–2149.

Hemrika W, Renirie R, Macedo-Ribeiro S, Messerschmidt A, Wever R. 1999. Heterologous expression of the vanadium-containing chloroperoxidase from *Curvularia inaequalis* in *Saccharomyces cerevisiae* and site-directed mutagenesis of the active site residues His496, Lys353, Arg360 and Arg490. *J. Biol. Chem.* **274**: 23820–23827.

Ishikawa K, Mihara Y, Gondoh K, Suzuki E, Asano Y. 2000. X-ray structures of a novel acid phosphatase from *Escherichia blattae* and its complex with the transition-state analog molybdate. *EMBO J.* **19**: 2412–2423.

Isupov MN, Dalby AR, Brindley AA, Izumi Y, Tanabe T, Murshudov GN, Littlechild JA. 2000. Crystal structure of dodecameric vanadium-dependent bromoperoxidase from the red algae *Corallina officinalis*. *J Mol. Biol.* **299**: 1035–1049.

Kravitz JY, Pecoraro VL, Carlson HA. 2005. Quantum mechanics calculations of the vanadium dependent chloroperoxidase, *J. Chem. Theory Comput.* **1**: 1265–1274.

Kupper FC, Carpenter LJ, McFiggans GB, Palmer CJ, Walte TJ, Boneberg EM, Woitsch S, Weiller M, Abela R, Grolimund D, Potin P, Butler A, Luther GW, Kroneck PMH, Meyer-Klaucke W, Feiters MC. 2008. Iodide accumulation provides kelp with an inorganic antioxidant impacting atmospheric chemistry. *Proc. Natl. Acad Sci. USA* **105**: 6954–6958.

Laturnus F, Svensson T, Wiencke C, Öberg G. 2004. Ultraviolet radiation affects emission of ozone-depleting substances by marine macroalgae: results from a laboratory incubation study. *Environ. Sci. Technol.* **38**: 6605–6609.

Leblanc C, Colin C, Cosse A, Delage L, La Barre S, Morin P, Fievet B, Voiseux C, Ambroise Y, Verhaeghe E, Amouroux D, Donard O, Tessier E, Potin P. 2006. Iodine transfers in the coastal marine environment: the key role of brown algae and of their vanadium-dependent haloperoxidases. *Biochemie* **88**: 1773–1785.

Littlechild J, Garcia-Rodriguez E, Dalby A, Isupov M. 2002. Structural and functional comparisons between vanadium haloperoxidase and acid phosphatase enzymes. *J. Mol. Recogn.* **15**: 291–296.

Macedo-Ribeiro S, Hemrika W, Renirie R, Wever R, Messerschmidt A. 1999. X-ray crystal structures of active site mutants of the vanadium-containing chloroperoxidase from the fungus *Curvularia inaequalis*. *J. Biol. Inorg. Chem.* **4**: 209–219.

Macedo-Ribeiro S, Renirie R, Wever R, Messerschmidt A. 2008. Crystal structure of a trapped phosphate intermediate in vanadium apochloroperoxidase catalyzing a dephosphorylation reaction. *Biochemistry* **47**: 929–934.

Makde RD, Mahajan SK, Kumar V. 2007. Structure and mutational analysis of the PhoN protein of *Salmonella typhimurium* provide insight into mechanistic details. *Biochemistry* **46**: 2079–2090.

Martínez V, De Cremer G, Roeffaers MBJ, Sliwa M, Mukulesh M, De Vos DE, Hofkens J, Sels BF. 2008. Exploration of single molecule events in a haloperoxidase and its biomimic: localization of halogenation activity. *J. Am. Chem. Soc.* **130**: 13192–13193.

McFiggans G, Coe H, Burgess R, Allan J, Cubison M, Alfarra MR, Saunders R, Saiz- Lopez A, Plane JMC, Wevill DJ, Carpenter LJ, Rickard AR, Monks PS. 2004. Direct evidence for coastal iodine particles from *Laminaria macroalgae*-linkage to emissions of molecular iodine. *Atm. Chem. Phys.* **4**: 701–713.

Messerschmidt A, Prade L, Wever R. 1997. Implications for the catalytic mechanism of the vanadium-containing enzyme chloroperoxidase from the fungus *Curvularia inaequalis* by X-ray structures of the native and peroxide form. *Biol. Chem.* **378**: 309–315.

Messerschmidt A, Wever R. 1996. X-ray structure of a vanadium-containing enzyme: chloroperoxidase from the fungus *Curvularia inaequalis*. *Proc. Natl. Acad. Sci. USA* **93**: 392–396.

Neuwald AF. 1997. An unexpected structural relationship between integral membrane phosphatases and soluble haloperoxidases, *Protein Sci.* **6**: 1764–1767.

O'Dowd D, Jimenez JL, Bahreini R, Flagan RC, Seinfeld JH, Hämeri K, Pirjola L, Kulmala M, Jennings GS, Hoffmann T. 2002. Marine aerosol formation from biogenic iodine emissions. *Nature* **417**: 632–636.

Ohshiro T, Hemrika W, Aibara T, Wever R, Izumi Y. 2002. Expression of the vanadium dependent bromoperoxidase gene from a marine macro-alga *Corallina pilulifera* in *Saccharomyces cerevisiae* and characterization of the recombinant enzyme. *Phytochemistry* **60**: 595–601.

Ohshiro T, Littlechild J, Garcia-Rodriguez E, Isupov MN, Iida Y, Kobayashi T, Izumi Y. 2004. Modification of halogen specificity of a vanadium-dependent bromoperoxidase. *Protein Sci.* **13**: 1566–1571.

Ohshiro T, Nakano S, Takahashi Y, Suzuki M, Izumi Y. 1999. Occurrence of bromoperoxidase in the marine green macro-alga, Ulvella lens, and emission of volatile brominated methane by the enzyme. *Phytochemistry* **52**: 1211–1215.

Ortiz-Bermúdez P, Hirth KC, Srebotnik E, Hammel KE. 2007. Chlorination of lignin by ubiquitous fungi has a likely role in global organochlorine. *Proc. Natl. Acad. Sci. USA* **104**: 3895–3900.

Palmer CJT, Anders L, Carpenter LJ, Küpper FC, McFiggans G. 2005. Iodine and halocarbon response of *Laminaria digitata* to oxidative stress and links to atmospheric new particle production. *Environ. Chem.* **2**: 282–290.

Pooransingh-Margolis N, Renirie R, Hasan Z, Wever R, Vega AJ, Polenova T. 2006. 51V Solid-state magic angle spinning NMR spectroscopy of vanadium chloroperoxidase. *J. Am. Chem. Soc.* **128**: 5190–5208.

Raugei S, Carloni P. 2006. Structure and function of vanadium haloperoxidases. *J. Phys. Chem. B* **110**: 3747–3758.

Renirie R, Pierlot C, Aubry JM, Hartog AF, Schoemaker HE, Alsters PL, Wever R. 2003. Vanadium chloroperoxidase as a catalyst for hydrogen peroxide disproportionation to singlet oxygen in mildly acidic environment. *Adv. Synth. Catal.* **345**: 849–858.

Renirie R, Dewilde A, Pierlot C, Wever R, Hober D, Aubry J-M. 2008. Bactericidal and virucidal activity of the alkalophilic P395D/L241V/T343A mutant of vanadium chloroperoxidase. *J Appl. Microbiol.* **105**: 264–270.

Renirie R, Pierlot C, Wever R, Aubry JM. 2009. Singlet oxygenation in microemulsion catalyzed by vanadium chloroperoxidase. *J. Mol. Catal. B Enzym.* **56**: 259–264.

Renirie R, Hemrika W, Piersma SR, Wever R. 2000a. Cofactor and substrate binding to vanadium-chloroperoxidase determined by UV-VIS spectroscopy and evidence for high affinity for pervanadate. *Biochemistry* **39**: 1133–1141.

Renirie R, Hemrika W, Wever R. 2000b. Peroxidase and phosphatase activity of active-site mutants of vanadium chloroperoxidase from the fungus *Curvularia inaequalis*, implications for the catalytic mechanism. *J. Biol. Chem.* **275**: 11650–11657.

Simpson WR, von Glasow R, Riedel K, Anderson P, Ariya P, Bottenheim J, Burrows J, Carpenter LJ, Friess U, Goodsite ME, Heard D, Hutterli M, Jacobi HW, Kaleschke L, Neff B, Plane J, Platt U, Richter A, Roscoe H, Sander R, Shepson P, Sodeau J, Steffen A, Wagner T, Wolff E. 2007. Halogens and their role in polar boundary-layer ozone depletion. *Atmos. Chem. Phys.* **7**: 4375–4418.

Soedjak HS, Walker JV, Butler A. 1995. Inhibition and inactivation of vanadium bromoperoxidases by the substrate hydrogen peroxide and further mechanistic studies. *Biochemistry* **34**: 12689–12696.

Stukey J, Carman GM. 1997. Identification of a novel phosphatase sequence motif. *Protein Sci.* **6**: 469–472.

Tanaka N, Dumay V, Liao QN, Lange AJ, Wever R. 2002. Bromoperoxidase activity of vanadate-substituted acid phosphatases from *Shigella flexneri and Salmonella enterica* ser. typhimurium. *Eur. J. Biochem.* **269**: 2162–2167.

Tanaka N, Hasan Z, Wever R. 2003. Kinetic characterization of the active site mutants Ser402Ala and Phe397His of vanadium chloroperoxidase from the fungus *Curvularia inaequalis. Inorg. Chem. Acta* **356**: 288–296.

Ten Brink HB, Schoemaker HE, Wever R. 2001. Sulfoxidation mechanism of vanadium bromoperoxidase from *Ascophyllum nodosum. Eur. J. Biochem.* **268**: 132–138.

Ten Brink HB, Tuynman A, Dekker HL, Hemrika W, Izumi Y, Oshiro T, Schoemaker HE, Wever R. 1998. Enantioselective sulfoxidation catalyzed by vanadium haloperoxidases. *Inorg. Chem.* **37**: 6780–6784.

Tromp MGM, Olafsson G, Krenn BE, Wever R. 1990. Some structural aspects of vanadium bromoperoxidase from *Ascophyllum nodosum. Biochim. Biophys. Acta* **1040**: 192–198.

Van de Velde F, Arends IWCE, Sheldon RA. 2000. Biocatalytic and biomimetic oxidations with vanadium. *J. Inorg. Biochem.* **80**: 81–89.

Van Schijndel JWPM, Barnett P, Roelse J, Vollenbroek EGM, Wever R. 1994. The stability and steady state kinetics of vanadium chloroperoxidase from the fungus *Curvularia inaequalis. Eur. J. Biochem.* **22**: 151–157.

Van Schijndel JWPM, Vollenbroek EGM, Wever R. 1993. The chloroperoxidase from the fungus *Curvularia inaequalis*: a novel vanadium enzyme. *Biochim. Biophys. Acta* **1161**: 249–256.

Vilter H. 1984. Peroxidases from Phaeophyceae—a vanadium (V) dependent peroxidase from *Ascophyllum nodosum. Phytochem.* **23**: 1387–1390.

Vollenbroek EGM, Simons LH, van Schijndel J, Barnett P, Balzar M, Dekker H, van der Linden C, Wever R. 1995. Vanadium chloroperoxidases occur widely in nature. *Biochem. Soc. Trans.* **23**: 267–271.

Waller MP, Buhl M, Geethalakshmi KR, Wang DQ, Thiel W. 2007. V-51 NMR chemical shifts calculated from QM/MM models of vanadium chloroperoxidase. *Chem. Eur. J.* **13**: 4723–4732.

Wever R, Hemrika W. 2001. Vanadium haloperoxidases. In: Messerschmidt A, Hubert R, Poulos T, Wieghardt K, editors. *Handbook of Metalloproteins*. John Wiley, Chichester, UK, pp. 1417–1428.

Wever R, Renirie R, Hasan Z. 2006. Vanadium in biology. In: Scott RA, Section editor. *Encyclopedia Inorg. Chem. Section Bioinorganic Chemistry, www.wiley.com/eic2.*

Wever R, Tromp MGM, Krenn BE, Marjani A, van Tol M. 1991. Brominating activity of the seaweed *Ascophyllum nodosum*: impact on the biosphere. *Environ. Sci. Technol.* **25**: 446–449.

Weyand M, Hecht HJ, Kiess M, Liaud MF, Vilter H, Schomburg D. 1999. X-ray structure determination of a vanadium-dependent haloperoxidase from *Ascophyllum nodosum* at 2.0 Å resolution. *J. Mol. Biol.* **293**: 595–611.

Zhang Y, Gascón J. 2008. QM/MM investigation of structure and spectroscopic properties of a vanadium-containing peroxidase. *J. Inorg. Biochem.* **102**: 1684–1690.

Zampella G, Fantucci P, Pecoraro VL, De Gioia L. 2006. Insight into the catalytic mechanism of vanadium haloperoxidases: DFT investigation of vanadium cofactor reactivity. *Inorg. Chem.* **45**: 7133–7143.

CHAPTER 24

OTHER HEME PEROXIDASES AND ENZYMES

The list of newly discovered peroxidases grows every year. It is no surprise that they can be found in all members of the plant and animal kingdoms. Following is a brief summary of some heme peroxidases not discussed earlier, and for comparative purposes, some heme enzymes that perform other functions. With heme as the central prosthetic group, how do different enzymes utilize it for different purposes? Axial imidazole orientation is important (Zarić et al., 2001). The imaginative experiments in two research groups deserve special mention. One is that of Paul Ortiz de Montellano. They probe the heme environment by use of a battery of chemical reagents, site-directed mutants, and analytical techniques (Ortiz de Montellano, 1987; Swanson and Ortiz de Montellano, 1991; Tschirret-Guth and Ortiz de Montellano, 1999). The other is that of John Dawson, who utilizes spectroscopic techniques (Dawson, 1988).

DI-HEME PEROXIDASES

Pseudomonas aeruginosa Peroxidase

This enzyme from a primitive organism contains two heme groups attached to the same protein chain. Intensive pioneering work was conducted by the team of Nils Ellfolk and Marjaana Rönnberg and coworkers (Ellfolk and Soininen, 1970; Ellfolk, 1979; Ellfolk et al., 1991; Rönnberg, 1976).

EPR spectral studies were conducted (Aasa et al., 1981; Ellfolk et al., 1983), showing the two hemes have very different redox potentials. The active form of the enzyme is half-reduced. From a series of rapid-scan spectral and kinetic studies, mechanistic aspects were discussed (Araiso et al., 1980; Rönnberg et al., 1981a, 1981b, 1984, 1985a, 1985b). The ferric-ferric form of the enzyme reacts slowly, but by use

Peroxidases and Catalases: Biochemistry, Biophysics, Biotechnology, and Physiology, Second Edition.
By H. Brian Dunford

of azurin or ferrocyanide it can be reduced to the active ferrous-ferric form. Hydrogen peroxide reacts with the ferrous heme, so Compound I has a $Fe^{IV}{=}O$, Fe^{III} structure. One-electron reduction produces an active Compound II with a $Fe^{III}{-}Fe^{III}$ structure, different from the inactive form with the same oxidation states.

Ligand binding to photoactivated enzyme was investigated (Greenwood and Gibson, 1989). A quantitative model for the mechanism of action was established. Measurements of rate constants for elementary steps and computer modeling of the mechanism were in agreement (Foote et al., 1992). Azurin complex formation with oxidized *Pseudomonas* cytochrome *c* peroxidase is followed by an intramolecular electron exchange reaction (Brittain and Greenwood, 1992).

The amino acid sequence of *Pseudomonas* cytochrome *c* peroxidase was determined from the gene nucleotide sequence (Ridout et al., 1995). The crystal structure rapidly followed (Fülöp et al., 1993, 1995). Two domains each have a heme *c* prosthetic group. The low potential heme has two histidine axial ligands, and the high potential heme has histidine and methionine. The hydrophobic domain interface has a bound calcium ion. "The structure suggests that, in the half-reduced functional form of the enzyme, the low potential heme has to shed His71 in order to make the enzyme catalytically competent" (Fülöp et al., 1995, p. 1225). The *Pseudomonas* enzyme has been examined electrochemically after depositing it on both graphite and modified gold electrodes (Becker et al., 2009).

Paracoccus Cytochrome *c* Peroxidases

Studies have been conducted on the diheme cytochrome *c* peroxidase from *Paracoccus pantotrophus* (Pauleta et al., 2004a, 2004b). Competitive binding was demonstrated between pseudoazurin and cytochrome c_{550}. Pseudoazurin was shown to be an electron donor. A resonance Raman study demonstrated changes in the heme structure in the half-reduced and fully oxidized enzyme. A strong saddling deformation of the heme is present in the active form of the enzyme which contains calcium ion, an effect seen with other peroxidases. In the active form the heme iron is five-coordinate high spin (Pauleta et al., 2008). The amino acid sequence of *Paracoccus denitrificans* has been determined (Hu et al., 1997).

Rhodobacter capsulatus Cytochrome *c* Peroxidase

The crystal structure of *Rhodobacter capsulatus* cytochrome *c* peroxidase has been determined and, combined with site-directed mutagenesis experiments, provides a detailed picture of the structure-function relationships of the enzyme (De Smet et al., 2001, 2006). The enzyme reacts fastest of the known di-heme peroxidases with its physiological substrate. A mechanism is proposed for all the di-heme bacterial peroxidases.

Pseudomonas nautica 617 Cytochrome *c* Peroxidase

This enzyme has a great deal of homology with *P. aeruginosa* cytochrome *c* peroxidase (Alves et al., 1999). Its natural substrates are hydrogen peroxide and cytochrome

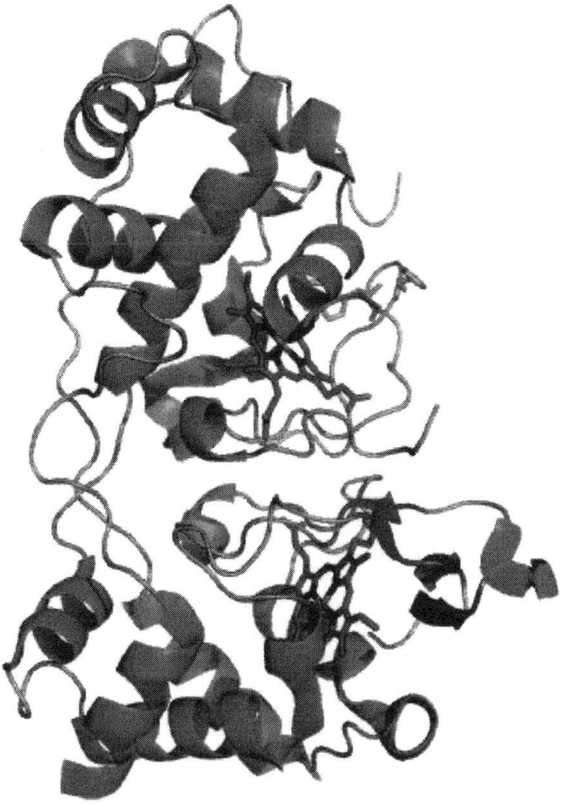

Figure 24.1. Structure of the di-heme peroxidase from *Pseudomonas nautica* 617.

c_{552}. Its crystal structure has been determined and a detailed discussion given of the two most plausible mechanisms of action. Ca^{2+} is required for its activation (Dias et al., 2004). A representation of its crystal structure is shown in Figure 24.1.

Other Di-Heme Peroxidases

Cellular location of di-heme peroxidases has been examined (Goodhew et al., 1990). Di-heme peroxidases have been found in *Paracoccus denitrificans* (Prazeres et al., 1995; Pettigrew et al., 2003), *Pleurotus ostreatus* (Y.-H. Han et al., 1996), *Pseudomonas denitrificans* (Coulson and Oliver, 1979), *Pseudomonas stutzeri* (Villalaín et al., 1984), and *Nitrosomonas europaea* (Arciero and Hooper, 1994; Shimizu et al., 2001). A lipoprotein cytochrome *c* peroxidase has been isolated from *Neisseria gonorrhoeae* (Turner et al., 2003).

PEROXIDASES EVERYWHERE YOU LOOK

Peroxidases homologous to mammalian peroxidases have been found in squid (Tomarev et al., 1993), in the ink gland of cuttlefish (Palumbo and Jackson, 1995), in *Drosphila*, called peroxidasin (Nelson et al., 1994), and in crayfish, called peroxinectin (Johansson and Söderhäll, 1988; Johnansson et al., 1995). Peroxidases are found in mosquito eggs (Q. Han et al., 2000), in *Flavobacterium meningosepticum* (Koga et al., 1999), and in the protozoan parasite *Trypanosoma cruzi*, the causative agent of Chagas disease (Wilkinson et al., 2002). Thioredoxin peroxidases are found in *Drosphila melanogaster* (Radyuk et al., 2001, 2003). *Thermomonospora fusca* contains a nonheme peroxidase (Svistunenko et al., 1999).

Among mammalian peroxidases, studies have been conducted on intestinal peroxidase (Kimura and Yamazaki, 1979; Ikeda-Saito and Kimura, 1990; Valoti et al., 1992; Chatterjee et al., 1993), lacrimal gland peroxidase (Mazumdar et al., 1996, 1997), and uterine peroxidase (Jellinck et al., 1991).

Peroxidases have been studied in African oil palm (Rodriguez et al., 2002; Sakharov and Sakharova, 2002), cucumber (Battistuzzi et al., 2001), the Mediterranean shrub *Euphorbia characias* (Medda et al., 2003; Mura et al., 2005), *Lupinus albus* (Hernández-Ruiz et al., 2000), potato (Espelie et al., 1986), Scots pine (Tarkka et al., 2001), sorghum (Dicko et al., 2006), and tobacco (Gazarian et al., 1996).

A Note on Peroxidase Nomenclature

What enzymes do is always discovered before one understands how they do it. The term cylooxygenase is a misnomer in the sense that the enzyme is a peroxidase, not an oxygenase. Like many other peroxidases, it produces free radicals that may be scavenged by dioxygen. Oxygen does not react directly with PGH synthase. Better to give a name that is noncommittal. My favorite is the Old Yellow Enzyme. Some noncommittal names are based on an absorbance maximum (cytochrome P450), on a substrate (manganese peroxidase, ascorbate peroxidase), or the name of one's wife: Stellacyanin. (My hat is off to Jack Peisach.) However, once a name is established, hopefully everyone understands its significance. With present day knowledge of peroxidases, the IUB numbering system has become increasingly irrelevant. Names are more likely to be used than numbers. A revised peroxidase nomenclature, similar to that accepted by cytochrome P450 workers, has been proposed (Dunford, 1999).

Chlorite Dismutase

An enzyme has been found that transforms chlorite into chloride and oxygen, an exciting new discovery. It appears to be a conventional peroxidase, but it has no sequence homologs in existing databases (Lee et al., 2008).

Artificial and Unusual Peroxidases

Unusual peroxidases have been created in the laboratory. Several microperoxidases have been produced from fragments of cytochrome *c*, which exhibit some activity. Antibody-heme complexes show peroxidase activity (Cochran and Schultz, 1990; Yin et al., 2004). Peroxidase activity was found in heme proteins derived from a designed combinatorial library (Moffet et al., 2000). Reaction of the Compound II species of the most active product reacted faster with 2,2′-azino-di-(3-ethylbenzthiazoline-6-sulfonic acid) (ABTS) than horseradish peroxidase, indicating a more accessible active site. Cytochrome *c* and cardiolipin complexes have peroxidase activity (Borisenko et al., 2008).

MYOGLOBINS

Now a brief look at other heme-containing proteins. The reduction potential of most myoglobins favors a stable ferrous state. Myoglobins reversibly bind dioxygen. They react with hydrogen peroxide to form a Compound II ($Fe^{IV}{=}O$) species but are incapable of forming Compound I, despite having an imidazole fifth ligand to the iron and a distal imidazole side chain. The ferrous deoxymyoglobin has five-coordinate iron and the ferric enzyme is six-coordinate with water as the sixth ligand. Myoglobins have different structures from the peroxidases and do not belong in the same family. The above paragraph summarizes some of the conventional wisdom about myoglobins. Some of these generalizations have been disproven with an unusual globin isolated from *Amphitrite ornata*, as described below. See also the concluding sentence about myoglobin.

Amphitrite ornata Dehaloperoxidase

Two isoenzymes were isolated from the marine worm *Amphitrite ornata* and one of them partially characterized (Chen et al., 1996). These enzymes can function as hemoglobins and as peroxidases (Nienhaus et al., 2008). Isoenzyme I converts trihalogenated phenols into dihalogenated quinones, and will oxidize bromo-, chloro-, and fluorophenols. The enzyme has an imidazole fifth ligand to iron which does not have imidazolate character (Franzen et al., 1998).

The full length coding sequence and crystal structure were obtained (LaCount et al., 2000). *Amphitrite ornata* dehaloperoxidase has a globin structure and has a common ancestry with the globins. The reducing substrate binding site is in the distal cavity. Changes in the positioning of the distal and proximal histidines compared to conventional globins are a major factor in the enhanced peroxidase activity.

The activity of the recombinant enzyme is midway between that of horseradish peroxidase and horse heart myoglobin. In its reaction with 2,4,6-tribromophenol (TBP), it requires TBP binding before peroxide addition. The TBP binding site above the heme is unique to peroxidases which usually react with substrates at the heme edge. Mutations showed that both distal and proximal histidines play important roles (Belyea et al., 2005).

Work on the *Amphitrite ornata* dehaloperoxidase continues in the laboratory of Stefan Franzen. Mutants abolishing distal and proximal histidines greatly decrease reactivity (Franzen et al., 2006a). Binding of carbon monoxide and nitric oxide to the ferrous enzyme and the effect of flash photolysis on geminate recombination have been examined (Nienhaus et al., 2006; Franzen et al., 2006b). The heme is activated at pH 6.5; at lower pH heme degradation occurs (Franzen et al., 2007).

Photodissociation of CO from the ferrous enzyme has been studied (Mikšovská et al., 2008). Compound I (called Compound ES) of the dehaloperoxidase has been detected (Feducia et al., 2009).

The above work on dehaloperoxidases led to another look at horse heart myoglobin, in which dehaloperoxidase activity was detected (Osborne et al., 2007). Conversion of trichlorophenol to a dichlorobenoquinone was demonstrated in a reaction involving myoglobin Compound II.

HEMOGLOBIN

Hemoglobin deserves the title of an honorary enzyme because of its influence on discussions of multi-subunit enzyme mechanisms. The famed extrema in mechanisms of subunit interactions in cooperative binding of oxygen: the Monod–Wyman Changeux model and Daniel Koshland's induced fit model both can describe the behavior of oxygen binding to hemoglobin, and neither can be proven to be right. A combination mechanism may be operative. It was pointed out that a total of 44 different states of hemoglobin are possible and "if all states were present a much larger number of parameters would be required for any realistic description" (Eigen, 1967, p. 342). This led to the famous cartoon showing that one can always fit the experimental data for a complex system with enough parameters, but it is not possible to attach any physical significance to their value (see Fig. 10.1).

CYTOCHROME *C* OXIDASE

Cytochrome *c* oxidase was discovered by Otto Warburg and initially was referred to as the respiratory enzyme. The reaction of the enzyme, which is highly exothermic, is the following:

$$O_2 + 4H^+ + 4e^- \longrightarrow 2H_2O$$

Electrons are supplied by heme-containing cytochrome *c*. The reaction occurs in four steps so the intermediates include superoxide or hydroperoxyl radical $HO_2^{\bullet}$ and hydrogen peroxide H_2O_2. Unlike peroxidases, which thrive on hydrogen peroxide, cytochrome *c* oxidase never releases it to the environment, but makes sure it is reduced all the way to water. Cytochrome *c* oxidase contains heme *a* as prosthetic group which has a long hydrocarbon side chain to anchor itself into a hydrophobic environment. This complicated enzyme is also a proton pump.

OXYGENASES

The oxygenases react directly with dioxygen. Thus peroxidases, including prostaglandin synthases, are not oxygenases; they can generate free radicals that then react with dioxygen but that is a different mechanism. The oxygenases were defined in the classic work of Howard Mason and Osamu Hayaishi who used oxygen isotopes to follow reaction pathways. They further defined monoxygenases, which utilize one oxygen atom of dioxygen, the other going to water; and dioxygenases that utilize both oxygen atoms. The most famous monoxygenase is the heme enzyme cytochrome P450.

Cytochrome P450

The cytochromes P450 are named after the wavelength maximum of their ferrous-CO absorbance. They are monooxygenases, that sometimes convert aromatic hydrocarbons into water-soluble carcinogens. They have an accessible hydrophobic distal binding site and do not have the built-in mechanism of peroxidases to attain higher oxidation states. Instead they rely on a complicated external system to deliver electrons obtained from NADH or NADPH. The electrons are transferred through a flavodehydrogenase and an iron-sulfur protein. Fe^{III} is reduced to Fe^{II}, dioxygen is bound, followed by O_2 cleavage leading to a formal $Fe^{V}{=}O$ species (Compound I) that carries out the oxygenation reaction. The elusive Compound I of a P450 has been detected (Newcomb et al., 2006; Sheng et al., 2008). Authoritative reviews on cytochromes P450 have been edited by Paul Ortiz de Montellano (Ortiz de Montellano, 2005).

Nitric Oxide Synthase

Relaxants that act upon blood vessels play an important role in regulating blood pressure. In 1980 it was discovered that the endothelium layer of blood vessels produced a hitherto unrecognized relaxant labeled endothelium-derived relaxing factor (EDFR; Furchgott and Zawadzki, 1980). Seven years later EDFR was identified as the diatomic free radical gas, nitric oxide (•NO) (Ignarro et al., 1987; Palmer et al., 1987). In 1998 the Nobel Prize in Physiology or Medicine was awarded to Furchgott, Ignarro, and Murad for their pioneering work on discovering the physiological role of nitric oxide.

The substrate for •NO production is the amino acid L-arginine. The overall reaction is

$$\text{L-arginine} + O_2 + e^- \longrightarrow \text{L-citrulline} + {}^{\bullet}NO + H_2O$$

Nitric oxide synthase is a dimeric enzyme. Each monomer has heme *b* as its prosthetic group and a thiolate fifth ligand attached to iron. Dioxygen is the oxidizing substrate, so the enzyme has many similarities to the cytochrome P450s. Each monomer consists of three regions: a reductase portion that binds NADPH, FAD, and FMN; a central portion that binds Ca^{2+} and calmodulin; and an oxidase portion where arginine and dioxygen are bound. NADPH is normally a two-electron donor but the flavins FAD and FMN can modulate flow to one electron at a time. Nitric oxide synthases are also found in the brain. Some are constitutive enzymes, others are inducible (Marletta, 1994). For a comprehensive review on nitric oxide and peroxynitrite see (Pacher et al., 2007).

The mechanism of action of nitric oxide synthase has been analyzed theoretically (de Visser and Tan, 2008; Cho et al., 2009).

Tryptophan Dioxygenase

Many dioxygenases contain nonheme iron. Tryptophan 2,3-dioxygenase is an example of a heme-containing dioxygenase. It catalyzes the conversion of L-tryptophan to formylkynurenine. Its crystal structure and that of mutants have been determined (Forouhar et al., 2007; Thackray et al., 2008). Indoleamine 2,3-dioxygenase can react with both D- and L-tryptophan. The results of a density functional theory (DFT) study of both tryptophan 2,3-dioxygenase and indoleamine 2,3-dioxygenase (Chung et al., 2008), and a recent experimental study (Chauhan et al., 2009), favor a mechanism in which direct electrophilic addition of substrate to heme dioxygen occurs. A multitude of oxygenase reactions occur in the oxidative degradation of tryptophan (Hayaishi, 1993, 2008).

HEME OXYGENASE

The first step in the degradation of heme is catalyzed by a heme oxygenase. There are three known isoenzymes of heme oxygenase, one induced by oxidative stress and the others constitutive. Crystal structures of heme oxygenase-1 have been published, both for the free enzyme and in complex with heme or biliverdin (Schuller et al., 1999; Sugishima et al., 2000, 2003; Lad et al., 2003, 2004). The heme oxidation products have been examined using resonance Raman spectroscopy (Takahashi et al., 1997), as has the ferric-NO complex (Tomita et al., 2001).

In mammals heme oxygenase catalyzes transformation of heme into biliverdin and carbon monoxide. The heme serves a dual role as both cofactor and substrate. Only the α-meso heme position is oxygenated in the enzyme reaction.

GUANYLYL CYCLASE

Soluble guanylyl cyclase is a dimeric enzyme with heme *b* loosely bound to each monomer. The nitric oxide generated by nitric oxide synthase binds to the heme iron as part of the activation process. The reaction catalyzed by guanylyl cyclase is the conversion of guanosine triphosphate to cyclic guanosine monophosphate. The cyclic product is apparently the final reagent in smooth muscle relaxation, the sexual overtones of which reverberate throughout society. The NO binding site in soluble guanylyl cyclase has been modeled in a computer simulation (Capece et al., 2008).

REFERENCES

Aasa R, Ellfolk N, Rönnberg M, Vänngård T. 1981. Electron paramagnetic resonance studies of *Pseudomonas* cytochrome *c* peroxidase. *Biochim. Biophys. Acta* **670**: 170–175.

Alves T, Besson S, Duarte LC, Pettigrew GW, Girio FMF, Devreese B, Vandenberghe I, Van Beeumen J, Fauque G, Moura I. 1999. A cytochrome *c* peroxidase from *Pseudomonas nautica* 617 active at high ionic strength: expression, purification and characterization. *Biochim. Biophys. Acta* **1434**: 248–259.

Araiso T, Rönnberg M, Dunford HB, Ellfolk N. 1980. The formation of the primary compound from hydrogen peroxide and *Pseudomonas* cytochrome *c* peroxidase. *FEBS Lett.* **118**: 99–102.

Arciero DM, Hooper AB. 1994. A di-heme cytochrome *c* peroxidase from *Nitrosomas europaea* catalytically active in both the oxidized and half-reduced states. *J. Biol. Chem.* **269**: 11878–11886.

Battistuzzi G, D'Onofrio M, Loschi L, Sola M. 2001. Isolation and characterization of two peroxidases from *Cucumis sativus*. *Arch. Biochem. Biophys.* **388**: 100–112.

Becker CF, Watmough NJ, Elliott SJ. 2009. Electrochemical evidence of multiple peroxidatic heme states of the diheme cytochrome *c* peroxidase of *Pseudomonas aeruginosa*. *Biochemistry* **48**: 87–95.

Belyea J, Gilvey LB, Davis MF, Godek M, Sit TL, Lommel SA, Franzen S. 2005. Enzyme function of the globin dehaloperoxidase from *Amphitrite ornata* is activated by substrate binding. *Biochemistry* **44**: 15637–15644.

Borisenko GG, Kapralov AA, Tyurin VA, Maeda A, Stoyanovsky DA, Kagan VE. 2008. Molecular design of new inhibitors of peroxidase activity of cytochrome c/cardiolipin complexes: fluorescent oxadiazole-derivatized cardiolipin. *Biochemistry* **47**: 13699–13710.

Brittain T, Greenwood C. 1992. Complex formation between the copper protein, azurin and the cytochrome *c* peroxidase of *Pseudomonas aeruginosa*. *J. Inorg. Biochem.* **48**: 71–77.

Capece L, Estrin DA, Marti MA. 2008. Dynamic characterization of the heme NO oxygen binding (HNOX) domain: insight into soluble guanylate cyclase allosteric transition. *Biochemistry* **47**: 9416–9427.

Chatterjee R, Bandyopadhyay U, Bhattacharyya D, Banerjee RK. 1993. Inhibition of intestinal peroxidase activity by nonsteroidal anti-inflammatory drugs. *Biochim. Biophys. Acta* **1161**: 168–176.

Chauhan N, Thackray SJ, Rafice SA, Eaton G, Lee M, Efimov I, Basran J, Jenkins PR, Mowat CG, Chapman SK, Raven EL. 2009. Reassessment of the reaction mechanism in the heme dioxygenases. *J. Am. Chem. Soc.* **131**: 4186–4187.

Chen YP, Woodin SA, Lincoln DE, Lovell CR. 1996. An unusual dehalogenating peroxidase from the marine terebellid polychaete *Amphitrite ornata*. *J. Biol. Chem.* **271**: 4609–4612.

Cho K-B, Carvajal MA, Shaik S. 2009. First half-reaction mechanism of nitric oxide synthase: the role of proton and oxygen coupled electron transfer in the reaction by quantum mechanics/molecular mechanics. *J. Phys. Chem. B.* **113**: 336–346.

Chung LW, Li X, Sugimoto H, Shiro Y, Morokuma K. 2008. Density functional theory study on a missing piece in understanding of heme chemistry: the reaction mechanism for indoleamine 2,3-dioxygenase and tryptophan 2,3-dioxygenase. *J. Am. Chem. Soc.* **130**: 12299–12309.

Cochran AG, Schultz PG. 1990. Peroxidase activity in an antibody-heme complex. *J. Am. Chem. Soc.* **112**: 9414–9415.

Coulson AFW, Oliver RIC. 1979. Isolation and properties of a cytochrome *c* peroxidase from *Pseudomonas denitrificans*. *Biochem. J.* **181**: 159–169.

Dawson JH. 1988. Probing structure-function relations in heme-containing oxygenases and peroxidases. *Science* **240**: 433–439.

De Smet L, Pettigrew G, Van Beeumen JJ. 2001. Cloning, overproduction and characterization of cytochrome *c* peroxidase from the purple phototrophic bacterium *Rhodobacter capsulatus*. *Eur. J. Biochem.* **268**: 6559–6568.

De Smet L, Savvides SN, Van Horen E, Pettigrew G, Van Beeumen JJ. 2006. Structural and mutagenesis studies on the cytochrome *c* peroxidase from *Rhodobacter capsulatus* provide new insights into structure-function relationships of bacterial di-heme peroxidases. *J. Biol. Chem.* **281**: 4371–4379.

de Visser SP, Tan LS. 2008. Is the bound substrate in nitric oxide synthase protonated or neutral and what is the active oxidant that performs substrate hydroxylation? *J. Am. Chem. Soc.* **130**: 12961–12974.

Dias JM, Alves T, Bonifácio C, Pereira AS, Trincão J, Bourgeois D, Moura I, Romão J. 2004. Structural basis for the mechanism of Ca^{2+} activation of the di-heme cytochrome *c* peroxidase from *Pseudomonas nautica* 617. *Structure* **12**: 961–973.

Dicko MH, Gruppen H, Hilhorst R, Voragen AGJ, van Berkel WJH. 2006. Biochemical characterization of the major sorghum grain peroxidase. *FEBS J.* **273**: 2293–2307.

Dunford HB. 1999. Peroxidase nomenclature. Presented at the *Fifth International Peroxidase Symoposium*. Mark Lagrimini, organizer. Deer Creek State Park Lodge, Ohio. July 17–21.

Eigen M. 1967. Kinetics of reaction control and information transfer in enzymes and nucleic acids. In: Claesson S, editor. *Nobel Symposium 5. Fast Reactions and Primary Processes in Chemical Kinetics*. Almqvist and Wiksell, Stockholm, pp. 334–367.

Ellfolk M (nee Rönnberg). 1979. Mechanism of action of *Pseudomonas* cytochrome *c* peroxidase. Academic Dissertation. Department of Biochemistry, University of Helsinki.

Ellfolk N, Rönnberg M, Aasa R, Andréasson L-E, Vänngård T. 1983. Properties and functions of the two hemes in *Pseudomonas* cytochrome *c* peroxidase. *Biochim. Biophys. Acta* **743**: 23–30.

Ellfolk N, Rönnberg M, Österlund K. 1991. Structural and functional features of *Pseudomonas* cytochrome *c* peroxidase. *Biochim. Biophys. Acta* **1080**: 68–77.

Ellfolk N, Soininen R. 1970. *Pseudomonas* cytochrome *c* peroxidase. *Acta Chem. Scand.* **24**: 2126–2136.

Espelie KE, Franceschi VR, Kolattukudy PE. 1986. Immunocytochemical localization and time course of appearance of an anionic peroxidase associated with suberization in wound-healing potato tuber tissue. *Plant Physiol.* **81**: 487–492.

Feducia J, Dumarieh R, Gilvey BGL, Smirnova T, Franzen S, Ghiladi RA. 2009. Characterization of dehaloperoxidase Compound ES and its reactivity with trihalophenols. *Biochemistry* **48**: 995–1005.

Foote N, Turner R, Brittain T, Greenwood C. 1992. A quantitative model for the mechanism of action of the cytochrome *c* peroxidase from *Pseudomonas aeruginosa*. *Biochem. J.* **283**: 839–843.

Forouhar F, Andersson JLR, Mowat CG, Vorobiev SM, Hussain A, Abashidze M, Bruckmann C, Thackray SJ, Seetharaman J, Tucker T, Xiao R, Ma L-C, Zhao L, Acton TB, Montelione GT, Chapman SK, Tong L. 2007. Molecular insights into substrate recognition by tryptophan 2,3-dioxygenase. *Proc. Natl. Acad. Sci. USA* **104**: 473–478.

Franzen S, Belyea J, Gilvey LB, Davis MF, Chaudhary CE, Sit TL, Lommel SA. 2006a. Proximal cavity, distal histidine, and substrate hydrogen-bonding mutations modulate the activity of *Amphitrite ornata* dehaloperoxidase. *Biochemistry* **45**: 9085–9094.

Franzen S, Gilvey LB, Belyea J. 2007. The pH dependence of the activity of dehaloperoxidase from *Amphitrite ornate*. *Biochim. Biophys. Acta* **1774**: 121–130.

Franzen S, Jasaitis A, Belyea J, Brewer SH, Casey R, MacFarlane IV AW, Stanley RJ, Vos MH, Martin J-L. 2006b. Hydrophobic distal pocket affects NO-heme geminate recombination dynamics in dehaloperoxidase and H64V myoglobin. *J. Phys. Chem. B* **110**: 14483–14493.

Franzen S, Roach MP, Chen YP, Dyer RB, Woodruff WH, Dawson JH. 1998. The unusual reactivities of *Amphitrite ornata* dehaloperoxidase and *Notomastus lobatus* chloroperoxidase do not arise from a histidine imidazolate proximal heme iron ligand. *J. Am. Chem. Soc.* **120**: 4658–4661.

Fülöp V, Little R, Thompson A, Greenwood C, Hajdu J. 1993. Crystallization and preliminary X-ray analysis of the di-haem cytochrome *c* peroxidase from *Pseudomonas aeruginosa*. *J. Mol. Biol.* **232**: 1208–1210.

Fülöp V, Ridout CJ, Greenwood C, Hajdu J. 1995. Crystal structure of the di-haem cytochrome *c* peroxidase from *Pseudomonas aeruginosa*. *Structure* **3**: 1225–1233.

Furchgott RF, Zawadzki JV. 1980. The obligatory role of endothelial cells in the relaxation of arterial smooth muscle by acetylcholine. *Nature* **288**: 373–376.

Gazarian IG, Lagrimini LM, George SJ, Thorneley RNF. 1996. Anionic tobacco peroxidase is active at extremely low pH: veratryl alcohol oxidation with a pH optimum of 1.8. *Biochem. J.* **320**: 369–372.

Goodhew CF, Wilson IBH, Hunter DJB, Pettigrew GW. 1990. The cellular location and specificity of bacterial cytochrome *c* peroxidases. *Biochem. J.* **271**: 707–712.

Greenwood C, Gibson QH. 1989. Ligand binding to cytochrome *c* peroxidase from *Pseudomonas aeruginosa*. *J. Biol. Chem.* **264**: 19022–19027.

Han Q, Li G, Li J. 2000. Purification and characterization of chorion peroxidase from *Aedes aegypti* eggs. *Arch. Biochem. Biophys.* **378**: 107–115.

Han Y-H, Shin K-S, Youn H-D, Hah YC, Kang S-O. 1996. Mode of action and active site of an extracellular peroxidase from *Pleurotus ostreatus*. *Biochem. J.* **314**: 421–426.

Hayaishi O. 1993. My life with tryptophan — never a dull moment. *Protein Sci.* **2**: 472–475.

Hayaishi O. 2008. From oxygenase to sleep. *J. Biol. Chem.* **283**: 19165–19175.

Ikeda-Saito M, Kimura S. 1990. Axial ligand coordination in intestinal peroxidase. *Arch. Biochem. Biophys.* **283**: 351–355.

Hernández-Ruiz J, Rodriguez-López JN, Garcia-Cánovas F, Acosta M, Arnao MB. 2000. Characterization of isoperoxidase-B2 inactivation in etiolated *Lupinus albus* hypocotyls. *Biochim. Biophys. Acta* **1478**: 78–88.

Hu W, Van Driessche G, Devreese B, Goodhew CF, McGinnity DF, Saunders N, Fulop V, Pettigrew GW, Van Beeuman J. 1997. Structural characterization of *Paracoccus denitrificans* cytochrome *c* peroxidase and assignment of low and high potential heme sites. *Biochemistry* **36**: 7958–7966.

Ignarro LG, Buga GM, Wood KS, Byrns RE, Chaudhuri G. 1987. Endothelium-derived relaxing factor produced and released from artery and vein is nitric oxide. *Proc. Natl. Acad. Sci. USA* **84**: 9265–9269.

Jellinck PH, Newbold RR, McLachlan JA. 1991. Characteristics of estrogen-induced peroxidase in mouse uterine luminal fluid. *Steroids* **56**: 162–167.

Johansson MW, Lind MI, Holmblad T, Thörnqvist P-O, Söderhäll K. 1995. Peroxinectin, a novel adhesion protein from crayfish blood. *Biochem. Biophys. Res. Commun.* **216**: 1079–1087.

Johansson MW, Söderhäll K. 1988. Isolation and purification of a cell adhesion factor from crayfish blood cells. *J. Cell Biol.* **106**: 1795–1803.

Kimura S, Yamazaki I. 1979. Comparisons between hog intestinal peroxidase and bovine lactoperoxidase: Compound I formation and inhibition by benzhydroxamic acid. *Arch. Biochem. Biophys.* **198**: 580–588.

Koga S, Ogawa J, Choi Y-M, Shimizu S. 1999. Novel bacterial peroxidase without catalase activity from *Flavobacterium meningosepticum:* purification and characterization. *Biochim. Biophys. Acta* **1435**: 117–126.

LaCount MW, Zhang E, Chen YP, Han K, Whitton MW, Lincoln DE, Woodin SA, Lebodia L. 2000. The crystal structure and amino acid sequence of dehaloperoxidase from *Amphitrite ornata* indicate common ancestry with globins. *J. Biol. Chem.* **275**: 18712–18716.

Lad L, Friedman J, Li H, Bhaskar B, Ortiz de Montellano PR, Poulos TL. 2004. Crystal structure of heme oxygenase-1 in a complex with biliverdin. *Biochemistry* **43**: 3793–3801.

Lad L, Schuller DJ, Shimizu H, Friedman J, Li H, Ortiz de Montellano PR, Poulos TL. 2003. Comparison of the heme-free and -bound crystal structures of human heme oxygenase-1. *J. Biol. Chem.* **278**: 7834–7843.

Lee AQ, Streit BR, Zdilla MJ, Abu-Omar MM, DuBois JL. 2008. Mechanism of and exquisite selectivity for O-O bond formation by the heme-dependent chlorite dismutase. *Proc. Natl. Acad. Sci. USA* **105**: 15654–15659.

Marletta MA. 1994. Nitric oxide synthase: aspects concerning structure and catalysis *Cell* **78**: 927–930.

Mazumdar A, Adak S, Chatterjee R, Banerjee RK. 1997. Mechanism-based inactivation of lacrimal-gland peroxidase by phenylhydrazine: a suicidal substrate to probe the active site. *Biochem. J.* **324**: 713–719.

Mazumdar A, Chatterjee R, Adak S, Ghosh A, Mondal C, Banerjee RK. 1996. Characterization of sheep lacrimal-gland peroxidase and its major physiological electron donor. *Biochem. J.* **314**: 413–419.

Meda R, Padiglia A, Longu S, Bellelli A, Arcovito A, Cavallo S, Pedersen JZ, Floris G. 2003. Critical role of Ca^{2+} ions in the reaction mechanism of *Euphorbia characias* peroxidase. *Biochemistry* **42**: 8909–8918.

Mikšovská J, Horsa S, Davis MF, Franzen S. 2008. Conformational dynamics associated with photodissociation of CO from dehaloperoxidase studied using photoacoustic calorimetry. *Biochemistry* **47**: 11510–11517.

Moffett DA, Certain LK, Smith AJ, Kessel AJ, Beckwith KA, Hecht MH. 2000. Peroxidase activity in heme proteins derived from a designed combinatorial library. *J. Am. Chem. Soc.* **122**: 7612–7613.

Mura A, Medda R, Longu S, Floris G, Rinaldi AC, Padiglia A. 2005. A Ca^{2+}/calmodulin binding peroxidase from *Euphorbia* latex: novel aspects of calcium-hydrogen peroxide cross-talk in the regulation of plant defences. *Biochemistry* **44**: 14120–14130.

Nelson RE, Fessler LI, Takagi Y, Blumberg B, Keene DR, Olson PF, Parker CG, Fessler JH. 1994. Peroxidasin: a novel enzyme-matrix protein of *Drosphila* development. *EMBO J.* **13**: 3438–3447.

Newcomb M, Zhang R, Chandrasena REP, Halgrimson JA, Horner JH, Makris TM, Sligar SG. 2006. Cytochrome P450 Compound I. *J. Am. Chem. Soc.* **128**: 4580–4581.

Nienhaus K, Deng P, Belyea J, Franzen S, Nienhaus GU. 2006. Spectroscopic study of substrate binding to the carbonmonoxy form of dehaloperoxidase from *Amphitrite ornate*. *J. Phys. Chem. B* **110**: 13264–13276.

Nienhaus K, Nickel E, Davis MF, Franzen S, Nienhaus GU. 2008. Determinants of substrate internalization in the distal pocket of dehaloperoxidase hemoglobin of *Amphitrite ornate*. *Biochemistry* **47**: 12985–12994.

Osborne RL, Coggins MK, Walla M, Dawson JH. 2007. Horse heart myoglobin catalyzes the H_2O_2 dependent oxidative dehalogenation of chlorophenols to DNA-binding radicals and quinones. *Biochemistry* **46**: 9823–9829.

Ortiz de Montellano PR. 1987. Control of the catalytic activity of prosthetic heme by the structure of hemoproteins. *Acc. Chem. Res.* **20**: 289–294.

Ortiz de Montellano PR, editor. 2005. *Cytochrome P450: Structure, Mechanism and Biochemistry*, 3rd edition. Kluwer Academic/Plenum Publishers, New York.

Pacher P, Beckman JS, Liaudet L. 2007. Nitric oxide and peroxynitrite in health and disease. *Physiol. Rev.* **87**: 315–424.

Palmer RMJ, Ferrige AG, Moncada S. 1987. Nitric oxide release accounts for the biological activity of endothelium-derived relaxing factor. *Nature* **327**: 524–526.

Palmer RMJ, Ashton DS, Moncada S. 1988. Vascular endothelial cells synthesize nitric oxide from L-arginine. *Nature* **333**: 664–666.

Palumbo A, Jackson IJ. 1995. Peroxidase activity in the ink gland of *Sepia officinalis* and partial nucleotide sequence of a candidate cDNA encoding the enzyme. *Biochim. Biophys. Acta* **1247**: 173–178.

Pauleta SR, Cooper A, Nutley M, Errington N, Harding S, Guerlesquin F, Goodhew CF, Moura I, Moura JJG, Pettigrew GW. 2004a. A copper protein and a cytochrome bind at the same site on bacterial cytochrome *c* peroxidase. *Biochemistry* **43**, 14566–14576.

Pauleta SR, Goodhew CF, Moura I, Pettigrew GW, Shelnutt JA. 2008. Calcium-dependent heme structure in the reduced forms of the bacterial cytochrome *c* peroxidase from *Paracoccus pantotrophus*. *Biochemistry* **47**, 5841–5850.

Pauleta SR, Guerlesquin F, Goodhew CF, Devreese B, Van Beeumen J, Pereira AS, Moura I, Pettigrew GW. 2004b. *Paracoccus pantotrophus* pseudoazurin is an electron donor to cytochrome *c* peroxidase. *Biochemistry* **43**, 11214–11225.

Pettigrew GW, Pauleta SR, Goodhew CF, Cooper A, Nutley M, Jumel K, Harding SE, Costa C, Krippahl L, Moura I, Moura J. 2003. Electron transfer complexes of cytochrome *c* peroxidase from *Paracoccus denitrificans* containing more than one cytochrome. *Biochemistry* **42**: 11968–11981.

Prazeres S, Moura I, Gilmour R, Pettigrew G, Ravi N, Huynh BH. 1995. Redox and spin-state control of the activity of a diheme cytochrome *c* peroxidase: spectroscopic studies.

In: La Mar G, editor. *Nuclear Magnetic Resonance of Paramagnetic Molecules*. Kluwer Academic, New York, pp. 141–163.

Radyuk SN, Klichko VI, Spinola B, Sohal RS, Orr WC. 2001. The peroxiredoxin gene family in *Drosphila melongaster*. *Free Rad. Biol. Med.* **31**: 1090–1100.

Radyuk SN, Sohal RS, Orr WC. 2003. Thioredoxin peroxidases can foster cytoprotection or cell death in response to different stressors: over- and under-expression of thiredoxin peroxidase in *Drosphila* cells. *Free Rad. Biol. Med.* **31**: 1090–1100.

Ridout CJ, James R, Greenwood C. 1995. Nucleotide sequence encoding the di-haem cytochrome c_{551} peroxidase from *Pseudomonas aeruginosa*. *FEBS Lett.* **365**: 152–154.

Rodriguez A, Pina DG, Yélamos B, Castillo Leon JJ, Zhadan GG, Villar E, Gavilanes F, Roig MG, Sakharov IYu, Shnyrov VL. 2002. *Eur. J. Biochem.* **269**: 2584–2590.

Rönnberg M. 1976. *Pseudomonas* cytochrome *c* peroxidase. XII. Product inhibition studies. *Acta Chem. Scand. B* **30**, 721–726.

Rönnberg M, Araiso T, Ellfolk N, Dunford HB. 1981a. The catalytic mechanism of *Pseudomonas* cytochrome *c* peroxidase. *Arch. Biochem. Biophys.* **207**: 197–204.

Rönnberg M, Araiso T, Ellfolk N, Dunford HB. 1981b. The reaction between reduced azurin and oxidized cytochrome *c* peroxidase from *Pseudomonas aeruginosa*. *Arch. Biochem. Biophys.* **207**: 197–204.

Rönnberg M, Ellfolk N, Dunford HB. 1984. The formation of the primary hydrogen peroxide compound (Compound I) of *Pseudomonas* cytochrome *c* peroxidase as a function of pH. *Acta Chem. Scand. B* **38**: 79–83.

Rönnberg M, Lambeir A-M, Ellfolk N, Dunford HB. 1985a. Kinetics of cyanide binding by half-reduced *Pseudomonas* cytochrome *c* peroxidase. *Biochim. Biophys. Acta* **828**: 67–72.

Rönnberg M, Lambeir A-M, Ellfolk N, Dunford HB. 1985b. A rapid-scan spectrophotometric and stopped-flow study of Compound I and Compound II of *Pseudomonas* cytochrome *c* peroxidase. *Arch. Biochem. Biophys.* **236**: 714–719.

Sakharov IYu, Sakharova IV. 2002. Extremely high stability of African palm tree peroxidase. *Biochim. Biophys. Acta* **1598**: 108–114.

Schuller DJ, Wilks A, Ortiz de Montellano PR, Poulos TL. 1999. Crystal structure of human heme oxygenase-1. *Nature Struct. Biol.* **6**: 860–867.

Sheng X, Horner JH, Newcomb M. 2008. Spectra and kinetic properties of the Compound I derivative of cytochrome P450 119. *J. Am. Chem. Soc.* **130**: 13310–13320.

Shimizu H, Schuller DJ, Lanzilotta WN, Sundaramoorthy M, Arciero DM, Hooper AB, Poulos TL. 2001. Crystal structure of *Nitrosomonas europaea* cytochrome *c* peroxidase and the structural basis for ligand switching in bacterial di-heme peroxidases. *Biochemistry* **40**: 13485–13490.

Sugishima M, Omata Y, Kakuta Y, Sakamoto H, Noguchi M, Fukuyama K. 2000. Crystal structure of rat heme oxygenase-1 in complex with heme. *FEBS Lett.* **471**: 61–66.

Sugishima M, Sakamoto H, Higashimoto Y, Noguchi M, Fukuyama K. 2003. Crystal structure of rat heme oxygenase-1 in complex with biliverdin-iron chelate. *J. Biol. Chem.* **278**: 32352–32358.

Svistunenko DA, Rob A, Ball A, Torres J, Symons MCR, Wilson MT, Cooper CE. 1999. The electron paramagnetic resonance characterization of a copper-containing extracellular peroxidase from *Thermomonospora fusca* BD25. *Biochim. Biophys. Acta* **1434**: 74–85.

Swanson BA, Ortiz de Montellano PR. 1991. Structure and absolute stereochemistry of the four *N*-phenylprotoporphyrin IX regioisomers isolated from phenylhydrazine-treated myoglobin. *J. Am. Chem. Soc.* **113**: 8146–8153.

Takahashi S, Matera KM, Fujii H, Zhou H, Ishikawa K, Yoshida T, Ikeda-Saito M, Rousseau DL. 1997. Resonance Raman spectroscopic characterization of α-hydroxyheme and verdoheme complexes of heme oxygenase. *Biochemistry* **36**: 1402–1410.

Tarkka MT, Nyman TA, Kalkkinen N, Raudaskoski M. 2001. Scots pine expresses short-root-specific peroxidase during development. *Eur. J. Biochem.* **268**: 86–92.

Thackray SJ, Bruckmann C, Anderson JLR, Campbell LP, Xiao R, Zhao L, Mowat CG, Forouhar F, Tong L, Chapman SK. 2008. Histidine 55 of tryptophan 2,3-dioxygenase is not an active site base but regulates catalysis by controlling substrate binding. *Biochemistry* **47**: 10677–10684.

Tomarev SI, Zinovieva RD, Weis VM, Chepelinsky AB, Piatigorsky J, McFall-Ngai MJ. 1993. Abundant mRNAs in the squid light organ encode proteins with a high similarity to mammalian peroxidases. *Gene* **132**: 219–226.

Tomita T, Haruta N, Aki M, Kitigawa T, Ikeda-Saito M. 2001. UV resonance Raman detection of a ligand vibration on ferric nitrosyl heme proteins. *J. Am. Chem. Soc.* **123**: 2666–2667.

Tschirret-Guth RA, Ortiz de Montellano PR. 1999. Trifluoromethyldiazirinphenyldiazenes: new hemoprotein active site probes. *J. Am. Chem. Soc.* **121**: 4731–4737.

Turner S, Reid E, Smith H, Cole J. 2003. A novel cytochrome *c* peroxidase from *Neisseria gonorrhoeae:* a lipoprotein from a Gram-negative bacterium. *Biochem. J.* **373**: 865–873.

Valoti M, Tipton KF, Sgaragli GP. 1992. Oxidative ring-coupling of tyrosine and its derivatives by purified rat intestinal peroxidase. *Biochem. Pharmacol.* **43**: 945–951.

Villalaín J, Moura I, Liu MC, Payne WJ, LeGall J, Xavier AV, Moura JJG. 1984. NMR and electron paramagnetic-resonance studies of a dihaem cytochrome from *Pseudomonas stutzeri* (ATCC11607) (cytochrome *c* peroxidase). *Eur. J. Biochem.* **141**: 305–312.

Wilkinson SR, Obado SO, Mauricio IL, Kelly JM. 2002. *Trypanosoma cruzi* expresses a plant-like ascorbate-dependent hemoperoxidase localized in the endoplasmic reticulum. *Proc. Natl. Acad. Sci. USA* **99**: 13453–13458.

Yin J, Mills JH, Schultz PG. 2004. A catalysis-based selection for peroxidase antibodies with increased activity. *J. Am. Chem. Soc.* **126**: 3006–3007.

Zarić SD, Popović DM, Knapp E-W. 2001. Factors determining the orientation of axially coordinated imidazoles in heme proteins. *Biochemistry* **40**: 7914–7928.

CHAPTER 25

APPLICATION OF PEROXIDASES

RON WEVER

INTRODUCTION

In this chapter potential applications and utilization of the two major classes of peroxidases are discussed. The first class is the vanadium peroxidases that contain the metal oxide vanadate. These enzymes are able to oxidize halides to their corresponding hypohalous acids. By virtue of their prosthetic group they are very resistant towards oxidative inactivation thus allowing more easy applications. This is in contrast to the second class, the heme-containing peroxidases. Due to the electron-rich nature of the heme prosthetic group they are easily oxidized and inactivated either by the substrate or by the products formed. The heme peroxidases, however, catalyze a much wider variety of reactions. Many applications of the vanadium enzymes and the heme peroxidases have been envisaged and consequently a wide range of patents has been published. With the exception of the use of heme peroxidases in analytical applications and diagnostics the number of commercial products that have been produced using peroxidases or containing peroxidases is still limited. In this chapter five major areas of (potential) applications of products and processes are discussed. These are the potential use of peroxidases as pharmaceutical or antimicrobial agents, the application in bleaching and detergents, their use in polymerization and depolymerization processes, their use in the insertion of oxygen in organic molecules and the oxidation of organic molecules, and analytical and medical applications. References are also made to the literature on patents; however, some of the more recent patents are refinements of previous ideas and will not be discussed in detail.

BACKGROUND INFORMATION

Many enzymes have made their way into everyday life. Examples are the use of proteases, lipases, and amylases in common household detergents. A variety of enzymes

Peroxidases and Catalases: Biochemistry, Biophysics, Biotechnology, and Physiology, Second Edition.
By H. Brian Dunford

are being used now in the industrial production process or conversions of a variety of products in the food, pharmaceutical, and chemical industries (Kirk et al., 2002). Examples of enzymes being used in the food industry are glucose isomerase and pectolytic enzymes. In the largest scale biocatalytic process (10^6 tons annually) the isomerase converts glucose to the sweeter fructose, resulting in high fructose corn syrup. The pectolytic enzymes are being used in the treatment of fruit juices to allow easier filtration, clarification, and concentration. Enzymes are also replacing processes in the chemical industry; examples are the use of nitril hydratase in the large-scale production of acryl amide (annual production about 5×10^4 tons) and in the production of antibiotics such as 6-aminopenicillic acid and cephalexine. Thanks to the development of recombinant enzyme technology the production of many enzymes on a very large and industrial scale is now possible allowing many uses in the detergent and fermentation industries. The availability of these enzymes has also triggered research in other directions. Lipase is such an example. It was originally produced for application in detergents but has found various applications in organic synthesis, for example, in the large-scale production of enantiomerically pure alcohols. The use of recombinant technology, protein engineering, and directed evolution to create enzymes with more specific properties has further revolutionized the development of industrial enzymes.

For successful application of enzymes in new industrial processes a number of requirements have to be met which may seriously delay or even prevent the introduction of an enzymatic process. Some of these are obvious; however, other requirements are not always considered a problem. Whether an enzyme-based process will penetrate a market successfully is determined by many factors. For example, the process should be competitive towards the existing process, be cheaper, be cleaner, and preferably produce less waste and consume less energy. An example is the cell- and enzyme-based production of the dye indigo (Berry et al., 2002) by Genencor. Indigo is widely used to give the characteristic blue color to denim. The enzyme-based process developed was clearly much cleaner than the original industrial process that uses aniline and cyanide. However, the cost of the indigo produced by the enzyme-based process was higher and therefore unable to compete with the existing process. To be successful it should also be possible to produce a recombinant enzyme on an industrial scale with a high operational stability and a large turnover number in terms of product produced or converted per unit time per molecule of enzyme. Finally if the enzyme contains a prosthetic group the incorporation of the prosthetic group should be easy and not require complex reactivation procedures. Unfortunately this is the case for most heme peroxidases. In general an apoenzyme can easily be produced but reactivation is not straightforward. An exception appears to be the reactivation of the *Coprinus cinereus* heme peroxidase (Cherry et al., 1999).

PEROXIDASES AS PHARMACEUTICAL AND/OR ANTIMICROBIAL AGENTS

Most peroxidases catalyze the two-electron oxidation of a halide (Cl^-, Br^-, I^-) and the pseudohalide (SCN^-) by H_2O_2 to hypohalous acids (Eq. 25.1).

$$H_2O_2 + X^- + H^+ \longrightarrow HOX + H_2O \tag{25.1}$$

The peroxidases catalyzing this reaction are generically named haloperoxidases. The oxidizing intermediate HOX formed by the vanadium haloperoxidases may be released from the active site and reacts aspecifically (Eq. 25.2) with a variety of organic compounds (RH) resulting in the formation of a diversity of halogenated compounds (RX).

$$HOX + R{-}H \longrightarrow RX + H_2O \qquad (25.2)$$

For the heme chloroperoxidase there is substantial evidence that an enzyme-bound halogenating enzyme intermediate is present that specifically interacts with substrates (Hofrichter and Ulrich, 2006). The nomenclature of the haloperoxidases is based on the most electronegative halide these enzymes can oxidize. The chloroperoxidases can oxidize Cl^- and Br^- and I^- and bromoperoxidases can oxidize Br^- and I^- and iodoperoxidases oxidize only I^-.

At more alkaline pH values, in particular at high concentrations of H_2O_2 and in the absence of suitable scavenger for HOX a competing reaction (Eq. 25.3) between H_2O_2 and HOX occurs.

$$H_2O_2 + HOX \longrightarrow {}^1O_2 + H_2O + X^- + H^+ \qquad (25.3)$$

As a result singlet oxygen is formed and some of the observed bactericidal effects described later may also be mediated by this activated oxygen molecule.

As outlined by Allen (1992), the H_2O_2-mediated oxidation of halides is thermodynamically favored but the change in free energy (ΔG) is a function of the halide concerned. The free energy change for chloride is smaller than for bromide and chloride is consequently more difficult to oxidize. The change in free energy declines rapidly with increasing pH and the ΔG of the reaction becomes nearly zero at pH 9. Consequently chloride is hardly oxidized above pH 8.5. In contrast both iodide and bromide are still oxidizable, with a relatively large change in free energy. The secondary reaction of HOX with hydrogen peroxide is more than sufficient exergonic for the generation of singlet oxygen but the reaction is favored at more alkaline pH. However, HOI reacts only in the alkaline region with hydrogen peroxide to generate singlet oxygen.

Several patents have appeared in which the bactericidal activity of peroxidases is claimed. A large portion of the patents deals with mammalian peroxidases (myeloperoxidases, eosinophil peroxidase, and lactoperoxidase). These enzymes catalyze the formation of the bactericidal products in vivo and are implicated in defense and attack mechanisms. On of the first patents in this area was that of Kessler and Rosenbaum (1984), in which the mammalian peroxidases were used for killing bacteria in the treatment of dental diseases in situ such as the oral cavity or as a denture cleaner (mouthwashes, toothpastes). An orally activated antimicrobial dentifrice was proposed (Montgomery, 1994) that includes SCN^- and lactoperoxidase. These inventions are based on the natural system present in saliva and in which the lactoperoxidase present in saliva generates HOSCN using hydrogen peroxide formed by metabolically active bacteria in the oral cavity (Eq. 25.4).

$$H_2O_2 + HSCN \longrightarrow HOSCN + H_2O \qquad (25.4)$$

Bacterial glycolysis is inhibited by oxidation of essential SH groups in bacterial glycolytic enzymes by the rather unstable hypothiocyanate (HOSCN) formed by the peroxidase. Bacterial glycolysis and particularly that of *Streptococcus mutans*, which is present in tooth plaque, generates acid and as a consequence of the localized pH drop in the bacterial biofilm tooth decay occurs, as a result of dissolution of hydroxyapatite. There are several patents on formulations of dentifrice claiming the use of lactoperoxidase in combination with other enzymes to combat tooth decay. In the Netherlands there is a toothpaste (Zendium) for sale in the supermarket which contains three enzymes: amyloglucosidase, glucose oxidase, and lactoperoxidase. The first two enzymes boost the generation of hydrogen peroxide from carbohydrates and the presence of lactoperoxidase should result in increase of the inhibitory $OSCN^-$ (Hoogendoorn et al., 1977). There is no scientific evidence that this system works in vivo (Kirstila et al., 1994). However, the toothpaste is effective in treating canker sores. The natural occurrence of lactoperoxidase in milk can be utilized in the preservation of several dairy products. By adding SCN^- and glucose oxidase for hydrogen peroxide production the antibacterial activity of lactoperoxidase is enhanced (Fuglsang et al., 1995) resulting in a longer shelf life.

The peroxidases myeloperoxidase, eosinophil peroxidase, lactoperoxidase, and the heme-containing chloroperoxidase were proposed for the treatment of infection and control of bacterial flora (Allen, 1992). It was suggested that compositions containing these enzymes might be highly useful in the therapeutic and prophylactic antiseptic treatment of human and animal subjects. The selectivity to inhibit growth appears to result from the selective binding to the microbes due to both hydrophobic properties and strong positive charge of some of the peroxidases. The target bound haloperoxidase catalyzes halide oxidation and facilitates the disproportionation of peroxide to singlet oxygen at the surface of the target microbe. Prophylactic and therapeutic applications of peroxidases (Pourtois et al., 1992) in the prevention and treatment of enveloped virus infections and, in particular, of herpes simplex and human immunodeficiency virus infections have also been proposed. Peroxidases include horseradish peroxidase, lactoperoxidase, and myeloperoxidase. It was also claimed that peroxidases can be applied to a localized area of the body (cream, gel, bandage) and be used as an oral dentifrice or an injectable composition. Since expression of the mammalian peroxidases in suitable hosts is very difficult and only possible in Chinese hamster ovarian cell lines application is unlikely. Growth media for mammalian cells are expensive and the peroxidases can only be produced for research purposes (Kooter et al., 1999). However, native lactoperoxidase may easily be recovered from whey, which is a side product of cheese production, and the enzyme is therefore relatively cheap.

A major problem in the application of heme-containing peroxidases is that in general the peroxidases are easily oxidatively inactivated by hydrogen peroxide at relatively low concentrations (0.5 to 10 mM). During catalysis also inactive peroxidase intermediates (Compound II or III) may accumulate that do not participate in the catalytic cycle. A solution to this problem is to control the hydrogen peroxide and keep it at a low level. However, this requires careful tuning of the reaction and this limits the usefulness of heme-containing peroxidases in many applications. By using directed evolution it is in principle possible to obtain mutants that are more resistant toward oxidative inactivation and this approach has been successful for the *Coprinus cinereus* peroxidase (Cherry et al., 1999). The native enzyme has been reported to have antimicrobial properties in the presence of an enhancing electron donor, usually an organic molecule or halides (Johansen, 1997).

The easy inactivation of heme peroxidases initiated screening of a variety of organisms for the presence of enzymes whose activity is not affected by high concentrations of peroxide and are not inhibited by the oxidative products being formed (Geigert et al., 1987). A number of terrestrial fungi, the so-called dematiaceous hyphomycetes present in the Death Valley, were screened for the presence of haloperoxidases with a high tolerance to alkaline pH and temperature and it was shown (Hunter-Cevera and Sotos, 1986; Hunter et al., 1990) that these fungi produced very stable haloperoxidases. Some of them had chloroperoxidase activity but were clearly not heme peroxidases (Liu et al., 1987). The properties of these enzymes resembled that of the vanadium bromoperoxidases that were discovered a few years earlier and it was subsequently proven (Van Schijndel et al., 1993) that the enzyme from the hyphomycete *Curvularia inaequalis* contained vanadate as the prosthetic group. The haloperoxidases that were secreted by several different fungi (Hunter et al., 1990) were later shown (Vollenbroek et al., 1995) to belong to the class of vanadium peroxidases. The bromo and chlorperoxidases are exceptionally stable; for example, they are able to withstand very high concentrations of hydrogen peroxide up to 0.5 M (Renirie et al., 2003), are very thermostable (Van Schijndel et al., 2004) and resistant towards organic solvents (DeBoer et al., 1987). These facts make them especially suitable for application.

After the discovery of vanadium haloperoxidases there was great activity in patenting the use of these enzymes in various applications. The patents were mainly focused on the production of HOCl or HOBr by these enzymes and their resulting antibacterial and bleaching properties. As outlined in Chapter 23 the bromoperoxidases essentially only oxidize halides but are also able to carry out enantioselective sulfoxidation of sulfides to sulfoxides. However, the rate of this reaction is slow (Ten Brink et al., 1998).

There are no reports that heme peroxidase substrates are oxidized by vanadium haloperoxidases. Surprisingly, it was found (Ten Brink et al., 2000) that the recombinant vanadium chloroperoxidase from the fungus *Curvularia inaequalis* catalyzes also the oxidation of 2,2′-azino-bis (3-ethylbenzthiazoline-6-sulfonic acid) (ABTS), a classical chromogenic heme peroxidase substrate in the presence of hydrogen peroxide with a turnover frequency of 11 s^{-1} at its pH optimum of 4.0. The turnover frequency of this reaction is at least a factor of 100 lower than that of lactoperoxidase. In addition, recombinant vanadium chloroperoxidase bleaches the industrial sulfonated azo dye, Chicago Sky Blue 6B, in the presence of hydrogen peroxide. No further studies have appeared on the subject.

The first patent (Barnett et al., 1995) from the Amsterdam group together with Unilever Research that appeared on these vanadium haloperoxidases was the use in an antimicrobial composition and also its production as a recombinant enzyme. According to the patent its use as a sanitizing agent provides hygiene benefits for hard-surface cleaning and fabric washing, but also provides hygiene and cleaning in industrial applications such as in hospitals and for cleaning and disinfecting medical equipment. Another application was in the dairy industry, for disinfecting milking equipment. The antimicrobial composition can also be used in deodorants in view of their ability to combat bacteria that cause malodor.

The antimicrobial activity and potential application as a surface disinfectant (Hansen et al., 2003) of vanadium haloperoxidase from *Curvularia verruculosa* has been evaluated in detail. This enzyme from C. *verruculosa* has a sequence homology of 96% with that of *C. inaqualis* studied by the Amsterdam group. The haloperoxidase system caused several-log-unit reductions in counts of bacteria, yeasts, and filamentous fungi. Also bacteria adhering to the surface of contact lenses were killed and it was concluded that the haloperoxidase system is an effective sanitizing system and has the potential for a vast range of applications. Also spores that withstand severe environmental stress and are difficult to inactivate were susceptible towards haloperoxidases in the presence of hydrogen peroxide, a surfactant and iodide (Danielsen and Christensen, 2003).

The virucidal and bactericidal properties of the peroxidase system were also evaluated at more alkaline pH conditions using a chloroperoxidase mutant from *C. inaequalis* (Hasan et al., 2006) that was obtained by direct evolution and had 100-fold higher k_{cat} in the formation of HOBr at pH 8 compared to the wild-type enzyme. Indeed it was shown (Renirie et al., 2008) that this mutant was far more bactericidal and virucidal than the wild-type enzyme.

After the first patent application (Barnett et al., 1995) several other patent applications on the expression and the use of vanadium enzymes as antimicrobial agents appeared mainly by investigators from Novozymes (formerly Novo Nordisk) and these are variants or further refinements (Fuglsang et al., 1997; Danielsen and Schneider, 2002) or claim other enzymes or peptides having haloperoxidase activity. An example is the inhibition of growth of microbial cells present in laundry and in industrial water lines; and a method of killing microbial cells present on human or animal skin (Johansen, 2001). Malodor from hygiene products such as diapers and the like can also be reduced (Kirk et al., 1999) by incorporation of a haloperoxidase and hydrogen peroxide. The peroxidase in minced horseradish roots is also capable of deodorizing the malodor of swine manure (Govere et al., 2007).

Synergistic Effects

When a haloperoxidase, hydrogen peroxide, a halide, and an ammonium salt or amino alcohol are present a synergistic antimicrobial effect occurs between the halide and the ammonium source (Johansen, 1999). This is probably due to the formation of chloramines that are still highly oxidizing, noncharged, and more easily diffuse through bacterial membranes (Thomas, 1979). The invention may be used for the preservation of food, beverages, cosmetics, deodorants, contact lens products, food ingredients or enzyme compositions or as a general disinfectant. Similarly it was shown (Johansen et al., 2002) that the oxidizing bleaching and antimicrobial activity is considerably improved when combined with carboxylic acids. A chemical basis for this observation has not been reported.

Use as an Antifouling Agent

A totally different application is the use of haloperoxidases as an additive in antifouling paint (Wever et al., 1995) to hinder the attachment of marine organisms. These antifouling paints are for use on the hull of ships in the marine environment to minimize drag and consequently minimize fuel consumption. Many antifouling paints based on cytotoxic effects of biocides or metal complexes are being banned now and the peroxidase-based antifouling may provide a safer alternative. Seawater contains about 1 mm Br^- and 500 mM Cl^- and if sufficient peroxide is present the antifouling paint will continuously generate HOX as a bactericidal agent (Fig. 25.1).

A limiting factor may be the concentration of hydrogen peroxide in seawater, which is present in concentrations ranging from 0.1 to 0.3 μM (Petasne and Zika, 1997). The hydrogen peroxide is generated by photooxidation processes of water initiated by the UV light of the sun. Also as a result of biological activity peroxide may be generated resulting in higher peroxide levels. The idea to combat fouling of surfaces by enzymes has its roots in the physiological role of the vanadium bromoperoxidase. In some seaweed the peroxidase is located extracellularly on the surface of the plant (Wever et al., 1991) and its possible role is to control colonization of the surfaces of the seaweed by generating the bactericidal HOBr. More recently it was shown that very low concentrations of HOBr inactivated bacterial homoserine lactones. These compounds (Fig. 25.2) play an important role in bacterial signaling systems and interference with these systems inhibits quorum sensing and bacterial biofilm formation, a first step in the fouling of surfaces (Borchardt et al., 2001).

Similarly, it was shown (Rasmussen et al., 2000; Kjelleberg and Steinberg, 2001) that some red macro-algae produced halogenated furanones (see Fig. 25.3) that are encapsulated in gland cells in the seaweed, which provides a mechanism for the delivery of the metabolites to the surface of the algae at concentrations that deter a wide range of prokaryote and eukaryote fouling organisms.

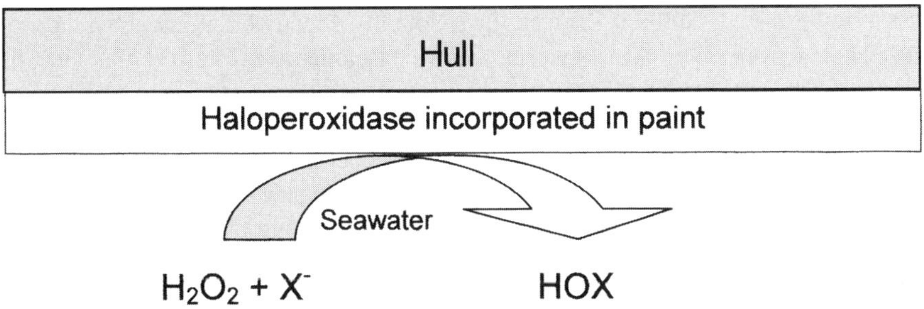

Figure 25.1. Schematic representation of the chloroperoxidase-containing antifouling paint coated on the hull of a ship and its principle. The enzyme should be bound to a suitable matrix and will in principle continuously release small amounts of hypohalous acid in the sea. Most of this will be in the form of HOBr since any HOCl formed will rapidly react with Br^- to yield HOBr.

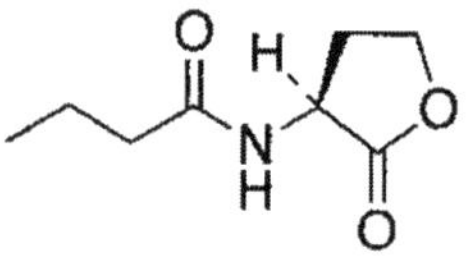

N-butanoyl-homoserine lactone (C_4-HSL)

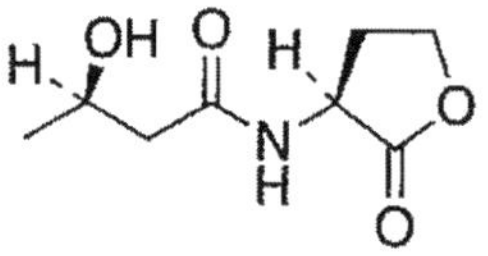

N-3-hydroxybutanoyl-homoserine lactone (3OHC_4-HSL)

Figure 25.2. Chemical structures of some of the homoserine lactones produced by bacteria.

Figure 25.3. Structures of halogenated furanones produced by *Delisae pulchra*. (a) 4-bromo-3-butyl-5-(dibromomethylene)-2(5H)-furanone. (b) 4-bromo-5-(bromomethylene)-3-butyl-2(5H)-furanone.

In the halogenation of the furanones a peroxidase is involved and halogenated furanones produced by the macroalga *Delisea pulchra* inhibit *N*-acyl-L-homoserine lactones-dependent gene expression by bacteria. These lactones are coregulatory ligands required for control of the expression of genes encoding virulence traits in many Gram-negative bacterial species. Thus, the effect of HOBr (Hansen et al., 2003; Renirie et al., 2008) is dual. HOBr is not only directly bactericidal but the oxidized halogen species also inactivate homoserine lactones and disturb communication signals between bacteria.

Initial tests using the peroxidase as an antifouling agent in marine paints were negative. However, the pH optimum of the chloroperoxidase from *C. inaequalis* is about 5.5 (Van Schijndel et al., 1994) and that of seawater about 8.2. This and the possible uses at higher pH values, for example, for washing purposes prompted the Amsterdam group and Novozymes, Denmark, to produce mutants that have an altered pH optimum. In the study by Novozymes (Svendsen and Jorgensen, 2002) the gene coding for the peroxidase from *C. verruculosa* was used to obtain a number of site-directed mutants. This enzyme has a sequence homology of 96% to that of *C. inaqualis*. Mutations were made within 10 Å around His496 and His404. His496 is the residue that ligates the vanadate cofactor and His404 may play a role as an acid-base group in catalysis. See Chapter 23 for details. One mutant was obtained

(D289L) that showed a pH shift in the brominating activity from 7 for the wild-type enzyme to pH 7.5. Most of the other mutants exhibited increased activity in the low pH range. Unfortunately very little detail is given in this patent regarding the kinetic parameters of these mutants. Directed evolution and high throughput screening at alkaline pH was used by the Amsterdam group (Hasan et al., 2006) to create mutants with a higher activity at pH 8. A triple mutant was obtained that showed a 100-fold increase in brominating activity at pH 8 (k_{cat}= 100 s^{-1}) and also a substantial increase in the chlorinating activity at pH 5 (k_{cat} = 36 s^{-1}) and the mutant is probably the best vanadium haloperoxidase known thus far. Whether these mutants are more effective in antifouling is not yet known. Interestingly, efforts are directed now to the inclusion of proteases and glycosylases in antifouling coatings (Kristensen et al., 2008) for use in the marine environment.

Use as a Paint Preservative

The use of haloperoxidases as a paint preservative in water-based paints has also been suggested (Wever et al., 1995). A similar idea using many oxidoreductases including laccases and peroxidases was patented later (Johansen and Fuglsang, 2006). The inventors also propose enzymatically preserved drilling fluids in the production of fossil oil and gas. The advantage would be substitution of conventional chemical biocides with enzymatic preservation systems.

APPLICATIONS IN BLEACHING AND DETERGENTS

Industrial enzymes in detergent formulations make up a substantial part of the biocatalyst market and it is likely that peroxidases with their degrading and bleaching properties will also be used in detergents in the near future. One of the first patents in which this novel lead was disclosed is that by Kirk et al., in 1989. It was claimed that an enzyme exhibiting peroxidase activity is capable of exerting a bleaching effect on fabrics. The main advantage would be that by using the detergent additives the amounts of peroxide or its precursors can be reduced and furthermore it is claimed that the overall performance of a detergent composition in which the bleaching enzymes are included is improved. A more specific application (Damhus et al., 1991) is that of the inhibition of dye transfer during washing or rinsing of fabrics by addition of enzymes exhibiting peroxidase activity to the wash liquor. Since heme peroxidase will oxidize or bleach the bleeding dye during the wash the peroxidases will inhibit the transfer of a textile dye from one dyed fabric to another during the washing process. This is most noticeable when white or light-colored fabrics are washed together with more intensely colored fabric. The heme peroxidases are derived from a strain of *Coprinus* or *Bacillus pumilis*. Also a microperoxidase, which is a hemopeptide may be used for this application (G. Pedersen et al., 1993). Similarly, these hemopeptides with peroxidase activity may be used for wastewater treatment and for pulp bleaching. However the wild-type peroxidase is rapidly inactivated under laundry conditions due to the high pH (10.5), high temperature (50°C), and high peroxide concentration (5 to 10 mM). The *Coprinus* peroxidase was therefore subjected to a directed evolution procedure and this led to most dramatic improvements in oxidative and thermal stability of the mutant (Pedersen et al., 1995; Cherry et al., 1999). A related application is the use of heme peroxidases in removing excess dye from newly dyed or printed textiles by using wash liquors containing peroxidases and suitable oxidants (Damhus et al., 2002).

Patent applications have also been filed in which the vanadium peroxidases or their derivatives are claimed as bleaching enzymes in washing liquids (Svendsen and Jorgensen, 2002) and to bleach stains present on fabrics (Verrips et al., 2000). As far as known the enzymatic bleach systems have not been applied in detergent so far.

Wool, wool fibers, or animal hair may also be treated with a haloperoxidase and a proteolytic enzyme (Mcdevitt and Winkler, 2000). The described method results in improved properties of the material, for example, shrink resistance, improved stretch, improved burst strength, and improved dyeing characteristics.

BIOTRANSFORMATIONS

The heme peroxidases catalyze a much wider variety of reactions than the vanadium enzymes. They efficiently catalyze the oxidation of a variety of substrates. In particular the chloroperoxidases isolated from the fungus *Caldariomyces fumago* is unusually versatile: it catalyzes not only the reactions typical of the classical peroxidases but also those of monoxygenases and it resembles cytochrome P450 in a number of aspects. Unlike P450, the chloroperoxidase catalyzes chlorination and bromination reactions. The enzyme is commercially available (e.g. from Sigma) and many research groups have therefore been able to explore the selective reactions that are catalyzed by this particular enzyme. Some of these transformations are difficult to carry out by conventional chemical synthesis and require aggressive chemicals that lead to the formation of a variety of undesired and toxic byproducts. The chloroperoxidase-catalyzed reactions have a promising biotechnological potential because they lead to regio- and enantioselective oxygenation of substrates and applications of chloroperoxidase in organic synthesis has been and will still be a research hotspot. Table 25.1 lists the various reactions catalyzed by the heme chloroperoxidase.

Not all of the reactions will be discussed in detail. For this the reader is referred to the following reviews (Colonna et al., 1999; Littlechild, 1999; Dembitsky, 2003; Hofrichter and Ullrich, 2006) and chapters in this book. Recently, a second heme thiolate enzyme has been discovered (Hofrichter and Ullrich, 2006) in the fungus *Agrocybe aegerita* and considering the wealth of fungal species it is likely that many more of these enzymes are present in nature. An example is a chloroperoxidase from *Rathayibacter biopuresis*. This enzyme can convert cephalexin to a halogenated cephalosporin antibiotic in a single step (Wong et al., 1996).

Of course it is also possible to use the vanadium enzymes in chlorination or bromination reactions of organic molecules (Leblanc et al., 2004). However, this swimming pool chemistry may have no large advantages except that the formation of HOBr or HOCl is controlled (Franssen et al., 1988) and as a result side reactions hardly occur. Interestingly the vanadium bromoperoxidase (Carter-Franklin and Butler,

TABLE 25.1. Reactions Catalyzed by the Chloroperoxidase from *C. fumago*

N-dealkylation	Okazaki and Guengeric (1993)
Oxidation of amino into nitroso groups	Corbett et al. (1978)
Enantioselective epoxidation of olefins	Lakner and Hager (1996); Colonna et al. (1993); Zaks and Dodds (1995)
Oxidative halogenation reactions	Geigert et al. (1983a, 1983b); Yamada et al. (1985); Wagenknecht and Woggn (1997)
Sulfoxidation reactions	Colonna et al. (1990); Casella et al. (1992)
Oxidation of indoles	Van Deurzen et al. (1996)
Propargylic hydroxylation	Miller et al. (1995); Hu and Hager (1998)
Oxidation of alcohols and aldehydes	Geigert et al. (1983a); Baciocchi et al. (1999)
Transformation of terpenes	Kaup et al. (2007)

2004) catalyzes the cyclization of some monoterpenes. The cyclization reaction only occurs in the presence of a large amount of organic cosolvent; in pure water only halohydrins and epoxides are formed. A pharmaceutically interesting halohydrin is also produced (Zhang et al., 1999) when indene is exposed to the bromoperoxidase from *Curvularia protuberata*. The nature of the prosthetic group of this enzyme has not been established but it is probably a vanadium enzyme. The recently discovered oxidation and halo-hydroxylation (Kaup et al., 2007) of monoterpenes adds another interesting reaction to the extensive catalytic potential of the heme chloroperoxidase. This reaction is particularly interesting since terpenes are cheap and abundantly available as a pool of chiral substances that can be transformed into valuable bioactive compounds. In this application the chloroperoxidase in the absence of halides catalyzes the oxidation of the terpene alcohols geraniol, nerol, and citronellol to the corresponding flavor aldehydes. The reaction appears to be superior to the existing chemical conversion or whole cells biocatalysis. In the presence of iodide or bromide, iodo- or bromohydrins are formed that by addition of sodium hydroxide are quantitatively converted into enantiopure terpene epoxides. Figure 25.4 gives an example. Halohydrin formation by heme chloroperoxidase has been reported before (Geigert et al., 1983b).

Other heme (Tuynman et al., 1998) and vanadium bromoperoxidases (Andersson et al., 1997) also catalyze the sulfoxidation of a variety of substrates. However, in

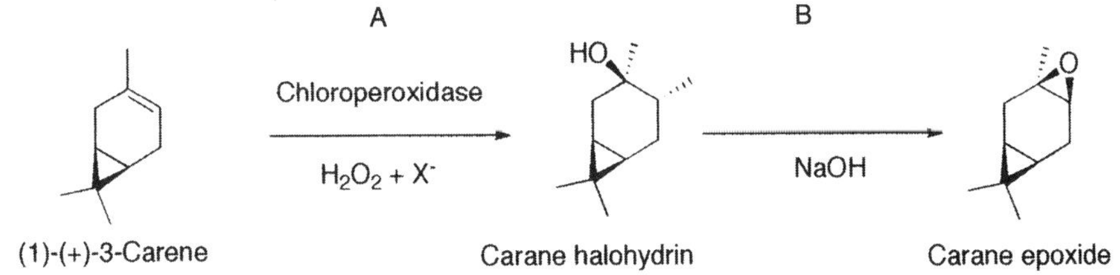

Figure 25.4. Chloroperoxidase catalyzed conversions of the monoterpene (1S)-(+)-3-carene into the carane halohydrin and formation of the enantiopure epoxide according to Kaup et al., 2007.

this the heme chloroperoxidase is the peroxidase of choice owing to its high enantioselectivity and versatility (Colonna et al., 1993, 1999). Unfortunately the activity of the heme chloroperoxidase is affected by the presence of cosolvents which are often present to solubilize the substrates (Van Deurzen et al., 1997a). Further the enzyme is irreversibly inactivated by hydrogen peroxide and the hydrogen peroxide concentration should be carefully controlled (Van Deurzen et al., 1997b). In contrast the vanadium bromoperoxidases are unaffected by the presence of cosolvents but the rate constant of the reaction is much lower (Andersson et al., 1997; Ten Brink et al., 1998). In sulfoxidation reactions catalyzed by both the heme and vanadium peroxidases the hydrogen peroxide concentration should be kept as low as possible; otherwise, the spontaneous oxidation of the sulfide by hydrogen peroxide to a racemic sulfoxide lowers the very high enantioselectivity of the reaction. An interesting application is also biodesulfurization of diesel (Ayala et al., 2000) to decrease sulfur emissions in the environment. Most fuels contain organosulfides and thiophenes and heme chloperoxidase catalyzes the oxidation of most organosulfur compounds found in straight-run diesel fuel. This oxidation allows the desulfurization of diesel fuel by distillation. The primary alcohol groups of carbohydrates may also be selectively oxidized in the presence of a peroxidase, a halide, hydrogen peroxide, and a nitroxyl radical mediator (Cui et al., 2006). Interestingly it has been reported that some heme peroxidases also break carbon-carbon double bonds (Tuynman et al., 1998).

In spite of the remarkable synthetic possibilities of the peroxidases a commercial-based process in which the heme chloroperoxidase is used has not been reported yet. A problem is that interesting substances for the chloroperoxidase have in general a low solubility in water resulting in limiting scaling up of the reaction. To add mixable organic solvents to water is an option but in general these solvents decrease activity and stability of most enzymes. A possibility to carry out reaction in hydrophobic organic solvents with low water content raises the problem of solubility of hydrogen peroxide in these solvents. As outlined these heme peroxidases are easily inactivated by excess peroxide. Directed evolution to improve the resistance towards oxidative inactivation may help to overcome this problem.

POLYMERIZATION REACTIONS AND WASTEWATER PURIFICATION

Heme peroxidases are able to oxidatively dehydrogenate a large number of organic compounds by one-electron oxidation steps. The radicals produced may dimerize or polymerize and the products formed are in general less soluble in water. This property may be used to remove carcinogenic aromatic amines and phenols from industrial aqueous effluent using cheap horseradish peroxidase (Klibanov et al., 1983; Nazari et al., 2007) that will oxidize phenols to phenoxy radicals. These radicals will react with other aromatic rings causing precipitation of toxic organic material. Also degradation of environmental pollutants is a suggested use, for example, the breakdown of perfluorooctanoic acid (Colosi et al., 2009) or to detoxify herbicides (Mougin et al., 1994). Peroxidases may also be used to remove color from bleach plant effluent (Paice and Jurasek, 1984). Similarly effluents of textile industries and wastewater containing discarded textile dyes may be treated with peroxidases. The oxidation of the azo dyes leads to bleaching, degradation, and polymerization facilitating downstream processing and improving wastewater quality (Shaffiqu et al., 2002; Matto and Husain, 2007).

Other applications in which a peroxidase is involved in polymerization processes are the preparation of a glue or binder from lignin sulphonate (Yde, 1994) using a peroxidase and hydrogen peroxide at alkaline pH to give high molecular weight polymers that may be used in the manufacture of wood composites. Its main advantage is that very high molecular weight lignin is prepared without the use of organic solvents. Similarly the treatment of pulp with phenol oxidizing enzymes derived from *Coprinus* or *Bacillus pumilllus* after completion and grinding and refining of logs gives paper or paperboard of improved strength (Hansen and Nielsen, 1993). Peroxidases may also be used in the tanning of hides, resulting in an increased degree of fixation (Ingvorsen et al., 1993).

Enzymatic modification of food proteins is of interest for the food industry. Modification with carbohydrate moieties will probably influence the gelling, emulsifying, and foaming properties of the protein. Addition of such modified proteins to food systems may have beneficial effects on their properties. Indeed it is possible to cross-link proteins via their tyrosines to ferulic acids (Oudgenoeg et al., 2001) using horseradish peroxidase and hydrogen peroxide. Subsequent cross-linking of feruloylated arabinoxylans and beta-casein by horseradish peroxidase has been demonstrated on a large scale (Boeriu et al., 2004). A somewhat related process (Maat and Roza, 1990) is the use of a peroxidase in dough to improve specific volume, staling resistance, and crumb structure of bread.

The use of a novel water-resistant, aqueous, and versatile phenolic adhesive or glue derived from algal materials has also been proposed (Vreeland and Grotkopp, 1996). Adhesive phenolic compounds derived from algae are activated with an enzyme catalyst or oxidizing agents and/or cross-linked with algal carbohydrate, other fibers, or proteins. The enzymes mentioned are horseradish peroxidase but surprisingly also vanadium bromoperoxidase. To date there is no evidence that the vanadium bromoperoxidase is directly oxidizing phenolic compounds. However, when bromide is present, HOBr is formed that may trigger further reactions. An improved biocatalytic process for the preparation of phenolic resins using soybean peroxidase has also been proposed. Soybean peroxidase has better temperature and solvent stability than horseradish peroxidase and is much more economical because it can be obtained from soybean hulls which are very inexpensive (Johnson et al., 1993).

Oxidative coupling may also be used in the coloring process of hair exposing it to soybean peroxidase, hydrogen peroxide, and aromatic compounds (Procter & Gamble, 1975). Lignin peroxidase may also have an application in cosmetic compositions suitable for skin and hair lightening (Belinky et al., 2006).

DEPOLYMERIZATION REACTIONS

Almost all terrestrial fixed carbon consists of lignocellulose, the principal structural component of plants. Lignin is a polymer in which the units are interconnected via diverse C—C and ether bonds that are not hydrolyzable under biological conditions. Further the three-dimensional structure is complex and nonrepeating. The complex polymeric structure of lignin makes it chemically very difficult to degrade and complicates the use of plant polysaccharides in paper manufacture and biofuel production. Lignocellulolytic filamentous fungi are able to oxidize lignin when they degrade lignocellulose using nonspecific oxidative mechanisms. Cleavage of alkylaromatic compounds occurs by abstraction of an electron from the ring to produce an unstable cation radical intermediate that spontaneously fragments. The lignin peroxidase is different from the classical peroxidases in the sense that the enzyme is a stronger oxidant. Further a tryptophane residue on the surface of the protein is involved in the long range electron transfer from the substrate to the enzyme. The manganese peroxidase that is present in many fungi plays a similar role. In this case, however, the peroxidase oxidizes Mn^{2+} present in the enzyme to Mn^{3+} and this metal ion after release from the enzyme is chelated to a physiological chelator. The diffusible strong oxidant may then generate other oxidants that can cleave nonphenolic structures (Hammel and Cullen, 2008). An obvious application is the enzymatic delignification and many patents have appeared; for example Crawford and Ramachandra (1993) but the major hurdle in application appears to be the large-scale production of lignin and Mn proxidases. However, it may also be possible to use soybean peroxidase (Pokora and Johnson, 1998).

ANALYTICAL APPLICATIONS

Peroxidases are used these days in a large number of diagnostics to determine the concentration of organic metabolites. The metabolite is converted by a specific enzyme to another product in a reaction in which hydrogen peroxide is generated in a stoichiometric amount. The peroxide is quantitatively determined using a peroxidase which in most cases is horseradish peroxidase, and a dye that upon oxidation gives a color change. The metabolites may be sugars (glucose, galactose, maltose, saccharase, or lactose; see, e.g. Tkáč et al., 2000) or other compounds, for example, phenol, cholesterol, urate, acetate, lactate, pyruvate, and L- and D-amino acids. The determination can be done in solution but nowadays the necessary enzymes are adsorbed or immobilized onto a convenient matrix such as filter paper. The glucose determination using these sticks is of great value especially for the determination of glucose in body fluids and for those suffering from diabetes. The peroxidase is also used in kits and immunoassays (ELISAs, monoclonal antibodies) in which the peroxidase is coupled to an antibody which will have a specific interaction with an antigen. The complex formed may be detected by a color reaction or chemiluminescence signal (Akhavan-Tafti, 1999) which gives high sensitivity. Thus the peroxidase acts as reporter molecule. Vanadium bromoperoxidase may also be used (Friedman et al., 1998). It has the advantage that unlike the classical peroxidases this enzyme does not inactivate during turnover, allowing detection of traces of an antigen. Peroxidases have also been incorporated into biosensors allowing easy determination of hydrogen peroxide (Shi et al., 2009). In particular soybean peroxidase is useful because of its high stability.

The vanadium bromo- and chloroperoxidases specifically oxidize halides and have a high affinity for halides. It is possible (Wever et al., 1995) to use these enzymes in a spectrophotometric assay for the determination of total halide and chloride concentrations of aqueous solutions. Although the procedure is convenient and much more sensitive (detection in the micromolar range) than the existing methods, it has the disadvantage that the enzymes required are not yet commercially available. A similar procedure for chloride determination was developed (Keener and Watwood, 2004) using 3,3′,5,5′-tetramethylbenzidine as a chromagenic scavenger for HOCl and heme chloroperoxidase.

MEDICAL APPLICATIONS

In the field of nuclear medicine there is considerable interest in the use of radioactive labeled proteins. Iodination of proteins is commonly used and commercial kits are available that use lactoperoxidase and glucose oxidase. Since the method is mild and results in stable attachment of the label it is particularly useful to label monoclonal antibodies. However, there are some disadvantages of using ^{125}I and ^{131}I in particular for patient applications. Bromination with ^{77}Br is an option and the advantage is that the binding strength of the C—Br bonds is stronger than that of the C—I bond. This bond strength is an important parameter for the in vitro stability of the radio-labeled compound. Catalytic halogenation by a haloperoxidase may be used and a procedure has been developed using the heme-containing bromoperoxidase from the algae *Penicillis capitatus* (Manthey et al., 1984). Care must be taken to control the hydrogen peroxide concentration since otherwise a sharp decrease in activity results. The vanadium bromoperoxidase from *Ascophyllum nodosum* has been used as an alternative (Lambert and Slegers, 1994). This enzyme is considerably more stable than its heme-containing counterpart. The method of enzymatic bromination (Rossin et al., 2007) is still being used in nuclear medicine and is a valuable tool for tumor detection using molecular imaging with positron emission tomography. Neidleman and Geigert (1986) also envisaged the use of peroxidases as cytotoxic agents if they can be delivered specifically to the site in the body where a cancer tumor is present. By coupling the enzymes to a specific monoclonal antibody the haloperoxidase could be directed to the growing cancer cells and by releasing their potent oxidants these cells could be selectively destroyed. This interesting idea has had no follow-up as far as known.

REFERENCES

Akhavan-Trafti H. 1999. Chemiluminescence compositions and methods. Patent WO/1999/014358.

Allen RC. 1992. Methods and compositions for the treatment of infection and control of flora using haloperoxidase. European Patent 500387.

Andersson M, Willetts A, Allenmark S. 1997. Asymmetric sulfoxidation catalyzed by a vanadium-containing bromoperoxidase. *J. Org. Chem.* **62**: 8455–8458.

Ayala M, Robledo NR, Lopez-Munguia A, Vazquez-Duhalt R. 2000. Substrate specificity and ionization potential in chloroperoxidase-catalyzed oxidation of diesel fuel. *Environ. Sci. Technol.* **34**: 2804–2809.

Baciocchi E, Manduchi L, Lanzalunga O. 1999. Prochiral selectivity and deuterium kinetic isotope effect in the oxidation of benzyl alcohol catalyzed by chloroperoxidase. *Chem. Commun.* 1715–1716.

Barnett P, Hondmann D, Simons LH, Ter Steeg PF, Wever R. 1995. Enzymatic antimicrobial compositions. Patent WO/1995/027046.

Beggs TS, Berry MJ, Davis PJ, Frenken LGJ, Verrips CT. 2001. Bleaching enzymes. US Patent 6218350.

Belinky P, Lasser H, Dosoretz C. 2006. Methods of producing lignin peroxidase and its use in skin and hair lightening. US Patent 0051305.

Berry A, Dodge TC, Pepsin M, Weyler W. 2002. Application of metabolic engineering to improve both the production and use of biotech indigo. *J. Ind. Microbiol. Biotechnol.* **28**: 127–133.

Boeriu CG, Oudgenoeg G, Spekking WTJ, Berendsen LBJM, Vancon L, Boumans H, Gruppen H, Van Berkel WJH, Laane C, Voragen AGJ. 2004. Horseradish peroxidase-catalyzed cross-linking of feruloylated arabinoxylans with beta-casein. *J. Agric. Food Chem.* **52**: 6633–6639.

Borchardt SA, Allain EJ, Michels JJ, Stearns GW, Kelly RF, McCoy WF. 2001. Reaction of acetylated homoserine lactone signaling molecules with oxidized halogen antimicrobials. *Appl. Environ. Microbiol.* **67**: 3174–3179.

Carter-Franklin JN, Butler A. 2004. Vanadium bromoperoxidase-catalyzed biosynthesis of halogenated marine natural products. *J. Am. Chem. Soc.* **126**: 15060–15066.

Casella L, Gullotti M, Ghezzi R, Poli S, Beringhelli T, Colonna S, Carrea G. 1992. Mechanism of enantioselective oxygenation of sulfides catalyzed by chloroperoxidase and horseradish peroxidase: spectral studies and characterization of enzyme substrate complexes. *Biochemistry* **31**: 9451–9459.

Cherry JP, Lamsa MH, Schneider P, Vind J, Svendson A, Jones A, Pedersen AH. 1999. Directed evolution of a fungal peroxidase. *Nature Biotechnol.* **17**: 379–384.

Colonna S, Gagereo N, Casella L, Carrea G, Pasta P. 1993. Enantioselective epoxidation of styrene derivatives by chloroperoxidase catalysis. *Tetrahedron Asymmetry* **4**: 1325–1330.

Colonna S, Gaggero N, Manfredi A, Casella L, Gulotti M, Carrea G, Pasta P. 1990. Enantioselective oxidations of sulfides catalyzed by chloroperoxidase. *Biochemistry* **29**: 10465–10468.

Colonna S, Gagereo N, Richelmi, Pasta P. 1999. Recent biotechnological development in the use of peroxidases. *Tibtech* **17**: 163–168.

Colosi LM, Pinto RA, Huang QG, Weber WJ. 2009. Peroxidase mediated degradation of perfluorooctanoic acid. *Environ. Toxic. Chem.* **28**: 264–271.

Corbett MD, Chipko BR, Baden DG. 1978. Chloroperoxidase-catalysed oxidation of 4-chloroaniline to 4-chloronitrosobenzene. *Biochem. J.* **175**: 353–360.

Crawford DL, Ramachandra M. 1993. Bacterial extracellular lignin peroxidase. *US Patent* 5200338.

Cui X, Cimecioglu LA, Shi Y. 2006. Process for the selective modification of carbohydrates by peroxidase catalyzed oxidation. US Patent 7138035.

Damhus T, Kirk O, Pedersen G, Venegas MG, Christensen BE, Schneider P. 1991. Dye transfer inhibition. Patent WO/1991/5839.

Damhus T, Vogt U, Schneider P. 2002. Process for removal of excess dye from printed or dyed fabric or yarn. US Patent 6409771.

Danielsen S, Christensen BE. 2003. Method for killing spores. Patent WO/2003/04735.

Danielsen S, Schneider P. 2002. Polypeptides having haloperoxidase activity. US Patent 641029.

DeBoer E, Plat H, Tromp MGM, Wever R, Franssen MCR, Van der Plas HC, Meijer EM, Schoemaker HE. 1987. Vanadium containing bromoperoxidase: an example of an oxidoreductase with high operational stability in aqeous and organic media. *Biotechnol. Bioeng.* **30**: 607–610.

Dembitsky VM. 2003. Oxidation, epoxidation and sulfoxidation reactions catalyzed by haloperoxidases. *Tetrahedron* **59**: 4701–4720.

Franssen MCR, Jansma JD, Van der Plas HC, De Boer E, Wever R. 1988. Enzymatic bromination of barbituric acid and some of its derivatives. *Bioorg. Chem.* **16**: 352–363.

Friedman AE, Grouix SF, Butler A. 1998. Use of vanadium bromoperoxidase as a signal-generating enzyme for chemiluminescent systems: test kits and analytical methods. US Patent 5811253.

Fuglsang C, Johanson C, Christgau S, Adler-Nissen J. 1995. Antimicrobial enzymes: applications and futures potential in the food industry. *Trends Food Sci. Techn.* **6**: 390–396.

Fuglsang C, Oxenboll KM, Halkier T, Berka RM, Cherry J. 1997. Haloperoxidases from *Curvularia verrucolosa* and nucleic acids encoding same. Patent WO/1997/004102.

Geigert J, Liu TE, N'timkulu T. 1987. Stable haloperoxidase method. US Patent 4707446.

Geigert J, Dalietos DJ, Neidleman SL, Lee TD, Wadsworth J. 1983a. Peroxide oxidation of primary alcohols to aldehydes by chloroperoxidase catalysis. *Biochem. Biophys. Res.* **114**: 1104–1108.

Geigert J, Neidleman SL, Dalietos DJ, de Witt SK. 1983b. Haloperoxidases: enzymatic synthesis of α,β-halohydrins from gaseous alkenes. *Appl. Environ. Microbiol.* **45**: 366–374.

Govere EM, Tonegawa M, Bruns MA, Wheeler EF, Kephart KB, Voigt JW, Dec J. 2007. Using minced horseradish roots and peroxides for the deodorization of swine manure: a pilot scale study. *Bioresource Techn.* **98**: 1191–1198.

Hammel KE, Cullen D. 2008. Role of fungal peroxidases in biological ligninolysis. *Curr. Opin. Plant. Biol.* **11**: 349–355.

Hansen TT, Nielsen PH. 1993. Production of paperboard of improved strength by treating pulp with a phenol oxidizing enzyme after completion of grinding and refining of logs. Patent WO/1993/23606.

Hansen EH, Albertsen L, Schafer T, Johansen C, Frisvad JC, Molin S, Gram L. 2003. *Curvularia haloperoxidase*: antimicrobial activity and potential application as a surface disinfectant. *Appl. Environ. Microbiol.* **69**: 4611–4617.

Hasan Z, Renirie R, Kerkman R, Ruijssenaars HJ, Hartog AF, Wever R. 2006. Laboratory-evolved vanadium chloroperoxidase exhibits 100-fold higher halogenating activity at alkaline pH: catalytic effects from first and second coordination sphere mutations. *J. Biol. Chem.* **281**: 9738–9744.

Hofrichter M, Ullrich R. 2006. Heme-thiolate haloperoxidases: versatile biocatalysts with biotechnological and environmental significance. *Appl. Microbiol. Biotechnol.* **71**: 276–288.

Hoogendoorn H, Piessens JP, Scholtens W, Stoddard LS. 1977. Hypothiocyanite ion, the inhibitor formed by the system lactoperoxidase-thiocyanate-hydrogen peroxide. *Caries Res.* **11**: 77–84.

Hu SH, Hager LP. 1999. Highly enantioselective propargylic hydroxylations catalyzed by chloroperoxidase. *J Am. Chem. Soc.* **121**: 872–873.

Hunter JC, Belt A, Sotos LS, Fonda ME. 1990. Fungal chloroperoxidase method. US Patent 4937192.

Hunter-Cevera JC, Sotos L. 1986. Screening for new enzyme in nature: haloperoxidase production by Death Valley dematiaceous hyphomyctes. *Microb. Ecol.* **12**: 121–127.

Ingvorsen K, Mollgaard A, Sorensen NH. 1993. Tanning of hides with added enzyme using bath containing oxidoreductase and having oxidizing agent added to increase degree of fixation. Patent WO/1993/12259.

Johansen C. 1997. Enzymatic compositions. Patent WO/1997/042825.

Johansen C. 1999. Antimicrobial composition containg haloperoxidase, hydrogen peroxide, a halide source and an ammonium source. Patent WO/1999/008531.

Johansen C. 2001. Enzymatic method for killing or inhibiting microbial cells at high pH. Patent WO/2001/0111969.

Johansen C, Fuglsang CC. 2006. Enzymatic preservation of water based paints. US Patent 7063970.

Johansen J, Skovgaard NM, Schneider P, Johansen JT. 2002. Use of haloperoxidase, peroxide and carboxylic acids. US Patent 137655.

Johnson MA, Pokora AR, Cyrus WL Jr. 1993. Biocatalytic oxidation using soybean peroxidase. US Patent 518895.

Kaup BA, Piantini U, Wüst M, Schrader J. 2007. Heme monoterpenes as novel substrates for oxidation and halo-hydroxylation with chloroperoxidase from *Caldariomyces fumago*. *Appl. Microbiol. Biotechnol.* **73**: 1087–1096.

Keener WK, Watwood ME. 2004. Chloride analysis using 3,3′,5,5′-tetramethylbenzidine and chloroperoxidase. *Anal. Biochem.* **334**: 406–408.

Kessler JH, Rosenbaum S. 1984. Bactericidal method. US Patent 447610.

Klibanov AM, Tu TM, Schott KP. 1983. Peroxidases-catalyzed removal of phenols from coal-conversion waste-waters. *Science* **221**: 259–260.

Kirk O, Borchert TV, Fuglsang CC. 2002. Industrial enzyme applications. *Curr. Opinion Biotech.* **13**: 345–351.

Kirk O, Damhus T, Gotfredsen SE. 1989. A detergent additive for bleaching fabric. Patent WO/1989/09813.

Kirk O, Johansen C, Hansen TT. 1999. Reduction of malodour. Patent WO/1999/008726.

Kirstila V, Lenander-Lumikari M, Tenovuo J. 1994. Effects of a lactoperoxidase-system-containing toothpaste on dental plaque and whole saliva in vivo. *Acta Odont. Scand.* **52**: 346–353.

Kjelleberg S, Steinberg P. 2001. Surface warfare in the sea. *Microbiol. Today* **28**, 134–135.

Kooter IM, Moguilevsky N, Bollen A, van der Veen LA, Otto C, Dekker HL, Wever R. 1999. The sulfonium ion linkage in myeloperoxidase: direct spectroscopic detection by isotopic labeling and effect of mutation. *J. Biol. Chem.* **274**: 26794–26802.

Kristensen JB, Meyer RL, Sogaard Laursen B, Shipovskov S, Besenbacher F, Besenbacher F, Horsmans Poulsen C. 2008. Antifouling enzymes and the biochemistry of marine settlement. *Biotechn. Adv.* **26**: 47–481.

Lakner FJ, Hager LP. 1996. Chloroperoxidase as enantioselective epoxidation catalyst: an efficient synthesis of (R)-(−)-mevalonolactone. *J. Org. Chem.* **61**: 3923–3925.

Lambert F, Slegers G. 1994. Preparation of Br-77 labeled proteins using bromoperoxidase. *Appl. Rad. Isotop.* **45**: 11–15.

Leblanc C, Colin C, Kloareg B, Delage L, Roeder VJ-MW, Potin P. 2004. Vanadium haloperoxidases from the brown alga *Laminaria digitata* and uses thereof. Patent WO/2004/078976.

Littlechild J. 1999. Haloperoxidases and their role in biotransformation reactions. *Curr. Opin. Chem. Biol.* **3**: 28–34.

Liu TE, N'timkulu T, Geigert J, Wolf B, Neidleman SL, Silva D, Hunter-Cevera JC. 1987. Isolation and characterization of a novel nonheme chloroperoxidase. *Biochem. Biophys. Res. Commun.* **142**: 329–333.

Maat J, Roza M. 1990. Bread improver compositions comprising cellulase and other enzyme-oxidase and/or peroxidase to improve specific volume, staling resistance and crumb structure. European Patent 396162.

Manthey JA, Hager LP, McElvany KD. 1984. Protein bromination by bromoperoxidase from *Penicillus capitatus*. *Methods Enzymol.* **107**: 439–445.

Matto M, Husain Q. 2007. Peroxide and redox mediators, decolorization of direct dyes by salt fractionated turnip proteins enhanced in the presence of hydrogen. *Chemosphere* **69**: 338–345.

Mcdevitt JP, Winkler J. 2000. A method for enzymatic treatment of wool. US Patent 6140109.

Miller VP, Tschirret-Guth RA, Ortiz de Montellano PR. 1995. Chloroperoxidase-catalyzed benzylic hydroxylation. *Arch. Biochem. Biophys.* **319**: 333–340.

Montgomery RE. 1994. Stabilized enzymatic antimicrobial compositions. Patent WO/1994/04127.

Mougin C, Laugero C, Asther M, Dubroca J, Frasse P, Asther M. 1994. Biotransformation of the herbicide atrazine by the white-rot fungus *phanerochaete-chrysosporium*. *Appl. Environ. Microbiol.* **60**: 705–708.

Nazari K, Esmaeili N, Mahmoudi A, Rahimi H, Moosavi-Movahedi AA. 2007. Peroxidative phenol removal from aqueous solutions using activated peroxidase biocatalyst. *Enzyme Microb. Technol.* **41**: 226–233.

Neidleman SL, Geigert J. 1986. *Biohalogenation: Principles, Basic Roles and Applications.* Ellis Horwood, Chicester, UK, pp. 140–155.

Okazaki O, Guengerich FP. 1993. Evidence for specific base catalysis in *N*-dealkylation reactions catalyzed by cytochrome P450 and chloroperoxidase. *J. Biol. Chem.* **268**: 1546–1552.

Oudgenoeg G, Hilhorst R, Piersma SR, Boeriu CG, Gruppen H, Hessing M, Voragen AGJ, Laane C. 2001. Peroxidase-mediated cross-linking of a tyrosine-containing peptide with ferulic acid. *J. Agric. Food. Chem.* **49**: 2503–2510.

Paice MG, Jurasek L. 1984. Peroxidase catalyzed color removal from bleach plant effluents. *Biotechn. Bioeng.* **26**: 477–480.

Pedersen G, Christensen BE, Schneider P. 1993. Hemopeptide having peroxidase activity for bleaching dyes. US Patent 5273896.

Pedersen AH, Vind J, Svendsen A, Cherry JR, Schneider P, Jensen B. 1995. H_2O_2 stable peroxidase variants. Patent WO/1995/10602.

Petasne RG, Zika RG. 1997. Hydrogen peroxide lifetimes in south Florida coastal and offshore waters. *Mar. Chem.* **56**: 15–25.

Pokora AR, Johnson MA. 1998. Process for treating lignocellulosic material with soybean peroxidase in the presence of peroxide. US Patent 5785811.

Pourtois M, Bollen A, Moguilhevsky N. 1992. Prophylactic and therapeutic applications of peroxidases. Patent WO/1992/01466.

Procter & Gamble. 1975. Hair dyeing premixes containing peroxidase enzymes stabilized with heme complexing agents. US Patent 3893803.

Rasmussen TB, Manefield M, Andersen JB, Eberl L, Anthoni U, Christophersen C, Steinberg P, Kjelleberg S, Givskov M. 2000. How *Delisea pulchra* furanones affect quorum sensing and swarming motility in *Serratia liquefaciens* MG1. *Microbiology* **146**: 3237–3244.

Renirie R, Dewilde A, Pierlot C, Wever R, Hober D, Aubry JM. 2008. Bactericidal and virucidal activity of the alkalophilic P395D/L241V/T343A mutant of vanadium chloroperoxidase. *J. Appl. Microbiol.* **105**: 264–270.

Renirie R, Pierlot C, Aubry JM, Hartog AF, Schoemaker HE, Alsters PL, Wever R. 2003. Vanadium chloroperoxidase as a catalyst for hydrogen peroxide disproportionation to singlet oxygen in mildly acidic environment. *Adv. Synth. Catal.* **345**: 849–858.

Rossin R, Berndorff D, Friebe M, Dinkelborg LM, Welch MJ. 2007. Small-animal PET of tumor angiogenesis using a Br-76-labeled human recombinant antibody fragment to the ED-B domain of fibronectin. *J. Nuclear Med.* **48**: 1172–1179.

Shaffiqu TS, Roy JJ, Nair RA, Abraham TE. 2002. Degradation of textile dyes mediated by plant peroxidases. *Appl. Biochem. Biotechnol.* **102**: 315–326.

Shi LH, Liu XQ, Niu WX, Li HJ, Han S, Chen J, Xu GB. 2009. Hydrogen peroxide biosensor based on direct electrochemistry of soybean peroxidase immobilized on single-walled carbon nanohorn modified electrode. *Biosens. Bioelectron.* **24**: 1159–1163.

Svendsen A, Jorgensen L. 1999. Variant vanadium containing haloperoxidases useful in bleaching microbial control or hard surface cleaning. Patent WO/1999/47651.

Svendsen A, Jorgensen L. 2002. Haloperoxidases with altered pH profiles. US patent 1041666.

Ten Brink HB, Dekker HL, Schoemaker HE, Wever R. 2000. Oxidation reactions catalyzed by vanadium chloroperoxidase from *Curvularia inaequalis*. *J. Inorg. Biochem.* **80**: 91–98.

Ten Brink HB, Tuynman A, Dekker HL, Hemrika W, Izumi Y, Oshiro T, Schoemaker HE, Wever R. 1998. Enantioselective sulfoxidation catalyzed by vanadium haloperoxidases. *Inorg. Chem.* **37**: 6780–6784.

Thomas EL. 1979. Myeloperoxidase, hydrogen peroxide, chloride antimicrobial system: nitrogen-chlorine derivatives of bacterial components in bactericidal action against *Escherichia coli*. *Infect Immun.* **23**: 522–531.

Tkáč J, Šturdík E, Gemeiner P. 2000. Novel glucose non-interference biosensor for lactose detection based on galactose oxidase-peroxidase with and without co-immobilised β-galactosidase. *Analyst* **125**: 1285–1289.

Tuynman A, Spelberg JL, Kooter IM, Schoemaker HE, Wever R. 2000. Enantioselective epoxidation and carbon-carbon bond cleavage catalyzed by *Coprinus cinereus* peroxidase and myeloperoxidase. *J. Biol. Chem.* **275**: 3025–3030.

Tuynman A, Vink MKS, Dekker HL, Schoemaker HE, Wever R. 1998. The sulphoxidation of thioanisole catalysed by lactoperoxidase and *Coprinus cinereus* peroxidase: evidence for an oxygen-rebound mechanism. *Eur. J. Biochem.* **258**: 906–913.

Van Deurzen MPJ, van Rantwijk F, Sheldon RA. 1996. Synthesis of substituted oxindoles by chloroperoxidase catalyzed oxidation of indoles. *J. Mol Catal. B.* **2**: 33–42.

Van Deurzen MPJ, Remkes IJ, van Rantwijk F, Sheldon RA. 1997a. Chloroperoxidase catalyzed oxidations in t-butyl alcohol/water mixtures. *J. Mol Catal. A.* **117**: 329–337.

Van Deurzen MPJ, Seelbach K, Van Rantwijk F, Kragl U, Sheldon RA. 1997b. Chloroperoxidase: use of a hydrogen peroxide-stat for controlling reactions and improving enzyme performance. *Biocat. Biotransform.* **15**: 1–16.

Van Schijndel JWPM, Barnett P, Roelse J, Vollenbroek EGM, Wever R. 1994. The stability and steady-state kinetics of vanadium chloroperoxidase from the fungus *Curvularia inaequalis*. *Eur. J. Biochem.* **22**: 151–157.

Van Schijndel JWPM, Vollenbroek EGM, Wever R. 1993. The chloroperoxidase from the fungus *Curvularia inaequalis*; a novel vanadium enzyme. *Biochim. Biophys. Acta* **1161**, 249–256.

Verrips CJ, Swarthoff T, Van der Helm M, Van der Heiden M, Hondmann DH, Smits A. 2000. Enzymatic bleach composition. US Patent 6107264.

Vollenbroek EGM, Simons LH, Van Schijndel J, Barnett P, Balzar M, Dekker H, van der Linden C, Wever R. 1995. Vanadium chloroperoxidases occur widely in nature. *Biochem. Soc. Trans.* **23**: 267–271.

Vreeland V, Grotkopp E. 1996. Aqueous algal-based phenolic type adhesives and glues. US Patent 5520727.

Wagenknecht HA, Woggon WD. 1997. Identification of intermediates in the catalytic cycle of chloroperoxidase. *Chem. Biol.* **4**: 367–372.

Wever R, Dekker HL, Van Schijndel JWPM, Vollenbroek EGM. 1995. Antifouling paint containing haloperoxidase and method to determine halide. Patent WO/1995/027009.

Wever R, Tromp MGM, Krenn BE, Marjani A, van Tol M. 1991. Brominating activity of the seaweed *Ascophyllum nodosum*: impact on the biosphere. *Environ. Sci. Technol.* **25**: 446–449.

Wong BL, Shen Y-Q, Chem Y-P. 1996. Enzymatic production of halogenated cephalosporin. Patent WO/1996/019569.

Yamada H, Itoh N, Izumi Y. 1985. Chloroperoxidase-catalyzed halogenation of trans-cinnamic acid and its derivatives. *J. Biol. Chem.* **260**: 11962–11969.

Yde B, 1994. Polymerisation of lignin comprises reaction with a peroxidase and hydrogen peroxide at alkaline pH to give high molecular weight polymers. Patent WO/1994/01488.

Zaks A, Dodds DR. 1995. Chloroperoxidase-catalyzed asymmetric oxidations substrate-specificity and mechanistic studies. *J. Am. Chem. Soc.* **117**: 10419–10424.

Zhang J, Roberge C, Reddy J, Connors N, Chartrain M, Buckland B, Greasham R. 1999. Bioconversion of indene to trans-2S,1S-bromoindanol and 1S, 2R-indene oxide by a bromoperoxidase/dehydrogenase preparation from *Curvularia protuberata* MF5400. *Enzyme Microb. Technol.* **24**: 86–95.

AUTHOR INDEX

Listed by chapter, parentheses indicate number of citations.

Peroxidases and Catalases: Biochemistry, Biophysics, Biotechnology, and Physiology, Second Edition.
By H. Brian Dunford

SUBJECT INDEX

Peroxidases and Catalases: Biochemistry, Biophysics, Biotechnology, and Physiology, Second Edition.
By H. Brian Dunford